EIGHTH EDITION

The Atmosphere

An Introduction to Meteorology

Frederick K. Lutgens
Edward J. Tarbuck

Illinois Central College

Illustrated by
Dennis Tasa

PRENTICE HALL
Upper Saddle River, New Jersey 07458

Library of Congress Cataloging-in-Publication Data

Lutgens, Frederick K.
    The atmosphere/Frederick K. Lutgens, Edward J. Tarbuck; illustrated by Dennis Tasa.--8th ed.
        p.cm
    Includes index.
    ISBN 0-13-087957-6
    1. Atmosphere. 2. Meteorology. 3. Weather. I. Tarbuck, Edward J. II. Title
    QC861.2 .L87 2001
    551.5--dc21

Executive Editor: Daniel Kaveney
Editorial Director: Paul F. Corey
Assistant Vice President of Production and Manufacturing: David W. Riccardi
Executive Managing Editor: Kathleen Schiaparelli
Assistant Managing Editor: Beth Sturla
Production Editor: Edward Thomas
Manufacturing Manager: Trudy Pisciotti
Manufacturing Buyer: Michael Bell
Assistant Editor: Amanda Griffith
Marketing Manager: Christine Henry
Marketing Assistant: Erica Clifford
Director of Creative Services: Paul Belfanti
Art Director: Joseph Sengotta
Art Manager: Gus Vibal
Art Editor: Grace Hazeldine
Photo Research Administrator: Melinda Reo
Image Permission Coordinator: Debbie Hewittson
Photo Researcher: Barbara Scott
Copy Editor: James Tully
Editorial Assistant: Margaret Ziegler
Text Composition: Molly Pike/Lido Graphics
Production Assistant: Nancy Bauer
Interior Designer: Donna Wickes
Cover Designer: Thomas Nery
Cover Photo: Mountaineer Wanda Rutkiewicz climbs a steep face of Gasherbrum II peak.
    Photo by Galen Rowell/Corbis.
Assistant Managing Editor, Science Media: Alison Lorber
Media Editor: Michael Banino

Prentice Hall © 2001, 1998, 1995, 1992, 1989, 1986, 1982, 1979 by Prentice-Hall, Inc.
Upper Saddle River, New Jersey 07458

Printed in the United States of America

10 9 8 7 6 5 4 3 2 1

ISBN 0-13-087957-6

Prentice-Hall International (UK) Limited, *London*
Prentice-Hall of Australia Pty. Limited, *Sydney*
Prentice-Hall Canada Inc., *Toronto*
Prentice-Hall Hispanoamericana, S.A., *Mexico*
Prentice-Hall of India Private Limited, *New Delhi*
Prentice-Hall of Japan, Inc., *Tokyo*
Pearson Education Asia Pte. Ltd.
Editora Prentice-Hall do Brasil, Ltda., *Rio de Janeiro*

Brief Contents

Contents

CHAPTER 3
Temperature 56

CHAPTER 4
Moisture and Atmospheric Stability 86

CHAPTER 5
Forms of Condensation and Precipitation 118

CHAPTER 9
Weather Patterns   236

CHAPTER 10
Thunderstorms and Tornadoes   268

CHAPTER 11
Hurricanes   299

CHAPTER 12
Weather Analysis and Forecasting   319

CHAPTER 13
Air Pollution   347

CHAPTER 14
The Changing Climate   367

CHAPTER 15
World Climates 390

CHAPTER 16
Optical Phenomena
of the Atmosphere 413

APPENDIX A

APPENDIX B

APPENDIX C

APPENDIX D

APPENDIX E

APPENDIX F

Preface

There are few aspects of the physical environment that influence our daily lives more than the phenomena we collectively call *weather*. Newspapers, magazines, and television stations regularly report a wide range of weather events as major news stories—an obvious reflection of people's interest and curiosity about the atmosphere. We also face important environmental problems related to the atmosphere. Such issues as air pollution, ozone depletion, and global warming require our attention. A basic meteorology course can take advantage of our interest and curiosity about the weather as well as our desire to understand the impact that people have on the atmospheric environment.

The Atmosphere: An Introduction to Meteorology, Eighth Edition, is designed to meet the needs of students who enroll in such a course. It is our hope that the knowledge gained by taking a class and using this book will encourage many to actively participate in bettering the environment, and others may be sufficiently stimulated to continue their study of meteorology. Equally important, however, is our belief that a basic understanding of the atmosphere and its processes will greatly enhance appreciation of our planet and thereby enrich the reader's life.

In addition to being informative and up-to-date, a major goal of *The Atmosphere* is to meet the need of beginning students for a readable and user-friendly text, a book that is a highly usable "tool" for learning basic meteorological principles and concepts.

Distinguishing Features

Readability

The language of this book is straightforward and *written to be understood*. Clear, readable discussions with a minimum of technical language are the rule. The frequent headings and subheadings help students follow discussions and identify the important ideas presented in each chapter. In the present edition, improved readability was achieved by examining chapter organization and flow, and writing in a more personal style. Large portions of the text were substantially rewritten in an effort to make the material more understandable.

Illustrations and Photographs

Meteorology is highly visual. Therefore, photographs and artwork are a very important part of an introductory book. *The Atmosphere, Eighth Edition*, contains dozens of new high-quality photographs that were carefully selected to aid understanding, add realism, and heighten the interest of the reader.

The illustrations in each new edition of *The Atmosphere* keep getting better and better. In the eighth edition more than 100 pieces of line art are new or revised. The new art illustrates ideas and concepts more clearly and realistically than ever before. Dennis Tasa, a gifted artist and respected science illustrator, carried out the art program.

Focus on Learning

New to the eighth edition. To assist student learning, every chapter now concludes with a *Chapter Summary*. When a chapter has been completed, five useful devices help students review. First, the *Chapter Summary* recaps all the major points. Next the *Vocabulary Review* provides a checklist of key terms with page references. Learning the language of meteorology helps students learn the material. This is followed by the *Review Questions* section, which helps students examine their knowledge of significant facts and ideas. In most chapters, *Problems*, with a quantitative orientation, follow the review questions. Most problems require only basic mathematical skills and allow students to enhance their understanding by applying skills and principles explained in the chapter. Each chapter closes with some suggested Web sites for further exploration. Moreover, students are reminded to visit the *all-new* Web site for *The Atmosphere, Eighth Edition* (*http://www.prenhall.com/lutgens*). It contains many excellent opportunities for review and exploration.

Environmental Issues and Atmospheric Hazards

Many of the serious environmental issues that face humanity are related to the atmosphere. This new edition includes up-to-date treatment of air pollution, ozone depletion, global warming, and more.

Because atmospheric hazards adversely affect millions of people worldwide every day, coverage of this topic has been expanded. At appropriate places throughout the book students will have an opportunity to learn more about atmospheric hazards. Two entire chapters (Chapter 10, "Thunderstorms and Tornadoes" and Chapter 11, "Hurricanes") focus almost entirely on hazardous weather. In addition, a number of the book's special-interest boxes are devoted to a broad variety of atmospheric hazards including heat waves, winter storms, floods, dust storms, drought, mudslides, and lightning.

Maintaining a Focus on Basic Principles

Although many topical issues are treated in the eighth edition of *The Atmosphere*, it should be emphasized that the main focus remains the same as that of its predecessors—to foster a basic understanding of the atmospheric environment. In keeping with this aim, the organization of the text remains intentionally traditional. Following an overview of the atmosphere in Chapter 1, the next 10 chapters are devoted to a presentation of the major elements and concepts of meteorology. Chapter 12, on weather analysis, follows and serves to reinforce and apply many of the concepts presented in the preceding chapters. Chapter 13 is devoted to the important issue of air pollution.

The text concludes with two chapters on climate (Chapters 14 and 15), and one devoted to optical phenomena (Chapter 16). Chapter 14, "The Changing Climate," explores a topic that is the focus of much public interest as well as scientific research: Is global climate changing, and, if so, in what ways? How are people causing or contributing to these changes? The discussions in Chapter 14 have been carefully and thoroughly revised and updated to reflect the fast-changing nature of this sometimes controversial subject.

More About the Eighth Edition

In addition to adding chapter summaries, strengthening the atmospheric hazards theme, and making substantial changes to the photography and art programs, much more is new to the eighth edition.

This latest edition of *The Atmosphere* represents a thorough revision. *Every* part of the book was examined carefully with the dual goals of keeping topics current and improving the clarity of text discussions. Based on feedback from reviewers and students, we believe we have succeeded.

Supplements

The authors and publisher have been pleased to work with a number of talented people to produce an excellent supplements package. This package includes the traditional supplements that students and instructors have come to expect from authors and publishers.

Internet Support. Authored by Kenneth G. Pinzke, of Belleville Area College, this new site, specific to the text, contains numerous review exercises (from which students get immediate feedback), exercises to expand one's understanding of weather and climate, and resources for further exploration. This Web site provides an excellent platform from which to start using the Internet for the study of meteorology. Please visit the site at *http://www.prenhall.com/lutgens*.

Instructor's Manual with Tests (0-13-088514-2). Written by Don Yow, University of South Carolina, Columbia, the *Instructor's Manual* is intended as a resource for both new and experienced instructors. It includes a variety of lecture outlines, additional source materials, teaching tips, advice about how to integrate visual supplements (including the Web-based resources), and various other ideas for the classroom. A test item file provides instructors with a wide variety of test questions.

PH Custom Test. Available for both Macintosh (0-13-089249-1) and Windows (0-13-088512-6), and based on the powerful testing technology developed by Engineering Software Associates, Inc. (ESA), *Prentice Hall Custom Test* allows instructors to create and tailor exams to their own needs. With the online testing program, exams can also be administered online and data can then be automatically transferred for evaluation. The comprehensive desk reference guide is included along with online assistance.

Transparency Set (0-13-088513-4). More than *150* full-color acetates of illustrations from the text are available free of charge to qualified adopters.

Science on the Internet: A Student's Guide (0-13-028253-7). This unique resource helps students locate and explore the myriad geoscience resources on the World Wide Web. It also provides an overview of the Web, general navigation strategies, and brief student activities. It is available *free* when packaged with the text.

Rand McNally Atlas of World Geography (0-13-959339-X). This atlas includes 126 pages of up-to-date, accurate regional maps and 20 pages of illustrated world information tables. It is available *free* when packaged with *The Atmosphere, Eighth Edition*. Please contact your local Prentice Hall representative for details.

Digital Files (0-13-088500-2). All the maps and figures from the text, and many of the photographs, are available digitally on a CD-ROM. These files are ideal for those instructors who use PowerPoint or a comparable presentation software for their classes, or for professors who create text-specific Web sites for their students.

Prentice Hall–New York Times Themes of the Times Supplements for Geography. This supplement reprints significant articles from the *New York Times*. The Geography *New York Times* article is available to students free of charge. Please contact your local Prentice Hall representative for details.

WebCT/Prentice Hall. The WebCT Course Management System equips faculty members with easy-to-use tools to create sophisticated Web-based educational programs. Prentice Hall provides the content, which is specifically crafted to accompany *The Atmosphere, Eighth Edition*. Please contact your local Prentice Hall representative for more details.

For the Laboratory

Exercises in Atmospheric Science (0-13-861006-1). Written by Gregory J. Carbone. This lab manual, which complements Lutgens and Tarbuck's *The Atmosphere, Eighth Edition*, offers students an opportunity to review important ideas and concepts through problem solving, simulations, and guided thinking. This revised edition features an upgraded graphics program, and five computer-based simulations and tutorials that help students better learn key concepts. The lab manual is available at a reduced price when packaged with the text.

Acknowledgments

Writing a college textbook requires the talents and cooperation of many individuals. Working with Dennis Tasa, who is responsible for all of the outstanding illustrations, is always special for us. We not only value his outstanding artistic talents and imagination, but also his friendship. We are also grateful to Professor Ken Pinzke at Belleville Area College. In addition to his many helpful suggestions regarding the manuscript, Ken developed the Web site and prepared the chapter summaries. Ken is an important part of our team and a valued friend as well.

Students remain our most effective critics. Their comments and suggestions continue to help us maintain our focus on readability and understanding.

Special thanks goes to those colleagues who prepared in-depth reviews. Their critical comments and thoughtful input helped guide our work and clearly strengthened the text. We wish to thank:

Mark R. Anderson, *University of Nebraska-Lincoln*
Greg Carbone, *University of South Carolina*
Colleen E. Keen, *Minnesota State University*
Scott Kirsch, *University of Memphis*
John A. Knox, *Valparaiso University*
Arlene Laing, *University of South Florida*
Scott M. Robeson, *Indiana State University*
Robert Rohli, *Louisiana State University*
Steven Rutledge, *Colorado State University*
Anthony J. Vega, *Clarion University*

We also want to acknowledge the team of professionals at Prentice Hall. Thanks to Editorial Director Paul F. Corey. We sincerely appreciate his continuing strong support for excellence and innovation. Thanks also to our Executive Editor Dan Kaveney. His strong communication skills and energetic style contributed greatly to the project. The production team, led by Ed Thomas, as always, has done an outstanding job. They are true professionals with whom we are very fortunate to be associated.

Frederick K. Lutgens

Edward J. Tarbuck

Introduction to the Atmosphere

Travelers in northern Iowa battle blowing snow and white-out conditions in January 1999. Such winter weather events are to be expected in the central plains. *(Associated Press photo)*

Figure 1–1 North America has a great variety of weather. Severe weather events are frequent and costly occurrences. In January 1998 an ice storm of historic proportions caused enormous damage in New England and southeastern Canada. Nearly five days of freezing rain left millions without electricity—some for as long as a month. *(Photo by Syracuse Newspapers/The Image Works)*

Weather influences our everyday activities, our jobs, and our health and comfort. Many of us pay little attention to the weather unless we are inconvenienced by it or when it adds to our enjoyment of outdoor activities. Nevertheless, there are few other aspects of our physical environment that affect our lives more than the phenomena we collectively call the weather.

On January 25, 2000, North Carolina and nearby states experienced a record-breaking winter storm. Up to 51 centimeters (20 inches) of snow fell on a region that only occasionally receives modest snows. Hundreds of thousands were without power, and more than 1 million people were trapped in their homes because streets and highways were impassable. Chilling cold engulfed the region. This was the third major weather event to hit the area since mid-1999.

In September of that year, Hurricane Floyd brought flooding rains, damaging winds, and rough seas to a large portion of the Atlantic seaboard. More than 2.5 million people evacuated their homes from Florida, north to the Carolinas, and beyond. It was the largest evacuation in U.S. history. Torrential rains created devastating floods that caused an estimated $6 billion in damages and cost 77 lives. Just a few months earlier North Carolina and nearby mid-Atlantic states were suffering from a heat wave and drought—the worst in more than a century at some locations. Crops were ruined, and many reservoirs dropped to dangerously low levels.

These memorable weather events serve to illustrate the fact that the United States has the greatest variety of weather of any country in the world (chapter opening photo and Figure 1–1). Severe weather events such as tornadoes, flash floods, and intense thunderstorms, as well as hurricanes and blizzards, are collectively more frequent and more damaging in the United States than in any other nation. Beyond its direct impact on the lives of individuals, the weather has a strong effect on the world economy, by influencing agriculture, energy use, water resources, transportation, and industry.

Weather clearly influences our lives a great deal. Yet it is also important to realize that people influence the atmosphere and its behavior as well. There are, and will continue to be, significant political and scientific decisions to make involving these impacts. Answers to questions regarding air pollution and its control and the effects of various emissions on global climate and the atmosphere's ozone layer are important examples. So there is a need for increased awareness and understanding of our atmosphere and its behavior.

Weather and Climate

Acted on by the combined effects of Earth's motions and energy from the Sun, our planet's formless and invisible envelope of air reacts by producing an infinite variety of weather, which in turn creates the basic pattern of global climates. Although not identical, weather and climate have much in common.

Weather is constantly changing, sometimes from hour to hour and at other times from day to day. It is a term that refers to the state of the atmosphere at a given time and place. Whereas changes in the weather are continuous and sometimes seemingly erratic, it is nevertheless possible to arrive at a generalization of these variations. Such a description of aggregate weather conditions is termed **climate**. It is based on observations that have been accumulated over many decades. Climate is often defined simply as "average weather," but this is an inadequate definition. In order to more accurately portray the character of an area, variations and extremes must also be included, as well as the probabilities that such departures will take place. For example, it is not only necessary for farmers to know the average rainfall during the growing season, but it is also important to know the frequency of extremely wet and extremely dry years. Thus, climate is the sum of all statistical weather information that helps describe a place or region.

Maps similar to the one in Figure 1–2 are familiar to everyone who checks the weather report in the morning newspaper or on a local television station. In addition to showing predicted high temperatures for the day, this map shows other basic weather information about cloud cover, precipitation, and fronts.

Suppose you were planning a vacation trip to an unfamiliar place. You would probably want to know what kind

of weather to expect. Such information would help as you selected clothes to pack and could influence decisions regarding activities you might engage in during your stay. Unfortunately, weather forecasts that go beyond a few days are not very dependable. Thus, it would not be possible to get a reliable weather report about the conditions you are likely to encounter during your vacation.

Instead, you might ask someone who is familiar with the area about what kind of weather to expect. "Are thunderstorms common?" "Does it get cold at night?" "Are the afternoons sunny?" What you are seeking is information about the climate, the conditions that are typical for that place. Another useful source of such information is the great variety of climate tables, maps, and graphs that are available. For example, the map in Figure 1–3 shows the average percentage of possible sunshine in the United States for the month of November, whereas the graph in Figure 1–4 shows average daily high and low temperatures for each month, as well as extremes for New York City.

Such information could no doubt help as you planned your trip. But it is important to realize that *climate data cannot predict the weather*. Although the place may usually (climatically) be warm, sunny, and dry during the time

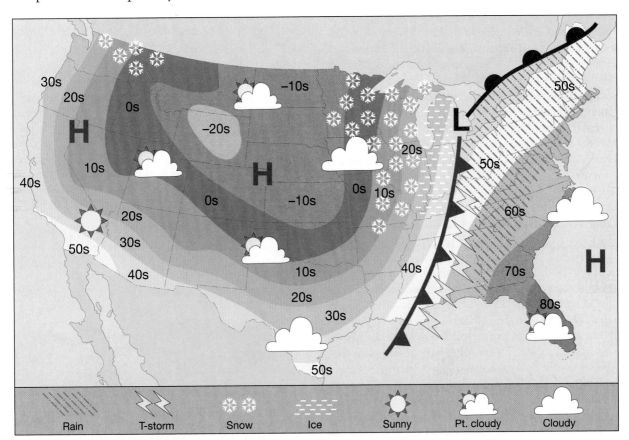

Figure 1–2 A typical newspaper weather map for a day in late December. The temperatures shown on the map are the highs forecast for the day.

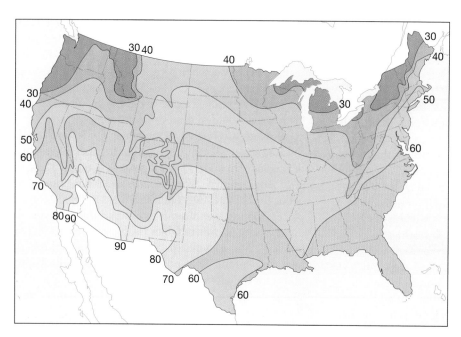

Figure 1–3 Mean percentage of possible sunshine for November. Southern Arizona is clearly the sunniest area. By contrast, parts of the Pacific Northwest receive less than half as much possible sunshine. Climate maps such as this one are based on many years of data.

of your planned vacation, you may actually experience cool, overcast, and rainy weather!

The nature of both weather and climate is expressed in terms of the same basic **elements**, those quantities or properties that are measured regularly. The most important are (1) the temperature of the air, (2) the humidity of the air, (3) the type and amount of cloudiness, (4) the type and amount of precipitation, (5) the pressure exerted by the air, and (6) the speed and direction of the wind. These elements constitute the variables by which weather patterns and climatic types are depicted. Although you will study these elements separately at first, keep in mind that they are very much interrelated. A change in one of the elements often produces changes in the others.

Atmospheric Hazards: Assault by the Elements

Natural hazards are a part of living on Earth. Every day they adversely affect literally millions of people worldwide and are responsible for staggering damages. Some, such as earthquakes and volcanic eruptions, are geological. But a greater number are related to the atmosphere.

Occurrences of severe weather have a fascination that ordinary weather phenomena cannot provide. A spectacular lightning display generated by a severe thunderstorm can elicit both awe and fear. Of course, hurricanes and tornadoes attract a great deal of much-deserved attention (Figure 1–5). A single tornado outbreak or hurricane can cause billions of dollars in property damage, much human suffering, and many deaths. In a typical year the United

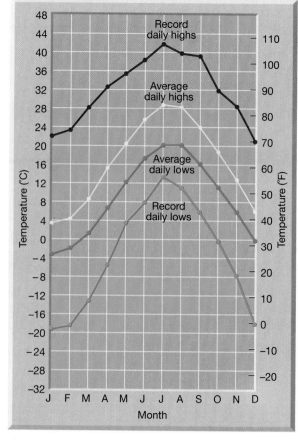

Figure 1–4 Graph showing daily temperature data for New York City. In addition to the average daily maximum and minimum temperatures for each month, extremes are also shown. As this graph shows, there can be significant departures from the average.

Figure 1–5 Tornadoes are intense and destructive local storms of short duration that cause many deaths each year. Many people have incorrect perceptions of weather dangers and are unaware of the relative differences of weather threats to human life. For example, they are awed by the threat of tornadoes and plan accordingly on how to respond (e.g. tornado-awareness week each spring), but they fail to realize that lightning or winter storms are greater threats. *(Photo by Alan R. Moller/Tony Stone Images)*

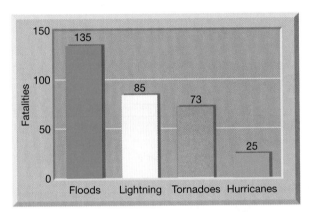

Figure 1–6 Annual average number of storm-related deaths in the United States, 1966–95.

States experiences thousands of violent thunderstorms, hundreds of floods and tornadoes, and several hurricanes. Tornadoes take an average of 73 lives each year, whereas hurricanes are responsible for about one-third as many deaths. Surprisingly (to many), lightning and flash floods are responsible for more deaths (Figure 1–6).

Of course, there are other atmospheric hazards that adversely affect us. Some are storm-related, such as blizzards, hail, and freezing rain. Others are not the direct result of a storm. Heat waves, cold waves and drought are important examples. In some years, the loss of human life due to excessive heat or bitter cold exceeds that caused by all other weather events combined. Moreover, although severe storms and floods usually generate more attention, droughts can be just as devastating and carry an even bigger price tag.

From 1988 through 1999 the United States experienced 38 weather-related disasters in which overall damages and costs reached or exceeded $1 billion. The combined costs of these 38 events exceeded $170 billion! Table 1–1 lists the 10 "billion-dollar weather disasters" that occurred in 1998 and 1999. As you can see, although hurricanes, tornadoes, and floods stand out, ice storms, drought, heat waves, and hail also figure prominently.

At appropriate places throughout this book you will have an opportunity to learn about atmospheric hazards. Two entire chapters (Chapter 10 Thunderstorms and Tornadoes and Chapter 11 Hurricanes) focus almost entirely

Table 1–1 Billion-Dollar Weather Events 1998–1999			
Event	**Date**	**Damage/costs**	**Fatalities**
Hurricane Floyd	September 1999	$6 billion	77
Eastern drought and heat wave	Summer 1999	$1 billion	256
Oklahoma–Kansas Tornadoes	May 1999	$1 billion	55
Arkansas–Tennessee Tornadoes	January 1999	$1.3 billion	17
Texas flooding	October–November 1998	$1 billion	31
Hurricane Georges	September 1998	$5.9 billion	16
Hurricane Bonnie	August 1998	$1 billion	3
Southern drought and heat wave	Summer 1998	$6–9 billion	200
Minnesota severe storms/hail	May 1998	$1 billion	1
Northeast ice storm	January 1998	$1.4 billion	16

Source: National Climatic Data Center

on hazardous weather. In addition, a number of the book's special-interest boxes are devoted to a broad variety of atmospheric hazards, including heat waves, winter storms, floods, dust storms, drought, mudslides, and lightning. To alert you to these discussions, each of the boxes that examines an atmospheric hazard is identified by the same small icon you see at the beginning of this section.

Every day our planet experiences an incredible assault by the atmosphere, so it is important to develop an awareness and understanding of these significant weather events.

The Atmosphere: A Part of the Earth System

A view of Earth from space affords us a unique perspective of our planet (Figure 1–7). At first it may strike us that Earth is a fragile-appearing sphere surrounded by the blackness of space. It is, in fact, just a speck of matter in a vast universe. As we look more closely, it becomes apparent that Earth is much more than the land upon which we live. Indeed, when seen from space, Earth's most conspicuous features are not the continents, but the swirling clouds suspended above the surface and the vast global ocean. This view serves to emphasize the importance of air and water to our planet.

From such a vantage point, we can also appreciate why Earth's physical environment is traditionally divided into three major parts: the solid Earth, the water portion of our planet, and Earth's gaseous envelope. In addition, our physical environment supports a vast array of life forms that constitute a fourth major part of our planet.

Earth's Four Spheres

On a human scale, Earth is huge. Its surface area occupies 500,000,000 square kilometers (193 million square miles). As indicated, we divide this vast planet into four independent parts. Because each part loosely occupies a shell around Earth, we call them spheres. The four spheres include the *atmosphere* (gaseous envelope), the *lithosphere* (solid Earth), the *hydrosphere* (water portion), and the *biosphere* (life). It is important to remember that these spheres are not separated by well-defined boundaries; rather, each sphere is intertwined with all of the others. In addition, each of Earth's four major spheres can be thought of as being composed of numerous interrelated parts.

The Atmosphere. Earth is surrounded by a life-giving gaseous envelope called the **atmosphere**. This thin blanket of air is an integral part of the planet. It not only provides the air that we breathe but also acts to protect us

Figure 1–7 Africa and Arabia are prominent in this image of Earth taken from *Apollo 17*. The tan cloud-free zones over the land coincide with major desert regions. The band of clouds across central Africa is associated with a much wetter climate that in places sustains tropical rain forests. The dark blue of the oceans and the swirling cloud patterns remind us of the importance of the oceans and the atmosphere. Antarctica, a continent covered by glacial ice, is visible at the South Pole. *(Courtesy of NASA/Science Source/Photo Researchers, Inc.)*

from the dangerous radiation emitted by the Sun. The energy exchanges that continually occur between the atmosphere and Earth's surface and between the atmosphere and space produce the effects we call *weather*. If, like the Moon, Earth had no atmosphere, our planet would not only be lifeless but many of the processes and interactions that make the surface such a dynamic place could not operate.

The Lithosphere. The solid Earth is divided into three principal regions: the dense inner sphere called the core, the less dense mantle, and the crust, which is the light and very thin outer skin of Earth. The term **lithosphere** refers to the rigid outer layer of Earth that includes the crust and uppermost part of the mantle. (The term lithosphere is sometimes used in reference to the entire solid planet.) Beneath the rigid rocks that compose the lithosphere, the rocks become weak and are able to slowly flow in response to the uneven distribution of heat deep within Earth.

The Hydrosphere. The **hydrosphere** is the water portion of our planet. This dynamic mass of liquid is continuously on the move, from the oceans to the air, to the land, and back again. The global ocean is obviously the most prominent feature of the hydrosphere, blanketing nearly 71 percent of Earth's surface and accounting for about 97 percent of Earth's water (Figure 1–8). However, the hydrosphere also includes the water found in streams, lakes, glaciers, and underground.

Although water outside of the ocean constitutes just a tiny fraction of the total, it is still very important because water is essential to life on Earth. To many people's surprise, glacial ice is Earth's largest reservoir of freshwater, accounting for about 85 percent of the planet's supply.

The Biosphere. The **biosphere** includes all life on Earth and penetrates parts of the lithosphere, hydrosphere, and atmosphere. Although plants and animals depend on the physical environment for the basics of life, it should be emphasized that organisms do more than just respond to their environment. Indeed, through countless interactions, life forms help *maintain* and *alter* their physical environment. Without life, the makeup and nature of the solid Earth, hydrosphere, and atmosphere would be very different.

Earth's Spheres Interact

A *system* is a group of interacting or interdependent parts that form a complex whole. Most of us hear and use the term frequently. We may service our car's cooling *system*, make use of the city's transportation *system*, and are

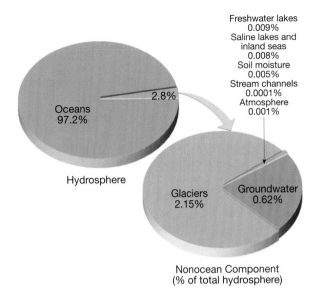

Figure 1–8 Distribution of Earth's water. Obviously most of Earth's water is in the oceans. Glacial ice represents about 85 percent of all the water *outside* the oceans. When only *liquid freshwater* is considered, more than 90 percent is groundwater.

affected by an approaching weather *system*. Further, we know that Earth is just a small part of a large system known as the *solar system*.

As we study Earth, it becomes clear that our planet can be viewed as a system with many separate but interacting parts or subsystems. The lithosphere, hydrosphere, atmosphere, and biosphere and all of their components can be studied separately. However, the parts are not isolated. Each is related in some way to the others to produce a complex and continuously interacting whole that we call the *Earth system*.

One way to think of Earth as a system is to imagine the four spheres—atmosphere, lithosphere, hydrosphere, and biosphere—as huge reservoirs of material and energy, some of which is continually moving from one sphere to another. Thus, any change in one part of the Earth system results in changes in one or more of the other parts.

The energy to drive this system comes primarily from two sources. Solar energy drives external processes that occur at or above Earth's surface. External processes, such as weather and climate, ocean circulation, and erosion are driven by energy from the Sun. Earth's interior is the second source of energy. Heat remaining from when our planet formed, and heat that is continuously generated by radioactive decay, powers the internal processes that produce volcanoes, earthquakes, and mountains.

The Carbon Cycle

To help illustrate the movement of material and energy from one sphere to another in the Earth system, take a brief look at the *carbon cycle* (Figure 1–9). Pure carbon is relatively rare in nature. It is found predominantly in two minerals: diamond and graphite. Most carbon is bonded chemically to other elements to form compounds such as carbon dioxide, calcium carbonate, and the hydrocarbons found in coal and petroleum. Carbon is also the basic building block of life as it readily combines with hydrogen and oxygen to form the fundamental organic compounds that compose living things.

In the atmosphere, carbon is found mainly chemically bonded to oxygen to form carbon dioxide (CO_2). As you will see later, carbon dioxide plays a very important role in the heating of the atmosphere. Because many of the processes that operate on Earth involve carbon dioxide, this gas is cycled in and out of the atmosphere at a rather rapid rate (Figure 1–9). For example, through the process of photosynthesis, plants absorb carbon dioxide from the atmosphere in order to produce the essential organic compounds needed for their growth. Animals that eat these plants (or consume other animals that eat plants) use these organic compounds as a source of energy and, through the process of respiration, return carbon dioxide to the atmosphere. (Plants also return some CO_2 to the atmosphere via respiration.) Further, when trees die and decay or are burned as is done on a large scale in the tropical rain forests, this biomass is oxidized, and carbon dioxide is released into the atmosphere.

Over long periods of geologic time, considerable biomass is buried with sediment. Under the right conditions, these carbon-rich deposits do not completely decompose but rather are converted to fossil fuels—coal, petroleum, or natural gas. Eventually some of this biomass is recovered (mined or pumped from a well) and burned to run factories and fuel our transportation system. One result of fossil-fuel combustion is the release of huge quantities of CO_2 into the atmosphere. Certainly one of the most active parts of the carbon cycle is the movement of CO_2 from the atmosphere to the biosphere and back again.

Carbon also moves from the lithosphere and hydrosphere to the atmosphere and back again. For example, volcanic activity early in Earth's history is thought to be the source of much of the carbon dioxide found in the atmosphere (see Box 1–1). One way that carbon dioxide makes its way back to the hydrosphere and then to the solid Earth is by first combining with water (rain water or

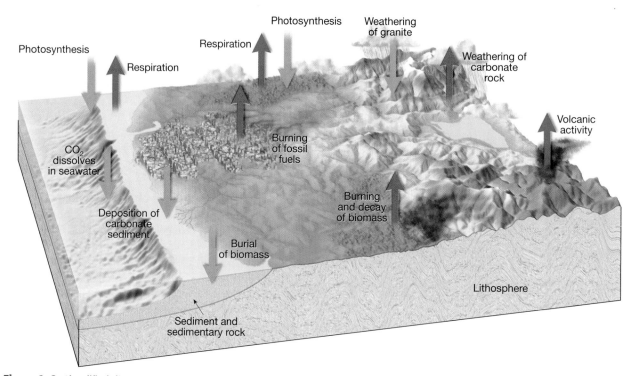

Figure 1–9 Simplified diagram of the carbon cycle, with emphasis on the flow of carbon between the atmosphere and the hydrosphere, lithosphere and biosphere. The colored arrows show whether the flow of carbon is into or out of the atmosphere.

soil moisture) to form a very weak acid called carbonic acid (H_2CO_3). Carbonic acid then attacks the rocks that compose the lithosphere (such as granite). One product of this chemical weathering of solid rock is the soluble carbonate ion (HCO_3^-), which is carried by streams to the ocean. Here water-dwelling organisms extract this dissolved material to produce shells of calcium carbonate ($CaCO_3$). When the organisms die, these hard parts settle to the ocean floor as marine sediment and become sedimentary rock. In fact, the lithosphere is Earth's largest depository of carbon, where it is a constituent of a variety of rocks, the most abundant being the sedimentary rock limestone (Figure 1–9). Eventually the limestone may be exposed at Earth's surface, where natural chemical weathering will cause the carbon stored in the rock to be released to the atmosphere as CO_2.

In summary, our environment is highly integrated and characterized by continuous interaction as air comes in contact with rock, rock with water, and water with air. Moreover, the biosphere, the totality of life forms on our planet, is associated with each of the three physical realms and is an equally integral part of Earth. The interplay and interactions among the spheres of Earth's environment are uncountable.

Composition of the Atmosphere

In the days of Aristotle, air was believed to be one of four fundamental substances that could not be further subdivided into constituent components. The other three substances were fire, earth (soil), and water. Even today the term **air** is sometimes used as if it were a specific gas, which, of course, it is not. The envelope of air that surrounds our planet is a *mixture* of many discrete gases, each with its own physical properties, in which varying quantities of tiny solid and liquid particles are suspended.

Major Components

The composition of air is not constant; it varies from time to time and from place to place (see Box 1–1). If the water vapor, dust, and other variable components were removed from the atmosphere, we would find that its makeup is very stable up to an altitude of about 80 kilometers (50 miles).

As you can see in Figure 1–10 and Table 1–2, two gases—nitrogen and oxygen—make up 99 percent of the volume of clean, dry air. Although these gases are the most plentiful components of the atmosphere and are of great significance to life on Earth, they are of little or no importance in affecting weather phenomena. The remaining 1 percent of dry air is mostly the inert gas argon (0.93 percent) plus tiny quantities of a number of other gases.

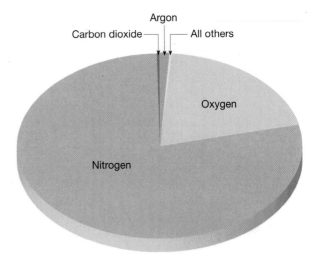

Figure 1–10 Proportional volume of gases composing dry air. Nitrogen and oxygen obviously dominate.

Carbon Dioxide

Carbon dioxide, although present in only minute amounts (0.036 percent), is nevertheless a meteorologically important constituent of air. Carbon dioxide is of great interest to meteorologists because it is an efficient absorber of energy emitted by Earth and thus influences the heating of the atmosphere. Although the proportion of carbon dioxide in the atmosphere is relatively uniform, its percentage has been rising steadily for more than a century. This rise is attributed to the burning of ever-increasing quantities of fossil fuels, such as coal and oil. Much of this additional carbon dioxide is absorbed by the waters of the ocean or is used by plants, but nearly half remains in the air. Estimates project that by sometime in the second half of the twenty-first century, carbon dioxide levels will be twice as high as they were early in the twentieth century.

Table 1–2 Principal gases of dry air

Constituent	Percent by Volume	Concentration in Parts Per Million (PPM)
Nitrogen (N_2)	78.084	780,840.0
Oxygen (O_2)	20.946	209,460.0
Argon (Ar)	0.934	9,340.0
Carbon dioxide (CO_2)	0.036	360.0
Neon (Ne)	0.00182	18.2
Helium (He)	0.000524	5.24
Methane (CH_4)	0.00015	1.5
Krypton (Kr)	0.000114	1.14
Hydrogen (H_2)	0.00005	0.5

Box 1–1 Earth's Atmosphere Evolves

Earth's atmosphere is unlike that of any other body in the solar system. No other planet is as hospitable or exhibits the same life-sustaining mixture of gases as Earth. Today the air you breathe is a stable mixture of 78 percent nitrogen, 20 percent oxygen, about 1 percent argon (an inert gas), and trace gases like carbon dioxide and water vapor. But our planet's original atmosphere, several billion years ago, was far different.

Earth's very earliest atmosphere probably was swept into space by the *solar wind*, a vast stream of particles emitted by the Sun. As Earth slowly cooled, a more enduring atmosphere formed. The molten surface solidified into a crust, and gases that had been dissolved in the molten rock were gradually released, a process called *outgassing*. Outgassing continues today from hundreds of active volcanoes worldwide. Thus, geologists hypothesize that Earth's original atmosphere was made up of gases similar to those released in volcanic emissions today: water vapor, carbon dioxide, nitrogen, and several trace gases.

As the planet continued to cool, the water vapor condensed to form clouds, and great rains commenced. At first the water evaporated in the hot air before reaching the ground, or quickly boiled away upon contacting the surface, just like water sprayed on a hot grill. This accelerated the cooling of Earth's crust. When the surface had cooled below water's boiling point (100°C, or 212°F), torrential rains slowly filled low areas, forming

the oceans. This reduced not only the water vapor in the air but also the amount of carbon dioxide, for it became dissolved in the water. What remained was a nitrogen-rich atmosphere.

If Earth's primitive atmosphere resulted from volcanic outgassing, we have a problem, because volcanoes do not emit free oxygen. Where did the very significant percentage of oxygen in our present atmosphere (20 percent) come from?

The major source of oxygen is green plants (Figure 1–A). Put another way, *life itself* has strongly influenced the composition of our present atmosphere. Plants did not just adapt to their environment; they actually influenced it, dramatically altering the composition of the entire planet's atmosphere by using carbon dioxide and releasing oxygen. This is a good example of how Earth operates as a giant system in which living things interact with their environment.

How did plants come to alter the atmosphere? The key is the way in which plants create their own food. They employ *photosynthesis*, in which they use light energy to synthesize food sugars from carbon dioxide and water. The process releases a waste gas—oxygen. Those of us in the animal kingdom rely on oxygen to metabolize our food, and we in turn exhale carbon dioxide as a waste gas. The plants use this carbon dioxide for more photosynthesis, and so on, in a continuing system.

The precise impact of the increased carbon dioxide is difficult to predict, but most atmospheric scientists believe that it will bring about a warming of the lower atmosphere and thus trigger global climate change. The role of carbon dioxide in the atmosphere and its possible effect on climate are examined in more detail in Chapter 14.

Variable Components

Air includes many gases and particles that vary significantly from time to time and place to place. Important examples include water vapor, dust particles, and ozone. Although usually present in small percentages, they can have significant effects on weather and climate.

Water Vapor The amount of water vapor in the air varies considerably, from practically none at all up to about 4 percent by volume. Why is such a small fraction of the

atmosphere so significant? Certainly the fact that water vapor is the source of all clouds and precipitation would be enough to explain its importance. However, water vapor has other roles. Like carbon dioxide, it has the ability to absorb heat energy given off by Earth, as well as some solar energy. It is therefore important when we examine the heating of the atmosphere.

When water changes from one state to another (see Figure 4–2, p. 88), it absorbs or releases heat. This energy is termed *latent heat*, which means "hidden" heat. As we shall see in later chapters, water vapor in the atmosphere transports this latent heat from one region to another, and it is the energy source that helps drive many storms.

Aerosols The movements of the atmosphere are sufficient to keep a large quantity of solid and liquid particles suspended within it. Although visible dust sometimes clouds

Figure 1–A Life has strongly influenced the composition of our atmosphere. The primary source of the abundant oxygen in Earth's atmosphere is photosynthesis by green plants. *(Photo by Pat O'Hara/DRK Photo)*

The first life forms on Earth, probably bacteria, did not need oxygen. Their life processes were geared to the earlier, oxygenless atmosphere. Even today, many *anaerobic* bacteria thrive in environments that lack free oxygen. Later, primitive plants evolved that used photosynthesis and released oxygen. Slowly the oxygen content of Earth's atmosphere increased. The geologic record of this ancient time suggests that much of the first free oxygen did not remain free, because it combined with (oxidized) other substances dissolved in water, especially iron. Iron has tremendous affinity for oxygen, and the two elements combine to form iron oxides (rust) at any opportunity.

Then, once the available iron satisfied its need for oxygen, substantial quantities of oxygen accumulated in the atmosphere. By the beginning of the Paleozoic era, about 4 billion years into Earth's existence (after seven-eighths of Earth's history had transpired) the fossil record reveals abundant ocean-dwelling organisms that require oxygen to live. Hence, the composition of Earth's atmosphere has evolved together with its life forms, from an oxygenless envelope to today's oxygen-rich environment.

the sky, these relatively large particles are too heavy to stay in the air for very long. Still, many particles are microscopic and remain suspended for considerable periods of time. They may originate from many sources, both natural and human made, and include sea salts from breaking waves, fine soil blown into the air, smoke and soot from fires, pollen and microorganisms lifted by the wind, ash and dust from volcanic eruptions, and more. Collectively, these tiny solid and liquid particles are called **aerosols**.

Aerosols are most numerous in the lower atmosphere near their primary source, Earth's surface. Nevertheless, the upper atmosphere is not free of them, because some dust is carried to great heights by rising currents of air, and other particles are contributed by meteoroids that disintegrate as they pass through the atmosphere.

From a meteorological standpoint, these tiny, often invisible particles can be significant. First, many act as surfaces on which water vapor may condense, an important function in the formation of clouds and fog. Second, aerosols can absorb or reflect incoming solar radiation. Thus, when an air-pollution episode is occurring or when ash fills the sky following a volcanic eruption, the amount of sunlight reaching Earth's surface can be measurably reduced. Finally, aerosols contribute to an optical phenomenon we have all observed—the varied hues of red and orange at sunrise and sunset (Figure 1–11).

Ozone Another important component of the atmosphere is **ozone**. It is a form of oxygen that combines three oxygen atoms into each molecule (O_3). Ozone is not the same as the oxygen we breathe, which has two atoms per molecule (O_2). There is very little ozone in the atmosphere, and its distribution is not uniform. In the lowest portion of the atmosphere, ozone represents less

Figure 1–11 Dust in the air can cause sunsets to be especially colorful. *(Photo by Steve Elmore/The Stock Market)*

than one part in 100 million. It is concentrated well above the surface in a layer called the *stratosphere*, between 10 and 50 kilometers (6 and 31 miles).

In this altitude range, oxygen molecules (O_2) are split into single atoms of oxygen (O) when they absorb ultraviolet radiation emitted by the Sun. Ozone is then created when a single atom of oxygen (O) and a molecule of oxygen (O_2) collide. This must happen in the presence of a third, neutral molecule that acts as a *catalyst* by allowing the reaction to take place without itself being consumed in the process. Ozone is concentrated in the 10 to 50 kilometer height range because a crucial balance exists there: The ultraviolet radiation from the Sun is sufficient to produce single atoms of oxygen, and there are enough gas molecules to bring about the required collisions.

The presence of the ozone layer in our atmosphere is crucial to those of us who dwell on Earth. The reason is that ozone absorbs the potentially harmful ultraviolet (UV) radiation from the Sun. If ozone did not filter a great deal of the ultraviolet radiation, and if the Sun's UV rays reached the surface of Earth undiminished, our planet would be uninhabitable for most life as we know it. Thus, anything that reduces the amount of ozone in the atmosphere could affect the well-being of life on Earth. Just such a problem exists and is described in the next section.

Ozone Depletion— A Global Issue

The loss of ozone high in the atmosphere as a consequence of human activities is a serious global-scale environmental problem. For nearly a billion years Earth's ozone layer has protected life on the planet. However, over the past half century, people have unintentionally placed the ozone layer in jeopardy by polluting the atmosphere. The offending chemicals are known as chlorofluorocarbons (CFCs). They are versatile compounds that are chemically stable, odorless, nontoxic, noncorrosive, and inexpensive to produce. Over the decades many uses were developed for CFCs, including as coolants for airconditioning and refrigeration equipment, cleaning solvents for electronic components, propellants for aerosol sprays, and the production of certain plastic foams.

No one worried about how CFCs might affect the atmosphere until three scientists, Paul Crutzen, F. Sherwood Rowland, and Mario Molina, studied the relationship. In 1974 they alerted the world when they reported that CFCs were probably reducing the average concentration of ozone in the stratosphere. In 1995 these scientists were awarded the Nobel Prize in chemistry for their pioneering work.

They discovered that because CFCs are practically inert (that is, not chemically active) in the lower atmosphere, a portion of these gases gradually makes its way to the ozone layer, where sunlight separates the chemicals into their constituent atoms. The chlorine atoms released this way, through a complicated series of reactions, have the net effect of removing some of the ozone.

The Ozone Hole

Although ozone depletion by CFCs occurs worldwide, measurements have shown that ozone concentrations take an especially sharp drop over Antarctica during the Southern Hemisphere spring (September and October). Later, during November and December, the ozone concentration recovers to more normal levels (Figure 1–12). Between the early 1980s, when it was discovered, and the late 1990s, this well-publicized *ozone hole* intensified and grew larger. By 1999 it covered an area more than twice the size of the continental United States (Figure 1–13).

The hole is caused in part by the relatively abundant ice particles in the south polar stratosphere. The ice boosts the effectiveness of CFCs in destroying ozone, thus causing a greater decline than would otherwise occur. The zone of maximum depletion is confined to the Antarctic region by a swirling upper-level wind pattern. When this

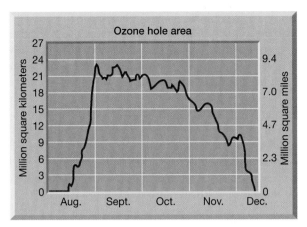

Figure 1–12 Changes in the size of the 1999 Antarctic ozone hole. It began to form in late August and was well developed in September and October. The ozone hole persisted through November and disappeared in December. At its maximum, the area of the ozone hole exceeded 21 million square kilometers (8.2 million square miles). For comparison, the area of the continental United States is about 9.4 million square kilometers (3.7 million square miles).

vortex weakens during the late spring, the ozone-depleted air is no longer restricted, and mixes freely with air from other latitudes where ozone levels are higher.

A few years after the Antarctic ozone hole was discovered, scientists detected a similar but smaller ozone thinning in the vicinity of the North Pole during spring and early summer. When this pool breaks up, parcels of ozone-depleted air move southward over North America, Europe, and Asia.

Effects of Ozone Depletion

Because ozone filters out most of the UV radiation in sunlight, a decrease in its concentration permits more of these harmful wavelengths to reach Earth's surface. Scientists in New Zealand discovered that during the decade of the 1990s, damaging UV radiation gradually increased as concentrations of stratospheric ozone decreased. By 1999 peak sunburning UV levels in New Zealand were about 12 percent higher than at the beginning of the decade. What are the effects of the increased ultraviolet radiation? Each 1 percent decrease in the concentration of stratospheric ozone increases the amount of UV radiation that reaches Earth's surface by about 2 percent. Therefore, because ultraviolet radiation is known to induce skin cancer, ozone depletion seriously affects human health, especially among fair-skinned people and those who spend considerable time in the Sun.

The fact that up to a half million cases of these cancers occur in the United States annually means that ozone depletion could ultimately lead to many thousands of additional cases each year.[*] In addition to upping the risk of skin cancer, an increase in damaging UV radiation can negatively impact the human immune system as well as promote cataracts, a clouding of the eye lens that reduces vision and may cause blindness if not treated.

The effects of additional UV radiation on animal and plant life are also important. There is serious concern that crop yields and quality will be adversely affected. Some scientists also fear that increased UV radiation in the Antarctic will penetrate the waters surrounding the continent and impair or destroy the microscopic plants, called phytoplankton, that represent the base of the food chain. A decrease in phytoplankton, in turn, could reduce the population of copepods and krill that sustain fish, whales, penguins, and other marine life in the high latitudes of the Southern Hemisphere.

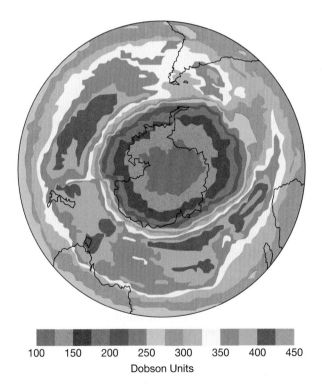

100   150   200   250   300   350   400   450
Dobson Units

Figure 1–13 This satellite image shows ozone distribution in the Southern Hemisphere in October 1998. The "hole" is that region where ozone concentrations are less than 220 Dobson units. The 1998 Antarctic ozone hole was the largest ever observed, extending over about 26 million square kilometers (10 million square miles). In 1999 the hole was somewhat smaller. *(Data from NOAA)*

[*]For more on this, see Box 2–2 "The Ultraviolet Index."

Montreal Protocol

What has been done to protect the atmosphere's ozone layer? Realizing that the risks of not curbing CFC emissions were difficult to ignore, an international agreement known as the Montreal Protocol was concluded under the auspices of the United Nations in late 1987. The treaty specified a 50 percent reduction in CFC production by the end of the century. More than 40 nations eventually endorsed the proposal. The treaty represented a positive international response to the ozone problem.

When subsequent evidence showed that atmospheric ozone levels were dropping more rapidly than had been predicted, a stronger response was necessary. In June 1990 a new agreement was reached. The updated protocol called for a complete phaseout of CFCs by early in the twenty-first century. Although some nations called for more rapid reductions, this agreement was nevertheless viewed by many as a major step forward in environmental diplomacy.

Although relatively strong action has been taken, CFC levels in the atmosphere will not drop rapidly. Once in the atmosphere, CFC molecules can take many years to reach the ozone layer and, once there, can remain active for decades. This does not promise a near-term reprieve for the ozone layer. Even after CFC levels begin to decline, the ozone layer probably will not cleanse itself of most of the pollutants until sometime in the second half of the twenty-first century.

Probing the Atmosphere

Scientific study of the atmosphere began in the seventeenth century as instruments were developed to measure different elements (see Box 1–2). The instruments provided data that helped observers formulate laws applying to the atmosphere. In 1593 Galileo invented an early version of the thermometer, and in 1643 Torricelli built the first barometer (for measuring air pressure). By 1661 Robert Boyle discovered the basic relationship between pressure and volume in a gas. During the eighteenth century, instruments were improved and standardized, and extensive data collection began. The acquisition of such data was fundamental to the study of physical processes and the development of explanations about atmospheric phenomena.

It became obvious to those studying the atmosphere that gathering data from only ground-level sites significantly limited understanding. In the late 1700s the only data for conditions at high altitudes came from observations made in the mountains. In 1752 Benjamin Franklin, using a kite, made his famous discovery that lightning is an electrical discharge. Not many years later kites were

being used to observe temperatures above the surface. In the late eighteenth century, manned balloons were used in an attempt to investigate the composition and properties of the "upper" atmosphere. Although several manned ascents were attempted over the years, they were dangerous undertakings and seldom exceeded heights of 5 to 8 kilometers (3 to 5 miles). Unmanned balloons, on the other hand, could rise to higher altitudes, but there was no assurance that the instruments carried aloft would be recovered. Notwithstanding the difficulties and dangers, considerable data were gathered on the nature of the air above.

Today balloons continue to play a significant role in the systematic investigation of the atmosphere. Giant balloons are launched regularly, primarily for research purposes (Figure 1–14a). Such balloons can stay aloft for extended periods and represent an important means of carrying monitoring instruments into the region of the atmosphere known as the stratosphere.

Since the late 1920s balloons have carried aloft **radiosondes**. These lightweight packages of instruments are fitted with radio transmitters that send back data on temperature, pressure, and relative humidity in the lower portions of the atmosphere (Figure 1–14b). Radiosondes are sent aloft twice daily from an extensive network of stations worldwide. The data that they supply are essential for making accurate weather forecasts.

Other important means for exploring the atmosphere include rockets and airplanes. After World War II, rockets revolutionized the study of the upper atmosphere. Prior to this time, knowledge of the atmosphere beyond about 30 kilometers (20 miles) came almost exclusively from indirect ground-based measurements. Airplanes also play a significant role in atmospheric studies. High-flying aircraft are capable of reaching portions of the stratosphere. Others are designed to measure such relatively small-scale yet complex phenomena as cloud systems or to fly directly into hurricanes to monitor their current state of development.

Among the methods of studying the atmosphere that are best known to the general public and most useful to atmospheric scientists are weather radar and satellites. Today, when we watch a television weather report, we expect to see satellite images that show moving cloud patterns and radar displays that depict the intensity and regional extent of precipitation (Figure 1–15). Recent technological advances greatly enhance the value of weather radar for the purpose of storm detection, warning, and research. Meteorological satellites give us a perspective of the atmosphere that is unique and invaluable. For example, they provide images that allow us to

(a)

(b)

Figure 1–14 Exploring the atmosphere using balloons. (a) Giant balloons such as this one are filled with helium and carry instrument packages high into the atmosphere. (b) A lightweight package of instruments, the radiosonde, is carried aloft by a small weather balloon. *(a) Courtesy of University Corporation for Atmospheric Research/National Center for Atmospheric Research/National Science Foundation; (b) (AP Photo)*

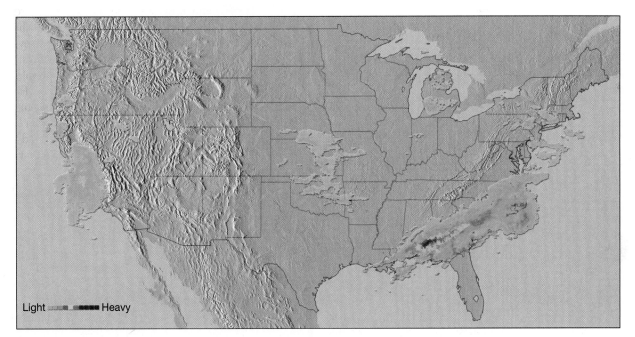

Light ▨▨▨ ▨▨▨ Heavy

Figure 1–15 Radar images such as this composite of the United States show the distribution and intensity of precipitation.

Box 1–2 The Nature of Scientific Inquiry

All science is based on the assumption that the natural world behaves in a consistent and predictable manner. The overall goal of science is to discover the underlying patterns in the natural world and then to use this knowledge to make predictions about what should or should not be expected to happen given certain facts or circumstances.

The development of new scientific knowledge involves some basic, logical processes that are universally accepted. To determine what is occurring in the natural world, scientists collect scientific facts through observation and measurement (Figure 1–B). These data are essential to science and serve as the springboard for the development of scientific theories.

Hypothesis

Once facts have been gathered and principles have been formulated to describe a natural phenomenon, investigators try to explain how or why things happen in the manner observed. They can do this by constructing a tentative (or untested) explanation, which we call a scientific *hypothesis.* Often, several hypotheses are advanced to explain the same facts. Next, scientists think about what will occur or be observed if a hypothesis is correct and devise ways or methods to test the accuracy of predictions drawn from the hypothesis.

If a hypothesis cannot be tested, it is not scientifically useful, no matter how interesting it might seem. Testing usually involves making observations, developing models, and performing experiments. What if test results do not turn out as expected? One possibility is that there were errors in the observations or experiments. Of course, another possibility is that the hypothesis is not valid. Before rejecting the hypothesis, the tests may be

Figure1–B An automated weather station. The meteorologist is downloading the accumulated data into a laptop computer. *(Photo by David Parker/Science Photo Library/Photo Researchers, Inc.)*

repeated or new tests may be devised. The more tests, the better. The history of science is littered with discarded hypotheses. One of the best known is the idea that Earth was at the center of the universe, a proposal

study the distribution of clouds and the circulation patterns that they reveal (Figure 1–16). Moreover, they let us see the structure and determine the speed of weather systems over the oceans and other regions where observations are scanty or nonexistent. One obvious benefit is that storms can be detected early and tracked with great precision.

Height and Structure of the Atmosphere

To say that the atmosphere begins at Earth's surface and extends upward is obvious. However, where does the atmosphere end and outer space begin? There is no sharp boundary; the atmosphere rapidly thins as you

that was supported by the apparent daily motion of the Sun, Moon, and stars around Earth.

Theory

When a hypothesis has survived extensive scrutiny and when competing hypotheses have been eliminated, a hypothesis may be elevated to the status of a scientific *theory*. In everyday language we may say "that's only a theory." But a scientific theory is a well-tested and widely accepted view that scientists agree best explains certain observable facts. It is not enough for scientific theories to fit only the data that are already at hand. Theories must also fit additional observations that were not used to formulate them in the first place. Put another way, theories should have predictive power.

Scientific theories, like scientific hypotheses, are accepted only provisionally. It is always possible that a theory that has withstood previous testing may eventually be disproved. As theories survive more testing, they are regarded with higher levels of confidence. Theories that have withstood extensive testing, as, for example, the theory of plate tectonics and the theory of evolution, are held with a very high degree of confidence.

Scientific Methods

The process just described, in which scientists gather facts through observations and formulate scientific hypotheses and theories, is called the *scientific method*. Contrary to popular belief, the scientific method is not a standard recipe that scientists apply in a routine manner

to unravel the secrets of our natural world. Rather, it is an endeavor that involves creativity and insight. Rutherford and Ahlgren put it this way: "Inventing hypotheses or theories to imagine how the world works and then figuring out how they can be put to the test of reality is as creative as writing poetry, composing music, or designing skyscrapers."[*]

There is not a fixed path that scientists always follow that leads unerringly to scientific knowledge. Nevertheless, many scientific investigations involve the following steps: (1) the collection of scientific facts through observation and measurement; (2) the development of one or more working hypotheses to explain these facts; (3) development of observations and experiments to test the hypothesis; and (4) the acceptance, modification, or rejection of the hypothesis based on extensive testing. Other scientific discoveries represent purely theoretical ideas, which stand up to extensive examination. Still other scientific advancements have been made when a totally unexpected happening occurred during an experiment. These serendipitous discoveries are more than pure luck; for as Louis Pasteur said, "In the field of observation, chance favors only the prepared mind." Scientific knowledge is acquired through several avenues, so it might be best to describe the nature of scientific inquiry as the methods of science rather than the scientific method.

[*]F. James Rutherford and Andrew Ahlgren, *Science for All Americans* (New York: Oxford University Press, 1990), p. 7.

---

travel away from Earth, until there are too few gas molecules to detect.

Pressure Changes

To understand the vertical extent of the atmosphere, let us examine the changes in atmospheric pressure with height. Atmospheric pressure is simply the weight of the air above. At sea level the average pressure is slightly more than 1000 millibars. This corresponds to a weight of slightly more than 1 kilogram per square centimeter (14.7 pounds per square inch). Obviously the pressure at higher altitudes is less (Figure 1–17).

One-half of the atmosphere lies below an altitude of 5.6 kilometers (3.5 miles). At about 16 kilometers (10 miles) 90 percent of the atmosphere has been traversed, and above 100 kilometers (62 miles) only

0.00003 percent of all the gases composing the atmosphere remain.

At an altitude of 100 kilometers the atmosphere is so thin that the density of air is less than could be found in the most perfect artificial vacuum at the surface. Nevertheless, the atmosphere continues to even greater heights. The truly rarefied nature of the outer atmosphere is described very well by Richard Craig:

The earth's outermost atmosphere, the part above a few hundred kilometers, is a region of extremely low density. Near sea level, the number of atoms and molecules in a cubic centimeter of air is about 2×10^{19}; near 600 km, it is only about 2×10^7, which is the sea-level value divided by a million million. At sea level, an atom or molecule can be expected, on the average, to move about 7×10^{-6} cm before colliding with

Figure 1–16 Satellites are invaluable tools for tracking storms and gathering atmospheric data. This is a satellite image of a storm over the British isles. (Courtesy of European Space Agency/Science Photo Library/Photo Researchers, Inc.)

another particle; at the 600-km level, this distance, called the "mean free path," is about 10 km. Near sea level, an atom or molecule, on the average, undergoes about 7×10^9 such collisions each second; near 600 km, this number is reduced to about 1 each minute.[°]

The graphic portrayal of pressure data (Figure 1–17) shows that the rate of pressure decrease is not constant. Rather, pressure decreases at a decreasing rate with an increase in altitude until, beyond an altitude of about 35 kilometers, the decrease is slight.

Put another way, data illustrate that air is highly compressible—that is, it expands with decreasing pressure and becomes compressed with increasing pressure. Consequently, traces of our atmosphere extend for thousands of kilometers beyond Earth's surface. Thus, to say where the atmosphere ends and outer space begins is arbitrary and, to a large extent, depends on what phenomenon one is studying. It is apparent that there is no sharp boundary.

In summary, data on vertical pressure changes reveal that the vast bulk of the gases making up the atmosphere is very near Earth's surface and that the gases gradually merge with the emptiness of space. When compared with the size of the solid Earth, with its radius of about 6400 kilometers (4000 miles), the envelope of air surrounding our planet is indeed very shallow.

[°]Richard Craig. The Edge of Space: Exploring the Upper Atmosphere (New York: Doubleday & Company, Inc. 1968), p. 130.

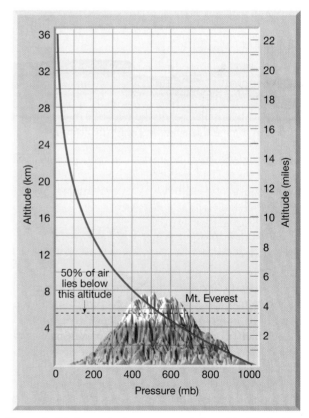

Figure 1–17 Atmospheric pressure variation with altitude. The rate of pressure decrease with an increase in altitude is not constant. Rather, pressure decreases rapidly near Earth's surface and more gradually at greater heights.

Figure 1–18 Temperatures drop with an increase in altitude in the troposphere. Therefore, it is possible to have snow on a mountaintop and warmer, snow-free lowlands below. Mount Kerkeslin, Jasper National Park, Alberta, Canada. *(Photo by Carr Clifton/Minden Pictures)*

Temperature Changes

By the early twentieth century much had been learned about the lower atmosphere. The upper atmosphere was partly known from indirect methods. Data from balloons and kites had revealed that the air temperature dropped with increasing height above Earth's surface. This phenomenon is felt by anyone who has climbed a high mountain, and is obvious in pictures of snow-capped mountaintops rising above snow-free lowlands (Figure 1–18).

Although measurements had not been taken above a height of about 10 kilometers (6 miles), scientists believed that the temperature continued to decline with height to a value of absolute zero (–273°C) at the outer edge of the atmosphere. In 1902, however, the French scientist Leon Philippe Teisserenc de Bort refuted the notion that temperature decreases continuously with an increase in altitude. In studying the results of more than 200 balloon launchings, Teisserenc de Bort found that the temperature stopped decreasing and leveled off at an altitude between 8 and 12 kilometers (5 and 7.5 miles). This surprising discovery was at first doubted, but subsequent data-gathering confirmed his findings. Later, through the use of radiosondes and rocket-sounding techniques, the temperature structure of the atmosphere

up to great heights became clear. Today the atmosphere is divided vertically into four layers on the basis of temperature (Figure 1–19).

Troposphere The bottom layer in which we live, where temperature decreases with an increase in altitude, is the **troposphere**. The term was coined in 1908 by Teisserenc de Bort and literally means the region where air "turns over," a reference to the appreciable vertical mixing of air in this lowermost zone.

The temperature decrease in the troposphere is called the **environmental lapse rate**. Its average value is 6.5°C per kilometer (3.5°F per 1000 feet), a figure known as the *normal lapse rate*. It needs to be emphasized, however, that the environmental lapse rate is not a constant, but rather can be highly variable and must be regularly measured using radiosondes. It can vary during the course of a day with fluctuations of the weather, as well as seasonally and from place to place. Sometimes shallow layers where temperatures actually increase with height are observed in the troposphere. When such a reversal occurs, a *temperature inversion* is said to exist.[*]

The temperature decrease continues to an average height of about 12 kilometers (7.5 miles). Yet the thickness of the troposphere is not the same everywhere. It reaches heights in excess of 16 kilometers (10 miles) in the tropics, but in polar regions, it is more subdued, extending to 9 kilometers (5.5 miles) or less (Figure 1–20). Warm surface temperatures and highly developed thermal mixing are responsible for the greater vertical extent of the troposphere near the equator. As a result, the environmental lapse rate extends to great heights; and despite relatively high surface temperatures below, the lowest tropospheric temperatures are found aloft in the tropics and not at the poles.

The troposphere is the chief focus of meteorologists because it is in this layer that essentially all important weather phenomena occur. Almost all clouds and certainly all precipitation, as well as all our violent storms, are born in this lowermost layer of the atmosphere. There should be little wonder why the troposphere is often called the "weather sphere."

Stratosphere Beyond the troposphere lies the **stratosphere**; the boundary between the troposphere and the stratosphere is known as the **tropopause**. Below the tropopause, atmospheric properties are readily transferred by large-scale turbulence and mixing, but above it, in the stratosphere, they are not. In the stratosphere, the temperature at first remains nearly

[*]Temperature inversions are described in greater detail in Chapter 13.

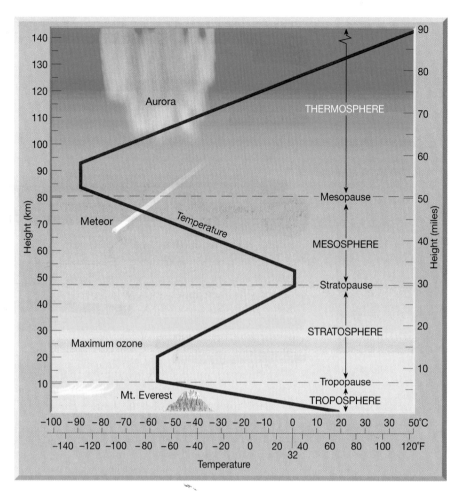

Figure 1–19 Thermal structure of the atmosphere.

constant to a height of about 20 kilometers (12 miles) before it begins a rather sharp increase that continues until the **stratopause** is encountered at a height of about 50 kilometers (30 miles) above Earth's surface. Higher temperatures occur in the stratosphere because it is in this layer that the atmosphere's ozone is concentrated. Recall that ozone absorbs ultraviolet radiation from the Sun. Consequently, the stratosphere is heated. Although the maximum ozone concentration exists between 15 and 30 kilometers (9 and 19 miles), the smaller amounts of ozone above this height range absorb enough UV energy to cause the higher observed temperatures (see Figure 1–19).

Mesosphere In the third layer, the **mesosphere**, temperatures again decrease with height until at the **mesopause**, some 80 kilometers (50 miles) above the surface, the temperature approaches –90°C (–130°F).

Thermosphere The fourth layer extends outward from the mesopause and has no well-defined upper limit. It is the **thermosphere**, a layer that contains only a minute fraction of the atmosphere's mass. In the extremely rarefied air of this outermost layer, temperatures again increase, owing to the absorption of very shortwave, high-energy solar radiation by atoms of oxygen and nitrogen.

Temperatures rise to extremely high values of more than 1000°C (1800°F) in the thermosphere. But such temperatures are not comparable to those experienced near Earth's surface. Temperature is defined in terms of the average speed at which molecules move. Because the gases of the thermosphere are moving at very high speeds, the temperature is very high. But the gases are so sparse that collectively they possess only an insignificant quantity of heat. For this reason, the temperature of a satellite orbiting Earth in the thermosphere is determined chiefly by the amount of solar radiation it absorbs and not by the high temperature of the almost nonexistent surrounding air. If an astronaut inside were to expose his or her hand, the air in this layer would not feel hot.

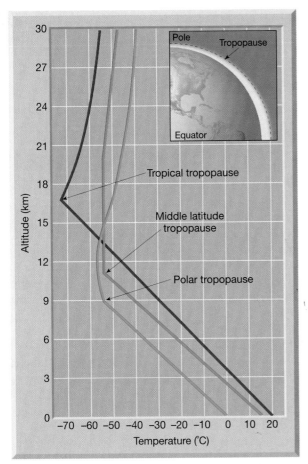

Figure 1–20 Differences in height of the tropopause. The variation in the height of the tropopause, as shown on the small inset diagram, is greatly exaggerated.

Vertical Variations in Composition

In addition to the layers defined by vertical variations in temperature, other layers, or zones, are also recognized in the atmosphere. Based on composition, the atmosphere is often divided into two layers: the homosphere and the heterosphere. From Earth's surface to an altitude of about 80 kilometers (50 miles), the makeup of the air is uniform in terms of the proportions of its component gases. That is, the composition is the same as that shown earlier in Table 1–2. (To compare this with Mars, see Box 1–3.) This lower uniform layer is termed the *homosphere*, the zone of homogeneous composition.

In contrast, the rather tenuous atmosphere above 80 kilometers is not uniform. Because it has a heterogeneous composition, the term *heterosphere* is used. Here the gases are arranged into four roughly spherical shells, each with a distinctive composition. The lowermost layer is dominated by molecular nitrogen (N_2); next, a layer of atomic oxygen (O) is encountered, followed by a layer dominated by helium (He) atoms, and finally a region consisting of hydrogen (H) atoms. The stratified nature of the gases making up the heterosphere varies according to their weights. Molecular nitrogen is the heaviest, and so it is lowest. The lightest gas, hydrogen, is outermost.

The Ionosphere

Located in the altitude range between 80 to 400 kilometers (50 to 250 miles), and thus coinciding with the lower portions of the thermosphere and heterosphere, is an electrically charged layer known as the **ionosphere**. Here molecules of nitrogen and atoms of oxygen are readily ionized as they absorb high-energy shortwave solar energy. In this process, each affected molecule or atom loses one or more electrons and becomes a positively charged ion, and the electrons are set free to travel as electric currents.

Although ionization occurs at heights as great as 1000 kilometers (620 miles) and extends as low as perhaps 50 kilometers (30 miles), positively charged ions and negative electrons are most dense in the range of 80 to 400 kilometers (50 to 250 miles). The concentration of ions is not great below this zone because much of the short-wavelength radiation needed for ionization has already been depleted. In addition, the atmospheric density at this level results in a large percentage of free electrons being swiftly captured by positively charged ions. Beyond the 400-kilometer (250-mile) upward limit of the ionosphere, the concentration of ions is low because of the extremely low density of the air. Because so few molecules and atoms are present, relatively few ions and free electrons can be produced.

The electrical structure of the ionosphere is not uniform. It consists of three layers of varying ion density. From bottom to top, these layers are called the D, E, and F layers, respectively. Because the production of ions requires direct solar radiation, the concentration of charged particles changes from day to night, particularly in the D and E zones. That is, these layers weaken and disappear at night and reappear during the day. The uppermost, or F layer, on the other hand, is present both day and night. The density of the atmosphere in this layer is very low, and positive ions and electrons do not meet and recombine as rapidly as they do at lesser heights, where density is higher. Consequently, the concentration of ions and electrons in the F layer does not change rapidly, and the layer, although weak, remains through the night.

As best we can tell, the ionosphere has little impact on our daily weather. But this layer of the atmosphere is the site of one of nature's most interesting spectacles, the auroras (Figure 1–21). The **aurora borealis** (northern lights) and its Southern Hemisphere counterpart, the

Box 1–3 Martian Weather

Mars has probably evoked more interest and speculation among people than any other planet. This curiosity stems mainly from this planet's accessibility to observation. All other planets within telescopic range have their surfaces hidden by clouds, except for Mercury, whose nearness to the Sun makes viewing difficult. Through a telescope Mars appears as a reddish ball interrupted by some dark regions that change in intensity during the Martian year. Huge dust storms periodically obscure the planet's surface, and brilliant white polar caps of frozen carbon dioxide (dry ice) seasonally advance and retreat during the Martian year, which lasts 687 Earth days.

Figure 1–C is a photograph that looks as though it could have been taken on Earth. However, the bleak dune- and boulder-covered surface is not a desert scene from some remote part of our planet, but rather is a view of the surface of Mars. In addition to dunes, Mars has other Earth-like surface features, such as volcanic peaks and canyons. Other similarities include a day that is just slightly longer than an Earth day (24.6 hours) and seasons that occur because the axis is tilted about the same as Earth's.

Although some people may dream of someday visiting or even inhabiting this planetary neighbor, it is clearly a forbidding place. For example, although afternoon maximum temperatures at the equator may exceed 10°C (50°F), nighttime minimums can plunge to –50°C (–58°F). During the Martian winter, temperatures can fall to –80°C (–112°F) in the moderate middle latitudes and approach –130°C (–202°F) at the poles.

The air on Mars is mostly carbon dioxide (about 95 percent), with nitrogen and argon accounting for most of the remainder. One amazing feature of the Martian climate is the dramatic change in atmospheric pressure that occurs with the seasons. When winter comes to the polar regions, temperatures are so low that between 25 and 30 percent of the atmospheric carbon dioxide turns directly into a solid and collects at the surface, causing the winter polar cap to grow. Because the amount of carbon dioxide in the Martian air is reduced, the force exerted by the weight of the air (that is, the air pressure) drops. When temperatures rise in spring, the carbon dioxide returns to the atmosphere as a gas and air pressure rises.

Dust storms are familiar weather events on Mars. It is not unusual for the planet to have more than 100 small storms every Martian year. Most last only a few days and can sometimes be detected by telescope as bright spots on the planet's reddish disk. For these storms to develop, wind speeds must exceed about 200 kilometers (125 miles) per hour. It takes such high winds to create these events because the density of the Martian atmosphere is so low. Therefore, tornado- or hurricane-like winds are necessary to create what a strong breeze can do on Earth. Occasionally, enormous dust storms can obscure large portions of the Martian surface.

Figure 1–C Like the other planets in our solar system, the atmosphere of Mars is very different from that of Earth. The Martian atmosphere is only 1 percent as dense as that of Earth and is composed primarily of carbon dioxide with a very small amount of water vapor. Although the atmosphere of Mars is very thin, extensive dust storms do occur, sometimes creating dunes that resemble those on Earth. *(Courtesy of NASA)*

Figure 1–21 Aurora borealis (northern lights) as seen from Alaska. The same phenomenon occurs toward the South Pole, where it is called the aurora australis (southern lights). *(Photo by Michio Hoshino/Minden Pictures)*

aurora australis (southern lights), appear in a wide variety of forms. Sometimes the displays consist of vertical streamers in which there can be considerable movement. At other times the auroras appear as a series of luminous expanding arcs or as a quiet glow that has an almost foglike quality.

The occurrence of auroral displays is closely correlated in time with solar-flare activity and, in geographic location, with Earth's magnetic poles. Solar flares are massive magnetic storms on the Sun that emit enormous amounts of energy and great quantities of fast-moving atomic particles. As the clouds of protons and electrons from the solar storm approach Earth, they are captured by its magnetic field, which in turn guides them toward the magnetic poles. Then, as the ions impinge on the ionosphere, they energize the atoms of oxygen and molecules of nitrogen and cause them to emit light—the glow of the auroras. Because the occurrence of solar flares is closely correlated with sunspot activity, auroral displays increase conspicuously at times when sunspots are most numerous.

Chapter Summary

- *Weather* refers to the state of the atmosphere at a given time and place. It is constantly changing, sometimes from hour to hour and other times from day to day. *Climate* is an aggregate of weather conditions, the sum of all statistical weather information that helps describe a place or region. The nature of both weather and climate is expressed in terms of the same basic *elements*, those quantities or properties measured regularly. The most important elements are (1) air temperature, (2) humidity, (3) type and amount of cloudiness, (4) type and amount of precipitation, (5) air pressure, and (6) the speed and direction of the wind.

- Earth's four spheres include the *atmosphere* (gaseous envelope), the *lithosphere* (solid Earth), the *hydrosphere* (water portion), and the *biosphere* (life). Each sphere is composed of many interrelated parts and is intertwined with all other spheres. The energy exchanges that continually occur between the atmosphere and Earth's surface, and between the atmosphere and space, produce the effects we call weather.

- A *system* is a group of interacting, interdependent parts that form a complex whole. The *Earth system* involves the intricate and continuous interactions among the lithosphere, hydrosphere, atmosphere, and biosphere. The two primary sources of energy that power this system are (1) *solar energy* that drives the external processes that occur at, or above, Earth's surface, and (2) *Earth's interior*, heat remaining from when the planet formed and heat that is continuously generated by radioactive decay. The *carbon cycle* is one example that illustrates the movement of material and energy from one sphere to another in the Earth system.

- Air is a mixture of many discrete gases and its composition varies from time to time and place to place. After water vapor, dust, and other variable components are removed, two gases, *nitrogen* and *oxygen*, make up 99 percent of the volume of the remaining clean, dry air. *Carbon dioxide*, although present in only minute amounts (0.036 percent), is an efficient absorber of energy emitted by Earth and thus influences the heating of the atmosphere. Because of the rising level of carbon dioxide in the atmosphere during the past century attributed to the burning of ever increasing quantities of fossil fuels, it is possible that a warming of the lower atmosphere will trigger global climate change.

- The variable components of air include *water vapor, dust particles*, and *ozone*. Like carbon dioxide, water vapor can absorb heat given off by Earth as well as some solar energy. When water vapor changes from one state to another, it absorbs or releases heat. In the atmosphere, water vapor transports this *latent* ("hidden") *heat* from one place to another, and it is the energy source that helps drive many storms. *Aerosols* (tiny solid and liquid particles) are meteorologically important because these often invisible particles act as surfaces on which water can condense and are also absorbers and reflectors of incoming solar radiation. *Ozone*, a form of oxygen that combines three oxygen atoms into each molecule (O_3), is a gas concentrated in the 10 to 50 kilometer height range in the atmosphere that absorbs the potentially harmful ultraviolet (UV) radiation from the Sun. Over the past half century, people have placed Earth's ozone layer in jeopardy by polluting the atmosphere with chlorofluorocarbons (CFCs) which remove some of the gas. Ozone concentrations take an especially sharp drop over Antarctica during the Southern Hemisphere spring (September and October). Furthermore, scientists have also discovered a similar but smaller ozone thinning near the North Pole during spring and early summer. Because ultraviolet radiation is known to produce skin cancer, ozone depletion seriously affects human health, especially among fair-skinned people and those who spend considerable time in the Sun. The *Montreal Protocol*, concluded under the auspices of the United Nations, represents a positive international response to the ozone problem.

- *Balloons* play a significant role in the systematic investigation of the atmosphere by carrying *radiosondes* (lightweight packages of instruments that send back data on temperature, pressure, and relative humidity) into the lower atmosphere. Rockets, airplanes, satellites, and weather radar are also among the methods used to study the atmosphere.

- No sharp boundary to the upper atmosphere exists. The atmosphere simply thins as you travel away from Earth until there are too few gas molecules to detect. The change that occurs in atmospheric pressure (the weight of the air above) depicts the vertical extent of the atmosphere. One-half of the atmosphere lies below an altitude of 5.6 kilometers (3.5 miles), and 90 percent lies below 16 kilometers (10 miles). Traces of atmosphere extend for thousands of kilometers beyond Earth's surface.

- Atmospheric temperature drops with increasing height above Earth's surface. Using temperature as the basis, the atmosphere is divided into four layers. The temperature decrease in the *troposphere*, the bottom layer in which we live, is called the *environmental lapse rate*. Its average value is 6.5°C per kilometer, a figure known as the *normal lapse rate*. The environmental lapse rate is not a constant and must be regularly measured using radiosondes. A *temperature inversion*, where temperatures increase with height, is sometimes observed in shallow layers in the troposphere. The thickness of the troposphere is generally greater in the tropics than in polar regions. Essentially all important weather phenomena occur in the troposphere. Beyond the troposphere lies the *stratosphere*; the boundary between the troposphere and stratosphere is known as the *tropopause*. In the stratosphere, the temperature at first remains constant to a height of about 20 kilometers (12 miles) before it begins a sharp increase due to the absorption of ultraviolet radiation from the Sun by ozone. The temperatures continue to increase until the *stratopause* is encountered at a height of about 50 kilometers (30 miles). In the *mesosphere*, the third layer, temperatures again decrease with height until the *mesopause*, some 80 kilometers (50 miles) above the surface. The fourth layer, the *thermosphere*, with no well-defined upper limit, consists of extremely rarefied air. Temperatures here increase with an increase in altitude.

- Besides layers defined by vertical variations in temperature, the atmosphere is often divided into two layers based on composition. The *homosphere* (zone of homogeneous composition), from Earth's surface to an altitude of about 80 kilometers (50 miles), consists of air that is uniform in terms of the proportions of its component gases. Above 80 kilometers, the *heterosphere* (zone of heterogenous composition) consists of gases arranged into four roughly spherical shells, each with a distinctive composition. With increasing altitudes, the four layers consist of molecular nitrogen (N_2), atomic oxygen (O), helium (He) atoms, and hydrogen (H) atoms respectively. The stratified nature of the gases in the heterosphere varies according to their weights, with the outermost gas, hydrogen, being the lightest.

- Occurring in the altitude range between 80 and 400 kilometers (50–250 miles) is an electrically charged layer known as the *ionosphere*. Here molecules of nitrogen and atoms of oxygen are readily ionized as they absorb high-energy, shortwave solar energy. Three layers of varying ion density make up the ionosphere. Auroras (the *aurora borealis*, northern lights, and its Southern Hemisphere counterpart the *aurora australis*, southern lights) occur within the ionosphere. Auroras form as clouds of protons and electrons ejected from the Sun during solar-flare activity enter the atmosphere near Earth's magnetic poles and energize the atoms of oxygen and molecules of nitrogen, causing them to emit light—the glow of the auroras.

Vocabulary Review

Review your understanding of important terms in this chapter by defining and explaining the importance of each term listed here. Terms are listed in alphabetical order. Page references indicate where the term is introduced and defined.

aerosols (p. 11)
air (p. 9)
atmosphere (p. 6)
aurora australis (p. 23)
aurora borealis (p. 21)
biosphere (p. 7)
climate (p. 3)
elements of weather and climate (p. 4)
environmental lapse rate (p. 19)
hydrosphere (p. 7)
ionosphere (p. 21)

lithosphere (p. 7)
mesopause (p. 20)
mesosphere (p. 20)
ozone (p. 11)
radiosonde (p. 14)
stratopause (p. 20)
stratosphere (p. 19)
thermosphere (p. 20)
tropopause (p. 19)
troposphere (p. 19)
weather (p. 2)

Review Questions

1. Distinguish between the terms "weather" and "climate."
2. The following statements refer to either weather or climate. On the basis of your answer to question 1, determine which statements refer to weather and which refer to climate. (*Note:* One statement includes aspects of both weather and climate.)
   a. The baseball game was rained out today.
   b. January is Peoria's coldest month.
   c. North Africa is a desert.
   d. The high this afternoon was 25°C.
   e. Last evening a tornado ripped through Canton.
   f. I am moving to southern Arizona because it is warm and sunny.
   g. The highest temperature ever recorded at this station is 43°C.
   h. Thursday's low of –20°C is the coldest temperature ever recorded for that city.
   i. It is partly cloudy.
3. What are the basic elements of weather and climate?
4. List and briefly describe the four "spheres" that constitute our environment.

5. What are the major components of clean, dry air?
6. Outline the stages in the formation of Earth's atmosphere. (See Box 1–1)
7. What is responsible for the increasing carbon dioxide content of the air? What is one possible effect of increased carbon dioxide in the atmosphere?
8. Why are water vapor and dust important constituents of our atmosphere?
9. a. Why is ozone important to life on Earth?
   b. What are CFCs, and what is their connection to the ozone problem?
   c. What are the effects on human health of a decrease in the stratosphere's ozone?
10. What is a radiosonde?
11. How is a scientific hypothesis different from a scientific theory? (See Box 1–2.)
12. The atmosphere is divided vertically into four layers on the basis of temperature. List the names of these layers and their boundaries in order (from lowest to highest), and list as many characteristics of each as you can.

13. Why does the temperature increase in the stratosphere?
14. Why are temperatures in the thermosphere not strictly comparable to those experienced near Earth's surface?
15. Distinguish between the homosphere and the heterosphere.
16. What is the primary cause of auroral displays?

Problems

1. Refer to the newspaper-type weather map in Figure 1–2 to answer the following:
   a. Estimate the predicted high temperatures in central New York State and the northwest corner of Arizona.
   b. Where is the coldest area on the weather map? Where is the warmest?
   c. On this weather map H stands for the center of a region of high pressure. Does it appear as though high pressure is associated with precipitation or fair weather?
   d. Which is warmer—central Texas or central Maine? Would you "normally" expect this to be the case?

2. Refer to the graph in Figure 1–4 to answer the following questions about temperatures in New York City:
   a. What is the approximate average daily high temperature in January? In July?
   b. Approximately what are the highest and lowest temperatures ever recorded?

3. Refer to the graph in Figure 1–17 to answer the following:
   a. Approximately how much does the air pressure drop (in millibars) between the surface and 4 kilometers? (Use a surface pressure of 1000 mb.)
   b. How much does the pressure drop between 4 and 8 kilometers?
   c. Based on your answers to parts a and b, with an increase in altitude, air pressure decreases at a(n) (constant, increasing, decreasing) rate. Underline the correct answer.

4. If the temperature at sea level were 23°C, what would the air temperature be at a height of 2 kilometers under average conditions?

5. Use the graph of the atmosphere's thermal structure (Figure 1–19) to answer the following:
   a. What is the approximate height and temperature of the stratopause?
   b. At what altitude is the temperature lowest? What is the temperature at that height?

6. Answer the following questions by examining the graph in Figure 1–20.
   a. In which one of the three regions (tropics, middle latitudes, poles) is the *surface* temperature lowest?
   b. In which region is the tropopause encountered at the lowest altitude? The highest? What are the altitudes and temperatures of the tropopause in those regions?

7. a. On a spring day a middle-latitude city (about 40° north latitude) has a surface (sea-level) temperature of 10°C. If vertical soundings reveal a nearly constant environmental lapse rate of 6.5°C per kilometer and a temperature at the tropopause of –55°C, what is the height of the tropopause?
   b. On the same spring day a station near the equator has a surface temperature of 25°C, 15°C higher than the middle-latitude city mentioned in part a. Vertical soundings reveal an environmental lapse rate of 6.5°C per kilometer and indicate that the tropopause is encountered at 16 kilometers. What is the air temperature at the tropopause?

Atmospheric Science Online

The following are informative and interesting Internet sites that address topics related to those presented in the chapter:

Atmospheric Chemistry (Ozone) and Structure (NASA):
- **http://daac.gsfc.nasa.gov/CAMPAIGN_DOCS/ ATM_CHEM/ozone_atmosphere.html**

National Weather Service (NOAA):
- **http://www.nws.noaa.gov/**

For direct links to these sites and others, chapter objectives and reviews, quiz questions, and topical investigations that utilize Web resources, visit *The Atmosphere, Eighth Edition* Home Page at:
- **http://www.prenhall.com/lutgens**

Heating Earth's Surface and Atmosphere

Solar radiation provides more than 99.9 percent of the energy that heats Earth's surface. *(Photo by Paul Harris/Tony Stone Images)*

From our experiences, we know that the Sun's rays feel hotter on a clear day than on an overcast day. After taking a barefoot walk on a sunny day, we realize that city pavement becomes much hotter than a grassy boulevard. A picture of a snow-capped mountain reminds us that temperature decreases with altitude. And we know that the fury of winter is always replaced by the newness of spring. You may not know, however, that these occurrences are manifestations of the same phenomenon that causes the blue color of the sky and the red color of a brilliant sunset. All such common occurrences are a result of the interaction of solar energy with Earth's atmosphere and its land–sea surface. That is the essence of this chapter.

Earth–Sun Relationships

Earth intercepts only a minute percentage of the energy given off by the Sun—less than one two-billionth. This may seem an insignificant amount until we realize that it is several hundred thousand times the electrical generating capacity of the United States. Solar radiation, in fact, represents more than 99.9 percent of the energy that heats our planet.

Solar energy is not distributed equally over Earth's land–sea surface. The amount of energy received varies with latitude, time of day, and season of the year. Contrasting images of polar bears on ice rafts and palm trees along a remote tropical beach serve to illustrate the extremes. It is the unequal heating of Earth that creates winds and drives the ocean's currents. These movements, in turn, transport heat from the tropics toward the poles in an unending attempt to balance energy inequalities. The consequences of these processes are the phenomena we call weather. If the Sun were "turned off," global winds and ocean currents would quickly cease. Yet as long as the Sun shines, the winds *will* blow and weather *will* persist. So to understand how the atmosphere's dynamic weather machine works, we must first know why different latitudes receive varying quantities of solar energy and why the amount of solar energy changes to produce the seasons. As we shall see, the variations in solar heating are caused by the motions of Earth relative to the Sun and by variations in Earth's land–sea surface.

Earth's Motions

Earth has two principal motions—rotation and revolution. **Rotation** is the spinning of Earth about its axis that produces the daily cycle of daylight and darkness. In the following chapter, we will examine the effects that this daily variation in solar heating has on the atmosphere.

The other motion of Earth, **revolution**, refers to its movement in orbit around the Sun. Hundreds of years

ago, most people believed that Earth was stationary in space. The reasoning was that, if Earth were moving, people would feel the movement of the wind rushing past them. Today we know that Earth is traveling at nearly 113,000 kilometers (70,000 miles) per hour in an elliptical orbit about the Sun. Why don't we feel the air rushing past us? The answer is that the atmosphere, bound by gravity to Earth, is carried along at the same speed as Earth.

The distance between Earth and Sun averages about 150 million kilometers (93 million miles). Because Earth's orbit is not perfectly circular, however, the distance varies during the course of a year. Each year, on about January 3, our planet is about 147 million kilometers (91 million miles) from the Sun, closer than at any other time. This position is called the **perihelion**. About six months later, on July 4, Earth is about 152 million kilometers (94 million miles) from the Sun, farther away than at any other time. This position is called the **aphelion**. Although Earth is closest to the Sun and thus receives more energy in January than in July, this difference plays only a minor role in producing seasonal temperature variations. As proof, consider that Earth is closest to the Sun during the cold Northern Hemisphere winter.

The Seasons

We know that it is colder in winter than in summer, but if variations in solar distance do not cause this seasonal temperature change, what does? We adjust to the continuous change in the duration of daylight that occurs throughout the year by planning our outdoor activities accordingly. The gradual but significant *change in day length* certainly accounts for some of the difference we notice between summer and winter. Furthermore, a gradual change in the angle of the noon Sun above the horizon is quite noticeable (Figure 2–1). At midsummer, the noon

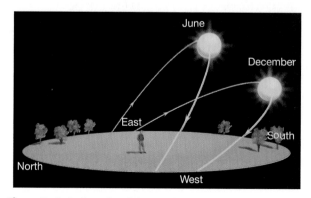

Figure 2–1 Daily paths of the Sun for June and December for an observer in the middle latitudes in the Northern Hemisphere. Notice that the angle of the Sun above the horizon is much greater in the summer than in the winter.

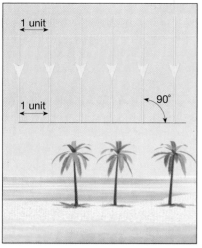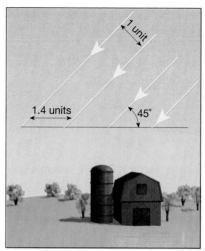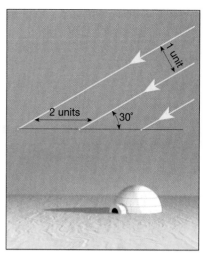

Figure 2–2 Changes in the Sun's angle causes variations in the amount of solar energy reaching Earth's surface. The higher the angle, the more intense the solar radiation.

Sun is seen high above the horizon. But as summer gives way to autumn, the noon Sun appears lower in the sky and sunset occurs earlier each evening.

The seasonal variation in the angle of the Sun above the horizon affects the amount of energy received at Earth's surface in two ways. First, when the Sun is directly overhead (at a 90° angle), the solar rays are most concentrated. The lower the angle, the more spread out and less intense is the solar radiation that reaches the surface. This idea is illustrated in Figure 2–2. You have probably experienced this when using a flashlight. If the beam strikes a surface perpendicularly, a small intense spot is produced. When the flashlight beam strikes the object at an oblique angle, however, the area illuminated is larger—and dimmer.

Second, and of lesser importance, the angle of the Sun determines the thickness of atmosphere that the rays must penetrate (Figure 2–3). When the Sun is directly overhead, the rays pass through a thickness of only 1 atmosphere. But rays entering at a 30° angle travel through twice this amount, and 5° rays travel through a thickness roughly equal to 11 atmospheres (Table 2–1). The longer the path, the greater is the chance that sunlight will be absorbed, reflected, or scattered by the atmosphere, all of which reduce the intensity at the surface. These same effects

Figure 2–3 Rays striking Earth at a low angle (toward the poles) must traverse more of the atmosphere than rays striking at a high angle (around the equator) and thus are subject to greater depletion by reflection and absorption.

Table 2–1 Distance radiation must travel through the atmosphere

Angle of Sun Above Horizon	Equivalent Number of Atmospheres Sunlight Must Pass Through
90° (Directly overhead)	1.00
80°	1.02
70°	1.06
60°	1.15
50°	1.31
40°	1.56
30°	2.00
20°	2.92
10°	5.70
5°	10.80
0° (At horizon)	45.00

account for the fact that we cannot look directly at the midday Sun, but we can enjoy gazing at a sunset.

It is important to remember that Earth has a spherical shape. Hence, on any given day, only places located along a particular latitude will receive vertical (90°) rays from the Sun. As we move either north or south of this location, the Sun's rays strike at an ever-decreasing angle. Thus, the nearer a place is situated to the latitude receiving the vertical rays of the Sun, the higher will be its noon Sun, and the more concentrated will be the radiation it receives.

In summary, the most important reasons for the variation in the amount of solar energy reaching a particular location are the seasonal changes in the angle at which the Sun's rays strike the surface and in the length of daylight.

Earth's Orientation

What causes the fluctuations in the Sun's angle and length of daylight that occur during the course of a year? They occur *because Earth's orientation to the Sun continually changes*. Earth's axis (the imaginary line through the poles around which Earth rotates) is not perpendicular to the plane of its orbit around the Sun, which is called the **plane of the ecliptic**. Instead, it is tilted 23 1/2° from the perpendicular, as shown in Figure 2–3. This is called the **inclination of the axis**. As we shall see, if the axis were not so inclined, we would have no seasonal changes. In addition, because the axis always remains pointed in the same direction (toward the North Star) as Earth journeys around the Sun, the orientation of Earth's axis to the Sun's rays is always changing (Figure 2–4).

For example, on one day in June each year, Earth's position in orbit is such that the Northern Hemisphere is "leaning" 23 1/2° toward the Sun (left in Figure 2–4). Six months later in December, when Earth has moved to the opposite side of its orbit, the Northern Hemisphere leans 23 1/2° away from the Sun (right in Figure 2–4). On days between these extremes, Earth's axis is leaning at amounts less than 23 1/2° to the rays of the Sun. This change in orientation causes the spot where the Sun's rays are vertical to make an annual migration from 23 1/2° north of the equator to 23 1/2° south of the equator. In turn, this migration causes the angle of the noon Sun to vary by up to 47° (23 1/2 + 23 1/2) for many locations during a year. A midlatitude city like New York, for instance, has a maximum noon Sun angle of 73 1/2° when the Sun's vertical rays have reached their farthest northward location in June and a minimum noon Sun angle of 26 1/2° six months later (Figure 2–5).

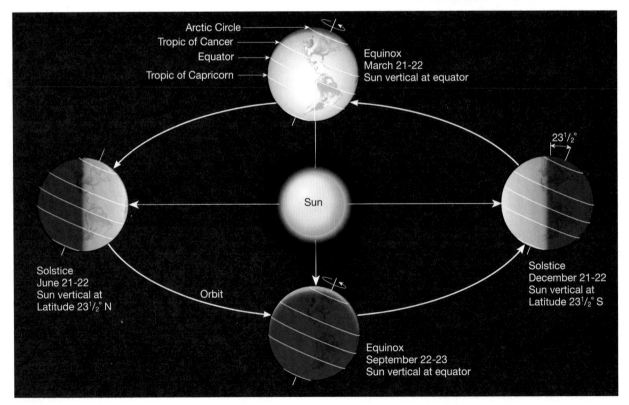

Figure 2–4 Earth–Sun relationships.

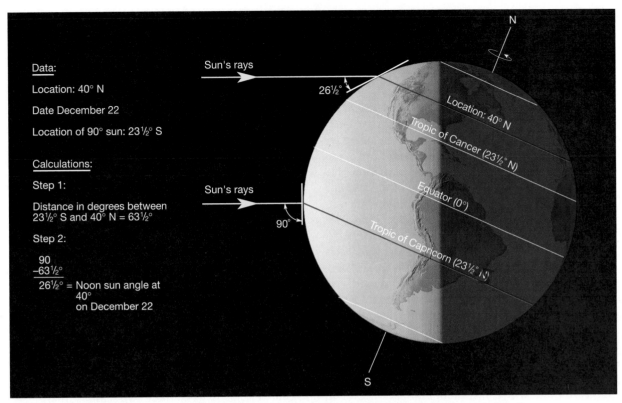

Figure 2–5 Calculating the noon Sun angle. Recall that on any given day, only one latitude receives vertical (90°) rays of the Sun. A place located 1° away (either north or south) receives an 89° angle at any location; a place 2° away, an 88° angle, and so forth. To calculate the noon Sun angle, simply find the number of degrees of latitude separating that location from the latitude that is receiving the vertical rays of the Sun. Then subtract that value from 90°. The example in this figure illustrates how to calculate the noon Sun angle for a city located at 40° north latitude on December 22 (winter solstice).

Solstices and Equinoxes

Historically, four days each year have been given special significance based on the annual migration of the direct rays of the Sun and its importance to the yearly cycle of weather. On June 21 or 22, Earth is in a position where the axis in the Northern Hemisphere is tilted 23 1/2° toward the Sun (Figure 2–4). At this time, the vertical rays of the Sun are striking 23 1/2° north latitude (23 1/2° north of the equator), a line of latitude known as the **Tropic of Cancer**. For people living in the Northern Hemisphere, June 21 or 22 is known as the **summer solstice**, the first "official" day of summer (see Box 2–1).

Six months later, on about December 21 or 22, Earth is in an opposite position, where the Sun's vertical rays are striking at 23 1/2° south latitude. This line is known as the **Tropic of Capricorn**. For those of us in the Northern Hemisphere, December 21 or 22 is the **winter solstice**, the first day of winter. However, at the same time in the Southern Hemisphere, people are experiencing just the opposite—the summer solstice.

The *equinoxes* occur midway between the solstices. September 22 or 23 is the date of the **autumnal equinox** in the Northern Hemisphere, and March 21 or 22 is the date of the **spring equinox** (also called the *vernal equinox*). On these dates, the vertical rays of the Sun strike along the equator (0° latitude), for Earth is in such a position in its orbit that the axis is tilted neither toward nor away from the Sun.

In addition, the length of daylight versus darkness is also determined by the position of Earth in its orbit. The length of daylight on June 21, the summer solstice in the Northern Hemisphere, is greater than the length of night. This fact can be established by examining Figure 2–6, which illustrates the **circle of illumination**—that is, the boundary separating the dark half of Earth from the lighted half. The length of daylight is established by comparing the fraction of a line of latitude that is on the lighted side of the globe with the fraction on the dark side. Notice that on June 21 (top left in Figure 2–6) all locations in the Northern Hemisphere experience longer periods of

Box 2–1 When Are the Seasons?

Have you ever been caught in a snowstorm around Thanksgiving, only to be told by the TV weatherperson that winter does not begin until December 21? Or perhaps you have endured several consecutive days of 100° temperatures only to discover that summer has not "officially" started? The idea of dividing the year into four seasons clearly originated from the Earth–Sun relationships discussed in this chapter (Table 2–A). This astronomical definition of the seasons defines winter (Northern Hemisphere) as the period from the winter solstice (December 21–22) to the spring equinox (March 21–22), and so forth. This is also the definition used most widely by the news media, yet it is not unusual for portions of the United States and Canada to have significant snowfalls weeks before the "official" start of winter (Figure 2–A).

Because the weather phenomena we normally associate with each season do not coincide well with the astronomical seasons, meteorologists prefer to divide the year into four 3-month periods based primarily on temperature. Thus, winter is defined as December, January, and February, the three coldest months of the year in the Northern Hemisphere. Summer is defined as the three warmest months, June,

Table 2–A Occurrence of the seasons in the Northern Hemisphere

Seasons	Astronomical Seasons	Climatological Seasons
Spring	March 21 or 22 to June 21 or 22	March, April, May
Summer	June 21 or 22 to September 22 or 23	June, July, August
Autumn	September 22 or 23 to December 21 or 22	September, October, November
Winter	December 21 or 22 to March 21 or 22	December, January, February

July, and August. Spring and autumn are the transition periods between these two seasons. Inasmuch as these four 3-month periods better reflect the temperatures and weather that we associate with the respective seasons, this definition of the seasons is more useful for meteorological discussions.

Figure 2–A Fall scene in the Adirondacks of upstate New York. *(Photo by Kim Heacox Photography/DRK Photo)*

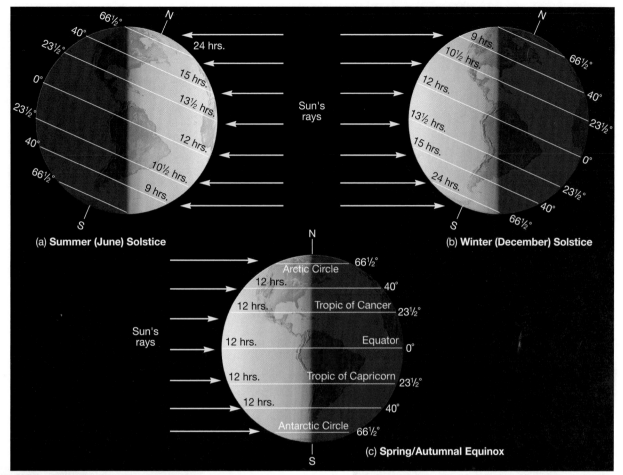

Figure 2–6 Characteristics of the solstices and equinoxes.

daylight than darkness. The opposite is true for the December solstice; then the length of darkness exceeds the length of daylight at all locations in the Northern Hemisphere. Again, for comparison, let us consider New York City (about 40° N). It has 15 hours of daylight on June 21 and only 9 hours on December 21.

Also note from Table 2–2 that the farther north you are of the equator on June 21, the longer the period of daylight. When you reach the Arctic Circle (66 1/2°N) the length of daylight is 24 hours. This is the "midnight Sun," which does not set for about six months at the North Pole (Figure 2–7).

Table 2–2	Length of daylight		
Latitude (Degrees)	Summer Solstice	Winter Solstice	Equinoxes
0	12 hr	12 hr	12 hr
10	12 hr 35 min	11 hr 25 min	12 hr
20	13 hr 12 min	10 hr 48 min	12 hr
30	13 hr 56 min	10 hr 04 min	12 hr
40	14 hr 52 min	9 hr 08 min	12 hr
50	16 hr 18 min	7 hr 42 min	12 hr
60	18 hr 27 min	5 hr 33 min	12 hr
70	2 mo	0 hr 00 min	12 hr
80	4 mo	0 hr 00 min	12 hr
90	6 mo	0 hr 00 min	12 hr

Figure 2–7 Multiple exposures of the midnight Sun in late June or July in high northern latitudes—Alaska, Scandinavia, Northern Canada, etc. *(Photo by Brian Stablyk/Tony Stone Images)*

During an equinox (meaning "equal night"), the length of daylight is 12 hours everywhere on Earth, for the circle of illumination passes directly through the poles, thus dividing the latitudes in half.

As a review of the characteristics of the summer solstice for the Northern Hemisphere, examine Figure 2–6 and Table 2–2 and consider the following facts:

1. The date of occurrence is June 21 or 22.
2. The vertical rays of the Sun are striking the Tropic of Cancer (23 1/2° north latitude).
3. Locations in the Northern Hemisphere are experiencing their longest length of daylight and highest Sun angle (opposite for the Southern Hemisphere).
4. The farther north you are of the equator, the longer the period of daylight until the Arctic Circle is reached, where the length of daylight becomes 24 hours long (opposite for the Southern Hemisphere).

The facts about the winter solstice are just the opposite. It should now be apparent why a midlatitude location is warmest in the summer. It is then that the days are longest and the angle of the Sun is highest.

In summary, *seasonal fluctuations in the amount of solar energy reaching various places on Earth's surface are caused by the migrating vertical rays of the Sun and the resulting variations in Sun angle and length of daylight.* These changes, in turn, cause the month-to-month variations in temperature observed at most locations outside the tropics. Figure 2–8 shows mean monthly temperatures for selected cities at different latitudes. Notice that places located more poleward experience larger temperature differences from summer to winter than do cities located nearer to the equator. Also notice that temperature minimums for Southern Hemisphere locations occur in July, just the opposite of places in the Northern Hemisphere.

All places situated at the same latitude have identical Sun angles and lengths of daylight. If the Earth–Sun relationships just described were the only controls of temperature, we would expect these places to have identical temperatures as well. Obviously, such is not the case. Although the angle of the Sun above the horizon and the length of daylight are the main controls of temperature, they are not the only controls, as we shall see in Chapter 3.

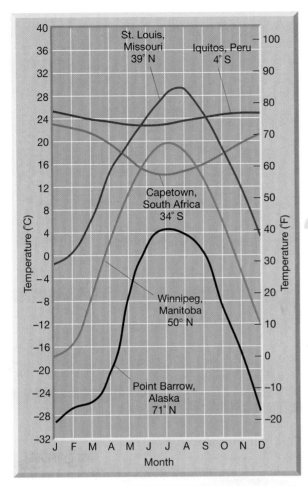

Figure 2–8 Mean monthly temperatures for six cities located at different latitudes. Note that Capetown, South Africa, experiences winter in June, July, and August.

Energy, Heat, and Temperature

The definition of **energy** is *the ability to do work*. We can think of work being done whenever an object is moved over some distance. Everyday examples include the chemical energy from gasoline that is released to power an automobile, electrical energy from a stove used to excite water molecules (boil water), and gravitational energy moving snow down a mountain slope in the form of an avalanche. These examples illustrate that energy takes many forms. Moreover, energy can change from one form to another. For example, the chemical energy in gasoline is first converted to heat energy in the engine of an automobile, which is then converted to energy of motion as the automobile speeds along. It is important to note that although energy can be converted from one form to another through ordinary chemical and physical processes, *energy cannot be created or destroyed* by such

processes. When energy is converted, the energy lost by one process *must* equal the energy gained by another.

Types of Energy

You are undoubtably familiar with some of the common forms of energy, such as heat, chemical, nuclear, radiant and gravitational energy. The various forms of energy can also be grouped into two major categories, *kinetic energy* and *potential energy*.

Kinetic Energy **Kinetic energy** can be thought of as *energy of motion*. A simple example of kinetic energy is the motion of a hammer when driving a nail. The faster the hammer is swung the greater will be its kinetic energy (energy of motion). In addition, a larger hammer (more massive) will possess more kinetic energy than a smaller one, provided they are swung at the same velocity. Therefore, it should be obvious that the winds associated with a hurricane possess much more kinetic energy than do light localized breezes. Hurricane-force winds are both larger in scale (more massive) and are traveling at a higher velocity.

Two forms of kinetic energy are particularly important in the study of the atmosphere, *heat energy* and *radiant energy*. All matter is composed of atoms or molecules that are constantly vibrating and therefore possess kinetic energy. The faster these atoms or molecules vibrate the more kinetic energy they contain. This type of kinetic energy is often referred to simply as *heat* or *heat energy*. Another important form of kinetic energy is *radiation* or *radiant energy*. This is the form of energy that travels from the Sun to Earth and is the *ultimate* source of most energy on our planet. Usually the word "radiation" conjures up thoughts of the Sun, or a nuclear power plant or possibly an atomic bomb. In reality, radiation is emitted constantly by all matter. Thus, in addition to the Sun, Earth's land–sea surface, the atmosphere and even you are emitting radiation.

Potential Energy As the name implies **potential energy** has the potential to do work. For example, large hailstones suspended by an updraft have potential energy because of their position. Should the updraft subside, these hailstones could do destructive work to someone's roof. Many substances, including wood, gasoline, and the food you eat, contain potential energy, which is capable of doing work.

Heat Energy Versus Temperature

The concepts of *heat* and *temperature* are often confused. The phrase "in the heat of the day" is one common expression in which the word "heat" is misused to describe the concept of temperature. *Heat* is a form of energy. Recall that all matter is composed of atoms or molecules that are constantly vibrating and therefore possess kinetic energy

(energy of motion). **Heat** or **heat energy** is defined as the *total* kinetic energy of all the atoms and molecules that make up a substance. By contrast, *temperature* refers to intensity—that is, the degree of "hotness." **Temperature** is a measure of the *average* kinetic energy of the individual atoms or molecules in a substance. When heat energy is added to a substance, its atoms move faster and faster and its temperature rises. When heat energy is removed, its atoms move slower and its temperature drops.

A cup of boiling water obviously has a higher *temperature* than a tub of lukewarm water, but the cup is small, so it does not contain as great a quantity of *heat* as the much larger tub. Far more ice would be melted in the tub of lukewarm water than in the cup of boiling water. The water temperature in the cup is higher, but the amount of heat energy is smaller. Thus, the quantity of heat depends on the mass of material considered, but the temperature does not. It is now easier to understand why the extremely thin air of the thermosphere (discussed in Chapter 1) can have a high temperature, yet contain little heat.

Although temperature and heat are distinct, they are nevertheless related. Certainly, the addition or subtraction of heat causes temperatures to increase or decrease. In addition, differences in temperature determine the direction of heat flow. When two bodies having different temperatures are in contact, *heat energy always moves from the higher-temperature body to the lower-temperature body*. For example, when you touch a hot stove, heat enters your hand because the stove is much warmer than your hand. By contrast, when you hold an ice cube, heat is transferred from your hand to the ice cube. Thus, if two objects of different temperature are in contact, the warmer object will become cooler and the cooler will become warmer until they reach the same temperature. Further, the greater the temperature difference, the greater will be the rate of heat flow from the hotter to the colder object.

Mechanisms of Energy Transfer

Three mechanisms of energy transfer are recognized: conduction, convection, and radiation (Figure 2–9). Although we present them separately, all three processes go on simultaneously in the atmosphere. In addition, these mechanisms operate to transfer heat between Earth's surface (both land and water) and the atmosphere.

Conduction

Conduction is familiar to most of us through our everyday experiences. Anyone who has attempted to pick up a metal spoon that was left in a hot pan is sure to realize that heat was conducted through the spoon. **Conduction**

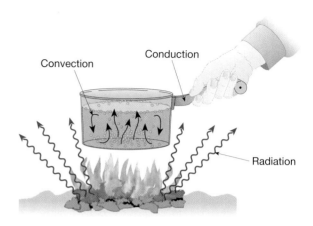

Figure 2–9 The three mechanisms of heat transfer: conduction, convection, and radiation.

is the transfer of heat through matter by molecular activity. The energy is transferred through collisions from one molecule to another, with the heat flowing from the higher temperature to the lower temperature. The ability of substances to conduct heat varies considerably. Metals are good conductors, as those of us who have touched a hot spoon have quickly learned. Air, in contrast, is a very poor conductor of heat. Consequently, conduction is important only between Earth's surface and the air immediately in contact with the surface. As a means of heat transfer for the atmosphere as a whole, conduction is the least significant and can be disregarded when considering most meteorological phenomena.

Objects that are poor conductors, like air, are called *insulators*. Most objects that are good insulators, such as cork, plastic foams, or goose down, contain many small air spaces. It is the poor conductivity of the trapped air that gives these materials their insulating value. Snow is also a poor conductor (good insulator). Like other insulators, fresh snow contains numerous air spaces that serve to retard the flow of heat. Thus, a wild animal will often burrow into a snowbank to escape the "cold." The snow, like a down-filled comforter, does not supply heat; it simply retards the loss of the animal's own body heat.

Convection

Much of the heat transport that occurs in the atmosphere is carried on by convection. **Convection** is the transfer of heat by mass movement or circulation within a substance. It can only take place in fluids (liquids like the ocean and gases like air) where the atoms and molecules are free to move about. In solids, such movement does not occur, so heat transfer through solids takes place by conduction.

The pan of water in Figure 2–9 illustrates the nature of simple convective circulation. Radiation from the fire warms the bottom of the pan, which conducts heat to the water near the bottom of the container. As the water is heated, it expands and thus becomes less dense than the water above. Because of this new buoyancy, the warmer water begins to rise. At the same time, cooler, denser water near the top of the pan sinks to the bottom, where it becomes heated. As long as the water is heated unequally—that is, from the bottom up—the water will continue to "turn over," producing a *convective circulation*.

In a similar manner, most of the heat acquired in the lowest layer of the atmosphere by way of radiation and conduction is transferred by convective flow. For example, on a warm, sunny day, a dark surface like an asphalt parking lot can heat the air directly above it more than the surrounding air. This warm, less dense air buoys upward, transporting heat aloft. These *thermals*, as they are called, are what hang glider pilots use to keep their crafts aloft. This type of convection not only transfers heat but also transports moisture that is in the air. The result is an increase in cloudiness that frequently can be observed on warm summer afternoons.

On a global scale, convection in the atmosphere creates a huge, worldwide air circulation. This is responsible for the redistribution of heat between hot equatorial regions and the frigid poles. This important process will be discussed in detail in Chapter 7.

In convective circulation, air moves both vertically and horizontally, so both vertical and horizontal heat transfer occurs. However, meteorologists use the term *convection* to describe *upward and downward* heat transfer.

By contrast, the term **advection** is used to denote the *horizontal* component of convective flow. (The common term for advection is "wind," a phenomenon we will examine closely in later chapters.) Those who reside in the midlatitudes often experience the effects of energy transfer by advection. For example, when frigid Canadian air invades the American Midwest in January, it brings bitterly cold winter weather. By contrast, the northward flow of air from the Gulf of Mexico is associated with warmth.

Radiation

The third mechanism of heat transfer is *radiation*. As shown in Figure 2–9, radiation travels out in all directions from its source. Unlike conduction and convection, which need a medium to travel through, radiant energy does not. Radiation is the only mechanism of energy transfer that travels through the vacuum of space. Thus, radiation is the heat-transfer mechanism by which solar energy reaches our planet.

Solar Radiation   As noted, the Sun is the ultimate source of energy that drives the weather machine. For this reason, we consider the nature of solar radiation in more detail. From our everyday experience, we know that the Sun emits light and heat as well as the rays that give us a suntan. Although these forms of energy constitute a major portion of the total energy that radiates from the Sun, they are only a part of a large array of energy called **radiation** or **electromagnetic radiation**. This array or spectrum of electromagnetic energy is shown in Figure 2–10.

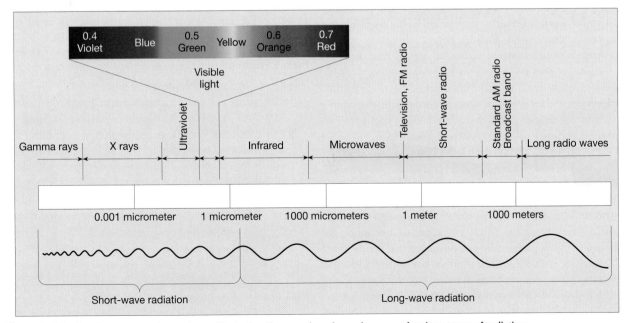

Figure 2–10 The electromagnetic spectrum, illustrating the wavelengths and names of various types of radiation.

All types of radiation, whether X-rays, radio, or heat waves, travel through the vacuum of space at 300,000 kilometers (186,000 miles) per second and only slightly slower through air. It helps to understand radiant energy by picturing ocean waves, or ripples made in a pond by tossing in a pebble. Not unlike these waves, *electromagnetic waves*, as waves of radiant energy are called, come in various sizes. For our purposes, the most important characteristic is their *wavelength*, or distance from one crest to the next. Radio waves have the longest wavelengths, ranging to tens of kilometers. Gamma waves are the shortest, being less than a billionth of a centimeter long.

Radiation is often identified by the effect that it produces when it interacts with an object. The retinas of our eyes, for instance, are sensitive to a range of wavelengths that we call **visible light**. We often refer to visible light as white light, for it appears "white" in color. It is easy to show, however, that white light is really an array of colors, each color corresponding to a specific wavelength. By using a prism, white light can be divided into the colors of the rainbow, from violet with the shortest wavelength, 0.4 micrometer (1 micrometer is 0.0001 centimeter), to red with the longest wavelength, 0.7 micrometer. Located adjacent to red, and having a longer wavelength, is **infrared** radiation, which we cannot see but can detect as heat. The closest invisible waves to violet are called **ultraviolet** rays. They are responsible for the sunburn that can occur after an intense exposure at the beach (see Box 2–2).

Although we divide radiant energy into categories based on our ability to perceive them, all forms of radiation behave in a similar manner. When any form of radiant energy is absorbed by an object, the result is an increase in molecular motion (kinetic energy) and a corresponding increase in temperature. One important difference among the various wavelengths of radiant energy is that shorter wavelengths are associated with greater energy. This accounts for the fact that relatively short, high-energy ultraviolet rays can damage human tissue more readily than similar exposure to longer wavelength radiation. The damage can be skin cancer and cataracts (see Box 2–2).

It is important to note that the Sun emits all of the forms of radiation shown in Figure 2–10, but in varying quantities. Over 95 percent of all solar radiation is emitted in wavelengths between 0.1 and 2.5 μm (micrometers) with much more emitted in certain wavelengths. Most radiant energy from the Sun is concentrated in the visible and near-visible parts of the electromagnetic spectrum. The narrow band of visible light, between 0.4 and 0.7 micrometer, represents over 43 percent of the total emitted. The bulk of the remainder lies in the infrared zone nearest to the visible band (49 percent) and ultraviolet (UV) section (7 percent). Less than 1 percent of solar radiation is emitted as X-rays, gamma rays, and radio waves.

Laws of Radiation

To obtain a better appreciation of how the Sun's radiant energy interacts with Earth's atmosphere and surface, you need a general understanding of the basic laws of radiation. The principles that follow were set forth by physicists during the late 1800s and early 1900s. Although their mathematics are beyond the scope of this book, the concepts themselves are easy to grasp:

1. *All objects above −273° Celsius (absolute zero) emit radiant energy.* Thus, not only do hot objects like the Sun continually emit energy, but Earth does as well, even its polar ice caps.

2. *Hotter objects radiate more total energy per unit area than do colder objects.* The Sun, which has a surface temperature of 6000 K (10,000°F), emits about 160,000 times more energy than does Earth, which has an average surface temperature of 288 K (59°F). This concept is called the *Stefan–Boltzman law* and is expressed mathematically in Box 2–3.

3. *The hotter the radiating body, the shorter is the wavelength of maximum radiation.* This law can be easily visualized if we examine something from our everyday experiences. For example, a very hot metal rod will emit visible radiation and produce a white glow. On cooling, it will emit more of its energy in longer wavelengths and will glow a reddish color. Eventually, no light will be given off, but if you place your hand near the rod, the still longer infrared radiation will be detectable as heat. The Sun (surface temperature of 6000 K) (10,000°F), radiates maximum energy at 0.5 micrometer, which is in the visible range (Figure 2–11).° The maximum radiation emitted from Earth occurs at a wavelength of 10 micrometers, well within the infrared (heat) range. Because the maximum Earth radiation is roughly 20 times longer than the maximum solar radiation, *it is often referred to as* long-wave radiation *whereas solar radiation is called* shortwave radiation. This concept is called the *Wien's displacement law* and is expressed mathematically in Box 2–3.

°Because the Sun is a gaseous body, it has no true surface. Most energy radiated from the Sun is emitted from the photosphere, the visible layer of the Sun.

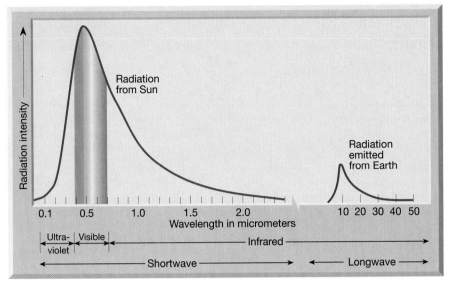

Figure 2–11 Comparison of the intensity of solar radiation and radiation emitted by Earth. Because of the Sun's high surface temperature, most of its energy is radiated at wavelengths shorter than 4 micrometers, with the greatest concentration in the visible range of the electromagnetic spectrum. Earth, in contrast, radiates most of its energy in wavelengths longer than 4 micrometers, primarily in the infrared band. Thus, we call the Sun's radiation *shortwave* and Earth's radiation *longwave*. *(After Tom L. McKnight, Physical Geography, ©1990, Prentice Hall, Inc.)*

4. *Objects that are good absorbers of radiation are also good emitters.* The perfect absorber (and emitter) is a theoretical object called a **blackbody**. The term "black" is misleading, for the object need not be black in color. However, a dull black surface is nearly a blackbody because it absorbs roughly 90 percent of the radiation striking it. Technically, a blackbody is any object that radiates, for every wavelength, the maximum intensity of radiation possible for that temperature. Earth's surface and the Sun approach being blackbodies (perfect radiators) because they absorb and radiate with nearly 100 percent efficiency for their respective temperatures. The principle that good absorbers of radiation are also good emitters is very important in understanding how the atmosphere is heated. For example, gases are selective absorbers and radiators. It turns out that the atmosphere is nearly transparent to (does not absorb) certain wavelengths of radiation, but is nearly opaque to (a good absorber of) others. Our experience tells us that the atmosphere is quite transparent to visible light emitted by the Sun; hence, it readily reaches Earth's surface.

Figure 2–9 depicts the various mechanisms of heat transfer. To summarize: A portion of the radiant energy generated by the campfire is absorbed by the pan. This energy is readily transferred through the metal container by the process of conduction. Conduction also increases the temperature of the water at the bottom. Once warmed, this layer of water moves upward and is replaced by cool water descending from above. Thus, convection currents that redistribute the newly acquired energy throughout the pan are established. Meanwhile, the camper is warmed by radiation emitted by the fire and the pan. Furthermore, because metals are good conductors, the camper's hand is likely to be burned if he or she does not use a potholder. Like this example, the heating of Earth's atmosphere involves the processes of conduction, convection, and radiation, all of which occur simultaneously.

The Fate of Incoming Solar Radiation

When radiation strikes an object there are usually three different results. First, some of the energy is *absorbed* by the object. Recall that when radiant energy is absorbed it is converted to heat, which causes an increase in temperature. Second, substances such as water and air are transparent to certain wavelengths of radiation. Such materials simply *transmit* this energy. Radiation that is transmitted does not contribute energy to the object. Third, some radiation may "bounce off" the object without being absorbed or transmitted. *Reflection* and *scattering* are responsible for redirecting incoming solar radiation. In summary, *radiation may be absorbed, transmitted, or redirected (reflected or scattered).*

Figure 2–12 shows the fate of incoming solar radiation averaged for the entire globe. Notice that the atmosphere is quite transparent to incoming solar radiation. On average, about 50 percent of the energy reaching the top of the atmosphere is absorbed at Earth's surface. Another

Box 2–2 Ultraviolet Index

Gong-Yuh Lin*

Most people welcome sunny weather. On warm days, when the sky is cloudless and bright, many spend a great deal of time outdoors "soaking up" the Sun (Figure 2–B). The goal is to develop a dark tan, one that sunbathers often describe as "healthy-looking." Ironically, there is strong evidence that too much sun (specifically, too much ultraviolet radiation) can lead to serious health problems including skin cancer and cataracts.

Since June 1994, the National Weather Service has issued the next-day ultraviolet index (UVI) for 58 cities to warn the public of potential health risks of exposure to the Sun. The UVI is determined by taking into

Figure 2–B Exposing sensitive skin to too much solar ultraviolet radiation has potential health risks. *(Photo by Robert Mort/Tony Stone Worldwide)*

30 percent is reflected back to space by the atmosphere, clouds, and reflective surfaces such as snow and water. The remaining 20 percent is absorbed by clouds and the atmosphere's gases.

What determines whether solar radiation will be transmitted to the surface, scattered, reflected back to space, or absorbed by the atmosphere? As we shall see, it depends greatly upon the *wavelength* of the energy being transmitted, as well as upon the size and nature of the absorbing or reflecting substance.

Reflection and Scattering

Reflection is the process whereby light bounces back from an object at the same angle at which it encounters a surface and with the same intensity (Figure 2–13a). By contrast, **scattering** produces a larger number of weaker rays, traveling in different directions. Although scattering disperses light both forward and backward (*backscattering*), more energy is dispersed in the forward direction (Figure 2–13b).

Reflection and Earth's Albedo Energy returns to space from Earth in two ways: reflection and emission (radiated back to space). The portion of solar energy that is reflected back to space leaves in the same short wavelengths in which it came to Earth. About 30 percent of the solar energy reaching the outer atmosphere is reflected back to space (Figure 2–12). Included in this figure is the amount sent skyward by backscattering. This energy is lost to Earth and does not play a role in heating the atmosphere.

account the predicted cloud cover and reflectivity of the surface, as well as the Sun angle and atmospheric depth for each forecast location. Because atmospheric ozone strongly absorbs ultraviolet radiation, the extent of the ozone layer is also considered. The UVI values lie on a scale from 0 to 15, with 15 representing the greatest risk.

The U.S. Environmental Protection Agency has established five exposure categories based on UVI values—Minimal, Low, Moderate, High, and Very High (Table 2–B). Precaution measures have been developed for each category. The public is advised to minimize outdoor activities when the UVI is High or Very High. Sunscreen with a sun protection factor (SPF) of 15 or higher is recommended for all exposed skin. This is especially important after swimming or while sunbathing, even on cloudy days with the UVI in the Low category.

Table 2–B shows the range of minutes to burn for the most susceptible skin type (pale or milky white) for each exposure category. Note that the exposure that results in sunburn varies from 30 minutes for the Minimal category to less than four minutes for the Very High category. It takes approximately five times longer to cause sunburn of the least susceptible skin type, brown to dark. The most susceptible skin type develops red sunburn, painful swelling, and skin peeling when exposed to excessive sunlight. By contrast, the least susceptible skin type rarely burns and shows very rapid tanning response.

°Professor Lin is a faculty member in the Department of Geography at California State University, Northridge.

Table 2–B The UV Index: Precaution measures and minutes to burn for the most susceptible skin type.

Exposure Category	UVI Value	Precaution Measures	Minutes to Burn
Minimal	0–2	Hat	30–60 minutes
Low	3–4	Hat, SPF 15+	15–20 minutes
Moderate	5–6	Hat, SPF 15+, shady areas	10–12 minutes
High	7–9	Hat, SPF 15+, shady areas, stay indoors 10 am to 4 pm	7–8.5 minutes
Very High	10–15	Stay indoors as much as possible, take other precautions when outdoors	4–6 minutes

The fraction of radiation that is reflected by a surface is called its **albedo**. Thus, the albedo for Earth as a whole (planetary albedo) is 30 percent. The albedo from place to place as well as from time to time, however, varies considerably, depending on the amount of cloud cover and particulate matter in the air, plus the angle of the Sun's rays and the nature of the surface. A lower Sun angle means that more atmosphere must be penetrated, thereby making the "obstacle course" longer and the loss of solar radiation greater.

Table 2–3 gives the albedo for various surfaces. Fresh snow and thick clouds have high albedos, reflecting much of their received radiation. Dark soil is not very reflective and thus absorbs much of the radiation it receives. In the case of a lake or the ocean, note that the angle at which the Sun's rays strike the water surface greatly affects its albedo.

The amount of light reflected from Earth's land–sea surface represents only about 5 percent of the total planetary albedo of 30 percent (Figure 2–12). Clouds are responsible for most of Earth's "brightness" as seen from space. This high reflectivity of clouds should not surprise anyone who has tried to drive on a foggy night with bright lights.

In comparison to Earth, the Moon, which is without clouds or an atmosphere, has an average albedo of only 7 percent. Even though a full Moon gives us a good bit of light on a clear night, the much brighter Earth would provide an astronaut on the Moon with far more light for an "Earth-lit" walk at night.

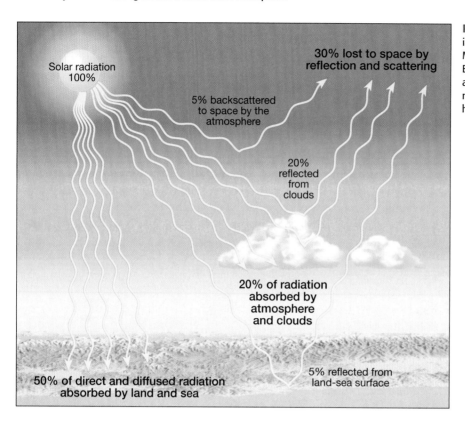

Figure 2–12 Average distribution of incoming solar radiation by percentage. More solar energy is absorbed by Earth's surface than by the atmosphere. Consequently, the air is not heated directly by the Sun, but is heated indirectly from Earth's surface.

Blue Skies and Red Sunsets Although incoming solar radiation travels in a straight line, small dust particles and gas molecules in the atmosphere scatter some of this energy in all directions. The result, called **diffused light**, explains how light reaches into the area beneath a shade tree, and how a room is lit in the absence of direct sunlight. Further, scattering accounts for the brightness and even the blue color of the daytime sky. In contrast, bodies like

the Moon and Mercury, which are without atmospheres, have dark skies and "pitch black" shadows, even during daylight hours. Overall, about one-half of solar radiation that is absorbed at Earth's surface arrives as scattered light.

To a large extent, the degree of scattering is determined by the size of the intervening gas molecules and dust particles. When light is scattered by very small particles, primarily gas molecules, it is distributed in all directions; however, more energy is scattered in the forward direction. The light that is backscattered is lost to space,

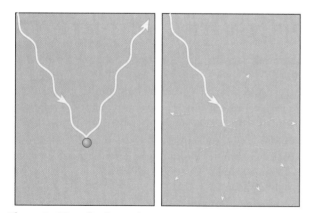

Figure 2–13 Reflection and scattering. (a) Reflected light bounces back from a surface at the same angle at which it strikes that surface and with the same intensity. (b) When a beam of light is scattered, it results in a larger number of weaker rays, traveling in all different directions. Usually more energy is scattered in the forward direction than is backscattered.

Table 2–3 Albedo (reflectivity) of various surfaces	
Surface	**Percent Reflected**
Fresh snow	80–85
Old snow	50–60
Sand (beach, desert)	20–30
Grass	20–25
Dry soil (plowed field)	15–25
Wet earth (plowed field)	10
Forest	5–10
Water (Sun near horizon)	50–80
Water (Sun near zenith)	3–5
Thick cloud	70–80
Thin cloud	25–30
Earth and atmosphere (overall total)	30

Box 2–3 Radiation Laws

Gregory J. Carbone*

All bodies radiate energy. Both the rate and the wavelength of radiation emission depend on the temperature of the radiating body.

Stefan–Boltzman Law

This law mathematically expresses the rate of radiation emission per unit area:

$$E = \sigma T^4$$

E, the rate of radiation emitted by a body, is proportional to the fourth power of the body's temperature (T). The Stefan–Boltzmann constant (σ) is equal to 5.67 $\times$ 10^{-8} W/m²K⁴. Compare the difference between the radiation emission from the Sun and Earth. The Sun, with an average temperature of 6000 K, emits 73,483,200 watts per square meter (Wm⁻²):

$$E = (5.67 \times 10^{-8} \text{ W/m}^2\text{K}^4)(6000 \text{ K})^4$$
$$= 73,483,200 \text{ W/m}^2$$

By contrast, Earth has an average temperature of only 288 K. If we round the value to 300 K, we have

$$E = (5.67 \times 10^{-8} \text{ W/m}^2 \text{ K}^4)(300 \text{ K})^4$$
$$= 459 \text{ W/m}^2$$

The Sun has a temperature that is approximately 20 times higher than Earth and thus emits approximately 160,000 times more radiation per unit area. This makes sense because $20^4 = 160,000$.

Wien's Displacement Law

This law describes mathematically the relationship between the temperature (T) of a radiating body and its wavelength of maximum emission (λ_{max}):

$$\lambda_{max} = C/T$$

Wien's constant (C) is equal to 2898 micrometers $\times$ K (2898 μK). If we use the Sun and Earth as examples, we find

$$\lambda_{max} (\text{Sun}) = \frac{2898 \text{ μK}}{6000 \text{ K}} = 0.483 \text{ μ}$$

and

$$\lambda_{max} (\text{Earth}) = \frac{2898 \text{ μK}}{300 \text{ K}} = 9.66 \text{ μm}$$

Note that the Sun radiates its maximum energy within the visible portion of the electromagnetic spectrum. The cooler Earth radiates its maximum energy in the infrared portion of the electromagnetic spectrum.

*Professor Carbone is a faculty member in the Department of Geography at the University of South Carolina.

---

but the remainder continues toward Earth's surface, where it interacts with other molecules that further diffuse it.

Gas molecules more effectively scatter the shorter wavelengths (blue and violet) of visible light than the longer wavelengths (red and orange). This fact, in turn, explains the blue color of the sky and the orange and red colors seen at sunrise and sunset (Figure 2–14). Remember, sunlight appears white, but it is composed of all colors. When the Sun is overhead, you can look in any direction away from the direct Sun and see predominantly blue light, which is the wavelength more readily scattered by the atmosphere.

Conversely, the Sun appears to have an orangish-to-reddish tint when viewed near the horizon (Figure 2–15). This is because solar radiation must travel through a greater thickness of atmosphere before it reaches your eyes (see Table 2–1). As a consequence, most of the blue and violet wavelengths will be scattered out, leaving light that consists mostly of reds and oranges. The reddish appearance of clouds during sunrise and sunset also results because the clouds are illuminated by light from which the blue color has been subtracted by scattering.

The most spectacular sunsets occur when large quantities of fine dust or smoke particles penetrate into the stratosphere. For three years after the great eruption of the Indonesian volcano Krakatau in 1883, brilliant sunsets occurred worldwide. The European summer that followed this colossal explosion was cooler than normal, a fact that has been attributed to the greater loss of radiation caused by backscattering.

Large particles associated with haze, fog, or smog scatter light more equally in all wavelengths. Because no color is predominant over any other, the sky appears white or gray on days when large particles are abundant. Scattering

Figure 2–14 At sunset, clouds often appear red because they are illuminated by sunlight in which most of the blue light has been lost due to scattering. *(Photo by Tom Ives)*

of sunlight by haze, water droplets, or dust particles makes it possible for us to observe bands (or rays) of sunlight called *crepuscular rays.* These bright fan-shaped bands are most commonly seen when the Sun shines through a break in the clouds, as shown in Figure 2–16. Crepuscular rays can also be observed around twilight when towering clouds cause alternating lighter and darker bands (light rays and shadows) to streak across the sky.

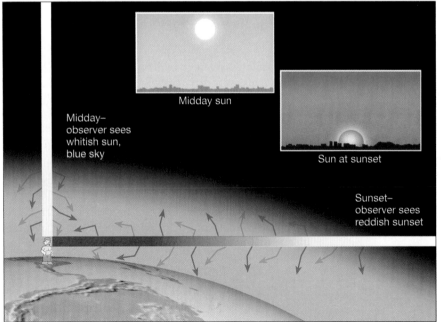

Figure 2–15 Short wavelengths (blue and violet) of visible light are scattered more effectively than are longer wavelengths (red, orange). Therefore, when the Sun is overhead, an observer can look in any direction and see predominantly blue light that was selectively scattered by the gases in the atmosphere. By contrast, at sunset, the path that light must take through the atmosphere is much longer. Consequently, most of the blue light is scattered before it reaches an observer. Thus, the Sun appears reddish in color.

Figure 2–16 Crepuscular rays are produced when haze scatters light. Crepuscular rays are most commonly seen when the Sun shines through a break in the clouds. *(Photo by Stephen Trimble/DRK Photo)*

In summary, the color of the sky gives an indication of the number of large or small particles present. Lots of small particles produce red sunsets, whereas large particles produce a white sky. Furthermore, the bluer the sky, the cleaner the air.

Absorption by Earth's Surface and Atmosphere

Although Earth's surface is a relatively good absorber (effectively absorbs most wavelengths of solar radiation) the atmosphere is not. This accounts for the fact that 50 percent of the solar radiation that reaches Earth is absorbed by Earth's land–sea surface, whereas only 20 percent of this energy is absorbed directly by the atmosphere (see Figure 2–12). The atmosphere is not as effective an absorber because gases are selective absorbers (and emitters) of radiation. As you will see in the following section, this fact greatly influences how the atmosphere is heated.

Freshly fallen snow is another example of a selective absorber. Anyone who has experienced the blinding reflection of sunlight from snow is aware that snow is a poor absorber of visible light (reflects up to 85 percent). Because of this, the temperature directly above a snow-covered surface is colder than it would otherwise be because much of the incoming radiation is reflected away. However, snow is a very good absorber of the infrared (heat) radiation that is emitted from Earth's surface. As the ground radiates heat upward, the lowest layer of snow absorbs this energy and reradiates most of it downward. Thus, the depth at which a winter's frost can penetrate into the ground is much less when the ground has a snow cover than in an equally cold region without snow. The statement "the ground is blanketed with snow" can be taken literally. Farmers who plant winter wheat desire a deep snow cover because it insulates their crops from bitter midwinter temperatures.

Radiation Emitted by Earth

Although we often talk about radiation in terms of the Sun, recall that *all* objects continuously emit radiation. Because Earth is much cooler than the Sun, it emits considerably less radiant energy. Furthermore, radiation emanating from Earth's surface and atmosphere is emitted at longer wavelengths than most solar radiation. Over 95 percent of Earth's radiation has wavelengths between 2.5 and 30 micrometers, placing it in the infrared (heat) band of the electromagnetic spectrum (Figure 2–17). Recall that the bulk of solar radiation is emitted in wavelengths shorter than 2.5 micrometers. This difference between incoming solar radiation and radiation emitted by Earth is very important to our understanding of how our atmosphere is heated.

Heating the Atmosphere

As stated earlier, gases are selective absorbers, meaning that they absorb strongly in some wavelengths, moderately in others, and only slightly in still others. When a gas molecule absorbs radiation, this energy is transformed into internal molecular motion (heat), which is detectable as a rise in temperature. *Thus, it is the gases that are the most effective absorbers of radiation that play the primary role in heating the atmosphere.*

At first glance, Figure 2–17 appears complicated; nevertheless, it is a useful aid to understanding how the atmosphere is heated. The upper portion shows that most (95 percent) incoming solar radiation has wavelengths between 0.1 and 2.5 micrometers (abbreviated μm) and includes the band of visible light (shown as the colors of the rainbow). The lower half of Figure 2–17 gives the absorptivity of the principal atmospheric gases. Note that nitrogen, the most abundant constituent in the atmosphere (78 percent), is a relatively poor absorber of incoming solar radiation

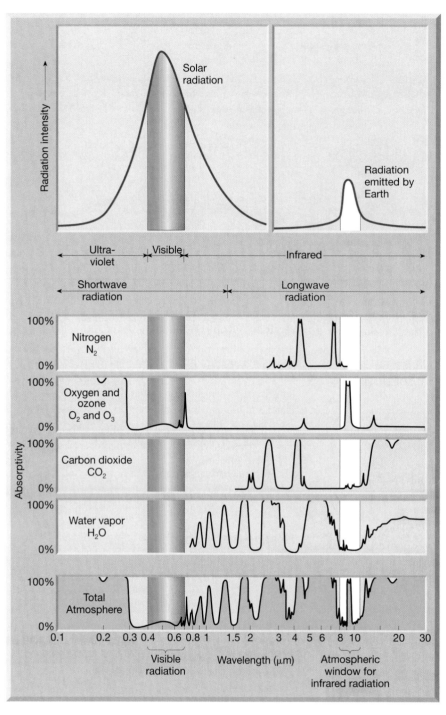

Figure 2–17 The absorptivity of selected gases of the atmosphere and the atmosphere as a whole. The atmosphere as a whole is quite transparent to solar radiation between 0.3 and 0.7 micrometer, which include the band of visible light. Most solar radiation falls in this range, explaining why a large amount of solar radiation penetrates the atmosphere and heats Earth's surface. Also, note that longwave infrared radiation in the zone between 8 and 11 micrometers can escape the atmosphere most readily. This zone is called the atmospheric window. (*Data from R. G. Fleagle and J. A. Businger,* An Introduction to Atmospheric Physics. *1963 by Academic Press*)

because it absorbs best in that part of the electromagnetic spectrum with wavelengths greater than 2.5 micrometers.

The only significant absorbers of incoming solar radiation are water vapor, oxygen, and ozone, which together accounts for most of the solar energy absorbed directly by the atmosphere. Oxygen and ozone are efficient absorbers of high-energy, shortwave radiation. Oxygen removes most of the shorter wavelength UV radiation high in the atmosphere, and ozone absorbs UV rays in the stratosphere between 10 and 50 kilometers (6-30 miles). The absorption of UV energy in the stratosphere accounts for the high temperatures experienced there. (The importance of the removal of harmful UV radiation by stratospheric ozone was discussed in Chapter 1.)

Looking at the bottom of Figure 2–17, you can see that for the atmosphere as a whole, none of the gases are effective absorbers of radiation that has wavelengths between 0.3 and 0.7 micrometers. This region of the spectrum corresponds to the visible light band, which constitutes nearly 50 percent of the energy radiated by the Sun. Because the atmosphere is a poor absorber of visible radiation, most of this energy is transmitted to Earth's surface. Thus, we say that *the atmosphere is nearly transparent to incoming solar radiation and that direct solar energy is not an effective "heater" of Earth's atmosphere.*

We can also see in Figure 2–17 that the atmosphere as a whole is a rather efficient absorber of longwave (infrared) radiation emitted by Earth (see bottom right of Figure 2–17). Water vapor and carbon dioxide are the principal absorbing gases, with water vapor absorbing about 60 percent of the radiation emitted by Earth. Therefore, water vapor accounts (more than any other gas) for the warm temperatures of the lower troposphere, where it is most highly concentrated.

Despite the fact that the atmosphere is a good absorber of radiation emitted by Earth, there exists a band of outgoing radiation having wavelengths between 8 and 11 micrometers to which the atmosphere is quite transparent. Notice from Figure 2–17 that water vapor and carbon dioxide do not absorb these wavelengths of energy. This zone is called the **atmospheric window** because radiation in this band is not readily absorbed by the atmosphere but passes through it to outer space. It should be pointed out, however, that a relatively small amount of Earth's radiation is lost directly to space through this window.

Because the atmosphere is largely transparent to solar (shortwave) radiation but more absorptive of the longwave radiation emitted by Earth, the atmosphere is heated from the ground up, instead of vice versa. This explains the general drop in temperature with increased altitude in the troposphere. The farther from the "radiator" (Earth's surface), the colder it gets. On average, the temperature drops 6.5°C for each kilometer increase in altitude, a figure known as the *normal lapse rate*. The fact that the atmosphere does not acquire the bulk of its energy directly from the Sun, but is heated by Earth's surface, is of utmost importance to the dynamics of the weather machine.

The "Greenhouse Effect"

If Earth had no atmosphere, it would experience an average surface temperature far below freezing. But the atmosphere warms the planet and makes Earth livable. The extremely important role the atmosphere plays in heating Earth's surface has been named the **greenhouse effect**.

As you saw earlier, cloudless air is largely transparent to incoming shortwave solar radiation and, hence, transmits it to Earth's surface. By contrast, a significant fraction of the longwave radiation emitted by Earth's land–sea surface is absorbed by water vapor, carbon dioxide, and other trace gases in the atmosphere. This energy heats the air and increases the rate at which it radiates energy, both out to space and back toward Earth's surface. The energy that is emitted back to the surface causes it to heat up more, which then results in greater emissions from the surface. This complicated game of "pass the hot potato" keeps Earth's average temperature 30°C (50°F) warmer than it would otherwise be (Figure 2–18). Without these absorptive gases in our atmosphere,

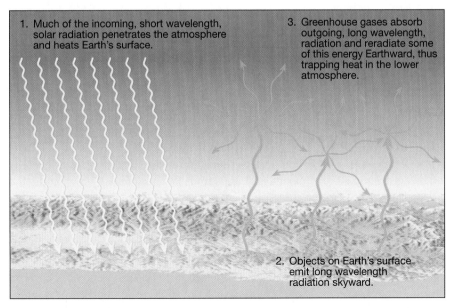

1. Much of the incoming, short wavelength, solar radiation penetrates the atmosphere and heats Earth's surface.

3. Greenhouse gases absorb outgoing, long wavelength, radiation and reradiate some of this energy Earthward, thus trapping heat in the lower atmosphere.

2. Objects on Earth's surface emit long wavelength radiation skyward.

Figure 2–18 The heating of the atmosphere. Most of the short-wavelength radiation from the Sun passes through the atmosphere and is absorbed by Earth's land–sea surface. This energy is then emitted from the surface as longer-wavelength radiation, much of which is absorbed by certain gases in the atmosphere. Some of the energy absorbed by the atmosphere will be reradiated Earthward. This so-called greenhouse effect is responsible for keeping Earth's surface much warmer than it would be otherwise.

Earth would not provide a suitable habitat for humans and other life-forms.

This natural phenomenon was named the *greenhouse effect* because it was once thought that greenhouses were heated in a similar manner. The glass in a greenhouse allows shortwave solar radiation to enter and be absorbed by the objects inside. These objects, in turn, radiate energy but at longer wavelengths, to which glass is nearly opaque. The heat, therefore, is "trapped" in the greenhouse. It has been shown, however, that air inside greenhouses attains higher temperatures than outside air mainly because greenhouses restrict the exchange of air between the inside and outside. Nevertheless, the term "greenhouse effect" remains (see Box 2–4).

The popular press frequently points to the greenhouse effect as the "villain" of the global warming problem. It is important to note that the greenhouse effect and global warming *are not* the same thing. Without the greenhouse effect Earth would be uninhabitable. We do have mounting evidence that human activity (particularly the release of carbon dioxide into the atmosphere) is responsible for a rise in global temperature (see Chapter 14). Thus, human activity seems to be enhancing an otherwise natural process (the greenhouse effect) to increase Earth's temperature. Nevertheless, to equate the greenhouse phenomenon, which makes life possible, with undesirable changes to our atmosphere caused by human activity is wrong.

Role of Clouds in Heating Earth

Clouds, like water vapor and carbon dioxide, are good absorbers of infrared radiation emitted by Earth. Thus, at night clouds play a key role in keeping the surface warm. A thick cloud cover will absorb most outward directed radiation and radiate much of that energy back to the surface. This explains why on clear, dry nights the surface cools considerably more than on cloudy or humid evenings. Furthermore, although desert areas often experience high daytime temperatures, they are also likely to experience cold nights, for they generally have cloudless skies.

During daylight hours the effect of clouds on heating Earth's surface depends on the type of clouds present. High thin clouds primarily transmit incoming solar radiation. At the same time these clouds absorb a portion of the outgoing infrared radiation emitted by Earth and radiate some of that energy back to the surface. The net effect is that high thin clouds warm the surface.

The impact of low thick clouds is opposite that of high clouds. Thick low clouds have a high albedo and therefore reflect a significant portion of incoming solar radiation back to space. Thus, because less solar radiation is transmitted, Earth's surface is cooler than it would otherwise be.

Whether a given cloud will cause the surface temperature to be higher or lower than when the sky is clear depends on several factors, including the time of day, the cloud's thickness, its height, and the nature of the particles that compose it. The balance between the cooling and warming effects of clouds is quite close; however, averaging the influence of all the clouds around the globe, cooling predominates.

Heat Budget

Worldwide, Earth's average temperature remains relatively constant, despite seasonal cold spells and heat waves. This stability indicates that a balance exists between the amount of incoming solar radiation and the amount of radiation emitted back to space; otherwise, Earth would be getting progressively colder or progressively warmer. The annual balance of incoming and outgoing radiation is called Earth's **heat budget**. The following examination of this budget provides a good review of the process just discussed.

Figure 2–19 illustrates Earth's heat budget. For simplicity we will use 100 units to represent the solar radiation intercepted at the outer edge of the atmosphere. You have already seen that, of the total radiation that reaches Earth, roughly 30 units (30 percent) are reflected back to space. The remaining 70 units are absorbed, 20 units within the atmosphere and 50 units by Earth's land–sea surface. How does Earth transfer this energy back to space?

If all of the energy absorbed by our atmosphere, land, and water were reradiated directly and immediately back to space, Earth's heat budget would be simple—100 units of radiation received and 100 units returned to space. In fact, this does happen *over time* (minus small quantities of energy that become locked up in biomass that may eventually become fossil fuel). What makes the heat budget complicated is the behavior of certain greenhouse gases, particularly water vapor and carbon dioxide. As you learned, these greenhouse gases absorb a large share of outward-directed infrared radiation and radiate much of that energy back to Earth. This "recycled" energy significantly increases the radiation received by Earth's surface. In addition to the 50 units received directly from the Sun, Earth's surface receives another 94 units from the atmosphere, bringing the total absorbed to 144 units (Table 2–4). A balance is maintained, however, because all 144 units are returned to the atmosphere and eventually lost to space (Table 2–4).

Earth's surface loses the 144 units mainly by emitting longwave radiation skyward. As Figure 2–19 illustrates, 102 units are emitted from Earth's surface and absorbed by the atmosphere. In addition, 12 units are transmitted

Box 2–4   Venus and the Runaway Greenhouse Effect

In describing planetary environments, the late Carl Sagan, one of the foremost experts on extraterrestrial life, called planet Earth "the Heaven of the solar system" and Venus "the Hell." Why should the environments of two planets that are nearly the same size and that are located in close proximity to one another be so dramatically different? The primary reason is the blistering temperature of the Venusian atmosphere, which is caused by a runaway greenhouse effect.

Recall that the greenhouse effect occurs when a planet's atmosphere is transparent to incoming solar energy, allowing it to penetrate to the surface. By contrast, certain gases in the planet's atmosphere can be very good absorbers of radiation emitted from the planet's surface. Some of the energy absorbed by these gases is radiated downward, causing the surface temperatures of these planets to be warmer than they would otherwise be.

Both Earth and Venus experience a greenhouse effect, so why do the surface temperatures of Venus reach 480°C (900°F) when much of Earth is hospitable? The answer is that carbon dioxide, which is a major contributor to the greenhouse effect, makes up 97 percent of the Venusian atmosphere, whereas our own atmosphere contains only 0.036 percent (Figure 2–C).

The primary source of carbon dioxide is outgassing during volcanic eruptions. Why does Earth, which has hundreds of active volcanoes, have an atmosphere that includes only a tiny percentage of carbon dioxide? The answer can be found in Earth's abundant plant life. Plants use carbon dioxide and water in photosynthesis to generate organic matter, while oxygen is released as a by-product. Thus, over millions of years, plant life has altered our atmosphere, making it carbon dioxide–poor and oxygen-rich.

In addition, carbon dioxide dissolved in seawater is used by a vast number of organisms to make their carbonate shells. These shells are eventually deposited as sediment on the ocean floor.

Consequently, huge quantities of carbon dioxide from Earth's atmosphere are continually being converted into organic matter and carbonate sediments. Apparently, Venus does not possess living organisms or a chemical mechanism capable of removing atmospheric carbon dioxide. Hence, our neighboring planet has a carbon dioxide concentration that has reached extreme proportions and a greenhouse effect to match.

Earth's atmosphere is unlikely to experience the extreme conditions that exist on Venus. Nonetheless, scientists are concerned that human activity is increasing carbon dioxide levels sufficiently to alter our climate. Through the burning of fossil fuels, we are converting increasingly larger amounts of oxygen and organic materials into carbon dioxide and water. In this way, we are altering Earth's atmosphere by rapidly undoing what has taken plant life millions of years to accomplish.

We should expect that further studies of Venus will provide insights into how our own atmosphere evolved and how it will respond to changes induced by human activity. These insights could keep our planet from becoming, as some have suggested, "hothouse Earth."

Figure 2–C Venus is shrouded in a hot, cloud-filled atmosphere composed mainly of carbon dioxide. *(Courtesy of NASA)*

through the atmosphere without being absorbed. (Recall that radiation between 8 and 11 micrometers escapes the troposphere most readily because water vapor and carbon dioxide do not absorb these wavelengths.) In addition, energy is carried from Earth's land–sea surface to the atmosphere by water molecules during the process of

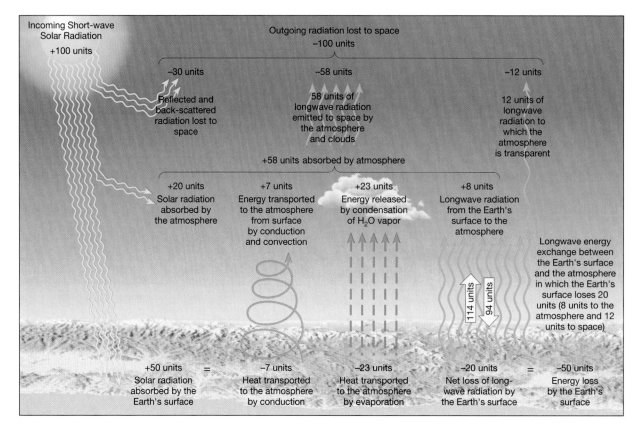

Figure 2–19 Heat budget of Earth and atmosphere. These estimates of the average global energy budget come from satellite observations and radiation studies. As more data are accumulated these numbers will be modified. *(Data from Kiehl, Trenberth, Liou, and others.)*

evaporation (23 units) and by conduction and convection (7 units).

The energy transferred to the atmosphere is reradiated skyward as well as toward the surface. However, a balance has been established, so that on average the atmosphere radiates back to space the same amount of energy as it receives. A careful examination of Table 2–4 and Figure 2–19 confirms that incoming shortwave radiation is in fact balanced by outgoing longwave radiation.

Latitudinal Heat Balance

Because the amount of incoming solar radiation is nearly equal to the amount of outgoing radiation for Earth as a whole, the average worldwide temperature remains constant. However, the balance of incoming and outgoing radiation that holds for the entire planet obviously is not maintained at each latitude. Averaged over the entire year, a zone around Earth between 36°N and 36°S *receives more solar radiation than is lost to space* (Figure 2–20a).

Not surprisingly, this zone, centered on the equator, includes the southern United States and the warm tropical areas of Earth. The opposite is true for higher latitudes, where *more heat is lost through longwave terrestrial radiation than is received.*

A conclusion that might be drawn is that the tropics should be getting hotter and the poles should be getting colder. But we know that is not happening. Instead, the atmosphere and the oceans act as giant thermal engines transferring surplus heat from the tropics poleward. In effect, it is this energy imbalance that drives the winds and the ocean currents.

Furthermore, the radiation budget of a given place fluctuates with changes in cloud cover, atmospheric composition, and, most importantly, Sun angle and length of daylight. Thus, areas of radiation surplus and deficit *migrate* seasonally as the Sun's angle and length of daylight change. Figure 2–20b shows the situation in December, when areas between 15° north and 70° south latitude experience a radiation surplus, whereas the rest of the

Table 2–4 Heat budget of Earth's surface, atmosphere, and overall planetary heat budget.*

Heat Budget of Earth's Surface			
Incoming		**Outgoing**	
Solar radiation	50	Earth radiation	114
Atmospheric radiation	94	Evaporation	23
		Conduction/Convection	7
Total	144	Total	144

Heat Budget of the Atmosphere			
Incoming		**Outgoing**	
Solar radiation	20	Radiation to space	58
Condensation	23	Radiation to surface	94
Earth radiation	102		
Conduction	7		
Total	152	Total	152

Planetary Heat Budget			
Incoming		**Outgoing**	
Solar radiation	100	Reflected and scattered	30
		Atmospheric radiation to space	58
		Earth radiation to space	12
Total	100	Total	100

*Units are in hundredths of the incoming solar radiation.

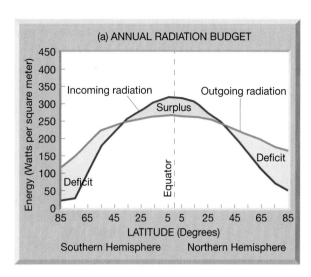

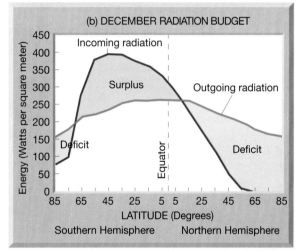

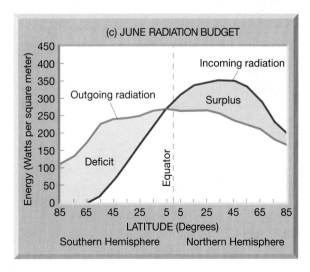

Figure 2–20 Latitudinal heat balance. (a) Averaged over the entire year, we see that equatorward from 36°, the amount of incoming solar radiation exceeds the loss from outgoing Earth radiation. The reverse is true for the middle and high (polar) latitudes, where losses from outgoing Earth radiation exceed gains from incoming solar radiation. (b) Because solar radiation migrates seasonally, during December, areas between 15° north and 70° south latitude experience a radiation surplus, whereas the rest of the world has a deficit. (c) By June, the area of radiation surplus has migrated far to the north.

Box 2–5 Solar Power

Nearly 95 percent of the world's energy needs are derived from fossil fuels, primarily oil, coal, and natural gas. Present estimates indicate that the amount of recoverable fossil fuels may equal 10 trillion barrels of oil, which at the present rate of consumption is enough to last 170 years. Of course, as world population soars, the rate of consumption will climb. Thus, reserves will eventually be in short supply. In the meantime, the environmental impact of burning huge quantities of fossil fuels will undoubtedly have an adverse effect on the environment. How can a growing demand for energy be met without radically altering the planet we inhabit? Although no clear answer has yet emerged, we must consider greater use of alternate energy sources, such as solar power.

The term *solar energy* generally refers to the direct use of the Sun's rays to supply energy for the needs of people. The simplest, and perhaps most widely used, *passive solar collectors* are south-facing windows. As shortwave sunlight passes through the glass, its energy is absorbed by objects in the room. These objects, in turn, radiate longwave heat that warms the air in the room. In the United States, we often use south-facing windows, along with better-insulated and more airtight construction, to reduce heating costs substantially.

More elaborate systems used for home heating involve an *active solar collector*. These roof-mounted devices are normally large, blackened boxes that are covered with glass. The heat they collect can be transferred to where it is needed by circulating air or fluids through pipes. Solar collectors are also used successfully to heat water for domestic and commercial needs. In Israel, for example, about 20 percent of all homes are equipped with some type of solar device.

Although solar energy is free, the necessary equipment and its installation are not. The initial cost of setting up a system, including a supplemental heating unit for times when solar energy is diminished (cloudy days and winter) or unavailable (nighttime), can be substantial. Nevertheless, over the long term, solar energy is economical in most parts of the United States and will become even more cost-efficient as the price of other fuels increases.

Research is currently underway to improve the technologies for concentrating sunlight. One method being examined uses mirrors that track the Sun and keep its rays focused on a receiving tower. A facility, with an array of 2000 mirrors, has been constructed near Barstow, California (Figure 2–D). Solar energy focused on the tower heats water in pressurized panels to over 500°C. The superheated water is then transferred to turbines, which turn electrical generators.

Another type of collector uses photovoltaic (solar) cells that convert the Sun's energy directly into electricity. A large experimental facility using photovoltaic cells is located near Hesperia, California, and supplies electricity to customers of Southern California Edison.

Recently, small rooftop photovoltaic systems have begun appearing in rural households of some Third-World countries, including the Dominican Republic, Sri Lanka, and Zimbabwe. These units are about the size of an open briefcase and use a battery to store electricity that is generated during the daylight hours. In the tropics, these small photovoltaic systems are capable of running a television or radio, plus a few light bulbs, for three to four hours. Although much cheaper than building conventional electric generators, these units are still too expensive for poor families. Consequently, an estimated 2 billion people in developing countries still lack electricity.

world has a radiation deficit. By contrast, Figure 2–20c shows the opposite situation in June, when areas north of about 5° south latitude have a radiation surplus. Because energy imbalances drive the general circulation of the atmosphere and the oceans, the regions of greatest energy transfer also migrate seasonally.

It should be of interest to those who live in the middle latitudes—in the Northern Hemisphere, from the latitude of New Orleans at 30°N to the latitude of Winnipeg, Manitoba, at 50°N—that most heat transfer takes place across this region. Consequently, much of the stormy weather experienced in the middle latitudes can be attributed to this unending transfer of heat from the tropics toward the poles. These processes are discussed in more detail in later chapters.

Figure 2–D Solar One, a solar installation used to generate electricity in the Mojave Desert near Barstow, California. *(Photo by Thomas Braise/The Stock Market)*

Chapter Summary

- Earth has two principal motions—*rotation* and *revolution*. Rotation is the spinning of Earth about its axis. Revolution refers to the movement of Earth in its orbit around the Sun.

- The two most important reasons for the variation in solar energy reaching a particular location are the seasonal changes in the angle at which the Sun's rays strike the surface and the length of daylight. The seasonal variation in the angle of the Sun affects where on Earth the solar rays are most concerned and the thickness of atmosphere the rays must penetrate.

- The four days each year given special significance based on the annual migration of the direct rays of the Sun and its importance to the yearly cycle of weather are (1) June 21/22, the *summer solstice* in the Northern Hemisphere, when the vertical rays of the Sun are striking 23 1/2° north latitude (*Tropic of Cancer*), (2) December 21/22, the *winter solstice* in the Northern Hemisphere, when the vertical rays of the Sun are striking 23 1/2° south latitude (*Tropic of Capricorn*), (3) September 22/23, the *autumnal equinox* in the Northern Hemisphere, when the vertical rays of the Sun strike the equator, and (4) March 21/22, the *spring, or vernal, equinox* in the Northern Hemisphere, when the vertical rays of the Sun also strike the equator.

- *Energy* is the ability to do work. The two major categories of energy are (1) *kinetic energy*, which can be thought of as energy of motion, and (2) *potential energy*, energy that has the capability to do work.

- *Heat* is a form of energy. By contrast, *temperature* refers to intensity—that is, the degree of "hotness." Temperature is a measure of the average kinetic energy of the individual atoms or molecules in a substance.

- The three mechanisms of energy transfer are (1) *conduction*, the transfer of heat through matter by molecular activity, (2) *convection*, the transfer of heat by mass movement or circulation within a substance, and (3) *radiation*, the transfer mechanism by which solar energy reaches our planet.

- *Radiation* or *electromagnetic radiation*, whether x-rays, visible light, heat waves, or radio, travels as various size waves through the vacuum of space at 300,000 kilometers per hour. Shorter wavelengths of radiation are associated with greater energy. The wavelength of visible light ranges from 0.4 micrometer (violet) to 0.7 micrometer (red). Although the Sun emits many forms of radiation, most of the energy is concentrated in the visible and near visible (infrared and ultraviolet) parts of the spectrum. The basic laws of radiation are (1) all objects above –273 °C emit radiant energy, (2) hotter objects radiate more total energy per unit area than colder objects, (3) the hotter the radiating body, the shorter is the wavelength of maximum radiation, and (4) objects that are good absorbers of radiation are also good emitters.

- Approximately 50 percent of the solar energy that strikes the top of the atmosphere reaches Earth's surface. About 30 percent is reflected back to space. The remaining 20 percent of the energy is absorbed by clouds and the atmosphere's gases. The wavelength of the energy being transmitted, as well as the size and nature of the absorbing or reflecting substance, determine whether solar radiation will be scattered, reflected back to space, or absorbed. The fraction of radiation reflected by a surface is called its *albedo*.

- Radiant energy that is absorbed heats Earth and eventually is reradiated skyward. Because Earth has a much lower surface temperature than the Sun, its radiation is in the form of longwave infrared radiation. Because the atmospheric gases, primarily water vapor and carbon dioxide, are more efficient absorbers of terrestrial (longwave) radiation, the atmosphere is heated from the ground up. The general drop in temperature with increased altitude in the troposphere (about 6.5°C/kilometer, a figure called the *normal lapse rate*) supports the fact that the atmosphere is heated from below. The transmission of shortwave solar radiation by the atmosphere coupled with the selective absorption of Earth radiation by atmospheric gases that results in the warming of the atmosphere is referred to as the *greenhouse effect*.

- Because of the annual balance that exists between incoming and outgoing radiation, called the *heat budget*, Earth's average temperature remains relatively constant, despite seasonal cold spells and heat waves.

- Although the balance of incoming and outgoing radiation holds for the entire planet, it is not maintained at each latitude. Averaged over the entire year, a zone around Earth between 36°N and 36°S receives more solar radiation than is lost to space. The opposite is true for higher latitudes, where more heat is lost through outgoing longwave radiation than is received. It is this energy imbalance between the low and high latitudes that drives the global winds and ocean currents, which in turn transfer surplus heat from the tropics poleward. Furthermore, the radiation balance of a given place fluctuates with changes in cloud cover, atmospheric composition, and most important, Sun angle and length of daylight. Thus, areas of radiation surplus and deficit migrate seasonally as the Sun angle and length of daylight change.

Vocabulary Review

advection (p. 37)
albedo (p. 41)
aphelion (p. 28)
atmospheric window (p. 47)
autumnal equinox (p. 31)
blackbody (p. 39)
circle of illumination (p. 31)
conduction (p. 36)
convection (p. 36)
diffused light (p. 42)
energy (p. 35)
greenhouse effect (p. 47)

heat (p. 36)
heat budget (p. 48)
heat energy (p. 36)
inclination of the axis (p. 30)
infrared (p. 38)
kinetic energy (p. 35)
perihelion (p. 28)
plane of the ecliptic (p. 30)
potential energy (p. 35)
radiation or electromagnetic
   radiation (p. 37)
reflection (p. 40)

revolution (p. 28)
rotation (p. 28)
scattering (p. 40)
spring equinox (p. 31)
summer solstice (p. 31)
temperature (p. 36)
Tropic of Cancer (p. 31)
Tropic of Capricorn (p. 31)
ultraviolet (p. 38)
visible light (p. 38)
winter solstice (p. 31)

Review Questions

1. Can the annual variations in Earth–Sun distance adequately explain seasonal temperature changes? Explain.

2. Why does the amount of solar energy received at Earth's surface change when the angle of the Sun changes?

3. List four (4) characteristics of the summer solstice for the Northern Hemisphere. For the Southern Hemisphere.

4. Distinguish between heat and temperature.

5. Describe the three basic mechanisms of energy transfer. Which mechanism is least important meteorologically?

6. What is the difference between convection and advection?

7. Compare visible, infrared, and ultraviolet radiation. For each, indicate whether it is considered short wavelength or long wavelength.

8. In what portion of the electromagnetic spectrum is most solar radiation concentrated?

9. Describe the relationship between the temperature of a radiating body and the wavelengths it emits.

10. Why does the daytime sky usually appear blue?

11. Why may the sky appear to have a red or orange tint near sunrise or sunset?

12. What factors influence albedo from time to time and from place to place?

13. Explain why the atmosphere is heated chiefly by reradiation from Earth's surface, rather than directly by solar radiation.

14. Which gases are the primary heat absorbers in the lower atmosphere? Which one is most influential in weather?

15. How does Earth's atmosphere act as a "greenhouse"?

16. What is responsible for absorbing the largest portion of incoming solar radiation?

17. By what mechanism is most Earth radiation lost to space?

18. What two phenomena are driven by the imbalance of heating that exists between the tropics and poles?

Problems

1. Refer to Figure 2–5 and calculate the noon Sun angle on June 21 and December 21 at 50° north latitude, 0° latitude (the equator), and 20° south latitude. Which of these latitudes has the greatest variation in noon Sun angle between summer and winter?

2. For the latitudes listed in problem 1, determine the length of daylight and darkness on June 21 and December 21 (refer to Table 2–2). Which of these latitudes has the largest seasonal variation in length of daylight? Which latitude has the smallest variation?

3. How would our seasons be affected if Earth's axis were not inclined 23 1/2° to the plane of its orbit, but were instead perpendicular?

4. Describe the seasons if Earth's axis were inclined 40°. Where would the tropics of Cancer and Capricorn be located? How about the Arctic and Antarctic circles?

5. Calculate the noon Sun angle at your location for the equinoxes and solstices.

6. If Earth had no atmosphere, its longwave radiation emission would be lost quickly to space, making the planet approximately 33 K cooler. Calculate the rate of radiation emitted (E), and the wavelength of maximum radiation emission (λ_{max}) for Earth at 255 K.

7. The intensity of solar radiation can be calculated using trigonometry, as shown in Figure 2–21. For simplicity, consider a solar beam of 1 unit width. The surface area over which the beam would be spread changes with Sun angle, such that

$$\text{Surface area} = \frac{1 \text{ unit}}{\sin (\text{Sun angle})}$$

Therefore, if the Sun angle at solar noon is 56°:

$$\text{Surface area} = \frac{1 \text{ unit}}{\sin 56°} = \frac{1 \text{ unit}}{0.829} = 1.206 \text{ units}$$

Using this method and your answers to problem 5, calculate the intensity of solar radiation (surface area) for your location at noon during the summer and winter solstices.

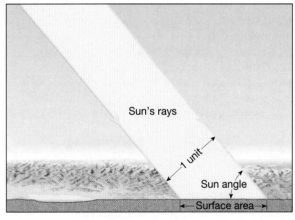

Figure 2–21 Calculating solar intensity.

Atmospheric Science Online

The following are informative and interesting Internet sites that address topics related to those presented in the chapter:

Earth's Radiation Balance (University of Wisconsin):
- **http://cimss.ssec.wisc.edu/wxwise/homerbe.html**

Surface Radiation Research Branch (NOAA):
- **http://titan.srrb.noaa.gov/**

For direct links to these sites and others, chapter objectives and reviews, quiz questions, and topical investigations that utilize Web resources, visit *The Atmosphere, Eighth Edition* Home Page at:
- **http://www.prenhall.com/lutgens**

Temperature

Tropical beach in the Maldives. These islands in the Indian Ocean are just 4.5 degrees north of the equator. Latitude is an important control of temperature. *(Photo by Paul Steel/The Stock Market)*

Figure 3–1 New Yorkers had to cope with frigid temperatures, strong winds, and lots of snow during this January blizzard in 1996. *(Photo by Kaz Chiba/Liaison International)*

Temperature is one of the basic elements of weather and climate. When someone asks what it is like outside, air temperature is often the first element we mention (Figure 3–1). From everyday experience, we know that temperatures vary on different time scales: seasonally, daily, sometimes even hourly. Moreover, we all realize that substantial temperature differences exist from one place to another. In Chapter 2, we learned how air is heated and examined the role of Earth–Sun relationships in causing temperature variations from season to season and from latitude to latitude. In Chapter 3, we will focus on several other aspects of this very important atmospheric property, including factors other than Earth–Sun relationships that act as temperature controls. We will also look at how temperature is measured and expressed and see that temperature data can be of very practical value to us all. Applications include calculations that are useful in evaluating energy consumption, crop maturity, and human comfort.

For the Record: Air Temperature Data

Temperatures recorded daily at thousands of weather stations worldwide provide much of the temperature data compiled by meteorologists and climatologists (see Box 3–1). Hourly temperatures are recorded by an observer or, more frequently, are obtained from a thermograph (a recording thermometer). At many locations, only the maximum and minimum temperatures are obtained. The **daily mean temperature** is determined by averaging the 24 hourly readings or by adding the maximum and minimum temperatures for a 24-hour period and dividing by 2. From the maximum and minimum, the **daily temperature range** is computed by finding the difference between these figures. Other data involving longer periods are also compiled:

1. The **monthly mean temperature** is calculated by adding together the daily means for each day of the month and dividing by the number of days in the month.
2. The **annual mean temperature** is an average of the 12 monthly means.
3. The **annual temperature range** is computed by finding the difference between the warmest and coldest monthly mean temperatures.

Mean temperatures are especially useful for making daily, monthly, or annual comparisons. It is common to hear a weather reporter state that "Last month was the warmest February on record" or "Today Omaha was 10 degrees warmer than Chicago." Temperature ranges are also useful statistics because they give an indication of extremes, a necessary part of understanding the weather and climate of a place or an area (see Box 3–2).

Box 3–1   The World's Hottest and Coldest Places

Most people living in the United States have experienced temperatures of 38°C (100°F) or more. When statistics for the 50 states are examined for the past century or longer, we find that every state has a maximum temperature record of 38°C or higher. Even Alaska has recorded a temperature this high. Its record was set June 27, 1915, at Fort Yukon, a town along the Arctic Circle in the interior of the state.

Maximum Temperature Records

Surprisingly, the state that ties Alaska for the "lowest high" is Hawaii. Panala, on the south coast of the big island, recorded 38°C on April 27, 1931. Although humid tropical and subtropical places like Hawaii are known for being warm throughout the year, they seldom experience maximum temperatures that surpass the low to mid-30s Celsius (90s Fahrenheit).

The highest accepted temperature record for the United States as well as the entire Western Hemisphere is 57°C (134°F). This long-standing record was set at Death Valley, California, on July 10, 1913. Summer temperatures at Death Valley are consistently among the highest in the Western Hemisphere. During June, July, and August, temperatures exceeding 49°C (120°F) are to be expected. Fortunately, Death Valley has few human summertime residents.

Why are summer temperatures at Death Valley so high? In addition to having the lowest elevation in the Western Hemisphere (53 meters below sea level), Death Valley is a desert. Although it is only about 300 kilometers (less than 200 miles) from the Pacific Ocean, mountains cut off the valley from the ocean's moderating influence and moisture. Clear skies allow a maximum of sunshine to strike the dry, barren surface. Because no energy is used to evaporate moisture as occurs in humid regions, all of the energy is available to heat the ground. In addition, subsiding air that warms by compression as it descends is also common to the region and contributes to its high maximum temperatures.

When maximum temperature records for other continents are examined (Table 3–A), we find situations similar to that in North America. That is, the highest temperatures have been recorded at locations that are arid or nearly so. The 58°C reading at Azizia, Libya, in North Africa's Sahara Desert, is the world record. Of course, Antarctica's highest recorded temperature of 15°C (59°F) does not come close to the others.

Minimum Temperature Records

Table 3–B presents minimum temperature records for each of the continents plus Greenland. The stations are ranked, with the lowest minimum at the top of the list. Remember that many other places have no doubt experienced equally low or even lower temperatures; they just were not officially recorded.

The temperature controls that make these places so frigid are predictable, and they should come as no surprise. We should expect extremely cold temperatures during winter in high-latitude places that lack the moderating influence of the ocean. Moreover, stations located on ice sheets and glaciers should be especially cold, as should stations positioned high in the mountains. Clearly *all* of these criteria apply to Vostok in

To examine the distribution of air temperatures over large areas, isotherms are commonly used. An **isotherm** is a line that connects points on a map that have the same temperature (*iso* = equal, *therm* = temperature). Therefore, all points through which an isotherm passes have identical temperatures for the time period indicated. Generally, isotherms representing 5° or 10° temperature differences are used, but any interval may be chosen. Figure 3–2 illustrates how isotherms are drawn on a map. Notice that most isotherms do not pass directly through the observing stations, because the station readings may not coincide with the values chosen for the isotherms. Only an occasional station temperature will be exactly the same as the value of the isotherm, so it is usually necessary to draw the lines by estimating the proper position between stations.

Isothermal maps are valuable tools because they clearly make temperature distribution visible at a glance. Areas of low and high temperatures are easy to pick out. In addition, the amount of temperature change per unit of distance, called the *temperature gradient*, is easy to visualize. Closely spaced isotherms indicate a rapid rate of temperature change, whereas more widely spaced lines indicate a more gradual rate of change. You can see this in Figure 3–2: The isotherms are closer in Colorado and Utah (steeper temperature gradient), whereas the isotherms are spread farther in Texas (gentler temperature gradient).

Table 3–A Maximum temperature records

Area	Highest (°C)	Highest (°F)	Place	Elevation (Meters)	Date
Africa	58	136	Azizia, Libya	114	Sept. 13, 1922
North America	57	134	Death Valley, Calif.	−53	July 10, 1913
Asia	54	129	Tirat Tsvi, Israel	−217	June 21, 1942
Australia	53	128	Cloncurry, Queensland	187	Jan. 16, 1889
Europe	50	122	Seville, Spain	8	Aug. 4, 1881
South America	49	120	Rivadavia, Argentina	203	Dec. 11, 1905
Oceania	42	108	Tuguegarao, Philippines	22	April 29, 1912
Antarctica	15	59	Vanda Station	8	Jan. 5, 1974

Table 3–B Minimum temperature records

Area	Lowest (°C)	Lowest (°F)	Place	Elevation (Meters)	Date
Antarctica	−89	−129	Vostok	3366	Aug. 24, 1960
Asia	−68	−90	Oymyakon, Russia	788	Feb. 6, 1933
Greenland	−66	−87	Northice	2307	Jan. 9, 1954
North America	−63	−81	Snag, Yukon, Canada	578	Feb. 3, 1947
Europe	−55	−67	Ust'Shchugor, Russia	84	Jan.°
South America	−33	−27	Sarmiento, Argentina	264	June 1, 1907
Africa	−24	−11	Ifrane, Morocco	1609	Feb. 11, 1935
Australia	−22	−8	Charlotte Pass, NSW	†	July 22, 1947°°

°Exact date unknown; lowest in 15-year period.
°°And earlier dates.
†Elevation unknown.

Antarctica and to Greenland's Northice station. Snag, in Canada's Yukon Territory, holds the record for North America, so no locations in the United States appear on the list. Nevertheless, Prospect Creek, located north of the Arctic Circle in the Endicott Mountains of Alaska, came close to the North American record on January 23, 1971, when the temperature plunged to −62°C (−80°F). In the lower 48 states, the record of −57°C (−70°F) was set in the mountains at Rogers Pass, Montana, on January 20, 1954.

Without isotherms, a map would be covered with numbers representing temperatures at tens or hundreds of places, which would make patterns difficult to see.

Why Temperatures Vary: The Controls of Temperature

The **controls of temperature** are factors that cause temperature to vary from place to place and from time to time. Chapter 2 examined the most important cause for temperature variation—differences in the receipt of solar radiation. Because variations in Sun angle and length of daylight depend on latitude, they are responsible for warm temperatures in the tropics and colder temperatures poleward. Of course, seasonal temperature changes at a given latitude occur as the Sun's vertical rays migrate toward and away from a place during the year.

But latitude is not the only control of temperature. If it were, we would expect all places along the same parallel to have identical temperatures. Such is clearly not the case. For instance, Eureka, California, and New York City are both coastal cities at about the same latitude, and both places have an annual mean temperature of 11°C (51.8°F). Yet New York City is 9.4°C (16.9°F) warmer than Eureka in July and 9.4°C (16.9°F) colder than Eureka in January.

Box 3–2 Atmospheric Hazard: Heat Waves: Deadly Events

A *heat wave* is a prolonged period of abnormally hot and usually humid weather that typically lasts from a few days to several weeks (Figure 3–A). The impact of heat waves on individuals varies greatly. The elderly are the most vulnerable because heat puts more stress on weak hearts and bodies. The poor, who often cannot afford air conditioning, also suffer disproportionally. Studies also show that the temperatures at which death rates increase varies from city to city. In Dallas, Texas, a temperature of 39°C (103°F) is required before the death rate climbs. In San Francisco, the key temperature is just 29°C (84°F).

Heat waves are deadly events.

The loss of human life by hot spells in summer exceeds that caused by all other weather events in the United States combined, including lightning, rainstorms, floods, hurricanes, and tornadoes. *Thus, the short-term extremes of summer weather conditions are the most severe weather events affecting human life in the United States.*°

This fact is reinforced when examining Table 3–C, which lists average annual weather-related deaths in Illinois. A

Figure 3–A Getting some relief during a heat wave in Washington, D.C. *(Photo by Mark Reinstein/The Image Works*

comparison of values reveals that the number of heat deaths far surpasses those of any other weather condition.

In July 1995 a brief but intense heat wave developed in the central United States. A total of 830 deaths were attributed to this severe five-day event, the worst in 50 years in the northern Midwest. The greatest loss of life

In another example, two cities in Ecuador, Quito and Guayaquil, are relatively close to one another, but the mean annual temperatures at these two cities differ by 12.2°C (22°F). To explain these situations and countless others, we must realize that factors other than latitude also exert a strong influence on temperature. In the next sections, we examine these other controls, which include:

1. Differential heating of land and water
2. Ocean currents
3. Altitude
4. Geographic position
5. Cloud cover and albedo

Land and Water

In Chapter 2, we saw that the heating of Earth's surface controls the heating of the air above it. Therefore, to understand variations in air temperature, we must understand the variations in heating properties of the different surfaces that Earth presents to the Sun—soil, water, trees, ice, and

so on. Different land surfaces reflect and absorb varying amounts of incoming solar energy, which, in turn, cause variations in the temperature of the air above. The greatest contrast, however, is not between different land surfaces, but between land and water (Figure 3–3). In side-by-side bodies of land and water, such as those shown in Figure 3–3, *land heats more rapidly and to higher temperatures than water, and it cools more rapidly and to lower temperatures than water.* Variations in air temperatures, therefore, are much greater over land than over water. Why do land and water heat and cool differently? Several factors are responsible.

1. An important reason that the surface temperature of water rises and falls much more slowly than the surface temperature of land is that *water is highly mobile.* As water is heated, convection distributes the heat through a considerably larger mass. Daily temperature changes occur to depths of 6 meters (20 feet) or more below the surface, and yearly, oceans and deep lakes experience temperature

Table 3–C Annual average weather-related deaths in Illinois

Event	Deaths
Tornado	5
Lightning	6
Winter storms/cold	4
Winds	1
Heavy rains/flood	4
Heat	74
Hail	0

occurred in Chicago, where there were 525 fatalities. The *Chicago Tribune* appropriately labeled the event "a citywide tragedy."

The severity of heat waves is usually greater in cities because of the *urban heat island* (see Box 3–4). Large cities do not cool off as much at night during heat waves as rural areas do, and this can be a critical difference in the amount of heat stress within the inner city (for more on heat stress see Box 4–3).

In addition to the tragic loss of life, the 1995 Midwest heat wave had many other impacts, including the following:

- Energy use vastly increased. This resulted in some power failures during peak stress hours and, of course, in substantially higher electric bills.
- Highways and railroads were damaged due to heat-induced heaving and buckling of roadway joints and rails.
- Many companies reported a substantial reduction in employee work efficiency.
- Shopping declined dramatically.
- In rural areas, livestock were affected. For example, on July 14 the *Wisconsin Journal* reported that 850 dairy cattle had died, major flocks of poultry were killed, and milk production was off by 25 percent.

Not everyone is adversely affected by a heat wave. In response to the 1995 event, tourism from Chicago to nearby (and cooler) Wisconsin increased 10 percent. As we would expect, sales of air conditioners in the Midwest rose as well—more than 50 percent over the previous year.

The 1995 heat wave in the upper Midwest was the most intense in half a century. It provided a sobering lesson by focusing attention on the need for more effective warning and response plans, especially in major urban areas where heat stress is greatest.

*Changnon, Stanley A., et al. "Impacts and Responses to the 1995 Heat Wave: A Call to Action," *Bulletin of the American Meteorological Society*, Vol. 77, No. 7, July 1996, p. 1504.

---

variations through a layer between 200 and 600 meters thick (650 and 2000 feet).

In contrast, heat does not penetrate deeply into soil or rock; it remains near the surface. Obviously, no mixing can occur on land because it is not fluid. Instead, heat must be transferred by the slow process of conduction. Consequently, daily temperature changes are small below a depth of 10 centimeters (4 inches), although some change can occur to a depth of perhaps 1 meter (3 feet). Annual temperature variations usually reach depths of 15 meters (50 feet) or less. Thus, as a result of the mobility of water and the lack of mobility in the solid Earth, a relatively thick layer of water is heated to moderate temperatures during the summer. On land, only a thin layer is heated but to much higher temperatures.

During winter, the shallow layer of rock and soil that was heated in summer cools rapidly. Water bodies, in contrast, cool slowly as they draw on the reserve of heat stored within. As the water surface cools, vertical motions are established. The chilled surface water, which is dense, sinks and is replaced by warmer water from below, which is less dense. Consequently, a larger mass of water must cool before the temperature at the surface will drop appreciably.

2. Because land surfaces are opaque, heat is absorbed only at the surface. This fact is easily demonstrated at a beach on a hot summer afternoon by comparing the surface temperature of the sand to the temperature just a few centimeters beneath the surface. Water, being more transparent, allows some solar radiation to penetrate to a depth of several meters.

3. The **specific heat** (the amount of heat needed to raise the temperature of 1 gram of a substance 1°C) is more than three times greater for water than for land. Thus, water requires considerably more heat to raise its temperature the same amount as an equal quantity of land.

4. Evaporation (a cooling process) from water bodies is greater than from land surfaces. Energy is

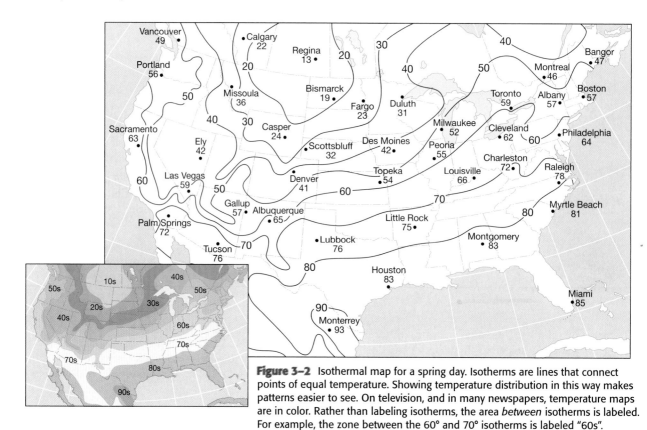

Figure 3–2 Isothermal map for a spring day. Isotherms are lines that connect points of equal temperature. Showing temperature distribution in this way makes patterns easier to see. On television, and in many newspapers, temperature maps are in color. Rather than labeling isotherms, the area *between* isotherms is labeled. For example, the zone between the 60° and 70° isotherms is labeled "60s".

required to evaporate water. When energy is used for evaporation, it is not available for heating.°

°Evaporation is an important process that is discussed more thoroughly in the section on "Water's Changes of State" in Chapter 4.

All these factors collectively cause water to warm more slowly, store greater quantities of heat energy, and cool more slowly than land.

It is interesting to compare monthly temperature data for two cities to demonstrate the moderating influence of a large water body and the temperature extremes

Figure 3–3 The differential heating of land and water is an important control of air temperatures. Temperatures over land rise to higher levels and fall to lower levels than over water. *(Photo by Warren Faidley/Weatherstock)*

associated with land. Vancouver, British Columbia, is located along the windward Pacific coast, whereas Winnipeg, Manitoba, is in a continental position far from the influence of water (Figure 3–4). Both cities are at about the same latitude and thus experience similar Sun angles and lengths of daylight. Winnipeg, however, has a mean January temperature that is 20°C (36°F) lower than Vancouver's. Conversely, Winnipeg's July mean is 2°C (3.6°F) higher than Vancouver's. Although their latitudes are nearly the same, Winnipeg, which has no water influence, experiences much greater temperature extremes than Vancouver, which does. The graph in Figure 3–4 clearly depicts the contrasting changes in monthly mean temperatures for marine and continental locations.

On a different scale, the moderating influence of water may also be demonstrated when temperature variations in the Northern and Southern hemispheres are compared. The views of Earth in Figure 3–5 show the uneven distribution of land and water over the globe. Water covers 61 percent of the Northern Hemisphere; land represents the remaining 39 percent. However, the figures for the Southern Hemisphere (81 percent water, 19 percent land) reveal why it is correctly called the *water hemisphere*. Between 45° north and 79° north latitude there is actually more land than water, whereas between 40° south and 65° south latitude there is almost no land to interrupt the oceanic and atmospheric circulation.

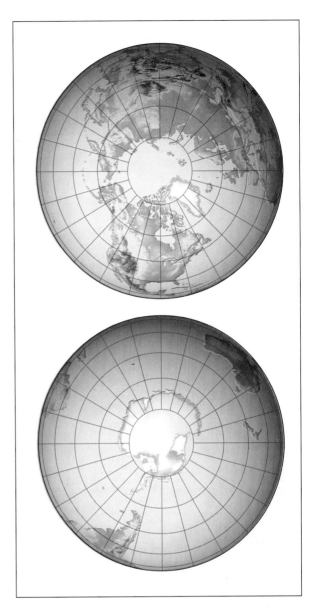

Figure 3–5 The uneven distribution of land and water between the Northern and Southern hemispheres. Almost 81 percent of the Southern Hemisphere is covered by the oceans—20 percent more than the Northern Hemisphere.

Table 3–1 portrays the considerably smaller annual temperature ranges in the water-dominated Southern Hemisphere compared with the Northern Hemisphere.

Ocean Currents

You probably have heard of the Gulf Stream, an important surface current in the Atlantic Ocean that flows northward along the eastern coast of the United States (Figure 3–6). Surface currents like this one are set in motion by

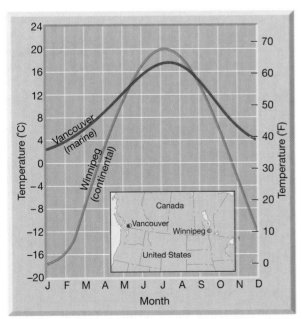

Figure 3–4 Mean monthly temperatures for Vancouver, British Columbia, and Winnipeg, Manitoba. Vancouver has a much smaller annual temperature range owing to the strong marine influence of the Pacific Ocean.

Table 3–1 Variation in mean annual temperature range (°C) with latitude

Latitude	Nothern Hemisphere	Southern Hemisphere
0	0	0
15	3	4
30	13	7
45	23	6
60	30	11
75	32	26
90	40	31

the wind. At the water surface, where the atmosphere and ocean meet, energy is passed from moving air to the water through friction. As a consequence, the drag exerted by winds blowing steadily across the ocean causes the surface layer of water to move. Thus, major horizontal movements of surface waters are closely related to the circulation of the atmosphere, which in turn is driven by the unequal heating of Earth by the Sun (Figure 3–7).°

°The relationship between global winds and surface ocean currents is examined in Chapter 7.

Figure 3–6 This satellite image of a portion of the North Atlantic shows the complexities of the Gulf Stream. It flows along the eastern coast of Florida and the Carolinas and northward into the North Atlantic. Reds and yellow denote warmer waters. The current transports heat from lower latitudes toward the North Pole. Meanders of the Gulf Stream pinch off to form eddies that may move about the ocean for up to two years before dissipating. *(Courtesy of National Oceanic and Atmospheric Administration)*

Surface ocean currents have an important effect on climate. It is known that for Earth as a whole, the gains in solar energy equal the losses to space of heat radiated from the surface. When most latitudes are considered individually, however, this is not the case. There is a net gain of energy in lower latitudes and a net loss at higher latitudes. Because the tropics are not becoming progressively warmer, nor the polar regions colder, there must be a large-scale transfer of heat from areas of excess to areas of deficit. This is indeed the case. *The transfer of heat by winds and ocean currents equalizes these latitudinal energy imbalances.* Ocean water movements account for about a quarter of this total heat transport, and winds the remaining three-quarters.

The moderating effect of poleward-moving warm ocean currents is well known. The North Atlantic Drift, an extension of the warm Gulf Stream, keeps wintertime temperatures in Great Britain and much of Western Europe warmer than would be expected for their latitudes (London is farther north than St. John's, Newfoundland). Because of the prevailing westerly winds, the moderating effects are carried far inland. For example, Berlin (52° north latitude) has a mean January temperature similar to that experienced at New York City, which lies 12° latitude farther south. The January mean at London (51° north latitude) is 4.5°C (8.1°F) higher than at New York City.

In contrast to warm ocean currents like the Gulf Stream, the effects of which are felt most during the winter, cold currents exert their greatest influence in the tropics or during the summer months in the middle latitudes. For example, the cool Benguela Current off the western coast of southern Africa moderates the tropical heat along this coast. Walvis Bay (23° south latitude), a town adjacent to the Benguela Current, is 5°C (9°F) cooler in summer than Durban, which is 6° latitude farther poleward but on the eastern side of South Africa away from the influence of the current (Figure 3–7). The east and west coasts of South America provide another example. Figure 3–8 shows monthly mean temperatures for Rio de Janeiro, Brazil, which is influenced by the warm Brazilian Current and Arica, Chile, which is adjacent to the cold Peruvian Current. Closer to home, because of the cold California current, summer temperatures in subtropical coastal southern California are lower by 6°C (10.8°F) or more compared to East Coast stations.

Altitude

The two cities in Ecuador mentioned earlier, Quito and Guayaquil, demonstrate the influence of altitude on mean temperature. Both cities are near the equator and relatively close to one another, but the annual mean temperature at Guayaquil is 25.5°C (77.9°F) compared with

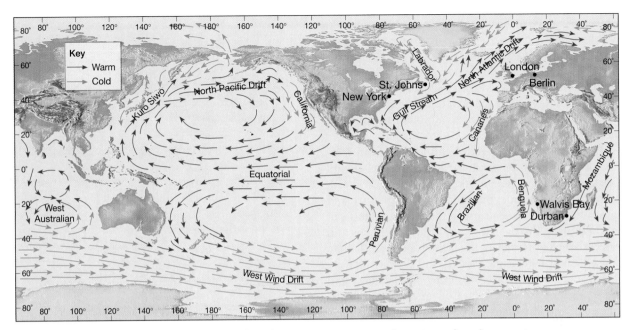

Figure 3–7 Major surface ocean currents. Poleward-moving currents are warm, and equatorward-moving currents are cold. Surface ocean currents are driven by global winds and play an important role in redistributing heat around the globe. Note that cities mentioned in the text discussion are shown on this map.

Quito's mean of 13.3°C (55.9°F). The difference may be understood when the cities' elevations are noted. Guayaquil is only 12 meters (39 feet) above sea level, whereas Quito is high in the Andes Mountains at 2800 meters (9200 feet) (Figure 3–9).

Recall that temperatures drop an average of 6.5°C per kilometer (3.5°F per 1000 feet) in the troposphere; thus, cooler temperatures are to be expected at greater heights.

Figure 3–8 Monthly mean temperatures for Rio de Janeiro, Brazil, and Arica, Chile. Both are coastal cities near sea level. Even though Arica is closer to the equator than Rio de Janeiro, its temperatures are cooler. Arica is influenced by the cold Peruvian Current, whereas Rio de Janeiro is adjacent to the warm Brazilian Current.

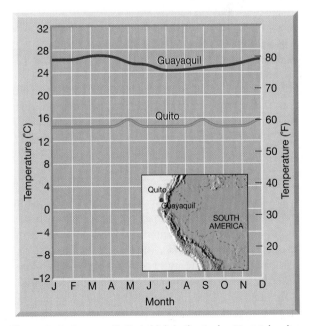

Figure 3–9 Because Quito is high in the Andes Mountains, it experiences much cooler temperatures than Guayaquil, which is near sea level.

Yet the magnitude of the difference is not totally explained by the normal lapse rate. If this figure were used, we would expect Quito to be about 18.2°C (32.7°F) cooler than Guayaquil, but the difference is only 12.2°C (22°F). The fact that high-altitude places, such as Quito, are warmer than the value calculated using the normal lapse rate results from the absorption and reradiation of solar energy by the ground surface.

In addition to the effect of altitude on mean temperatures, the daily temperature range also changes with variations in height. Not only do temperatures drop with an increase in altitude, but atmospheric pressure and density also diminish. Because of the reduced density at high altitudes, the overlying atmosphere absorbs and reflects a smaller portion of the incoming solar radiation. Consequently, with an increase in altitude, the intensity of solar radiation increases, resulting in relatively rapid and intense daytime heating. Conversely, rapid nighttime cooling is also the rule in high mountain locations. Therefore, stations located high in the mountains generally have a greater daily temperature range than do stations at lower elevations.

Geographic Position

The geographic setting can greatly influence the temperatures experienced at a specific location. A coastal location where prevailing winds blow from the ocean onto the shore (a *windward* coast) experiences considerably different temperatures than does a coastal location where prevailing winds blow from the land toward the ocean (a *leeward* coast). In the first situation, the windward coast will experience the full moderating influence of the ocean—cool summers and mild winters—compared to an inland station at the same latitude.

A leeward coastal situation, however, will have a more continental temperature regime because the winds do not carry the ocean's influence onshore. Eureka, California, and New York City, the two cities mentioned earlier, illustrate this aspect of geographic position (Figure 3–10). The annual temperature range at New York City is 19°C (34°F) greater than Eureka's.

Seattle and Spokane, both in the State of Washington, illustrate a second aspect of geographic position: mountains acting as barriers. Although Spokane is only about 360 kilometers (225 miles) east of Seattle, the towering Cascade Range separates the cities. Consequently, Seattle's temperatures show a marked marine influence, but Spokane's are more typically continental (Figure 3–11). Spokane is 7°C (12.6°F) cooler than Seattle in January and 4°C (7.2°F) warmer than Seattle

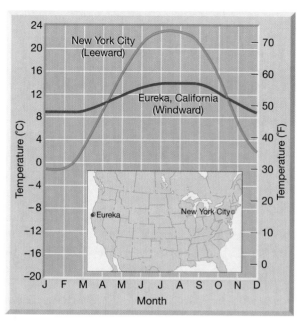

Figure 3–10 Monthly mean temperatures for Eureka, California, and New York City. Both cities are coastal and located at about the same latitude. Because Eureka is strongly influenced by prevailing winds from the ocean and New York City is not, the annual temperature range at Eureka is much smaller.

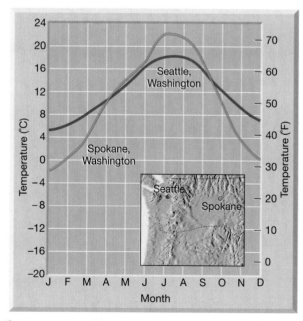

Figure 3–11 Monthly mean temperatures for Seattle and Spokane, Washington. Because the Cascade Mountains cut off Spokane from the moderating influence of the Pacific Ocean, its annual temperature range is greater than Seattle's.

in July. The annual range at Spokane is 11°C (nearly 20°F) greater than at Seattle. The Cascade Range effectively cuts off Spokane from the moderating influence of the Pacific Ocean.

Cloud Cover and Albedo

You may have noticed that clear days are often warmer than cloudy ones and that clear nights usually are cooler than cloudy ones. This demonstrates that cloud cover is another factor that influences temperature in the lower atmosphere. Studies using satellite images show that at any particular time about half of our planet is covered by clouds. Cloud cover is important because many clouds have a high albedo and therefore reflect a significant proportion of the sunlight that strikes them back to space. By reducing the amount of incoming solar radiation, daytime temperatures will be lower than if the clouds were absent and the sky were clear (Figure 3–12a). As was noted in Chapter 2, the albedo of clouds depends on the thickness of the cloud cover and can vary from 25 to 80 percent (see Table 2–3, Chapter 2).

At night, clouds have the opposite effect as during daylight. They absorb outgoing Earth radiation and emit a portion of it toward the surface (Figure 3–12b). Consequently, some of the heat that otherwise would have been lost remains near the ground. Thus, nighttime air temperatures do not drop as low as they would on a clear night. The effect of cloud cover is to reduce the daily temperature range by lowering the daytime maximum and raising the nighttime minimum. This is illustrated nicely in Figure 3–13.

The effect of cloud cover on reducing maximum temperatures can also be detected when monthly mean temperatures are examined for some stations. For example, each year, much of southern Asia experiences an extended period of relative drought during the cooler low-Sun period; it is then followed by heavy monsoon rains.° The graph for Rangoon, Myanmar (formerly Burma), illustrates this pattern (Figure 3–14). Notice that the highest monthly mean temperatures occur in April and May, before the summer solstice, rather than in July and August, as normally occurs at most stations in the Northern Hemisphere. Why?

The reason is that during the summer months when we would usually expect temperatures to climb, the extensive cloud cover increases the albedo of the region, which

°This pattern is associated with the monsoon circulation and is discussed in Chapter 7.

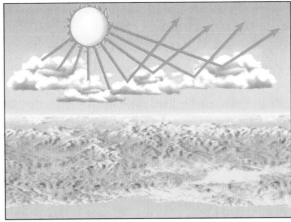

(a)

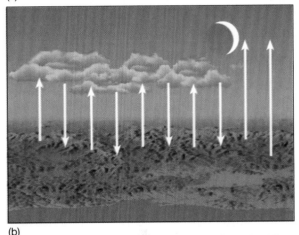

(b)

Figure 3–12 How clouds reduce the daily temperature range. (a) During daylight hours, clouds reflect solar radiation back to space. Therefore, the maximum temperature is lower than if the sky were clear. (b) At night, the minimum temperature will not fall as low because clouds retard the loss of heat.

reduces incoming solar radiation. As a result, the highest monthly mean temperatures occur in late spring when the skies are still relatively clear.

Cloudiness is not the only phenomenon that increases albedo and thereby reduces air temperature. We also recognize that snow- and ice-covered surfaces have high albedos. This is one reason why mountain glaciers do not melt away in the summer and why snow may still be present on a mild spring day. In addition, during the winter, when snow covers the ground, daytime maximums on a sunny day are cooler than they otherwise would be because energy that the land would have absorbed and used to heat the air is reflected and lost.

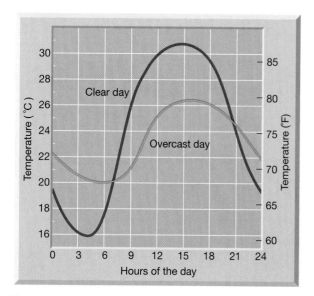

Figure 3–13 The daily cycle of temperature at Peoria, Illinois, for two July days. On the clear day, the maximum temperature was higher and the minimum temperature was lower than on the cloudy day.

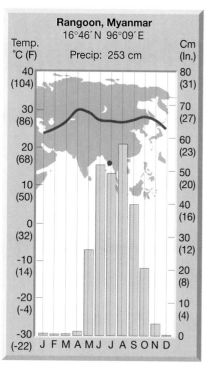

Figure 3–14 Monthly mean temperatures (curve) and monthly mean precipitation (bar graph) for Rangoon, Myanmar. The highest mean temperature occurs in April, just before the onset of heavy summer rains. The abundant cloud cover associated with the rainy period reflects back into space the solar energy that otherwise would strike the ground and raise summer temperatures.

World Distribution of Temperatures

Take a moment to study the two world isothermal maps (Figures 3–15 and 3–16). From hot colors near the equator to cool colors toward the poles, these maps portray sea-level temperatures in the seasonally extreme months of January and July. On these maps you can study global temperature patterns and the effects of the controlling factors of temperature, especially latitude, the distribution of land and water, and ocean currents. Like most isothermal maps of large regions, all temperatures on these world maps have been reduced to sea level to eliminate the complications caused by differences in altitude.

On both maps, the isotherms generally trend east and west and show a decrease in temperatures poleward from the tropics. They illustrate one of the most fundamental aspects of world temperature distribution: that the effectiveness of incoming solar radiation in heating Earth's surface and the atmosphere above it is largely a function of latitude.

Moreover, there is a latitudinal shifting of temperatures caused by the seasonal migration of the Sun's vertical rays. To see this, compare the color bands by latitude on the two maps. For example, on the January map, the "hot spots" of 30°C are *south* of the equator, but in July they have shifted *north* of the equator.

If latitude were the only control of temperature distribution, our analysis could end here, but this is not the case. The added effect of the differential heating of land and water is clearly reflected on the January and July temperature maps. The warmest and coldest temperatures are found over land—note the coldest area, a purple oval in Siberia, and the hottest areas, the deep orange ovals—all over land. Consequently, because temperatures do not fluctuate as much over water as over land, the north–south migration of isotherms is greater over the continents than over the oceans. In addition, it is clear that the isotherms in the Southern Hemisphere, where there is little land and where the oceans predominate, are much more regular than in the Northern Hemisphere, where they bend sharply northward in July and southward in January over the continents.

Isotherms also reveal the presence of ocean currents. Warm currents cause isotherms to be deflected toward the poles, whereas cold currents cause an equatorward bending. The horizontal transport of water poleward warms the overlying air and results in air temperatures that are higher than would otherwise be expected for the latitude. Conversely, currents moving toward the equator produce cooler than expected air temperatures.

Figures 3–15 and 3–16 show the seasonal extremes of temperature, so comparing them enables us to see

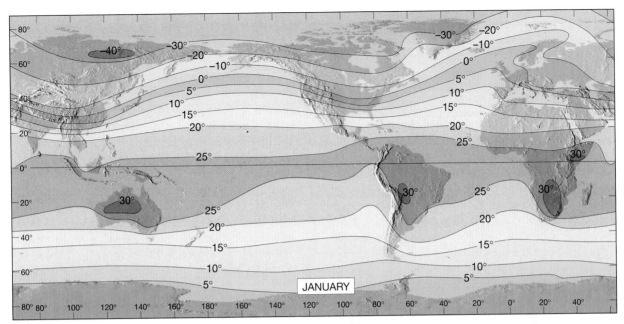

Figure 3–15 World mean sea-level temperatures in January in degrees Celsius.

the annual range of temperature from place to place. Comparing the two maps shows that a station near the equator has a very small annual range because it experiences little variation in the length of daylight and it always has a relatively high Sun angle. A station in the middle latitudes, however, experiences wide variations in Sun angle and length of daylight and hence large variations in temperature. Therefore, we can state that the annual temperature range increases with an increase in latitude (see Box 3–3).

Moreover, land and water also affect seasonal temperature variations, especially outside the tropics. A continental location must endure hotter summers and colder winters than a coastal location. Consequently, outside the tropics the annual range will increase with an increase in continentality.

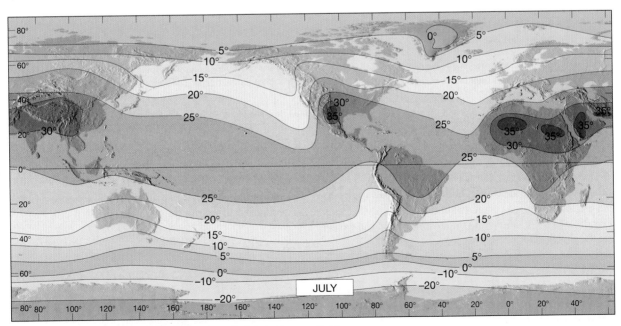

Figure 3–16 World mean sea-level temperatures in July in degrees Celsius.

Box 3-3 Latitude and Temperature Range

Gregory J. Carbone*

Latitude, because of its influence on Sun angle, is the most important temperature control. Figures 3–15 and 3–16 clearly show higher temperatures in tropical locations and lower temperatures in polar regions. The maps also show that higher latitudes experience a greater range of temperatures during the year than do lower latitudes. Notice also that the temperature gradient between the subtropics and the poles is greatest during the winter season. A look at two cities—San Antonio, Texas, and Winnipeg, Manitoba—illustrates how seasonal differences in Sun angle and day length account for these temperature patterns (Figure 3–B). Figure 3–C

shows the annual march of temperature for the two cities, while Figure 3–D illustrates the Sun angles for the June and December solstices.

San Antonio and Winnipeg are a fixed distance apart (approximately 20.5° latitude), so the difference in Sun angles between the two cities is the same throughout the year. However, in December, when the Sun's rays are least direct, this difference more strongly affects the intensity of solar radiation received at Earth's surface. Therefore, we expect both a greater difference in temperatures between the two stations during the winter and

Figure 3–B Winnipeg is more than 20 degrees of latitude north of San Antonio.

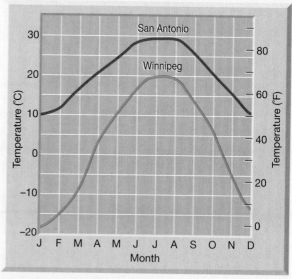

Figure 3–C The annual temperature range at Winnipeg is much greater than at San Antonio.

A classic example of the effect of latitude and continentality on annual temperature range is Yakutsk, Russia. This city in the heart of Siberia is just a few degrees south of the Arctic Circle and far from the influence of water. As a result, Yakutsk has an average annual temperature range of 62.2°C (112°F), among the greatest ranges in the world.

Cycles of Air Temperature

You know from experience that a rhythmic rise and fall of air temperature occurs almost every day. Your experience is confirmed by thermograph records like the one in Figure 3–17

(a thermograph is an instrument that continuously records temperature). The temperature curve reaches a minimum around sunrise (Figure 3–18). It then climbs steadily to a maximum between 2 P.M. and 5 P.M. The temperature then declines until sunrise the following day.

Daily Temperature Variations

The primary control of the daily cycle of air temperature is as obvious as the cycle itself: It is Earth's daily rotation, which causes a location to move into the circle of illumination for part of each day, and then out of it. As the Sun's angle increases throughout the morning, the intensity of

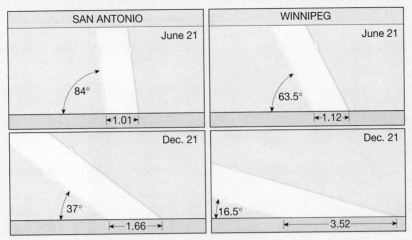

Figure 3–D A comparison of Sun angles (solar noon) for summer and winter solstices at two cities. The space covered by a 90° angle = 1.00.

a greater winter-to-summer difference in solar intensity at Winnipeg than at San Antonio, accounting for a greater annual temperature range at the more northerly station. Table 2–2, Chapter 2, shows that seasonal contrasts in day length also contribute to different temperature patterns at the two cities.

°Professor Carbone is a faculty member in the Department of Geography at the University of South Carolina.

sunlight also rises, reaching a peak at local noon and gradually diminishing in the afternoon.

Figure 3–19 shows the daily variation of incoming solar energy versus outgoing Earth radiation and the resulting temperature curve for a typical middle-latitude location at the time of an equinox. During the night, the atmosphere and the surface of Earth cool as they radiate heat away that is not replaced by incoming solar energy. The minimum temperature, therefore, occurs about the time of sunrise, after which the Sun again heats the ground, which, in turn, heats the air.

It is apparent that the time of highest temperature does not generally coincide with the time of maximum radiation. By comparing Figures 3–17 and 3–19, you can see that the curve for incoming solar energy is symmetrical with respect to noon, but the daily air temperature curves are not. The delay in the occurrence of the maximum until mid-to-late afternoon is termed the *lag of the maximum*.

Although the intensity of solar radiation drops in the afternoon, it still exceeds outgoing energy from Earth's surface for a period of time. This produces an energy surplus for up to several hours in the afternoon and contributes

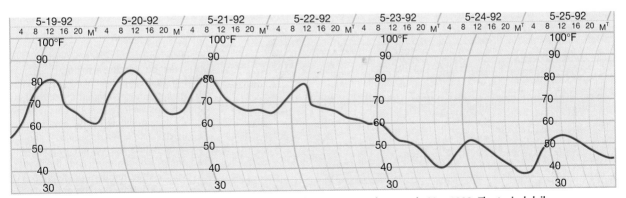

Figure 3–17 Thermograph of temperatures in Peoria, Illinois, during a seven-day span in May 1992. The typical daily rhythm, with minimums around sunrise and maximums in mid- to late afternoon, occurred on most days. The obvious exception occurred on May 23, when the maximum was reached at midnight and temperatures dropped throughout the day.

Figure 3–18 The minimum daily temperature usually occurs near the time of sunrise. As the ground and air cool during the nighttime hours, familiar early morning phenomena such as dew, frost, and ground fog may form. The area shown here is near Poplar Bluff, Missouri. *(Photo by Michael Collier/Stock Boston)*

substantially to the lag of the maximum. In other words, as long as the solar energy gained exceeds the rate of Earth radiation lost, the air temperature continues to rise. When the input of solar energy no longer exceeds the rate of energy lost by Earth, the temperature falls.

The lag of the daily maximum is also a result of the process by which the atmosphere is heated. Recall that air is a poor absorber of most solar radiation; consequently, it is heated primarily by energy reradiated from Earth's surface. The rate at which Earth supplies heat to the atmosphere through radiation, conduction, and other means, however, is not in balance with the rate at which the atmosphere radiates heat away. Generally, for a few hours after the period of maximum solar radiation, more heat is supplied to the atmosphere by Earth's surface than is emitted by the atmosphere to space. Consequently, most locations experience an increase in air temperature during the afternoon.

In dry regions, particularly on cloud-free days, the amount of radiation absorbed by the surface will generally be high. As a result, the time of the maximum temperature at these locales will often occur quite late in the afternoon. Humid locations, in contrast, will frequently experience a shorter time lag in the occurrence of their temperature maximum.

Magnitude of Daily Temperature Changes

The magnitude of daily temperature changes is variable and may be influenced by locational factors, or local weather conditions, or both (see Box 3–4). Three common examples illustrate this point. The first two relate to location and the third pertains to the influence of clouds.

1. Variations in Sun angle are relatively great during the day in the middle and low latitudes. However, points near the poles experience a low Sun angle all day. Consequently, the temperature change experienced during a day in the high latitudes is small.

2. A windward coast is likely to experience only modest variations in the daily cycle. During a typical 24-hour period the ocean warms less than 1°C. As a result, the air above it shows a correspondingly slight change in temperature. For example,

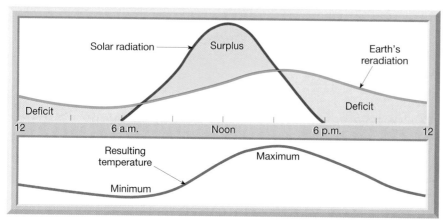

Figure 3–19 The daily cycle of incoming solar radiation, Earth's radiation, and the resulting temperature cycle. This example is for a midlatitude site around the time of an equinox. As long as solar energy gained exceeds outgoing energy emitted by Earth (zone labeled "Surplus"), the temperature rises. When outgoing energy from Earth exceeds the input of solar energy, temperature falls ("Deficit"). Note that the daily temperature cycle *lags* behind the solar radiation input by a couple of hours.

Eureka, California, a windward coastal station, consistently has a lower daily temperature range than Des Moines, Iowa, an inland city at about the same latitude. Annually, the daily range at Des Moines averages 10.9°C (19.6°F) compared with 6.1°C (11°F) at Eureka, a difference of 4.8°C (8.6°F).

3. As mentioned earlier, an overcast day is responsible for a flattened daily temperature curve (see Figure 3–13). By day, clouds block incoming solar radiation and so reduce daytime heating. At night, the clouds retard the loss of radiation by the ground and air. Therefore, nighttime temperatures are warmer than they otherwise would have been.

Although the rise and fall of daily temperatures usually reflect the general rise and fall of solar radiation, such is not always the case. For example, a glance back at Figure 3–17 reveals that on May 23, the maximum temperature occurred at midnight, after which temperatures fell throughout the day. If records for a station are examined for a period of several weeks, nonperiodic variations are seen. Obviously these are not Sun-controlled. Such irregularities are caused primarily by the passage of atmospheric disturbances (weather systems) that are often accompanied by variable cloudiness and winds that bring air having contrasting temperatures. Under these circumstances, the maximum and minimum temperatures may occur at any time of the day or night.

Annual Temperature Variations

Each year, the months of the highest and lowest temperatures do not coincide with the periods of maximum and minimum incoming solar radiation. Poleward of the tropics the greatest intensity of solar radiation occurs at the time of the summer solstice in June, yet the months of July and August are generally the warmest of the year in the Northern Hemisphere. Conversely, a minimum of solar energy is received in December at the time of the winter solstice, but January and February are usually colder.

The fact that the occurrence of annual maximum and minimum radiation does not coincide with the times of temperature maximums and minimums indicates that the amount of solar radiation received is not the only factor determining the temperature at a particular location. Recall that places equatorward of about 36° receive more solar radiation than is lost to space and that the opposite is true of more poleward regions. Based on this imbalance between incoming and outgoing radiation, any location in the southern United States, for example, should continue to get warmer late into autumn.

But this does not occur because more poleward locations begin experiencing a negative radiation balance shortly after the summer solstice. As the temperature contrasts become greater, the atmosphere and ocean currents "work harder" to transport heat from lower latitudes poleward.

Temperature Measurement

Thermometers are "meters of therms": They measure temperature. Thermometers measure temperature either mechanically or electrically.

Mechanical Thermometers

Most substances expand when heated and contract when cooled, so most common thermometers are based on this property. More precisely, they rely on the fact that different substances react to temperature changes differently.

The **liquid-in-glass thermometer** shown in Figure 3–20 is a simple instrument that provides relatively

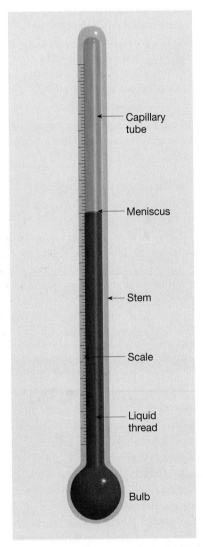

Figure 3–20 The main components of a liquid-in-glass thermometer.

Box 3–4 How Cities Influence Temperature: The Urban Heat Island

The most apparent human impact on climate is the modification of the atmospheric environment by the building of cities. The construction of every factory, road, office building, and house destroys microclimates and creates new ones of great complexity.

The most studied and well-documented urban climatic effect is the *urban heat island*. The term refers to the fact that temperatures within cities are generally higher than in rural areas. The heat island is evident when temperature data such as appear in Table 3–D are examined. As is typical, the data for Philadelphia show the heat island is most pronounced when minimum temperatures are examined. The magnitude of the temperature differences shown by Table 3–D is probably even greater than the figures indicate, because temperatures observed at suburban airports are usually higher than those in truly rural environments.

Table 3–D Average temperatures (°C) for suburban Philadelphia Airport and downtown Philadelphia (10-year averages).

	Airport	**Downtown**
Annual mean	12.8	13.6
Mean June max	27.8	28.2
Mean December max	6.4	6.7
Mean June min	16.5	17.7
Mean December min	–2.1	–0.4

Source: After H. Neuberger and J. Cahir, Principles of Climatology *(New York: Holt, Rinehart and Winston, 1969), 128.*

Figure 3–E, which shows the distribution of average minimum temperatures in the Washington, D.C., metropolitan area for the three-month winter period over a five-year span, also illustrates a well-developed heat island. The warmest winter temperatures occurred in the heart of the city, whereas the suburbs and surrounding countryside experienced average minimum temperatures that were as much as 3.3°C (6°F) lower. Remember that these temperatures are averages. On many clear, calm nights the temperature difference between the city center and the countryside was considerably greater, often 11°C (20°F) or more. Conversely, on many overcast or windy nights the temperature differential approached zero degrees.

Why are cities warmer? The radical change in the surface that results when rural areas are transformed into cities is a significant cause of the urban heat island. First, the tall buildings and the concrete and asphalt of the city absorb and store greater quantities of solar radiation than do the vegetation and soil typical of rural areas. In addition, because the city surface is impermeable, the runoff of water following a rain is rapid, resulting in a significant reduction in the evaporation rate. Hence, heat that once would have been used to convert liquid water to a gas now goes to increase further the surface temperature. At night, as both the city and countryside cool by radiative losses, the stone-like surface of the city gradually releases the additional heat accumulated during the day, keeping the urban air warmer than that of the outlying areas.

accurate readings over a wide temperature range. Its design has remained essentially unchanged ever since it was developed in the late 1600s. When temperature rises, the molecules of fluid grow more active and spread out (the fluid expands). Expansion of the fluid in the bulb is much greater than the expansion of the enclosing glass. As a consequence, a thin "thread" of fluid is forced up the capillary tube. Conversely, when temperature falls, the liquid contracts and the thread of fluid moves back down the tube toward the bulb. The movement of the end of this thread (known as the *meniscus*) is calibrated against an established scale to indicate the temperature.

The highest and lowest temperatures that occur each day are of considerable importance and are often obtained by using specially designed liquid-in-glass thermometers. Mercury is the liquid used in the **maximum thermometer**, which has a narrowed passage called a *constriction* in the bore of the glass tube just above the bulb (Figure 3–21). As the temperature rises, the mercury expands and is forced through the constriction. When the temperature falls, the constriction prevents a return of mercury to the bulb. As a result, the top of the mercury column remains at the highest point (maximum temperature attained during the measurement period). The

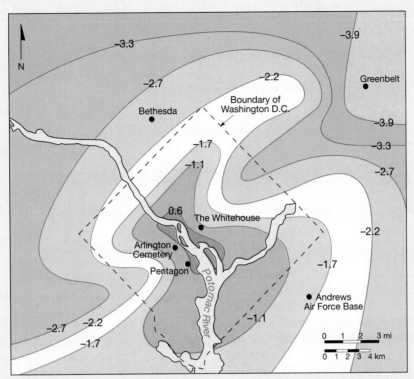

Figure 3–E The heat island of Washington, D.C., as shown by the average minimum temperatures (°C) during the winter season (December through February). The city center had an average minimum that was nearly 4°C higher than some outlying areas. *(After Clarence A. Woolum, "Notes from the Study of the Microclimatology of the Washington, D.C. Area for the Winter and Spring Seasons,"* Weatherwise, *17, no. 6 (1964), 264, 267.)*

A portion of the urban temperature rise is also attributed to waste heat from sources such as home heating and air conditioning, power generation, industry, and transportation. In addition, the "blanket" of pollutants over a city contributes to the heat island by absorbing a portion of the upward-directed long-wave radiation emitted by the surface and re-emitting some of it back to the ground.

instrument is reset by shaking or by whirling it to force the mercury through the constriction back into the bulb. Once the thermometer is reset, it indicates the current air temperature. A common thermometer used to measure human body temperature is an example of a maximum thermometer.

In contrast to a maximum thermometer that contains mercury, a **minimum thermometer** contains a liquid of low density, such as alcohol. Within the alcohol, and resting at the top of the column, is a small dumbbell-shaped index (Figure 3–21). As the air temperature drops, the column shortens and the index is pulled toward the bulb by the effect of surface tension with the meniscus. When the temperature subsequently rises, the alcohol flows past the index, leaving it at the lowest temperature reached. To return the index to the top of the alcohol column, the thermometer is simply tilted. Because the index is free to move, the minimum thermometer must be mounted horizontally; otherwise the index will fall to the bottom.

Another commonly used thermometer is the **bimetal strip**. As the name indicates, this thermometer consists of two thin strips of metal that are bonded together and have widely different expansion properties. When the temperature changes, both metals expand or contract, but

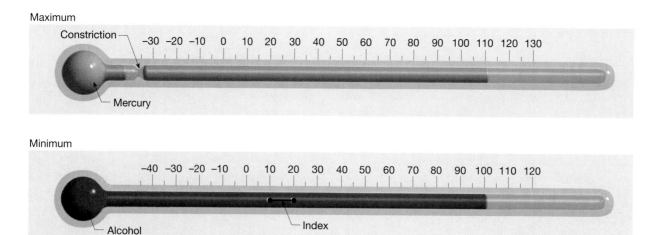

Figure 3–21   Maximum and minimum thermometers.

they do so unequally, causing the strips to curl. This change corresponds to the change in temperature.

The primary meteorological use of the bimetal strip is in the construction of a **thermograph**, an instrument that continuously records temperature. The changes in the curvature of the strip can be used to move a pen arm that records the temperature on a calibrated chart that is attached to a clock-driven, rotating drum (Figure 3–22). Although very convenient, thermograph records are generally less accurate than readings obtained from a mercury-in-glass thermometer. To obtain the most reliable values, it is necessary to check and correct the thermograph periodically by comparing it with an accurate, similarly exposed thermometer.

Electrical Thermometers

Some thermometers do not rely on differential expansion but instead measure temperature electrically.

A resistor is a small electronic part that resists the flow of electrical current. **A thermistor** (thermal resistor) is

Figure 3–23 This modern shelter contains an electrical thermometer called a *thermistor*. *(Photo by Bobbé Christopherson)*

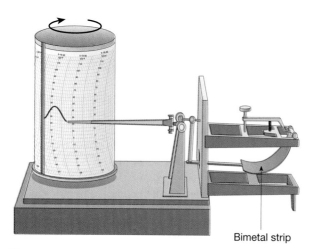

Figure 3–22 A common use of the bimetal strip is in the construction of a thermograph, an instrument that continuously records temperatures.

similar, but its resistance to current flow varies with temperature. As temperature increases, so does the resistance of the thermistor, reducing the flow of current. As temperature drops, so does the resistance of the thermistor, allowing more current to flow. The current operates a meter or digital display that is calibrated in degrees of temperature. The thermistor thus is used as a temperature sensor—an electrical thermometer.

Thermistors are rapid-response instruments that quickly register temperature changes. Therefore, they are commonly used in radiosondes where rapid temperature changes are often encountered. The National Weather Service also uses a thermistor system for ground-level readings. The sensor is mounted inside a shield made of louvered plastic rings and a digital readout is placed indoors (Figure 3–23).

Instrument Shelters

How accurate are thermometer readings? It depends not only on the design and quality of the instruments, but also where they are placed. Placing a thermometer

Figure 3–24 This traditional standard instrument shelter is white (for high albedo) and louvered (for ventilation). It protects instruments from direct sunlight and allows for the free flow of air. *(Courtesy of Qualimetrics, Inc.)*

in the direct sunlight will give a grossly excessive reading, because the instrument itself absorbs solar energy much more efficiently than the air. Placing a thermometer near a heat-radiating surface, such as a building or the ground, also yields inaccurate readings. Another way to assure false readings is to prevent air from moving freely around the thermometer.

So, where should a thermometer be placed to read air temperature accurately? The ideal location is an instrument shelter (Figure 3–24). The shelter is a white box that has louvered sides to permit the free movement of air through it, while shielding the instruments from direct sunshine, heat from the ground, and precipitation. Furthermore, the shelter is placed over grass whenever possible and as far away from buildings as circumstances permit. Finally, the shelter must conform to a standardized height so that the thermometers will be mounted at 1.5 meters (5 feet) above the ground.

Temperature Scales

In the United States, TV weather reporters give temperatures in degrees Fahrenheit. But scientists as well as most people outside of the United States use degrees Celsius. Scientists sometimes also use the Kelvin or absolute scale. What are the differences among these three temperature scales? To make quantitative measurement of temperature possible, it was necessary to establish scales. Such temperature scales are based on the use of reference points, sometimes called **fixed points**. In 1714, Gabriel Daniel **Fahrenheit**, a German physicist, devised the temperature scale that bears his name. He constructed a mercury-in-glass thermometer in which the zero point was the lowest temperature he could attain with a mixture of ice, water, and common salt. For his second fixed point, he chose human body temperature, which he arbitrarily set at 96°.

On this scale, he determined that the melting point of ice (the **ice point**) was 32° and the boiling point of water (the **steam point**) was 212°. Because Fahrenheit's original reference points were difficult to reproduce accurately, his scale is now defined by using the ice point and the steam point. As thermometers improved, average human body temperature was later shown to be 98.6°F.

In 1742, twenty-eight years after Fahrenheit invented his scale, Anders Celsius, a Swedish astronomer, devised a decimal scale on which the melting point of ice was set at 0° and the boiling point of water at 100°.* For many

---

*The boiling point referred to in the Celsius and Fahrenheit scales pertains to pure water at standard sea-level pressure. It is necessary to remember this fact, for the boiling point of water gradually decreases with altitude.

years it was called the *centigrade scale*, but it is now known as the **Celsius scale**, after its inventor.

Because the interval between the melting point of ice and the boiling point of water is 100 degrees on the Celsius scale and 180 degrees on the Fahrenheit scale, a Celsius degree (°C) is larger than a Fahrenheit degree (°F) by a factor of 180/100, or 1.8. So, to convert from one system to the other, allowance must be made for this difference in the size of the degrees. Also, conversions must be adjusted because the ice point on the Celsius scale is at 0° rather than at 32°. This relationship is shown graphically in Figure 3–25.

The Celsius–Fahrenheit relationship also is shown by the following formulas:

$$°F = (1.8 \times °C) + 32$$

or

$$°C = \frac{°F - 32}{1.8}$$

You can see that the formulas adjust for degree size with the 1.8 factor, and adjust for the different 0° points with the ±32 factor.

The Fahrenheit scale is best known in English-speaking countries, where its official use is declining along with the British system of weights and measures. In other parts of the world as well as in the scientific community, where the metric system is used, the Celsius temperature scale is also used.

For some scientific purposes, a third temperature scale is used, the **Kelvin** or **absolute scale**. On this scale, degrees Kelvin are called *Kelvins* (abbreviated K). It is similar to the Celsius scale because its divisions are exactly the same; there are 100 degrees separating the melting point of ice and the boiling point of water. However, on the Kelvin scale, the ice point is set at 273 and the steam point at 373 (see Figure 3–25). The reason is that the zero point represents the temperature at which all molecular motion is presumed to cease (called **absolute zero**). Thus, unlike the Celsius and Fahrenheit scales, it is not possible to have a negative value when using the Kelvin scale, for there is no temperature lower than absolute zero. The relationship between Kelvin and Celsius scales is easily written as follows:

$$°C = K - 273 \text{ or } K = °C + 273$$

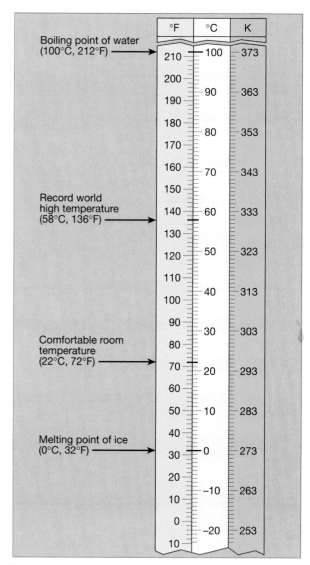

Figure 3–25 The three temperature scales compared.

Applications of Temperature Data

To make weather data more useful to people, many different applications have been developed over the years. In this section, we examine some commonly used practical applications. First, we will look at three indices that all have the term *degree-days* as part of their name: heating degree-days, cooling degree-days, and growing degree-days. The first two are relative measures that allow us to evaluate the weather-produced needs and costs of heating and cooling. The third is a simple index used by farmers to estimate the maturity of crops.

Heating Degree-Days

Developed by heating engineers early in the twentieth century, **heating degree-days** represent a practical method for evaluating energy demand and consumption. It starts from the assumption that heating is not required in a building when the daily mean temperature is 65°F (18.3°C) or higher.° Simply, each degree of temperature below 65°F is counted as 1 heating degree-day. Therefore, heating degree-days are determined each day by subtracting the daily mean below 65°F from 65°F. Thus, a day with a mean temperature of 50°F has 15 heating degree-days (65 – 50 = 15) and one with an average temperature of 65°F or higher has none.

The amount of heat required to maintain a certain temperature in a building is proportional to the total heating degree-days. This linear relationship means that doubling the heating degree-days usually doubles the fuel consumption. Consequently, a fuel bill will generally be twice as high for a month with 1000 heating degree-days as for a month with just 500. When seasonal totals are compared for different places, we can estimate differences in seasonal fuel consumption (Figure 3–26). For example, more than four times as much fuel is required to heat a building in Chicago (about 6200 total heating degree-days) than to heat a similar building in New Orleans (1400 heating degree-days). This statement is true, however, only if we assume that building construction and living habits in these areas are the same.

Each day, the previous day's accumulation is reported as well as the total thus far in the season. For reporting purposes, the heating season is defined as the period from July 1 through June 30. These reports often include a comparison with the total up to this date last year or with the long-term average for this date or both, and so it is a relatively simple matter to judge whether the season thus far is above, below, or near normal.

°Because the National Weather Service and the news media in the United States still compute and report degree-day information in Fahrenheit degrees, we will use degrees in Fahrenheit throughout this discussion.

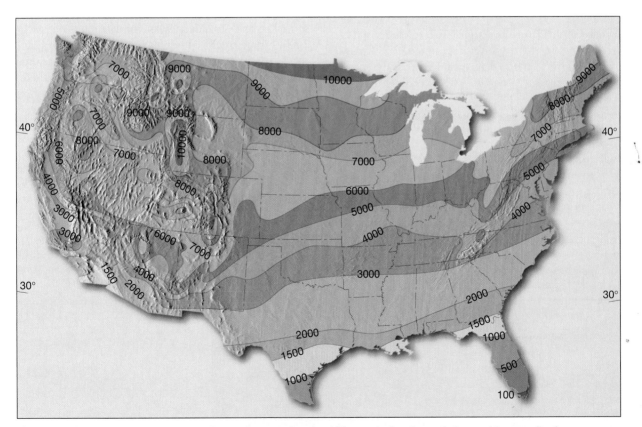

Figure 3–26 Mean annual total heating degree-days. Caution should be used when interpolating on this generalized map, particularly in mountainous regions. *(Courtesy of Environmental Data Service, NOAA)*

Table 3–2 Average annual cooling degree-days for selected cities

City	Cooling Degree-days
Miami, Fla.	4000
San Antonio, Tex.	3000
Tucson, Ariz.	2800
St. Louis, Mo.	1500
Washington, D.C.	1400
Los Angeles, Calif.	1200
Chicago, Ill.	1000
Boston, Mass.	700
Detroit, Mich.	700
Great Fall, Mont.	350
International Falls, Minn.	200
Seattle, Wash.	200

(*Source: National Oceanic and Atmospheric Administration.*)

Cooling Degree-Days

Just as fuel needs for heating can be estimated and compared by using heating degree-days, the amount of power required to cool a building can be estimated by using a similar index called the **cooling degree-day**. Because the 65°F base temperature is also used in calculating this index, cooling degree-days are determined each day by subtracting 65°F from the daily mean. Thus, if the mean temperature for a given day is 80°F, 15 cooling degree-days would be accumulated. Mean annual totals of cooling degree-days for selected cities are shown in Table 3–2. By comparing the totals for St. Louis and San Antonio, we can see that the fuel requirements for cooling a building in San Antonio are about twice as great as for a similar building in St. Louis. The "cooling season" is conventionally measured from January 1 through December 31. Therefore, when cooling degree-day totals are reported, the number represents the accumulation since January 1 of that year.

Although indices that are more sophisticated than heating and cooling degree-days have been proposed to take into account the effects of wind speed, solar radiation, and humidity, degree-days continue to be widely used.

Growing Degree-Days

Another practical application of temperature data is used in agriculture to determine the approximate date when crops will be ready for harvest. This simple index is called the **growing degree-day**.

The number of growing degree-days for a particular crop on any day is the difference between the daily mean temperature and the base temperature of the crop, which is the minimum temperature required for it to grow. For example, the base temperature for sweet corn is 50°F and for peas it is 40°F. Thus, on a day when the mean temperature is 75°F, the number of growing degree-days for sweet corn is 25 and the number for peas is 35.

Starting with the onset of the growth season, the daily growing degree-day values are added. Thus, if 2000 growing degree-days are needed for that crop to mature, it should be ready to harvest when the accumulation reaches 2000 (Figure 3–27). Although many factors important to plant growth are not included in the index, such as moisture conditions and sunlight, this system nevertheless serves as a simple and widely used tool in determining approximate dates of crop maturity.

Temperature and Comfort

A more familiar use of temperature data relates to human perception of temperature. Television weather reports use indices that attempt to portray levels of human comfort

Figure 3–27 Growing degree-days are used to determine the approximate date when crops will be ready for harvest. (*Photo by Inga Spence/Holt Studios International/Photo Researchers*)

and discomfort. Such indices are based on the fact that our sensation of temperature is often quite different from the actual air temperature recorded by a thermometer.

The human body is a heat generator that continually releases energy. Anything that influences the rate of heat loss from the body also influences our sensation of temperature, thereby affecting our feeling of comfort. Several factors control the thermal comfort of the human body, and certainly air temperature is a major one. Other environmental conditions are also significant, such as relative humidity, wind, and solar radiation.

Because evaporation is a cooling process, the evaporation of perspiration from skin is a natural means of

regulating body temperature. When the air is very humid, however, heat loss by evaporation is reduced. As a result, a hot and humid day feels warmer and more uncomfortable than a hot and dry day. This is the basis for a commonly used expression of summertime discomfort known as the *heat stress index* or simply the *heat index*. Because of its link to humidity, this index is the focus of a special-interest box in Chapter 4.

Wind is another significant factor affecting the sensation of temperature. A cold and windy winter day may feel much colder than the air temperature indicates. Box 3–5 focuses on this frequently used wintertime application known as *windchill*.

Chapter Summary

- Temperature is one of the basic elements of weather and climate. The *daily mean temperature* is determined by averaging the 24 hourly readings or by adding the maximum and minimum temperatures for a 24-hour period and dividing by two. The *daily temperature range* is computed by finding the difference between the maximum and minimum temperatures. Other temperature data involving longer periods include the *monthly mean temperature* (the sum of the daily means for each day of the month divided by the number of days in the month), the *annual mean temperature* (the average of the twelve monthly mean temperatures), and the *annual temperature range* (the difference between the warmest and coldest monthly mean temperatures).

- The *controls of temperature*—those factors that cause temperature to vary from place to place—are (1) differential heating of land and water; (2) ocean currents; (3) altitude; (4) geographic position; and (5) cloud cover and albedo.

- On maps illustrating the world distribution of temperature, *isotherms*, lines that connect points of equal temperature, generally trend east and west and show a decrease in temperature poleward. Moreover, the isotherms illustrate a latitudinal shifting of temperatures caused by the seasonal migration of the Sun's vertical rays and also reveal the presence of ocean currents. The north-south migration of isotherms is more pronounced over the continents because the temperatures do not fluctuate as much over water.

- The primary control of the daily cycle of air temperature is Earth's rotation. However, the magnitude of

these changes is variable and influenced by locational factors, local weather conditions, or both.

- As a consequence of the mechanism by which Earth's atmosphere is heated, the months of the highest and lowest temperatures do not coincide with the periods of maximum and minimum incoming solar radiation. In the Northern Hemisphere the greatest intensity of solar radiation occurs at the time of the summer solstice, yet the months of July and August are generally the warmest of the year. Conversely, in the Northern Hemisphere a minimum of solar energy is received in December at the time of the winter solstice, but January and February are usually colder.

- *Thermometers* measure temperature either mechanically or electrically. Most *mechanical thermometers* are based on the ability of a substance to expand when heated or contract when cooled. One type of mechanical thermometer, the liquid-in-glass thermometer, includes maximum thermometers, which use the liquid mercury, and minimum thermometers, which contain a liquid of low density, such as alcohol. A bimetal strip mechanical thermometer is frequently used in a thermograph, an instrument that continuously records temperature. *Electrical thermometers* use a thermistor (a thermal resistor) to measure temperature.

- Temperature scales use reference points, called *fixed points*. Three common temperature scales are (1) the *Fahrenheit scale*, which is defined by using the ice point (32°) and steam point (212°), (2) the *Celsius scale*, a decimal scale on which the melting point of ice is set at 0° and the boiling point of water at 100°, and

Box 3–5 Windchill: The Cooling Power of Moving Air

Most everyone is familiar with the wintertime cooling power of moving air (Figure 3–F). When the wind blows on a cold day, we realize that comfort would improve if the wind were to stop. A stiff breeze penetrates ordinary clothing and reduces its capacity to retain body heat while causing exposed parts of the body to chill rapidly. Not only is cooling by evaporation heightened in this situation, but the wind is also acting to carry heat away from the body by constantly replacing warmer air next to the body with colder air.

During the late 1930s and early 1940s, pioneering experiments performed in Antarctica by the polar scientist Paul Siple led to the development of the concept of *windchill*. Today, during cold weather, windchill normally is mentioned during weather reports. The figure that is actually being reported is the *windchill equivalent temperature* (WET for short). This is not an actual measured temperature, but a "feels like" expression that relates wind speed to the temperature we perceive.

The windchill charts (Tables 3–E for Celsius and 3–F for Fahrenheit) illustrate the combined effects of wind and temperature on the cooling rate of the human body. They translate the cooling power of a windy atmosphere to a temperature equivalent under nearly calm conditions. (Rather than using an absolute calm, a wind of about 6 kilometers per hour or 4 miles per hour is assumed because it represents the air motion that a person feels while walking briskly in calm air.)

For example, on a day when the temperature is –8°C and the wind speed is 30 kilometers per hour, the sensation of temperature—that is, the windchill equivalent temperature—would be reported as –25°C, or 17°C less than the actual air temperature. It should be emphasized that the temperature of the skin does not drop to –25°C. Skin temperature can fall no lower than the air temperature, which is –8°C in this example. What the WET does indicate is that any exposed skin will lose heat at a rate equal to the rate that occurs when the temperature is –25°C and the air is relatively calm.

By examining Tables 3–E and 3–F, it is clear that the cooling power of the wind increases as wind speed

Figure 3–F Strong winds make winter days seem much colder. *(Photo by Donald R. Winslow/Corbis-Bettman)*

rises and as temperature decreases. It should also be pointed out that in contrast to a cold and windy day, a calm and sunny day in winter often feels warmer than the thermometer reading. In this situation, the warm feeling is caused by the absorption of direct solar radiation by the body.

It is important to remember that the windchill equivalent temperature is only an *estimate* of human discomfort. The degree of discomfort felt by different people will vary because it is influenced by many factors. Even if clothing is assumed to be the same, individuals vary widely in their responses because of such factors as age, physical condition, state of health, and level of activity. Nevertheless, as a relative measure, WET is useful because it allows people to make more informed judgments regarding the potential harmful effects of wind and cold.

(3) the *Kelvin* or *absolute scale*, where the zero point represents the temperature at which all molecular motion is presumed to cease (called *absolute zero*), the ice point is set at 273, and the steam point at 373.

- Three common applications of temperature data are (1) *heating degree days*, where each degree of temperature below 65°F is counted as one heating degree day, (2) *cooling degree days*, which are determined by

Table 3–E Celsius windchill equivalent temperature (°C)

Actual Temperature (°C)	Wind Speed (KM/HR)								
	6	10	20	30	40	50	60	70	80
8	8	5	0	−3	−5	−6	−7	−7	−8
4	4	0	−5	−8	−11	−12	−13	−14	−14
0	0	−4	−10	−14	−17	−18	−19	−20	−21
−4	−4	−8	−15	−20	−23	−25	−26	−27	−27
−8	−8	−13	−21	−25	−29	−31	−32	−33	−34
−12	−12	−17	−26	−31	−35	−37	−39	−40	−40
−16	−16	−22	−31	−37	−41	−43	−45	−46	−47
−20	−20	−26	−36	−43	−47	−49	−51	−52	−53
−24	−24	−31	−42	−48	−53	−56	−58	−59	−60
−28	−28	−35	−47	−54	−59	−62	−64	−65	−66
−32	−32	−40	−52	−60	−65	−68	−70	−72	−73
−36	−36	−44	−57	−65	−71	−74	−77	−78	−79
−40	−40	−49	−63	−71	−77	−80	−83	−85	−86

Source: NOAA, National Weather Service.

Table 3–F Fahrenheit windchill equivalent temperature (°F)

Air Temperature (°F)	Wind Speed (Miles/HR)								
	5	10	15	20	25	30	35	40	45
40	37	28	22	18	15	13	11	10	9
35	32	22	16	11	8	5	3	2	1
30	27	16	9	4	0	−2	−4	−6	−7
25	22	10	2	−3	−7	−10	−12	−14	−15
20	16	4	−5	−10	−15	−18	−20	−22	−23
15	11	−3	−11	−17	−22	−25	−28	−29	−31
10	6	−9	−18	−25	−29	−33	−35	−37	−39
5	1	−15	−25	−32	−37	−41	−43	−45	−47
0	−5	−21	−32	−39	−44	−48	−51	−53	−55
−5	−10	−27	−38	−46	−52	−56	−59	−61	−62
−10	−15	−33	−45	−53	−59	−63	−67	−69	−70
−15	−20	−40	−52	−60	−66	−71	−74	−77	−78
−20	−26	−46	−58	−67	−74	−79	−82	−85	−86
−25	−31	−52	−65	−74	−81	−86	−90	−93	−94
−30	−36	−58	−72	−82	−89	−94	−98	−101	−102

Source: NOAA, National Weather Service.

subtracting 65°F from the daily mean, and (3) *growing degree days*, which are determined from the difference between the daily mean temperature and the base temperature of the crop, the minimum temperature required for it to grow.

- One familiar use of temperature data relates to human perception of temperature. The *heat stress index* (or *heat index*), a commonly used expression of summertime discomfort, links humidity and temperature to determine the thermal comfort of the human body. *Windchill*, a typical wintertime index, uses both wind and air temperature to calculate the human sensation of temperature.

Vocabulary Review

absolute zero (p. 78)
annual mean temperature (p. 57)
annual temperature range (p. 57)
bimetal strip (p. 75)
Celsius scale (p. 78)
controls of temperature (p. 59)
cooling degree-day (p. 80)
daily mean temperature (p. 57)
daily temperature range (p. 57)
Fahrenheit scale (p. 77)
fixed points (p. 77)
growing degree-days (p. 80)
heating degree-days (p. 79)

ice point (p. 77)
isotherm (p. 58)
Kelvin or absolute scale (p. 78)
liquid-in-glass thermometer (p. 73)
maximum thermometer (p. 74)
minimum thermometer (p. 75)
monthly mean temperature (p. 57)
specific heat (p. 61)
steam point (p. 77)
thermistor (p. 76)
thermograph (p. 76)
thermometers (p. 73)

Review Questions

1. How are the following temperature data calculated: daily mean, daily range, monthly mean, annual mean, and annual range?

2. What are isotherms and what is their purpose?

3. Why are summertime maximum temperatures so high in Death Valley, California? (See Box 3–1.)

4. **a.** State the relationship between the heating and cooling of land versus water.
   b. List and explain the factors that cause the difference between the heating and cooling of land and water.
   c. We are studying the atmosphere, so why are we concerned with the heating characteristics at Earth's surface?

5. How does the annual temperature range near the equator compare with the annual temperature ranges in the middle to high latitudes? Explain.

6. Three cities are at the same latitude (about 45°N). One city is located along a windward coast, another in the center of the continent, and the third along a leeward coast. Compare the annual temperature ranges of these cities.

7. Quito, Ecuador, is on the equator and is not a coastal city. It has an annual mean temperature of only 13°C. What is the likely cause for this low annual mean temperature?

8. How does the daily march of temperature on a completely overcast day compare with that on a cloudless, sunny day? Explain your answer.

9. Examine Figure 3–14 and explain why Rangoon's monthly mean temperature for April is higher than the July monthly mean.

10. Answer the following questions about world temperature distribution (you may wish to refer to the January and July isotherm maps).
    a. Isotherms generally trend east–west. Why?
    b. Isotherms bend (poleward, equatorward) over continents in summer. Underline the correct answer and explain.
    c. Isotherms shift north and south from season to season. Why?
    d. Where do isotherms shift most: over land or water? Explain.
    e. How do isotherms show ocean currents? How can you tell if the current is warm or cold?
    f. Why are the isotherms more irregular in the Northern Hemisphere than in the Southern Hemisphere?

11. Although the intensity of incoming solar radiation is greatest at local noon, the warmest part of the day is most often midafternoon. Why? Use Figure 3–19 to explain your answer.

12. List at least three factors that contribute to the urban heat island. (See Box 3–4.)

13. The magnitude of the daily temperature range can vary significantly from place to place and from time to time. List and describe at least three factors that might cause such variations.

14. Describe how each of the following thermometers works: liquid-in-glass, maximum, minimum, bimetal strip, and thermistor.

15. What is a thermograph? Which one of the thermometers listed in question 14 is commonly used in the construction of a thermograph?

16. In addition to having an accurate thermometer, which other factors must be considered to obtain a meaningful air temperature reading?

17. **a.** What is meant by the terms *steam point* and *ice point*?
    b. What values are given these points on each of the three temperature scales presented in this chapter?

18. Why is it impossible to have a negative value when using the Kelvin temperature scale?

19. When heating and cooling degree-day totals for different places are examined to compare fuel consumption, what important assumption is made?

20. How are growing degree-days calculated? For what purpose is this index used?

Problems

1. Refer to the thermograph record in Figure 3–17. Determine the maximum and minimum temperature for each day of the week. Use these data to calculate the daily mean and daily range for each day.

2. By referring to the world maps of temperature distribution for January and July (Figures 3–15 and 3–16), determine the approximate January mean, July mean, and annual temperature range for a place located at 60° north latitude, 80° east longitude, and a place located at 60° south latitude, 80° east longitude.

3. Calculate the annual temperature range for three cities in Appendix E. Try to choose cities with different ranges and explain these differences in terms of the controls of temperature.

4. Referring to Table 3–E in Box 3–5, determine equivalent temperatures under the following circumstances:
   a. Temperature = –12°C, wind speed = 20 km/hr.
   b. Temperature = –12°C, wind speed = 50 km/hr.

5. The mean temperature is 55°F on a particular day. The following day the mean drops to 45°F. Calculate the number of heating degree-days for each day. How much more fuel would be needed to heat a building on the second day compared with the first day?

6. Use the appropriate formula to convert the following temperatures:

$$20°C = \underline{\quad 68 \quad} °F$$
$$-25°C = \underline{\quad 248 \quad} K$$
$$59°F = \underline{\quad 15 \quad} °C$$

Atmospheric Science Online

The following are informative and interesting Internet sites that address topics related to those presented in the chapter:

Current U.S. Temperatures (Penn State University):
- **http://www.ems.psu.edu/wx/usstats/ tempstats.html**

National Climatic Data Center:
- **http://www.ncdc.noaa.gov/**

For direct links to these sites and others, chapter objectives and reviews, quiz questions, and topical investigations that utilize Web resources, visit *The Atmosphere Eighth Edition* Home Page at:
- **http://www.prenhall.com/lutgens**

Moisture and Atmospheric Stability

Hot air balloons. *(Photo by Randy Wells/Tony Stone Images)*

As you observe day-to-day weather changes, you might ask: Why is it generally more humid in the summer than in the winter? Why do clouds form on some occasions but not on others? Why do some clouds look thin and harmless whereas others form gray and ominous towers? Answers to these questions involve the role of water vapor in the atmosphere, the central theme of this chapter.

Movement of Water Through the Atmosphere

Water is everywhere on Earth—in the oceans, glaciers, rivers, lakes, the air, soil, and in living tissue. All of these "reservoirs" constitute Earth's hydrosphere. In all, the water content of Earth's hydrosphere is about 1.36 billion cubic kilometers (326 million cubic miles).

The increasing demands on this finite resource have led scientists to focus on the continuous exchange of water among the oceans, the atmosphere, and the continents (Figure 4–1). This unending circulation of Earth's water supply has come to be called the **hydrologic cycle** (or water cycle).

The hydrologic cycle is a gigantic system powered by energy from the Sun in which the atmosphere provides the vital link between the oceans and continents. Water from the oceans, and to a much lesser extent from the continents, evaporates into the atmosphere. Winds transport this moisture-laden air, often over great distances.

Complex processes of cloud formation eventually result in precipitation. The precipitation that falls into the ocean has ended its cycle and is ready to begin another by evaporating again. The water that falls on the continents, however, must still flow back to the oceans.

Once precipitation has fallen on land, a portion of the water soaks into the ground, some of it moving downward, then laterally, and finally seeping into lakes and streams or directly into the ocean. Much of the water that soaks in or runs off eventually finds its way back to the atmosphere. In addition to evaporation from the soil, lakes, and streams, some water that infiltrates the ground is absorbed by plants through their roots. They then release it into the atmosphere, a process called **transpiration**.

Figure 4–1 not only shows Earth's hydrologic cycle but also its *water balance*. The water balance is a quantitative view of the hydrologic cycle. Although the amount

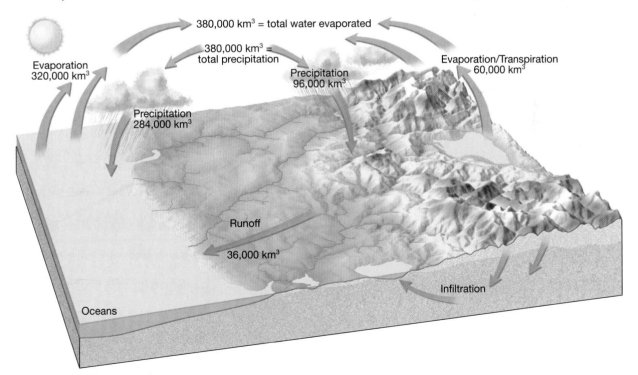

Figure 4–1 Earth's water balance. About 320,000 cubic kilometers of water are evaporated annually from the oceans, whereas evaporation from the land (including lakes and streams) contributes 60,000 cubic kilometers of water. Of this total of 380,000 cubic kilometers of water, about 284,000 cubic kilometers fall back to the ocean, and the remaining 96,000 cubic kilometers fall on Earth's land surface. Because 60,000 cubic kilometers of water leave the land through evaporation and transpiration, 36,000 cubic kilometers of water remain to erode the land during the journey back to the oceans.

of water vapor in the air is just a tiny fraction of Earth's total water supply, the absolute quantities that are cycled through the atmosphere in a year are immense, some 380,000 cubic kilometers (91,000 cubic miles). This is enough to cover Earth's surface uniformly to a depth of about 1 meter (3.3 feet). Estimates show that over North America, almost six times more water is carried within the moving currents of air than is transported by all the continent's rivers.

Because the total amount of water vapor in the entire global atmosphere remains about the same, the average annual precipitation over Earth must be equal to the quantity of water evaporated. However, for the continents, precipitation exceeds evaporation. Conversely, over the oceans, evaporation exceeds precipitation. Because the level of the world ocean is consistent, runoff from land areas must balance the deficit of precipitation over the oceans.

In summary, the hydrologic cycle depicts the continuous movement of water from the oceans to the atmosphere, from the atmosphere to the land, and from the land back to the sea. The movement of water through the cycle holds the key to the distribution of moisture over the surface of our planet and is intricately related to all atmospheric phenomena.

Water's Changes of State

Water vapor is an odorless, colorless gas that mixes freely with the other gases of the atmosphere. Unlike oxygen and nitrogen—the two most abundant components of the atmosphere—water can change from one state of matter to another (solid, liquid, or gas) at the temperatures and pressures experienced on Earth. (By contrast, nitrogen

will not condense to a liquid unless its temperature is lowered to –196°C [–371°F]). Because of this unique property water freely leaves the oceans as a gas and returns again as a liquid, producing the vital hydrologic cycle.

The process of changing state requires that heat be absorbed or released, as shown in Figure 4–2. The heat energy involved is often measured in calories. One **calorie** is the amount of heat required to raise the temperature of 1 gram of water 1°C (1.8°F). Thus, when 10 calories of heat are added to 1 gram of water, a 10°C (18°F) temperature rise occurs.

Under certain conditions, heat may be added to a substance without an accompanying rise in temperature. This situation occurs, for example, when ice changes to water. How can this be?

When heat is supplied to a glass of ice water the temperature of the ice water remains a constant 0°C (32°F) until all the ice has melted. If adding heat does not raise the temperature, then where does this heat go?

In this case, the added heat went to disrupt the internal crystalline structure of the ice cubes. Stated another way, the water molecules in the ice crystals were dislodged to form the noncrystalline substance liquid water. We call this process *melting*.

Because the heat used to melt ice does not produce a temperature change, it is referred to as **latent heat**. (Latent means *hidden*, like the latent fingerprints hidden at a crime scene.) This energy becomes stored in the liquid water and is not released as heat until the liquid returns to the solid state.

As we shall examine later, latent heat plays an important role in many atmospheric processes. In particular, the release of latent heat is an important source of energy for violent thunderstorms, tornadoes, and hurricanes.

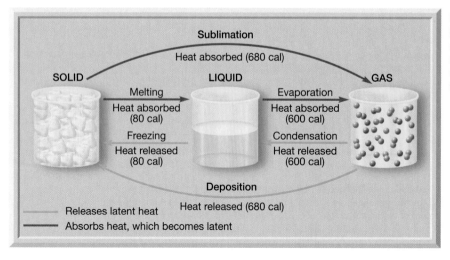

Figure 4–2 Change of state always involves an exchange of heat. The amounts of heat are expressed in calories and are shown here for the change of 1 gram of water from one state of matter to another.

Converting solid ice to liquid water requires heat. So does **evaporation**, the process of converting a liquid to a gas (vapor). It takes approximately 600 calories of energy to convert 1 gram of water to water vapor.° The energy absorbed by the water molecules during evaporation is used solely to give them the motion needed to escape the surface of the liquid and become a gas. This energy is referred to as the *latent heat of vaporization*. During the process of evaporation, it is the higher-temperature (faster-moving) molecules that escape the surface. As a result, the average molecular motion (temperature) of the remaining water is reduced—hence the common expression "evaporation is a cooling process." You have undoubtedly experienced this cooling effect on stepping dripping wet from a swimming pool or bathtub.

Condensation is the process wherein water vapor changes to the liquid state. For condensation to occur, the energetic water vapor molecules must release energy (*latent heat of condensation*) equivalent to what was absorbed during evaporation. This released energy plays an important role in producing violent weather. When condensation occurs in the atmosphere, it results in the formation of such phenomena as fog and clouds (Figure 4–3).

°The latent heat of vaporization depends upon temperature and varies from 569 cal/g to 629 cal/g as the temperature goes from 50°C to –50°C. For most purposes, a value of 600 cal/g is a good approximation.

Melting is the process by which a solid is changed to a liquid. As discussed, it requires the absorption of heat energy from the environment, approximately 80 calories of energy per gram of water. **Freezing**, the reverse process, releases these 80 calories per gram as the *latent heat of fusion*.

The last of the processes illustrated in Figure 4–2 are sublimation and deposition. **Sublimation** is the conversion of a solid directly to a gas without passing through the liquid state. Examples you may have observed include the gradual shrinking of unused ice cubes in a freezer and the rapid conversion of dry ice (frozen carbon dioxide) to wispy clouds that quickly disappear. "Freezer burn" is an example of ice sublimating from food left for long periods in a frost-free refrigerator. The food is not actually burned; it is simply dried out.

Deposition is used to denote the reverse process, the conversion of a vapor directly to a solid. This change occurs, for example, when water vapor is deposited as ice on solid objects such as grass or windows (Figure 4–4). These deposits are called *white frost* or *hoar frost* and are frequently referred to simply as *frost*. A household example of the process of deposition is the "frost" that accumulates in a freezer. As shown in Figure 4–2, sublimation and deposition involve an amount of energy equal to the total of the other two processes.

Figure 4–3 Condensation of water vapor generates phenomena such as clouds and fog. *(Photo by Jeremy Walker/Tony Stone Images)*

Figure 4–4 White frost on a window pane. *(Photo by Craig F. Bohren)*

Water in the Atmosphere

Water vapor constitutes only a small fraction of the atmosphere, varying from as little as one-tenth of 1 percent up to about 4 percent by volume. But the importance of water in the air is far greater than these small percentages would indicate. Indeed, scientists agree that *water vapor* is the most important gas in the atmosphere when it comes to understanding atmospheric processes.

Humidity is the general term used to describe the amount of water vapor in the air. Meteorologists employ several methods to express the water vapor content of the air, including (1) absolute humidity, (2) mixing ratio, (3) vapor pressure, (4) relative humidity, and (5) dew point. Two of these methods, *absolute humidity* and *mixing ratio*, are similar in that both are expressed as the quantity of water vapor contained in a specific amount of air.

Absolute Humidity and Mixing Ratio

Absolute humidity is *the mass of water vapor in a given volume of air* (usually as grams per cubic meter).

$$\text{Absolute humidity} = \frac{\text{mass of water vapor (grams)}}{\text{volume of air (cubic meters)}}$$

As air moves from one place to another, changes in pressure and temperature cause changes in its volume. When such volume changes occur, the absolute humidity also changes, even if no water vapor is added or removed. Consequently, it is difficult to monitor the water vapor content of a moving mass of air if absolute humidity is the index being used. Therefore, meteorologists generally prefer to employ mixing ratio to express the water vapor content of air.

The **mixing ratio** is *the mass of water vapor in a unit of air compared to the remaining mass of dry air*.

$$\text{mixing ratio} = \frac{\text{mass of water vapor (grams)}}{\text{mass of dry air (kilograms)}}$$

Because it is measured in units of mass (usually grams per kilogram), the mixing ratio is not affected by changes in pressure or temperature.†

Neither the absolute humidity nor the mixing ratio, however, can be easily determined by direct sampling. Therefore, other methods are also used to express the moisture content of the air. These include vapor pressure, relative humidity, and dew point.

Vapor Pressure and Saturation

Another measure of the moisture content of the air is obtained from the pressure exerted by water vapor. To understand how water vapor exerts pressure imagine a closed container half full of pure water and overlain by dry air, as shown in Figure 4–5a. Almost immediately some of the water molecules begin to leave the water surface and evaporate into the dry air above. The addition of water vapor into the air can be detected by a small increase in pressure (Figure 4–5b). This increase in pressure is the result of the motion of the water vapor molecules that were added to the air through evaporation. In the atmosphere, this pressure is called **vapor pressure**

---

†Another commonly used expression is "specific humidity," which is the mass of water vapor in a unit mass of air, including the water vapor. Because the amount of water vapor in the air rarely exceeds a few percent of the total mass of the air, the specific humidity of air is equivalent to its mixing ratio for all practical purposes.

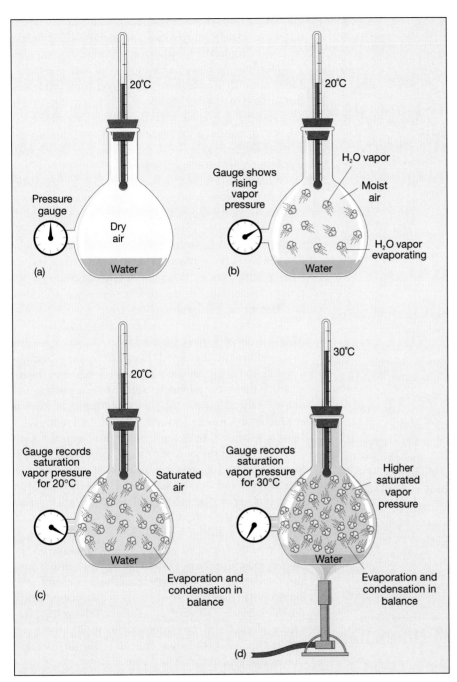

Figure 4–5 The relationship between vapor pressure and saturation. (a) Initial conditions—dry air at 20°C with no observable vapor pressure. (b) Evaporation generates measurable vapor pressure. (c) As more and more molecules escape from the water surface, the steadily increasing vapor pressure forces more and more of these molecules to return to the liquid. Eventually, the number of water-vapor molecules returning to the surface will balance the number leaving. At that point, the air is said to be saturated. (d) When the container is heated from 20° to 30°C, the rate of evaporation increases, causing the vapor pressure to increase until a new balance is reached.

and is defined as *that part of the total atmospheric pressure attributable to its water vapor content.*

Initially, many more molecules will leave the water surface (evaporate) than will return (condense). However, as more and more molecules evaporate from the water surface, the steadily increasing vapor pressure in the air above forces more and more water molecules to return to the liquid. Eventually a balance is reached in which the number of water molecules returning to the surface balances the number leaving. At that point, the air is said to have reached an equilibrium called **saturation** (Figure 4–5c). When air is saturated, the pressure exerted by the motion of the water vapor molecules is called the **saturation vapor pressure**.

Now suppose we were to disrupt the equilibrium by heating the water in our closed container, as illustrated in Figure 4–5d. The added energy would increase the rate at which the water molecules would evaporate from the

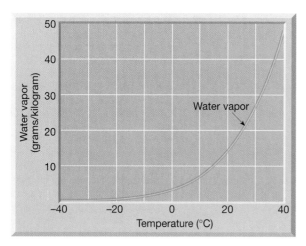

Figure 4–6 The amount of water vapor required to saturate 1 kilogram of dry air at various temperatures.

surface. This, in turn, would cause the vapor pressure in the air above to increase until a new equilibrium was reached between evaporation and condensation. Thus, we conclude that the saturation vapor pressure is temperature-dependent, such that, at higher temperatures, it takes more water vapor to saturate air (Figure 4–6). The amount of water vapor required for the saturation of 1 kilogram (2.2 pounds) of dry air at various temperatures is shown in Table 4–1. Note that for every 10°C (18°F) increase in temperature, the amount of water vapor needed for saturation almost doubles. Thus, roughly four times more water vapor is needed to saturate 30°C (86°F) air than 10°C (50°F) air.

The atmosphere behaves in much the same manner as our closed container. In nature, gravity, rather than a lid, prevents water vapor (and other gases) from escaping

into space. Also like our container, water molecules are constantly evaporating from liquid surfaces (such as lakes or cloud droplets) and other vapor molecules are arriving. However, in nature a balance is not always achieved. More often than not, more water molecules are leaving the surface of a water puddle than are arriving, causing what meteorologists call *net evaporation*. By contrast, during the formation of fog, more water molecules are condensing than are evaporating from the tiny fog droplets, resulting in *net condensation*.

What determines whether the rate of evaporation exceeds the rate of condensation (net evaporation) or vice versa? One of the major factors is the temperature of the surface water, which in turn determines how much motion (kinetic energy) the water molecules possess. At higher temperatures the molecules have more energy and can more readily escape. Thus, under otherwise similar conditions, because hot water has more energy it will evaporate faster than cold water.

The other major factor determining which will dominate, evaporation or condensation, is the vapor pressure in the air around the liquid. Recall from our example of a closed container that vapor pressure determines the rate at which the water molecules return to the surface (condense). When the air is dry (low vapor pressure) the rate at which water molecules return to the liquid phase is low. However, when the air around a liquid has reached the saturation vapor pressure the rate of condensation will be equal to the rate of evaporation. Thus, at saturation there is neither a net condensation nor a net evaporation. Therefore, all else being equal, net evaporation is greater when the air is dry (low vapor pressure) than when the air is humid (high vapor pressure).

In addition to vapor pressure and temperature, other factors exist in nature that affect rates of evaporation and condensation. Although these factors are of minimal importance to most of the processes operating at Earth's surface, they are significant in the atmosphere where clouds and precipitation form. Thus, we will revisit this idea when we consider the formation of clouds and precipitation in Chapter 5.

Relative Humidity

The most familiar and, unfortunately, the most misunderstood term used to describe the moisture content of air is relative humidity. **Relative humidity** *is a ratio of the air's actual water vapor content compared with the amount of water vapor required for saturation at that temperature (and pressure).* Thus, relative humidity indicates how near the air is to saturation, rather than the actual quantity of water vapor in the air (see Box 4–1).

Table 4–1 Saturation Mixing Ratio (at Sea-Level Pressure)	
Temperature **°C** **(°F)**	**Saturation Mixing Ratio** **(G/KG)**
–40 (–40)	0.1
–30 (–22)	0.3
–20 (–4)	0.75
–10 (14)	2
0 (32)	3.5
5 (41)	5
10 (50)	7
15 (59)	10
20 (68)	14
25 (77)	20
30 (86)	26.5
35 (95)	35
40 (104)	47

Box 4–1   Dry Air at 100 Percent Relative Humidity?

A common misconception relating to meteorology is the notion that air with a high relative humidity must have a greater water vapor content than air with a lower relative humidity. Frequently, this is not the case (Figure 4–A). To illustrate, let us compare a typical January day at International Falls, Minnesota, to one in the desert near Phoenix, Arizona. On this hypothetical day, the temperature in International Falls is a cold –10°C (14°F) and the relative humidity is 100 percent. By referring to Table 4–1, we can see that saturated –10°C (14°F) air has a water vapor content (mixing ratio) of 2 grams per kilogram (g/kg). By contrast, the desert air at Phoenix on this January day is a warm 25°C (77°F) and the relative humidity is just 20 percent. A look at Table 4–1 reveals that 25°C (77°F) air has a saturation

mixing ratio of 20 g/kg. Therefore, with a relative humidity of 20 percent, the air at Phoenix has a water vapor content of 4 g/kg (20 grams × 20 percent). Consequently, the "dry" air at Phoenix actually contains twice the water vapor as the "wet" air at International Falls.

This should make clear why places that are very cold are also very dry. The low water vapor content of frigid air (even when saturated) helps to explain why many arctic areas receive only meager amounts of precipitation and are sometimes referred to as "polar deserts." This also helps us understand why people frequently experience dry skin and chapped lips during the winter months. The water vapor content of cold air is low, even when compared to some hot, arid regions.

Figure 4–A Moisture content of hot air versus frigid air. Hot desert air with a low relative humidity generally has a higher water vapor content than frigid air with a high relative humidity. *(Top photo by E. J. Tarbuck, bottom photo by Matt Duvall)*

To illustrate, we see from Table 4–1 that at 25°C, air is saturated when it contains 20 grams of water vapor per kilogram of air. Thus, if the air contains 10 grams per kilogram on a 25°C day, the relative humidity is expressed as 10/20, or 50 percent. Further, if air with a temperature of 25°C had a water vapor content of 20 grams per kilogram, the relative humidity would be expressed as 20/20 or 100 percent. On those occasions when the relative humidity reaches 100 percent, the air is said to be saturated.

How Relative Humidity Changes

Because relative humidity is based on the air's water vapor content, as well as the amount of moisture required for saturation, it can be changed in either of two ways. First, relative humidity can be changed by the addition or removal of water vapor. Second, because the amount of moisture required for saturation is a function of air temperature, relative humidity varies with temperature. (Recall that the saturation vapor pressure is temperature dependent, such that,

at higher temperatures, it takes more water vapor to saturate air than at lower temperatures.)

Adding or Subtracting Moisture. Notice in Figure 4–7 that when water vapor is added to a parcel of air, its relative humidity increases until saturation occurs (100 percent relative humidity). What if even more moisture is added to this parcel of saturated air? Does the relative humidity exceed 100 percent? Normally, this situation does not occur. Instead, the excess water vapor condenses to form liquid water.

You may have experienced such a situation while taking a hot shower. The water leaving the shower is composed of very energetic (hot) molecules, which means that the rate of evaporation is high. As long as you run the shower, the process of evaporation continually adds water vapor to the unsaturated air in the bathroom. Therefore, if you stay in a hot shower long enough, the air eventually becomes saturated and the excess water vapor condenses on the mirror, window, tile, and other surfaces in the room.

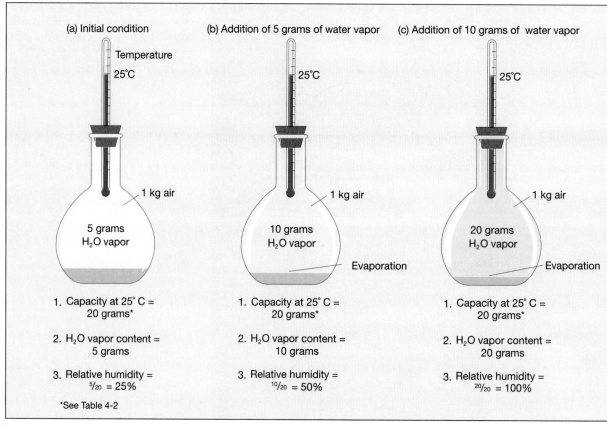

Figure 4–7 Relative humidity. At a constant temperature, the relative humidity will increase as water vapor is added to the air. Here, the capacity remains constant at 20 grams per kilogram and the relative humidity rises from 25 to 100 percent as the water vapor content increases.

In nature, moisture is added to the air mainly via evaporation from the oceans. However, plants, soil, and smaller bodies of water do make substantial contributions. Unlike your shower, however, the processes that add water vapor to the air generally do not operate at rates fast enough to cause saturation to occur directly. One exception is when you exhale on a cold winter day and "see your breath." What is happening is the warm moist air from your lungs mixes with the cold outside air, which has a very low saturation vapor pressure. Your breath has enough moisture to saturate a small quantity of cold outside air and the result is a miniature "cloud." Almost as fast as the "cloud" forms, it mixes with more of the dry outside air and quickly evaporates.

Changes With Temperature The second condition that affects relative humidity is air temperature (see Box 4–2). Examine Figure 4–8 carefully. Note that in Figure 4–8a, when air at 20°C contains 7 grams of water vapor per kilogram, it has a relative humidity of 50 percent. This can be verified by referring to Table 4–1. Here we can see that at 20°C, air is saturated when it contains 14 grams of water vapor per kilogram of air. Because the air in Figure 4–8a contains 7 grams of water vapor, its relative humidity is 7/14 or 50 percent.

How does cooling affect relative humidity? When the flask in Figure 4–8a is cooled from 20° to 10°C, as shown in Figure 4–8b, the relative humidity increases from 50 to 100 percent. We can conclude from this that when the water vapor content remains constant, a decrease in temperature results in an increase in relative humidity.

But there is no reason to assume that cooling would cease the moment the air reached saturation. What happens when the air is cooled below the temperature at which saturation occurs? Figure 4–8c illustrates this situation. Notice from Table 4–1 that when the flask is cooled to 0°C, the air is saturated at 3.5 grams of water vapor per kilogram of air. Because this flask originally contained 7 grams of water vapor, 3.5 grams of water vapor will condense to form liquid droplets that collect on the walls of the container. In the meantime, the relative humidity of the air inside remains at 100 percent. This raises an important concept. When air aloft is cooled below its saturation level, some of the water vapor condenses to form clouds. As clouds are made of liquid droplets, this moisture is no longer part of the *water vapor* content of the air.

Let us return to Figure 4–8 and see what would happen should the flask in Figure 4–8a be heated to 35°C. From Table 4–1, we see that at 35°C, saturation occurs at 35 grams of water vapor per kilogram of air. Consequently, by heating the air from 20° to 35°C, the relative humidity would drop from 7/14 (50 percent) to 7/35 (20 percent).

We can summarize the effects of temperature on relative humidity as follows. When the water vapor content of air remains at a constant level, a decrease in air temperature results in an increase in relative humidity and an increase in temperature causes a decrease in relative humidity.

Box 4–2 Humidifiers and Dehumidifiers

In summer, stores sell *dehumidifiers*. As winter rolls around, these same merchants feature *humidifiers*. Why do you suppose so many homes are equipped with both a humidifier and a dehumidifier? The answer lies in the relationship between temperature and relative humidity. Recall that if the water vapor content of air remains at a constant level, an increase in temperature lowers the relative humidity and a lowering of temperature increases the relative humidity.

During the summer months, warm, moist air frequently dominates the weather of the central and eastern United States. When hot and humid air enters a home, some of it circulates into the cool basement. As a result, the temperature of this air drops and its relative humidity increases. The result is a damp, musty-smelling basement. In response, the homeowner installs a dehumidifier to alleviate the problem. As air is drawn over the cold coils of the dehumidifier, water vapor condenses and collects in a bucket or flows down a drain. This process reduces the relative humidity and makes for a drier, more comfortable basement.

By contrast, during the winter months, outside air is cool and dry. When this air is drawn into the home, it is heated to room temperature. This process in turn causes the relative humidity to plunge, often to uncomfortably low levels of 40 percent or lower. Living with dry air can mean static electrical shocks, dry skin, sinus headaches, or even nose bleeds. Consequently, the homeowner may install a humidifier, which adds water to the air and increases the relative humidity to a more comfortable level.

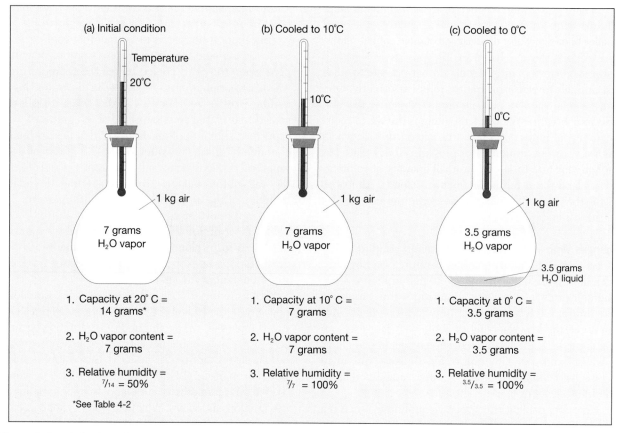

Figure 4–8 Relative humidity varies with temperature. When the water-vapor content (mixing ratio) remains constant, the relative humidity can be changed by increasing or decreasing the air temperature. In this example, when the temperature of the air in the flask was lowered from 20° to 10°C, the relative humidity increased from 50 to 100 percent. Further cooling (from 10° to 0°C) causes one-half of the water vapor to condense. In nature, cooling of air below its saturated mixing ratio generally causes condensation in the form of clouds, dew, or fog.

Natural Change in Relative Humidity

In nature there are three major ways that air temperatures change (over relatively short time spans) to cause corresponding changes in relative humidity. These are:

1. Daily changes in temperatures (daylight versus nighttime temperatures).
2. Temperature changes that result as air moves horizontally from one location to another.
3. Temperature changes caused as air moves vertically in the atmosphere.

The importance of the last two processes in creating weather will be discussed later. The effect of the typical daily temperature cycle on relative humidity is shown in Figure 4–9. Notice that during the warmer midday period, relative humidity reaches its lowest level, whereas the cooler evening hours are associated with higher relative humidities. In this example, the actual water vapor content

(mixing ratio) of the air remains unchanged; only the relative humidity varies. Now we can better understand why a high relative humidity does not necessarily indicate a high water vapor content.

Despite the previous example, we still describe air having a low relative humidity as being "dry" and vice versa. The use of the word "dry" in this context indicates that the air is far from being saturated. Thus, the rate of evaporation on a dry day is generally higher than on a humid day.

In summary, relative humidity indicates how near the air is to being saturated, whereas the air's mixing ratio denotes the actual quantity of water vapor contained in that air.

Dew Point Temperature

Another important measure of humidity is the dew point temperature. The **dew point temperature** or simply the **dew point** is the temperature to which a parcel of air

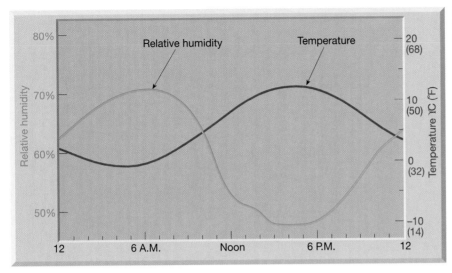

Figure 4–9 Typical daily variation in temperature and relative humidity during a spring day at Washington, DC.

would need to be cooled to reach saturation. Note that in Figure 4–8, unsaturated air at 20°C must be cooled to 10°C before saturation occurs. Therefore, 10°C would be the dew point temperature for this air. If the same air were cooled further, the air's saturation mixing ratio would be exceeded and the excess water vapor would condense, typically as dew, fog, or clouds. The term *dew point* stems from the fact that during nighttime hours, objects near the ground often cool below the dew point temperature and become coated with dew‡ (Figure 4–10).

Contrary to popular belief, white frost is not frozen dew. Rather, white frost (*hoar frost*) forms on occasions when saturation occurs at temperatures of 0°C (32°F)

‡Normally, we associate dew with grass. Because of transpiration by the blades of grass, the relative humidity on a calm night is much higher near the grass than a few inches above the surface. Consequently, dew forms on grass before it does on most other objects.

Figure 4–10 Condensation of dew on a spider web. *(Photo by Wolfgang Kaehler Photography)*

or below (a temperature called the *frost point*). Thus, frost forms when water vapor changes directly from a gas into a solid (ice), without entering the liquid state. This process, called *deposition*, produces delicate patterns of ice crystals often decorating windows during northern winters (see Figure 4–4).

Unlike relative humidity which is a measure of how near the air is to being saturated, dew point temperature is a measure of its *actual moisture* content. Because the dew point temperature is directly related to the amount of water vapor in the air, and because it is easy to determine, it is one of the most widely used measures of humidity.

Recall that the saturation vapor pressure is temperature-dependent and that for every 10°C (18°F) increase in temperature the amount of water vapor needed for saturation doubles. Therefore, relatively cold *saturated air* (0°C or 32°F) contains about half the water vapor of *saturated air* having a temperature of 10°C (50°F) and roughly one-fourth that of hot *saturated air* with a temperature of 20°C (68°F). Because the dew point is the temperature at which saturation occurs, we can conclude that high dew point temperatures equate to moist air and low dew point temperatures indicate dry air. More precisely, based on what we have learned about vapor pressure and saturation, we can state that for every 10°C (18°F) increase in the dew point temperature the air contains about twice as much water vapor. Therefore, we know that air over Fort Myers, Florida, with a dew point of 25°C (77°F) contains about twice the water vapor of air situated over St. Louis, Missouri, with a dew point of 15°C (59°F) and four times that of Tucson, Arizona, with a dew point of 5°C (41°F).

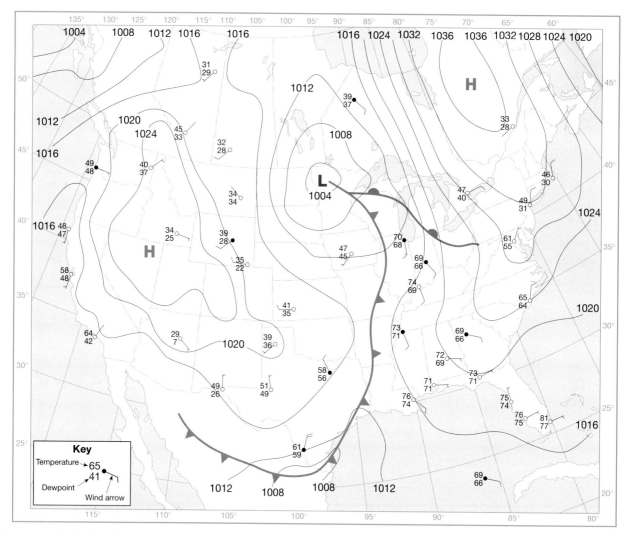

Figure 4–11 This simplified National Weather Service (NWS) surface map provides the air temperature, dew point and wind data for selected cities at 7 A.M., EST, October 6, 1998. Dew point temperatures above 60°F dominate the Southeast, indicating that this region is blanketed with humid air. By contrast, lower dew point temperatures are found in much of the remainder of the continent, which tells us that these areas are experiencing drier conditions.

Because the dew point temperature is a good measure of the amount of water vapor in the air, it is the measure of atmospheric moisture that appears on daily weather maps. Notice on the daily weather map in Figure 4–11 that many of the weather stations situated near the warm waters of the Gulf of Mexico have dew point temperatures that exceed 70°F (21°C). When the dew point exceeds 65°F (18°C) it is considered humid by most people, and air with a dew point 75°F (24°C) or higher is oppressive. Also notice in Figure 4–11 that although the Southeast is dominated by humid conditions (dew points above 60°F) most of the remainder of the country is experiencing comparatively cooler and dryer air.

Humidity Measurement

Aside from the fact that humidity is important meteorologically, many of us are interested in humidity because it influences comfort (see Box 4–3). Here we will look at the various ways that humidity is measured.

Absolute humidity and mixing ratio are difficult to measure directly, but if the relative humidity is known, they can be readily computed from a table or graph. How is relative humidity measured?

A variety of instruments, called **hygrometers**, can be used to measure relative humidity. One of the simplest hygrometers, a **psychrometer**, consists of two

Figure 4–12 Sling psychrometer. This instrument is used to determine relative humidity and dew point. The dry-bulb thermometer gives the current air temperature. The thermometers are spun until the temperature of the wet-bulb thermometer (covered with a cloth wick) stops declining. Then both thermometers are read and the data are used in conjunction with Tables C–1 and C–2 in Appendix C. *(Photo by E. J. Tarbuck)*

identical thermometers mounted side by side (Figure 4–12). One thermometer, called the *wet bulb*, has a thin muslin wick tied around the end.

To use the psychrometer, the cloth wick is saturated with water and a continuous current of air is passed over the wick, either by swinging the instrument freely in the air or by fanning air past it. As a result, water evaporates from the wick, absorbing heat energy from the thermometer to do so, and the temperature of the wet bulb drops. The amount of cooling that takes place is directly proportional to the dryness of the air. The drier the air, the greater the cooling. Therefore, the larger the difference between the wet- and dry-bulb temperatures, the lower the relative humidity; the smaller the difference, the higher the relative humidity. If the air is saturated, no evaporation will occur and the two thermometers will have identical readings.

Tables have been devised to obtain both the relative humidity and the dew-point temperature. (Please refer to Appendix C—Table C–1 and Table C–2.) All that is required is to record the air (dry-bulb) temperature and calculate the difference between the dry- and wet-bulb readings. The difference is known as the *depression of the wet bulb*. Assume, for instance, that the dry-bulb temperature is 20°C and that the wet-bulb reading after swinging or fanning is 14°C. To determine the relative humidity, find the dry-bulb temperature on the left-hand column of Table C–1 of Appendix C and the depression of the wet bulb across the top. The relative humidity is found where the two meet. In this example, the relative humidity is 51 percent. The dew point can be determined in the same way, using Table C–2. In this case, it would be 10°C.

Another instrument used for measuring relative humidity, the *hair hygrometer*, can be read directly without using tables. The hair hygrometer operates on the principle that hair changes length in proportion to changes in relative humidity. Hair lengthens as relative humidity increases and shrinks as relative humidity drops. People with naturally curly hair experience this phenomenon, for in humid weather their hair lengthens and hence becomes curlier. The hair hygrometer uses a bundle of hairs linked mechanically to an indicator that is calibrated between 0 and 100 percent. Thus, we need only glance at the dial to read directly the relative humidity. Unfortunately, the hair hygrometer is less accurate than the psychrometer. Furthermore, it requires frequent calibration and is slow in responding to changes in humidity, especially at low temperatures.

A different type of hygrometer is used in remote-sensing instrument packages, such as radiosondes, that transmit upper-air observations back to ground stations. The electric hygrometer contains an electrical conductor coated with a moisture-absorbing chemical. It works on the principle that electric current flow varies as the relative humidity varies. Most surface weather stations have converted from traditional hygrometers to electric hygrometers of the type used for upper-air observations.

Adiabatic Temperature Changes

So far we have considered some basic properties of water vapor and how its variability in the atmosphere is measured. We are now ready to examine the critical role that water vapor plays in our daily weather.

Recall that condensation occurs when water vapor is cooled enough to change to a liquid. Condensation may produce dew, fog, or clouds. Although each type of condensation is different, all require saturated air to form. As indicated earlier, saturation occurs either when sufficient water vapor is added to the air or, more commonly, when the air is cooled to its dew-point temperature.

Box 4–3 Atmospheric Hazard: Humidity and Heat Stress

During a heat wave in 1995 more than 500 heat-related fatalities occurred in the greater Chicago area. Although this was an exceptional event, the stress of high summer temperatures and exposure to the Sun claim about 175 American lives in an average year.

High humidity contributes significantly to the discomfort people feel during a heat wave. Why are hot, muggy days so uncomfortable? Humans, like other mammals, are warm-blooded creatures who maintain a constant body temperature regardless of the temperature of the environment. One of the ways the body prevents overheating is by perspiring or sweating. However, this process does little to cool the body unless the perspiration can evaporate. It is the cooling created by the evaporation of

perspiration that reduces body temperature. Because high humidity retards evaporation, people are more uncomfortable on a hot and humid day than on a hot and dry day.

Generally, temperature and humidity are the most important elements influencing summertime human comfort. Several indices combine these factors to establish the level or degree of comfort or discomfort. One index widely used by the National Weather Service was developed by R. G. Steadman and is called the *heat stress index*, or simply the *heat index*. It is a measure of *apparent temperature*, the air temperature that an individual perceives. It indicates how "hot" an average person feels given various combinations of temperature and relative humidity (Table 4–A).

TABLE 4-A: Heat Index

Temperature	Relative Humidity									
	10%	20%	30%	40%	50%	60%	70%	80%	90%	100%
80° F	75	77	78	79	81	82	85	86	88	91
85° F	80	82	84	86	88	90	93	97	102	108
90° F	85	87	90	93	96	100	106	113	122	*
95° F	90	93	96	101	107	114	124	136	*	*
100° F	95	99	104	110	120	132	144	*	*	*
105° F	100	105	113	123	135	149	*	*	*	*
110° F	105	112	123	137	150	*	*	*	*	*
115° F	111	120	135	151	*	*	*	*	*	*

Danger Category

- No discomfort
- I. Caution
- II. Extreme caution
- III. Danger
- IV. Extreme danger
- Not observed

Data from National Weather Service.
* Conditions not normally observed.

Heat near Earth's surface is readily exchanged between the ground and the air above. As the ground loses heat in the evening (radiation cooling), dew may condense on the grass and fog may form in the air near the surface. Thus, surface cooling that occurs after sunset accounts for some condensation. However, cloud formation often takes place

during the warmest part of the day. Clearly some other mechanism must operate aloft that cools air sufficiently to generate clouds.

The process that generates most clouds is easily visualized. Have you ever pumped up a bicycle tire with a hand pump and noticed that the pump barrel became

For example, we can see from Table 4–A that if the air temperature is 90°F and the relative humidity is 60 percent, it should feel like 100°F. Note that as the relative humidity increases, the apparent temperature, and thus heat stress, increases as well. Further, when the relative humidity is low, the apparent temperature can have a value that is less than the actual air temperature.

To advise the public on the potential danger from heat stress, the National Weather Service uses the apparent temperature to determine the level of human discomfort, as shown in Table 4–B. This method categorizes the impact that various apparent temperatures will have on the well-being of individuals. It is important to note that factors such as the length of exposure to direct sunlight, the wind speed, and the general health of the individual greatly affect the amount of stress a person will experience. Further, while a period of hot, humid weather in New Orleans might be tolerable, a similar event in Minneapolis, Minnesota, would tax that population. This is because hot and humid weather is more taxing on people who live where it is less common than it is on people who live where prolonged periods of heat and humidity are the rule.

Table 4–B Apparent temperature and associated human discomfort

Danger Category	Apparent Temperature (°F)	Heat Syndrome
I. Caution	80 to 90	Fatigue *possible* with prolonged exposure and physical activity.
II. Extreme caution	90 to 106	Sunstroke, heat cramps, and heat exhaustion *possible* with prolonged exposure and physical activity.
III. Danger	106 to 130	Sunstroke, heat cramps, or heat exhaustion *likely*. Heatstroke *possible* with prolonged exposure and physical activity.
IV. Extreme danger	Greater than 130	Heatstroke or sunstroke *imminent*.

Source: *National Weather Service.*

very warm? When you applied energy to compress the air, the motion of the gas molecules increased and the temperature of the air rose. Conversely, if you allow air to escape from a bicycle tire, it expands; the gas molecules move less rapidly and the air cools. You have probably felt the cooling effect of the propellant gas expanding as you applied hair spray or spray deodorant.

The temperature changes just described, in which heat was neither added nor subtracted, are called **adiabatic temperature changes**. They result when air is compressed or allowed to expand. In summary, *when air is allowed to expand, it cools; when air is compressed, it warms.*

Adiabatic Cooling and Condensation

To simplify the following discussion it helps if we imagine a volume of air enclosed in a thin elastic cover. Meteorologists call this imaginary volume of air a **parcel**. Typically, we consider a parcel to be a few hundred cubic meters in volume, and we assume that it acts independently of the surrounding air. It is also assumed that no heat is transferred into, or out of, the parcel. Although highly idealized, over short time spans, a parcel of air behaves in a manner much like an actual volume of air moving vertically in the atmosphere. In nature, sometimes the surrounding air infiltrates a vertically moving column of air, a process called **entrainment**. For the following discussion we will assume no mixing of this type is occurring.

Anytime a parcel of air moves upward, it passes through regions of successively lower pressure. As a result, ascending air expands and it cools adiabatically. Unsaturated air cools at a constant rate of 1°C for every 100 meters of ascent (5.5°F per 1000 feet). Conversely, descending air comes under increasing pressure and is compressed and heated 1°C for every 100 meters of descent (Figure 4–13). This rate of cooling or heating applies only to vertically moving unsaturated air and is known as the **dry adiabatic rate** ("dry" because the air is unsaturated).

If a parcel of air rises high enough, it will eventually cool to its dew point. Here the process of condensation begins. The altitude at which a parcel reaches saturation and cloud formation begins is called the **lifting condensation level**.

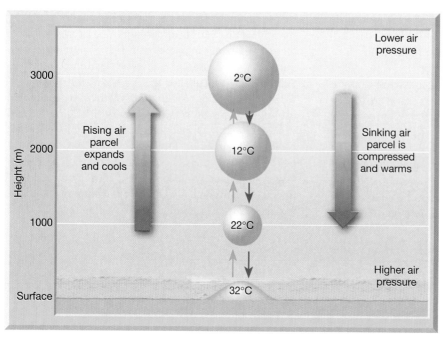

Figure 4–13 Whenever an unsaturated parcel of air is lifted it expands and cools at the *dry adiabatic rate* of 10°C per 1000 meters. Conversely, when air sinks, it is compressed and heats at the same rate.

At the lifting condensation level an important thing happens: The latent heat that was absorbed by the water vapor when it evaporated is liberated. Although the parcel will continue to cool adiabatically, the release of this latent heat slows the rate of cooling. In other words, when a parcel of air ascends above the lifting condensation level, the rate of cooling is reduced because the release of latent heat partially offsets the cooling due to expansion. This slower rate of cooling caused by the release of latent heat is called the **wet adiabatic rate** of cooling ("wet" because the air is saturated).

Because the amount of latent heat released depends on the quantity of moisture present in the air (generally between 0 and 4 percent), the wet adiabatic rate varies from 0.5°C per 100 meters for air with a high moisture content to 0.9°C per 100 meters for air with a low moisture content. Figure 4–14 illustrates the role of adiabatic

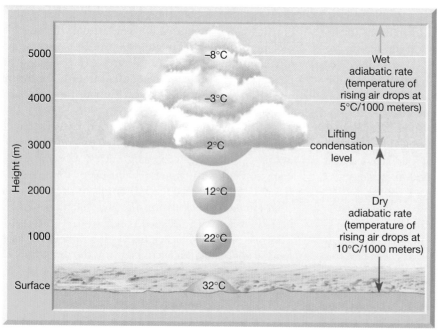

Figure 4–14 Rising air expands and cools at the dry adiabatic rate of 10°C per 1000 meters until the air reaches the dew point and condensation (cloud formation) begins. As air continues to rise, the latent heat released by condensation reduces the rate of cooling. The wet adiabatic rate is therefore always less than the dry adiabatic rate.

cooling in the formation of clouds. Note that from the surface up to the lifting condensation level, the air cools at the faster dry adiabatic rate. The slower wet adiabatic rate commences at the point where condensation begins.

Lifting Processes

To review, when air rises, it expands and cools adiabatically. If air is lifted sufficiently, it will eventually cool to its dew point temperature, saturation will occur, and clouds will develop. But why does air rise on some occasions and not on others?

It turns out that, in general, the tendency is for air to resist vertical movement. Therefore, air located near the surface tends to stay near the surface, and air aloft tends to remain aloft. Exceptions to this rule, as we shall see, include conditions in the atmosphere that give air sufficient buoyancy to rise without the aid of outside forces. In many situations, however, when you see clouds forming there is some mechanical phenomenon at work that forces the air to rise (at least initially).

We will be looking at four mechanism to rise. These are:

1. *Orographic lifting*—air is forced mountainous barrier.
2. *Frontal wedging*—warmer, less dense air, is forced over cooler, denser air.
3. *Convergence*—a pileup of horizontal air flow results in upward movement.
4. *Localized convective lifting*—unequal surface heating causes localized pockets of air to rise because of their buoyancy.

Orographic Lifting

Orographic lifting occurs when elevated terrains, such as mountains, act as barriers to the flow of air (Figure 4–15a). As air ascends a mountain slope, adiabatic cooling often generates clouds and copious precipitation. In fact, many of the rainiest places in the world are located on windward mountain slopes (see Box 4–4).

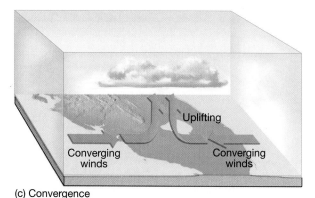

(a) Orographic lifting

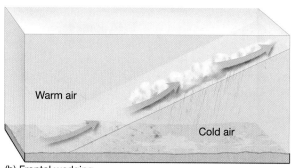

(b) Frontal wedging

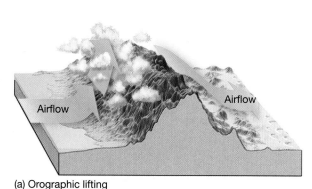

(c) Convergence

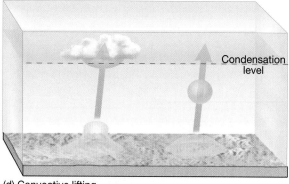

(d) Convective lifting

Figure 4–15 Four processes that lift air. (a) *Orographic lifting:* Air is forced over a topographic barrier. (b) *Frontal wedging:* Cold, dense air displaces warm, less dense air along their boundary. (c) *Convergence:* When surface air converges, it increases in height to allow for the decreased area it occupies. (d) *Localized convective lifting:* Unequal heating of Earth's surface causes pockets of air to be warmed more than the surrounding air.

Box 4–4 Precipitation Records and Mountainous Terrain

Many of the rainiest places in the world are located on windward mountain slopes. Typically, these rainy areas occur because the mountains act as a barrier to the general circulation. Thus, the prevailing winds are forced to ascend the sloping terrain, thereby generating clouds and often abundant precipitation. A station at Mt. Waialeale, Hawaii, for example, records the highest average annual rainfall in the world, some 1234 centimeters (486 inches). The station is located on the windward (northeast) coast of the island of Kauai at an elevation of 1569 meters (5148 feet). Incredibly, only 31 kilometers (19 miles) away lies sunny Barking Sands, with annual precipitation that averages less than 50 centimeters (20 inches).

The greatest recorded rainfall for a single 12-month period occurred at Cherrapunji, India, where an astounding 2647 centimeters (1042 inches), over 86 feet, fell. Cherrapunji, which is located at an elevation of 1293 meters (4309 feet), lies just north of the Bay of Bengal in an ideal location to receive the full effect of India's wet summer monsoon. Most of this rainfall occurred in the summer, particularly during the month of July, when a record 930 centimeters (366 inches) fell. For comparison, 10 times more rain fell in a single month at Cherrapunji, India, than falls in an average year at Chicago.

Because mountains can be sites of abundant precipitation, they are frequently very important sources of water. This is true for many dry locations in the western United States. Here the snow pack, which accumulates high in the mountains during the winter, is a major source of water for the summer season when precipitation is light and demand is great (Figure 4–B). Reservoirs in the Sierra Nevada, for example, accumulate and store spring runoff, which is then delivered to cities such as Los Angeles by way of an extensive network of canals. The record for greatest annual snowfall in the United States goes to the Mount Baker ski area north of Seattle, Washington, where 2896 centimeters (1140 inches) of snow were measured during the winter of 1998–1999.

Figure 4–B Heavy snow pack along Trail Ridge Road, in Colorado's Rocky Mountain National Park. *(Photo by Henry Lansford)*

In addition to providing lift, mountains remove additional moisture in other ways. By slowing the horizontal flow of air, they cause convergence and retard the passage of storm systems. Moreover, the irregular topography of mountains enhances the differential heating that causes some localized convective lifting. These combined effects account

Figure 4–16 Rain shadow desert. The arid conditions in California's Death Valley can be partially attributed to the adjacent mountains, which orographically remove the moisture from air originating over the Pacific. *(Photo by James E. Patterson/James Patterson Collection)*

for the generally higher precipitation associated with mountainous regions compared with surrounding lowlands.

By the time air reaches the leeward side of a mountain, much of its moisture has been lost. If the air descends, it warms adiabatically, making condensation and precipitation even less likely. As shown in Figure 4–15a, the result can be a **rain shadow desert** (see Box 4–5). The Great Basin Desert of the western United States lies only a few hundred kilometers from the Pacific Ocean, but it is effectively cut off from the ocean's moisture by the imposing Sierra Nevada (Figure 4–16). The Gobi Desert of Mongolia, the Takla Makan of China, and the Patagonia Desert of Argentina are other examples of deserts that exist because they are on the leeward sides of mountains.

Frontal Wedging

If orographic lifting were the only mechanism that forced air aloft, the relatively flat central portion of North America would be an expansive desert instead of the nation's breadbasket. Fortunately, this is not the case.

In central North America, masses of warm and cold air collide producing a **front**. Here the cooler, denser air acts as a barrier over which the warmer, less dense air rises. This process, called **frontal wedging**, is illustrated in Figure 4–15b.

It should be noted that weather-producing fronts are associated with storm systems called *middle-latitude cyclones*. Because these storms are responsible for producing a high percentage of the precipitation in the middle latitudes, we will examine them closely in Chapter 9.

Convergence

We saw that the collision of contrasting air masses forces air to rise. In a more general sense, whenever air in the lower troposphere flows together, lifting results. This phenomenon is called **convergence**. When air flows in from more than one direction, it must go somewhere. As it cannot go down, it goes up (Figure 4–15c). This, of course, leads to adiabatic cooling and possibly cloud formation.

Convergence can also occur whenever an obstacle slows or restricts horizontal air flow (wind). We saw earlier that mountains slow winds and cause convergence. Further, when air moves from a relatively smooth surface, such as the ocean, onto an irregular landscape its speed is reduced. The result is a pileup of air (convergence). This is similar to what happens when people leave a well-attended sporting event and pileup results at the exits. When air converges, the air molecules do not simply squeeze closer together (like people); rather, there is a net upward flow.

The Florida peninsula provides an excellent example of the role that convergence can play in initiating cloud development and precipitation. On warm days, the airflow is from the ocean to the land along both coasts of Florida. This leads to a pileup of air along the coasts and general convergence over the peninsula. This pattern of air movement and the uplift that results is aided by intense solar heating of the land. The result is that the peninsula of Florida experiences the greatest frequency of mid-afternoon thunderstorms in the United States (Figure 4–17).

More importantly, convergence as a mechanism of forceful lifting is a major contributor to the weather

Box 4–5   Rain Shadow Deserts and Chinooks

A good way to demonstrate the role that orographic lifting plays in the formation of rain shadow deserts is to examine a simplified, hypothetical situation. Consider air being forced over a 3000-meter-high mountain range, as shown in Figure 4–C. As the unsaturated air ascends the windward side of the mountain, it cools at the rate of 1°C per 100 meters (dry adiabatic rate) until it reaches the dew-point temperature of 20°C. Because the dew-point temperature is reached at 1000 meters, this marks the lifting condensation level and the height of the cloud base.

From the cloud base to the top of the mountain, water vapor within the rising air is continually condensing to form cloud droplets. Recall that condensation releases latent heat that was originally absorbed when the water evaporated. Also, recall that latent heat partially offsets the effect of adiabatic cooling to produce a slower rate of cooling, called the *wet adiabatic rate*.

In our hypothetical example, as the air rises above the lifting condensation level, it cools at the rate of 0.5°C per 100 meters (wet adiabatic rate). By the time the air reaches the top of our mountain, its temperature will have dropped to 10°C. Because condensation is still occurring, the relative humidity at the top of the mountain is 100 percent and the dew-point temperature is the same as the air temperature, that is, 10°C.

For simplicity, we will assume that the air that was forced to the top of the mountain is cooler than the surrounding air and hence begins to flow down the leeward slope of the mountain. As the air descends, it is compressed and *heated* at the dry adiabatic rate. (We are assuming that all of the water vapor that condensed as the air was forced up the mountain fell as precipitation.) Upon reaching the base of the mountain range, the temperature of the descending air has risen to 40°C, or 10°C warmer than the temperature at the base of the mountain on the windward side. The higher temperature on the leeward side is the result of the latent heat that was released during condensation as the air ascended the windward slope of the mountain range.

Now we will examine the moisture content of the air arriving on the leeward side of the mountain. Let us assume that (1) all of the moisture that condensed fell as precipitation; (2) no moisture was added to the air as it descended; and (3) as the air flowed down the mountain slope, its dew point temperature (10°C) remained unchanged.

Using Table 4–1 and the fact that the air temperature is 40°C and the dew point is 10°C, we can approximate the relative humidity on the leeward side of the mountain. From Table 4–1, we see that when saturated, 40°C air contains 47 grams of water vapor per kilogram of air. However, because this 40°C air has a dew point of 10°C, it contains only 7 grams of water vapor per kilogram of air. Therefore, this parcel of air is 7/47 or about 15 percent saturated.

Two reasons account for the low relative humidity commonly observed on leeward mountain slopes. First, water is extracted from air in the form of precipitation on the windward side. Second, the air on the leeward

associated with middle-latitude cyclones and hurricanes. The low-level horizontal airflow associated with these systems is inward and upward around their centers. These important weather producers will be covered in more detail later, but for now remember that convergence near the surface results in a general upward flow.

Localized Convective Lifting

On warm summer days, unequal heating of Earth's surface may cause pockets of air to be warmed more than the surrounding air. For instance, air above a paved parking lot will be warmed more than the air above an adjacent wooded park. Consequently, the parcel of air above the parking lot, which is warmer (less dense) than the surrounding air, will be buoyed upward. These rising parcels of warmer air are called *thermals*. Birds such as hawks and eagles use these thermals to carry them to great heights where they can gaze down on unsuspecting prey. People have learned to employ these rising parcels using hang gliders as a way to "fly."

The phenomenon that produces rising thermals is called **localized convective lifting**. When these warm parcels of air rise above the lifting condensation level, clouds form, which on occasion produce mid-afternoon rain showers. The height of clouds produced in this

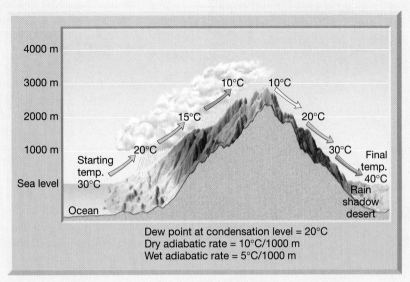

Dew point at condensation level = 20°C
Dry adiabatic rate = 10°C/1000 m
Wet adiabatic rate = 5°C/1000 m

Figure 4–C Orographic lifting and the formation of rain shadow deserts.

side is warmer than the air on the windward side. Recall that without the addition of water vapor, whenever temperature rises, relative humidity drops. It should be emphasized that the hypothetical example provided here is only a rough approximation of conditions in the natural world. Most often only a small percentage (sometimes none) of the moisture that condenses to form clouds actually falls as precipitation. Therefore, the contrast in temperature and humidity (windward versus leeward) in our example represents the extreme case. Nevertheless, warmer and dryer conditions are the rule for the leeward side of a mountainous barrier as compared to the windward side. A classic example is

provided by California's Sierra Nevada. Here, giant sequoias and Douglas firs are found on the well-watered western (windward) slopes. By contrast, a rain shadow desert, which includes Owens Valley and Death Valley, is found leeward of this mountain barrier.

In addition to producing rain shadow deserts, this orographic setting may also generate warm, dry winds that go by a variety of names depending on locale. For example, *chinook* winds occasionally move down the east slopes of the Rockies where they are known to abruptly warm the adjacent western plains by as much as 10°C (18°F), or more, during the winter months. Chinooks and related winds will be addressed again in Chapter 7.

fashion is somewhat limited, for instability caused solely by unequal surface heating is confined to, at most, the first few kilometers of the atmosphere. Also, the accompanying rains, although occasionally heavy, are of short duration and widely scattered.

Although localized convective lifting by itself is not a major producer of precipitation, the added buoyancy that results from surface heating contributes significantly to the lifting initiated by the other mechanisms. It should also be remembered that while the other mechanisms force air to rise, convective lifting occurs because the air is warmer (less dense) than the surrounding air and rises for the same reasons as a hot air balloon.

The Critical Weathermaker: Atmospheric Stability

When air rises, it cools and eventually produces clouds. Why do clouds vary so much in size, and why does the resulting precipitation vary so much? The answers are closely related to the *stability* of the air.

Recall that a parcel of air can be thought of as having a thin flexible cover that allows it to expand but prevents it from mixing with the surrounding air (picture a hot-air balloon). If this parcel were forced to rise, its temperature would decrease because of expansion. By comparing the

Figure 4–17 Southern Florida viewed from the space shuttle. On warm days, airflow from the Atlantic Ocean and Gulf of Mexico onto the Florida peninsula generates many mid-afternoon thunderstorms. *(Photo by NASA/Media Services)*

parcel's temperature to that of the surrounding air, we can determine its stability. If the parcel were *cooler* than the surrounding environment, it would be more dense; and if allowed to do so, it would sink to its original position. Air of this type, called **stable air**, resists vertical movement.

If, however, our imaginary rising parcel were *warmer* and hence less dense than the surrounding air, it would continue to rise until it reached an altitude where its temperature equalled that of its surroundings. This is exactly how a hot-air balloon works, rising as long as it is warmer and less dense than the surrounding air (Figure 4–18). This type of air is classified as **unstable air**. In summary, stability is a property of air that describes its tendency to remain in its original position (stable), or to rise (unstable).

Types of Stability

The stability of the atmosphere is determined by measuring the air temperature at various heights. Recall that this measure is called the *environmental lapse rate*. Do not confuse the environmental lapse rate with adiabatic temperature changes. The environmental lapse rate is the actual temperature of the atmosphere, as determined from observations made by radiosondes and aircraft. Adiabatic temperature changes are changes in temperature that a parcel of air would experience if it moved vertically through the atmosphere.

To illustrate how the stability of the atmosphere is determined consider a situation in which the prevailing environmental lapse rate is 5°C per 1000 meters (Figure 4–19).

Under this condition, when the air at the surface has a temperature of 25°C, the air at 1000 meters will be 5°C cooler, or 20°C, whereas the air at 2000 meters will have a temperature of 15°C, and so forth.

Examine Figure 4–19 and note that the air at the surface appears to be less dense than the air at 1000 meters, for it is 5°C warmer. However, if the air near the surface were to rise to 1000 meters, it would expand and cool at the dry adiabatic rate of 1°C per 100 meters. Therefore, on reaching 1000 meters, the temperature of the rising parcel would have dropped from 25°C to 15°C, a total of 10°C. Being 5°C cooler than its environment, it would be more dense and would tend to sink to its original position. Thus, we say that the air near the surface is *potentially cooler* than the air aloft and therefore it will not rise on its own.

By similar reasoning, if the air at 1000 meters subsided, adiabatic heating would increase its temperature 10°C by the time it reached the surface, making it warmer than the surrounding air; thus, its buoyancy would cause it to return to its original position. The air just described is stable and resists vertical movement.

With that background we will now look at three fundamental conditions of the atmosphere: absolute stability, absolute instability, and conditional instability.

Absolute Stability. Stated quantitatively, **absolute stability** prevails when *the environmental lapse rate is less than the wet adiabatic rate*. Figure 4–20 depicts this situation by

Figure 4–18 Hot air balloons. *(Photo by Jim Sulley/The Image Works)*

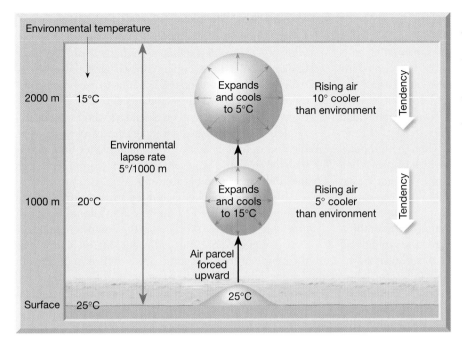

Figure 4–19 In a stable atmosphere, as an unsaturated parcel of air is lifted, it expands and cools at the dry adiabatic rate of 10°C per 1000 meters. Because the temperature of the rising parcel of air is lower than that of the surrounding environment, it will be heavier and, if allowed to do so, will sink to its original position.

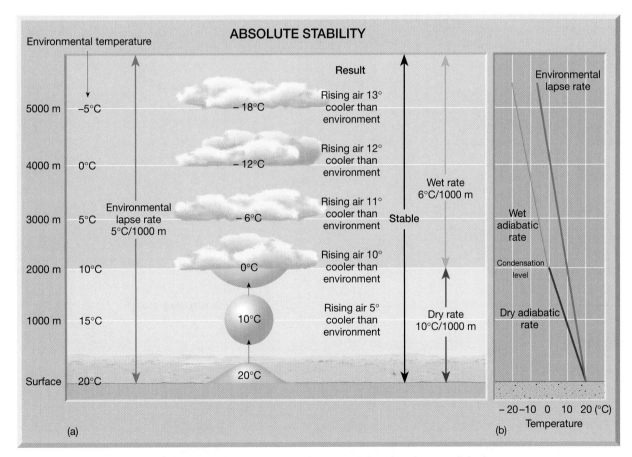

Figure 4–20 *Absolute stability* prevails when the environmental lapse rate is less than the wet adiabatic rate. (a) The rising parcel of air is always cooler and heavier than the surrounding air, producing stability. (b) Graphic representation of the conditions shown in part (a).

using an environmental lapse rate of 5°C per 1000 meters and a wet adiabatic rate of 6°C per 1000 meters. Note that at 1000 meters, the temperature of the surrounding air is 15°C and that the rising parcel of air has cooled by expansion to 10°C and so is the denser air. Even if this stable air were forced above the condensation level, it would remain cooler and denser than its environment and would have a tendency to return to the surface.

The most stable conditions occur when the temperature in a layer of air actually increases with altitude. When such a reversal occurs, a *temperature inversion* exists. Many circumstances can create temperature inversions. Temperature inversions frequently occur on clear nights as a result of radiation cooling at Earth's surface. Under these conditions, an inversion is created because the ground and the air next to it will cool more rapidly than the air aloft.

Temperature inversions also occur in winter when warm air from the Gulf of Mexico invades the cold, snow-covered surface of the midcontinent. Anytime warmer air overlies cooler air, the resulting layer is extremely stable and resists appreciable vertical mixing. Because of this, temperature inversions are responsible for trapping pollutants in a narrow zone near Earth's surface. This idea will be explored more fully in Chapter 13.

Absolute Instability. At the other extreme, a layer of air is said to exhibit **absolute instability** when *the environmental lapse rate is greater than the dry adiabatic rate.* As shown in Figure 4–21, the ascending parcel of air is always warmer than its environment and will continue to rise because of its own buoyancy. Absolute instability occurs most often during the warmest months and on clear days when solar heating is intense. Under these conditions, the lowermost layer of the atmosphere is heated to a much higher temperature than the air aloft. This results in a steep environmental lapse rate and a very unstable atmosphere.

Instability produced mainly by strong surface heating is generally confined to the first few kilometers of the atmosphere. Above this height, the environmental lapse rate assumes a more "normal" value. Stated another way, the temperature drops more slowly with altitude, creating a more stable temperature regime. Consequently, clouds produced by surface heating lack great vertical height and thus rarely produce violent weather.

Conditional Instability. A more common type of atmospheric instability is called **conditional instability**. This situation prevails when *moist air has an environmental lapse rate between the dry and wet*

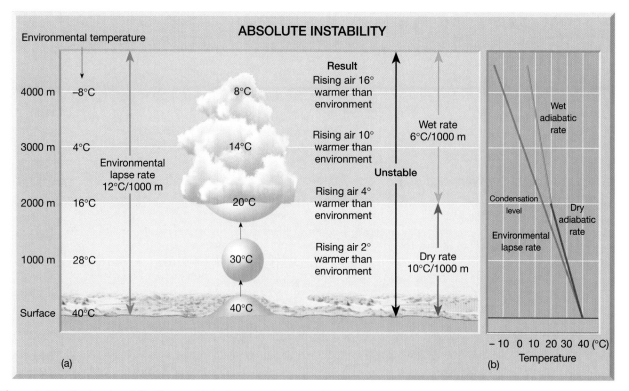

Figure 4–21 *Absolute instability* illustrated by using an environmental lapse rate of 12°C per 1000 meters. (a) The rising air is always warmer than the surrounding air, and therefore it is less dense and rises (unstable). (b) Graphic representation of the conditions shown in part (a).

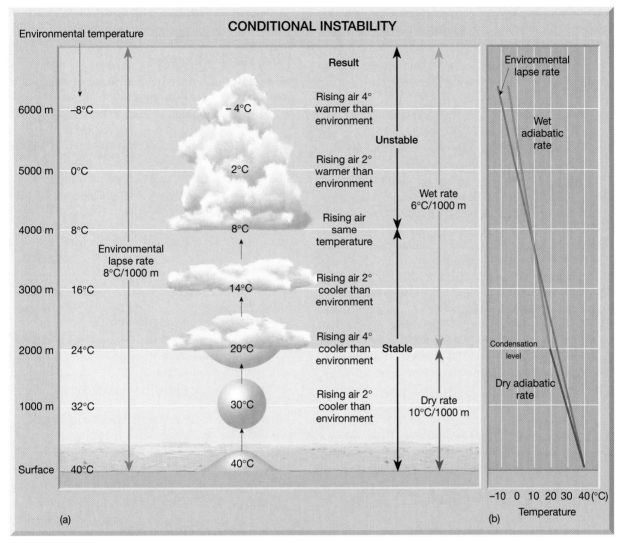

CONDITIONAL INSTABILITY

Figure 4–22 *Conditional instability* illustrated by using an environmental lapse rate of 8°C per 1000 meters, which lies between the dry adiabatic rate and the wet adiabatic rate. (a) The rising parcel of air is cooler than the surrounding air below 4000 meters (stable) but warmer above 4000 meters (unstable). (b) Graphic representation of the conditions shown in part (a).

adiabatic rates (between about 5° and 10°C per 1000 meters). Simply stated, the atmosphere is said to be conditionally unstable when it is *stable* with respect to an *unsaturated* parcel of air, but *unstable* with respect to a *saturated* parcel of air. Notice in Figure 4–22 that the rising parcel of air is cooler than the surrounding air for the first 4000 meters. With the release of latent heat above the lifting condensation level, the parcel becomes warmer than the surrounding air. From this point along its ascent, the parcel will continue to rise without an outside force.

Thus, conditional instability depends on whether or not the rising air is saturated. The word "conditional" is used because the air must be forced upward before it

reaches the level where it becomes unstable and rises on its own.

In summary, the stability of air is determined by measuring the temperature of the atmosphere at various heights (environmental lapse rate). In simple terms, a column of air is deemed unstable when the air near the bottom of this layer is significantly warmer (less dense) than the air aloft, indicating a steep environmental lapse rate. Under these conditions, the air actually turns over because the warm air below rises and displaces the colder air aloft. Conversely, the air is considered to be stable when the temperature decreases gradually with increasing altitude. The most stable conditions occur during a temperature inversion when the temperature actually increases with height. Under these conditions, there is very little vertical air movement.

Stability and Daily Weather

From the previous discussion, we can conclude that stable air resists vertical movement and that unstable air ascends freely because of its own buoyancy. How do these facts manifest themselves in our daily weather?

When stable air is forced aloft, the clouds that form are widespread and have little vertical thickness compared with their horizontal dimension. Precipitation, if any, is light to moderate. In contrast, clouds associated with unstable air are towering and are usually accompanied by heavy precipitation. Thus, we can conclude that on a dreary and overcast day with light drizzle, stable air was forced aloft. Conversely, on a day when cauliflower-shaped clouds appear to be growing as if bubbles of hot air were surging upward, we can be relatively certain that the atmosphere is unstable (Figure 4–23).

As noted earlier, the most stable conditions occur during a temperature inversion when temperature increases with height. In this situation, the air near the surface is cooler and heavier than the air aloft, and therefore little vertical mixing occurs between the layers. Because pollutants are generally added to the air from below, a temperature inversion confines them to the lowermost layer, where their concentration will continue to increase until the temperature inversion dissipates. Widespread fog is another sign of stability. If the layer containing fog were mixing freely with the "dry" layer above, evaporation would quickly eliminate the foggy condition.

How Stability Changes

Recall that the higher (steeper) the environmental lapse rate, the more rapidly the temperature drops with increasing altitude. Therefore, any factor that causes air near the surface to become warmed in relation to the air aloft increases instability. The opposite is also true; any factor that causes the surface air to be chilled results in the air becoming more stable. Also recall that the most stable conditions occur when the temperature in a layer of air actually increases with altitude.

Instability is enhanced by the following:

1. Intense solar heating warming the lowermost layer of the atmosphere
2. The heating of an air mass from below as it passes over a warm surface
3. General upward movement of air caused by processes such as orographic lifting, frontal wedging, and convergence
4. Radiation cooling from cloud tops

Stability is enhanced by the following:

1. Radiation cooling of Earth's surface after sunset
2. The cooling of an air mass from below as it traverses a cold surface
3. General subsidence within an air column

Figure 4–23 Cauliflower-shaped clouds provide evidence of unstable conditions in the atmosphere. *(Photo by Tom Bean/DRK Photo)*

Note that most processes that alter stability result from temperature changes caused by horizontal or vertical air movement, although daily temperature changes are important too. In general, any factor that increases the environmental lapse rate renders the air more unstable, whereas any factor that reduces the environmental lapse rate increases the air's stability.

Temperature Changes and Stability

As stated earlier, on a clear day when there is abundant surface heating, the lower atmosphere often becomes warmed sufficiently to cause parcels of air to rise. After the Sun sets, surface cooling generally renders the air stable again.

Similar changes in stability occur as air moves horizontally over a surface having markedly different temperatures. In the winter, warm air from the Gulf of Mexico moves northward over the cold, snow-covered Midwest. Because the air is cooled from below, it becomes more stable, often producing widespread fog.

The opposite occurs when wintertime polar air moves southward over the open waters of the Great Lakes. Although we would die of hypothermia in a few minutes if we fell into these cold waters (perhaps 5°C or 41°F), compared to the frigid air temperatures, these waters are warm. Recall from our discussion of vapor pressure that the temperature of the water surface and the vapor pressure of the surrounding air determine the rate at which water will evaporate. Because the water of the Great Lakes is comparatively warm and because the polar air is dry (low vapor pressure) the rate of evaporation will be high. The moisture and heat added to the frigid polar air from the water below are enough to make it unstable and generate the clouds that produce heavy snowfalls on the downwind shores of these lakes (called "lake-effect snows"—see Chapter 8).

Radiation Cooling from Clouds. On a smaller scale, the loss of heat by radiation from cloud tops during evening hours adds to their instability and growth. Unlike air, which is a poor radiator of heat, cloud droplets emit energy to space nearly as well as does Earth's surface. Towering clouds that owe their growth to surface heating lose that source of energy at sunset. After sunset, however, radiation cooling at their tops steepens the lapse rate near the top of the cloud and can lead to additional upward flow of warmer parcels from below. This process is believed responsible for producing nocturnal thunderstorms from clouds whose growth prematurely ceased at sunset.

Vertical Air Movement and Stability

Vertical movements of air also influence stability. When there is a general downward airflow, called **subsidence**, the upper portion of the subsiding layer is heated by compression, more so than the lower portion. Usually the air near the surface is not involved in the subsidence and so its temperature remains unchanged. The net effect is to stabilize the air, for the air aloft is warmed in relation to the surface air. The warming effect of a few hundred meters of subsidence is enough to evaporate the clouds found in any layer of the atmosphere. Thus, one sign of subsiding air is a deep blue, cloudless sky. Subsidence can also produce a temperature inversion aloft. The most intense and prolonged temperature inversions and associated air pollution episodes are caused by subsidence, a topic discussed more fully in Chapter 13.

Upward movement of air generally enhances instability, particularly when the lower portion of the rising layer has a higher moisture content than the upper portion, which is usually the situation. As the air moves upward, the lower portion becomes saturated first and cools at the lesser wet adiabatic rate. The net effect is to increase the lapse rate within the rising layer. This process is especially important in producing the instability associated with thunderstorms. In addition, recall that conditionally unstable air can become unstable if lifted sufficiently.

In summary, the role of stability in determining our daily weather cannot be overemphasized. The air's stability, or lack of it, determines to a large degree whether clouds develop and produce precipitation and whether that precipitation will come as a gentle shower or a violent downpour. In general, when stable air is forced aloft, the associated clouds have little vertical thickness, and precipitation, if any, is light. In contrast, clouds associated with unstable air are towering and are frequently accompanied by heavy precipitation.

Chapter Summary

- The unending circulation of Earth's water supply is called the *hydrologic cycle* (or water cycle). The cycle illustrates the continuous movement of water from the oceans to the atmosphere, from the atmosphere to the land, and from the land back to the sea.

- *Water vapor*, an odorless, colorless gas, can change from one state of matter (solid, liquid or gas) to another at the temperatures and pressures experienced on Earth. The heat energy involved in the change of state of water is often measured in *calories*. The processes involved in changes of state include *evaporation* (liquid to gas), *condensation* (gas to liquid), *melting* (solid to liquid), *freezing* (liquid to solid), *sublimation* (solid to gas), and *deposition* (gas to solid). During each change, *latent* (hidden, or stored) *heat* energy is either absorbed or released.

- *Humidity* is the general term used to describe the amount of water vapor in the air. The methods used to express humidity quantitatively include (1) *absolute humidity*, the mass of water vapor in a given volume of air, (2) *mixing ratio*, the mass of water vapor in a unit of air compared to the remaining mass of dry air, (3) *vapor pressure*, that part of the total atmospheric pressure attributable to its water vapor content, (4) *relative humidity*, the ratio of the air's actual water vapor content compared with the amount of water vapor required for saturation at that temperature, and (5) *dew point*, the temperature to which a parcel of air would need to be cooled to reach saturation. When air is *saturated*, the pressure exerted by the water vapor, called the *saturation vapor pressure*, produces a balance between the number of water molecules leaving the surface of the water and the number returning. Because the saturation vapor pressure is temperature-dependent, at higher temperatures, more water vapor is required for saturation to occur.

- Relative humidity can be changed in two ways, (1) by changing the amount of moisture in the air, or (2) by changing the air's temperature. Adding moisture to the air while keeping the temperature constant increases the relative humidity. Removing moisture lowers the relative humidity. When the water vapor content of air remains at a constant level, a decrease in air temperature results in an increase in relative humidity and an increase in temperature causes a decrease in relative humidity. In nature there are three major ways that air temperatures change to cause corresponding changes in relative humidity; (1) daily (daylight versus nighttime) changes in temperature, (2) temperature changes that result as air moves horizontally from one location to another, and (3) changes caused as air moves vertically in the atmosphere.

- An important concept related to relative humidity is the *dew-point temperature* (or simply *dew point*), which is the temperature to which a parcel of air would need to be cooled to reach saturation. Unlike relative humidity which is a measure of how near the air is to being saturated, dew point temperature is a measure of the air's actual moisture content. High dew point temperatures equate to moist air and low dew point temperatures indicate dry air. Because the dew point temperature is a good measure of the amount of water vapor in the air, it is the measure of atmospheric moisture that appears on daily weather maps.

- A variety of instruments, called *hygrometers*, can be used to measure relative humidity. One of the simplest hygrometers, a *psychrometer*, consists of two identical thermometers mounted side by side. One thermometer, called the wet bulb thermometer, has a thin muslin wick tied around the bulb. After spinning or fanning air past the instrument and noting the difference between the dry- and wet-bulb readings (known as the depression of the wet bulb), tables are consulted to determine the relative humidity. A second instrument, the *hair hygrometer*, can be read directly without using tables.

- When air is allowed to expand, it cools. When air is compressed, it warms. Temperature changes produced in this manner, in which heat is neither added nor subtracted, are called *adiabatic temperature changes*. The rate of cooling or warming of vertically moving unsaturated ("dry") air is 1°C for every 100 meters (5.5°F per 1000 feet), the *dry adiabatic rate*. At the *lifting condensation level* (the altitude where the parcel of air has reached saturation and cloud formation begins), latent heat is released and the rate of cooling is reduced. The slower rate of cooling, called the *wet adiabatic rate* of cooling ("wet" because the air is saturated) varies from 0.5°C per 100 meters for air with a high moisture content to 0.9°C per 100 meters for air with a low moisture content.

- When air rises, it expands and cools adiabatically. If air is lifted sufficiently high, it will eventually cool to its dew-point temperature, and clouds will develop. Four mechanisms that cause air to rise are (1) *orographic lifting*, where air is forced to rise over a mountainous barrier, (2) *frontal wedging*, where warmer, less dense air is forced over cooler, denser air along a *front*, (3) *convergence*, a pileup of horizontal air flow resulting in an upward flow, and (4) *localized convective lifting*, where unequal surface heating causes localized pockets of air to rise because of their buoyancy.

- When air rises, it cools and can eventually produce clouds. *Stable air* resists vertical movement, whereas *unstable air* rises because of its buoyancy. The stability of air is determined by knowing the *environmental lapse rate*, the temperature of the atmosphere at various heights. The three fundamental conditions of the atmosphere are (1) *absolute stability*, when the environmental lapse rate is less than the wet adiabatic rate, (2) *absolute instability*, when the environmental lapse rate is greater than the dry adiabatic rate, and (3) *conditional instability*, when moist air has an environmental lapse rate between the dry and wet adiabatic rates. In general, when stable air is forced aloft, the associated clouds have little vertical thickness, and precipitation, if any, is light. In contrast, clouds associated with unstable air are towering and frequently accompanied by heavy rain.

- Any factor that causes air near the surface to become warmed in relation to the air aloft increases the air's instability. The opposite is also true; any factor that causes the surface air to be chilled results in the air becoming more stable. Most processes that alter atmospheric stability result from temperature changes caused by horizontal or vertical air movements, although daily temperature changes are important too. Changes in stability occur as air moves horizontally over a surface having a markedly different temperature than the air. Furthermore, *subsidence* (a general downward airflow) generally stabilizes the air, while upward air movement enhances instability.

Vocabulary Review

absolute humidity (p. 90)
absolute instability (p. 110)
absolute stability (p. 108)
adiabatic temperature changes (p. 101)
calorie (p. 88)
condensation (p. 89)
conditional instability (p. 110)
convergence (p. 105)
deposition (p. 89)
dew point (p. 96)
dry adiabatic rate (p. 101)
entrainment (p. 101)
evaporation (p. 89)
freezing (p. 89)
front (p. 105)
frontal wedging (p. 105)
humidity (p. 90)
hydrologic cycle (p. 87)
hygrometer (p. 98)

latent heat (p. 88)
lifting condensation level (p. 101)
localized convective lifting (p. 106)
melting (p. 89)
mixing ratio (p. 90)
orographic lifting (p. 103)
parcel (p. 101)
psychrometer (p. 98)
rain shadow desert (p. 105)
relative humidity (p. 92)
saturation (p. 92)
saturation vapor pressure (p. 91)
stable air (p. 108)
sublimation (p. 89)
subsidence (p. 113)
transpiration (p. 87)
unstable air (p. 108)
vapor pressure (p. 90)
wet adiabatic rate (p. 102)

Review Questions

1. Describe the movement of water through the hydrologic cycle.

2. The quantity of water lost to evaporation over the oceans is not equaled by precipitation. Why, then, does the sea level not drop?

3. Summarize the processes by which water changes from one state to another. Indicate whether heat is absorbed or liberated.

4. After reviewing Table 4–1, write a generalization relating temperature and the amount of water vapor needed to saturate the air.

5. How do absolute humidity and mixing ratio differ? What do they have in common? How is relative humidity different from absolute humidity and the mixing ratio?

6. Refer to Figure 4–9 and then answer the following questions.

a. When is relative humidity highest during a typical day? When is it lowest?

b. At what time of day would dew most likely form?

c. Write a generalization relating changes in air temperature to changes in relative humidity.

7. If temperature remains unchanged and if the mixing ratio decreases, how will relative humidity change?

8. How much more water vapor does a mass of air having a dew point temperature of 24°C (75°F) contain than air having a dew point temperature of 4°C (39°F)?

9. Explain the principle of the psychrometer; the hair hygrometer.

10. What are the disadvantages of the hair hygrometer? Does this instrument have any advantages over the psychrometer?

11. What name is given to the processes whereby the temperature of the air changes without the addition or subtraction of heat?

12. At what rate does unsaturated air cool when it rises through the atmosphere?

13. Why does air expand as it moves upward through the atmosphere?

14. Explain why air warms whenever it sinks.

15. Why does the adiabatic rate of cooling change when condensation begins? Why is the wet adiabatic rate not a constant figure?

16. The contents of an aerosol can are under very high pressure. When you push the nozzle on such a can, the spray feels cold. Explain.

17. How do orographic lifting and frontal wedging act to force air to rise?

18. Explain why the Great Basin area of the western United States is dry. What term is applied to such a situation?

19. How is localized convective lifting different from the other three processes that cause air to rise?

20. How does stable air differ from unstable air?

21. Explain the difference between the environmental lapse rate and adiabatic cooling.

22. How is the stability of air determined?

23. Write a statement relating the environmental lapse rate to stability.

24. What weather conditions would lead you to believe that air is unstable?

25. List four ways instability can be enhanced.

26. List three ways stability can be enhanced.

Problems

1. Using Table 4–1, answer the following:

a. If a parcel of air at 25°C contains 10 grams of water vapor per kilogram of air, what is its relative humidity?

b. If a parcel of air at 35°C contains 5 grams of water vapor per kilogram of air, what is its relative humidity?

c. If a parcel of air at 15°C contains 5 grams of water vapor per kilogram of air, what is its relative humidity?

d. If the temperature of the parcel of air in part c dropped to 5°C, how would its relative humidity change?

e. If 20°C air contains 7 grams of water vapor per kilogram of air, what is its dew point?

2. Using the standard tables (Appendix C Tables C–1 and C–2), determine the relative humidity and dew-point temperature if the dry-bulb thermometer reads 22°C and the wet-bulb thermometer reads 16°C. How would the relative humidity and dew point change if the wet-bulb reading were 19°C?

3. If unsaturated air at 20°C were to rise, what would its temperature be at a height of 500 meters? If the dew-point temperature at the lifting condensation level were 11°C, at what elevation would clouds begin to form?

4. Using Figure 4–24, answer the following. (Hint: Read Box 4–5)

a. What is the elevation of the cloud base?

b. What is the temperature of the ascending air when it reaches the top of the mountain?

c. What is the dew-point temperature of the rising air at the top of the mountain? (Assume 100 percent relative humidity.)

d. Estimate the amount of water vapor that must have condensed (in grams per kilogram) as the air moved from the cloud base to the top of the mountain.

e. What will the temperature of the air be if it descends to point B? (Assume that the moisture that condensed fell as precipitation on the windward side of the mountain.)

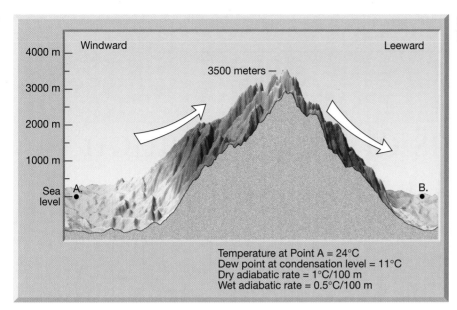

Figure 4–24 Orographic lifting problem.

Temperature at Point A = 24°C
Dew point at condensation level = 11°C
Dry adiabatic rate = 1°C/100 m
Wet adiabatic rate = 0.5°C/100 m

f. What is the approximate capacity of the air to hold water vapor at point B?

g. Assuming that no moisture was added or subtracted from the air as it traveled downslope, estimate the relative humidity at point B.

h. What is the *approximate* relative humidity at point A? (Use the dew-point temperature at the condensation level for the surface dew point.)

i. Give *two* reasons for the difference in relative humidity between points A and B.

j. Needles, California, is situated on the dry leeward side of a mountain range similar to the position of point B. What term describes this situation?

Atmospheric Science Online

The following are informative and interesting Internet sites that address topics related to those presented in the chapter:

Current U.S. Dewpoint Statistics (Penn State University):
• **http://www.ems.psu.edu/wx/usstats/dewpstats. html**

Weather Calculator (National Weather Service, El Paso):
• **http://nwselp.epcc.edu/elp/wxcalc.html**

For direct links to these sites and others, chapter objectives and reviews, quiz questions, and topical investigations that utilize Web resources, visit *The Atmosphere Eighth Edition* Home Page at:
• **http://www.prenhall.com/lutgens**

Forms of Condensation and Precipitation

Steam fog over a lake in Maine. *(Photo by Sara Gray/Tony Stone Images)*

Figure 5–1 People caught in a downpour. *(Photo by Mary Fulton/Liaison Agency, Inc.)*

Clouds, fog, rain, snow, and sleet are among the most observable weather phenomena (Figure 5–1). This chapter will provide a basic understanding of each. In addition to learning the basic scheme for classifying and naming clouds, you will learn that the formation of an average raindrop involves complex processes requiring water from roughly a million cloud droplets. Can scientists make rain by triggering these processes? Can modern weather-modification technology increase the precipitation that falls from a cloud? Can we control weather events such as fog and frost?

Condensation

As you learned earlier, condensation occurs when water vapor changes to a liquid. The result of this process may be dew, fog, or clouds. Although each type of condensation is different, all form when two conditions are met.

First, for any form of condensation to occur, the air must be *saturated*. Saturation occurs either when the air is cooled below its dew point, or when sufficient water vapor is added to the air.

Second, there generally must be a *surface* on which the water vapor can condense. When dew forms, objects at or near the ground, like blades of grass, serve this purpose. When condensation occurs in the air above the ground, tiny particles known as **condensation nuclei** serve as surfaces on which water vapor condenses. Nuclei are important because, if they are absent, a relative humidity well in excess of 100 percent is necessary to produce clouds. Condensation nuclei include microscopic dust, smoke, and salt particles, all of which are profuse in the lower atmosphere. Consequently, in the troposphere, the relative humidity seldom exceeds 100 percent by more than 1 or 2 percent.

Particles that are the most effective cloud condensation nuclei are **hygroscopic**, which means they are water-absorbent. Some familiar food items, such as crackers and cereal, are also hygroscopic, which is why they quickly absorb moisture when exposed to humid air and become stale. Some of the most common hygroscopic condensation nuclei are minute crystals of sulfate and nitrate compounds. Hygroscopic nuclei are introduced into the atmosphere mainly as a by-product of combustion (burning), from such sources as forest fires, automobiles, and coal-burning furnaces. In addition, salt from breaking ocean waves and some particles found in ordinary dust can serve as cloud condensation nuclei.

Condensation Aloft and Cloud Formation

Clearly the most important cloud-forming process is *adiabatic cooling that occurs as air ascends*. To review, any time a parcel of air ascends, it passes through regions of successively lower pressure. As a result, rising air expands and is cooled adiabatically. At a height called the *lifting condensation level*, the ascending parcel has cooled to its dewpoint temperature, and further ascent causes condensation.

Recall that condensation occurs on tiny particles (condensation nuclei). Initially, the growth rate of cloud droplets is rapid. Growth diminishes in a short time, however, because the available water vapor is quickly consumed by the large number of competing droplets. The result is the formation of a cloud consisting of billions of tiny water droplets, all so small that they remain suspended in air. Even in very moist air, the growth of cloud droplets by additional condensation is slow. Furthermore, the immense size difference between cloud droplets and raindrops (it takes about a million cloud droplets to form a single raindrop) suggests that condensation alone is not responsible for the formation of drops large enough to fall as rain. We will investigate this point later.

Clouds

Clouds are one form of condensation. They are best described as *visible aggregates of minute droplets of water or tiny crystals of ice*. In addition to being prominent and sometimes spectacular features in the sky, clouds are of continual interest to meteorologists because they provide a visible indication of what is going on in the atmosphere.

Anyone who observes clouds finds a bewildering variety of these white and gray masses streaming across the sky. Once the basic classification scheme for clouds is known, however, most of the confusion vanishes.

Cloud Classification

Prior to the beginning of the nineteenth century, there were no generally accepted names for clouds. In 1803, Luke Howard, an English naturalist, published a cloud classification that met with great success and subsequently served as the basis of our present-day system.

Clouds are classified on the basis of two criteria: *form* and *height* (Figure 5–2). Three basic cloud forms are recognized:

- **Cirrus** clouds are high, white, and thin. They are separated or detached and form delicate veil-like patches or extended wispy fibers and often have a feathery appearance. (*Cirrus* is a Latin word meaning "curl" or "filament.")
- **Cumulus** clouds consist of globular individual cloud masses. Normally, they exhibit a flat base and appear as rising domes or towers. Such clouds are frequently described as having a cauliflowerlike structure.
- **Stratus** clouds are best described as sheets or layers (strata) that cover much or all of the sky. Although there may be minor breaks, there are no distinct individual cloud units.

All clouds have one of these three basic forms or combinations or modifications of them.

Looking at the second aspect of cloud classification, height, three levels are recognized: high, middle, and low. **High clouds** normally have bases above 6000 meters (20,000 feet); **middle clouds** generally occupy heights from 2000 to 6000 meters; **low clouds** form below 2000 meters (6500 feet). These altitudes are not hard and fast. They vary somewhat by season of the year and by latitude. At high (poleward) latitudes or during cold winter months, high clouds generally occur at lower altitudes. Further, some clouds extend vertically to span more than one height range. These are called **clouds of vertical development**.

Definite weather patterns can be associated with specific clouds or combinations of clouds, so it is important to become familiar with cloud characteristics.

Cloud Descriptions

Ten basic cloud types are recognized internationally. We describe them below, and summarize them in Table 5–1.

High Clouds. Three cloud types make up the family of high clouds (above 6000 meters [20,000 feet]). They are *cirrus*, *cirrostratus*, and *cirrocumulus*. Because of the low temperatures and small quantities of water vapor present at high altitudes, all high clouds are thin and white and made up primarily of ice crystals.

Cirrus are detached clouds composed of white, delicate icy filaments. Winds aloft often cause these fibrous ice trails to bend or curl. As shown in Figure 5–3a, cirrus clouds with hooked filaments are called "mares' tails" (see Box 5–1).

Cirrostratus is a transparent, whitish cloud veil of fibrous or sometimes smooth appearance that may cover much or all of the sky. This cloud is easily recognized when it produces a halo around the Sun or Moon (Figure 5–3b). On occasions, cirrostratus may be so thin and transparent that the clouds are barely discernible. With the approach of a warm front, cirrostratus clouds generally thicken and grade into middle-level altostratus clouds.

Box 5–1 Aircraft Contrails and Cloudiness

You have undoubtedly seen a *contrail* (from *conden-sation trail*), in the wake of an aircraft flying on a clear day (Figure 5–A). Contrails are produced by jet aircraft engines that expel large quantities of hot, moist air. As this air mixes with the frigid air aloft, a streamlined cloud is produced. Because it often takes a few seconds for sufficient cooling to occur, the contrail usually forms a short distance behind the aircraft.

Why do contrails occur on some occasions and not on others? Contrails form under the same conditions as any other cloud—that is, when the air reaches saturation and condensation nuclei exist in sufficient numbers. Most contrails form when the exhaust gases add sufficient water vapor to the air to cause saturation. Further, it has been demonstrated that the exhaust gases of aircraft engines supply abundant sulfate molecules that serve as nuclei to promote the development of contrails.

Contrails typically form above 9 kilometers (6 miles) where air temperatures are a frigid –50°C (–58°F), or colder. Thus, it is not surprising that contrails are com-posed of minute ice crystals. Most contrails have a very short life span. Once formed, these streamlined clouds mix with surrounding cold, dry air and ultimately evapo-rate. However, if the air aloft is near saturation, contrails may survive for long periods. Under these conditions, the upper airflow usually spreads the streamlike clouds into broad bands of clouds called *contrail cirrus*.

With the increase in air traffic during the last few decades, an overall increase in cloudiness has been recorded, particularly near major transportation hubs. This is most evident in the American Southwest where aircraft contrails persist in otherwise cloudless, or mostly clear skies.

In addition to the added cloud cover associated with increasing levels of jet aircraft traffic, other less noticeable effects are of concern. Research is currently underway to assess the impact of contrail-produced cirrus clouds on the planet's heat budget. Recall from Chapter 3 that high thin clouds are effective transmitters of solar radiation (most radiation reaches the surface), but are good absorbers of outgoing infrared radiation emitted from Earth's surface. As a consequence, high cirrus clouds tend to have an over-all warming effect. However, research indicates that most contrails differ markedly from typical cirrus clouds, which are generated under quite different conditions. Although the results of these studies are inconclusive, it appears that human-induced cirrus clouds may actually lead to surface cooling rather than warming. Much more research will be needed to predict with accuracy the impact of contrails on climate change.

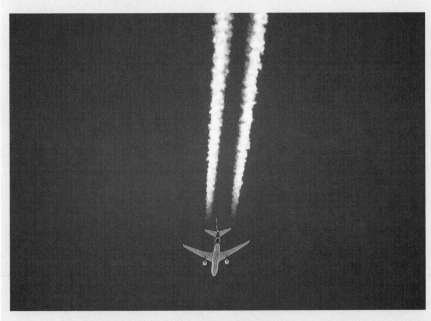

Figure 5–A Aircraft contrails. Condensation trails produced by jet aircraft often spread out to form broad bands of cirrus clouds. *(Photo by J. F. Towers/The Stock Market)*

Figure 5–2 Classification of clouds according to height and form. *(After Ward's Natural Science Establishment, Inc., Rochester, N.Y.)*

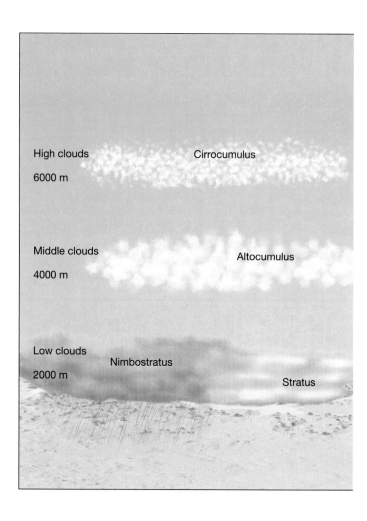

Cirrocumulus clouds appear as white patches composed of very small cells or ripples (Figure 5–3c). Most often the masses that make up these clouds have an apparent width similar to that of the Sun. Furthermore, these small globules, which may be merged or separate, are often arranged in a regular pattern. This pattern is commonly called "mackerel sky" because of the similarity to the pattern formed by fish scales.

High clouds are generally not precipitation makers. However, when cirrus clouds give way to cirrocumulus clouds that cover even more of the sky, they may warn of impending stormy weather. The following mariner's saying is based on this observation: *Mackerel scales and mares' tails make lofty ships carry low sails.*

Figure 5–3 Three basic cloud types make up the family of high clouds: (a) cirrus, (b) cirrostratus, and (c) cirrocumulus. *(Photos (a) and (c) by E. J. Tarbuck, (b) by A. and J. Verkaik/The Stock Market)*

(a)

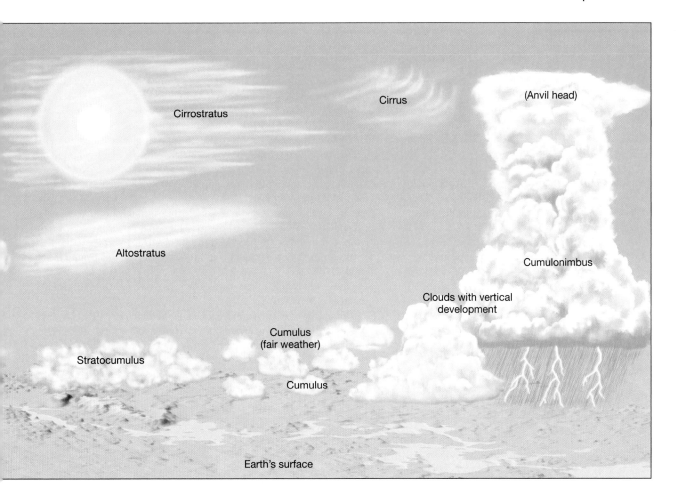

(b)

(c)

Table 5–1 Basic cloud types

Cloud Family and Height	Cloud Type	Characteristics
High clouds—above 6000 m (20,000 ft)	Cirrus	Thin, delicate, fibrous ice-crystal clouds. Sometimes appear as hooked filaments called "mares' tails" (cirrus uncinus; Figure 5–3a).
	Cirrostratus	Thin sheet of white ice-crystal clouds that may give the sky a milky look. Sometimes produces halos around the Sun and Moon (Figure 5–3b).
	Cirrocumulus	Thin, white ice-crystal clouds. In the form of ripples or waves, or globular masses all in a row. May produce a "mackerel sky." Least common of high clouds (Figure 5–3c).
Middle clouds—2000–6000 m (6500 to 20,000 ft)	Altocumulus	White to gray clouds often made up of separate globules; "sheepback" clouds (Figure 5–4a).
	Altostratus	Stratified veil of clouds that is generally thin and may produce very light precipitation. When thin, the Sun or Moon may be visible as a "bright spot," but no holes are produced (Figure 5–4b).
Low clouds—below 2000 m (6500 ft)	Stratus	Low uniform layer resembling fog but not resting on the ground. May produce drizzle.
	Stratocumulus	Soft, gray clouds in globular patches or rolls. Rolls may join together to make a continuous cloud.
	Nimbostratus	Amorphous layer of dark gray clouds. One of the chief precipitation-producing clouds (Figure 5–5).
Clouds of vertical development	Cumulus	Dense, billowy clouds often characterized by flat bases. May occur as isolated clouds or closely packed (Figure 5–6).
	Cumulonimbus	Towering cloud, sometimes spreading out on top to form an "anvil head." Associated with heavy rainfall, thunder, lightning, hail, and tornadoes (Figure 5–7).

Middle Clouds. Clouds that appear in the middle altitude range (2000 to 6000 meters [6500 to 20,000 feet]) have the prefix *alto* as part of their name. There are two types: *altocumulus* and *altostratus*.

Altocumulus tend to form in large patches composed of rounded masses or rolls that may or may not merge (Figure 5–4a). Because they are generally composed of water droplets rather than ice crystals, the individual cells usually have a more distinct outline. Altocumulus are most easily confused with two other cloud types: cirrocumulus (which are smaller and less dense) and stratocumulus (which are larger).

Altostratus is the name given to a formless layer of grayish clouds covering all or a large portion of the sky. Generally, the Sun is visible through these clouds as a bright spot, but with the edge of its disc not discernible (Figure 5–4b). However, unlike cirrostratus clouds, altostratus do not produce halos. Infrequent precipitation in the form of light snow or drizzle may accompany these clouds. Altostratus clouds are commonly associated with warm fronts. As the front approaches, the clouds thicken into a dark gray layer of nimbostratus that is capable of producing copious rainfall.

Low Clouds. There are three members of the family of low clouds (below 2000 meters [6500 feet]): *stratus*, *stratocumulus*, and *nimbostratus*.

Stratus is a uniform layer that frequently covers much of the sky and, on occasion, may produce light precipitation. When stratus clouds develop a scalloped bottom that appears as long parallel rolls or broken globular patches, they are called *stratocumulus* clouds.

Nimbostratus clouds derive their name from the Latin *nimbus*, "rain cloud," and *stratus*, "to cover with a layer" (Figure 5–5). As the name implies, nimbostratus clouds are one of the chief precipitation producers. Nimbostratus clouds form in association with stable conditions. We might not expect clouds to grow or persist in stable air, yet cloud growth of this type is common when air is forced to rise, as along a front or near the center of a cyclone where converging winds cause air to ascend. Such forced ascent of stable air leads to the formation of a stratified cloud layer that is large horizontally compared to its thickness. Precipitation associated with nimbostratus clouds is generally light to moderate but of long duration and widespread.

Clouds of Vertical Development. Some clouds do not fit into any one of the three height categories. Such clouds have their bases in the low height range and extend upward into the middle or high altitudes; they are referred to as *clouds of vertical development*. Vertically developed clouds are all closely related and are associated with unstable air. There are two types, *cumulus* and *cumulonimbus*.

(a)   (b)

Figure 5–4 Two forms of clouds are generated in the middle-altitude range. (a) Altocumulus tend to form in patches composed of rolls or rounded masses. (b) Altostratus occur as grayish sheets covering a large portion of the sky. When visible, the Sun appears as a bright spot through these clouds. *(Photos by E. J. Tarbuck)*

Cumulus clouds are individual masses that develop into vertical domes or towers, the tops of which often resemble cauliflower. Cumulus clouds most often form on clear days when unequal surface heating causes parcels of air to rise convectively above the lifting condensation level (Figure 5–6). This level is often apparent to an observer because the flat cloud bottoms define it.

On days when cumulus clouds are present, we usually notice an increase in cloudiness into the afternoon as solar heating intensifies. Furthermore, because cumulus clouds rarely produce appreciable precipitation, and because they form on "sunny" days, they are often called "fair weather clouds."

Although cumulus clouds are associated with fair weather, they may, under the proper circumstances, grow dramatically in height. Once upward movement is triggered, acceleration is powerful, and clouds with great vertical extent

are formed. As the cumulus enlarges, its top leaves the low height range and it is called a *cumulus congestus*. Finally, when the cloud becomes even more towering and rain begins to fall, it becomes a cumulonimbus.

Cumulonimbus are dark, dense, billowy clouds of considerable vertical extent in the form of huge towers (Figure 5–7). In its later stages of development, the upper part of a cumulonimbus turns to ice and appears fibrous. Furthermore, the tops of these clouds frequently spread out in the shape of an anvil. Cumulonimbus towers extend from a few hundred meters above the surface upward to 12 kilometers (7 miles), or, on rare occasions, 20 kilometers (12 miles). These huge towers produce heavy precipitation with accompanying lightning and thunder and occasionally hail. We will consider the development of these important weather producers in Chapter 10, which considers thunderstorms and tornadoes.

Figure 5–5 Nimbostratus clouds are one of the chief precipitation producers. These dark gray layers often exhibit a ragged-appearing base. *(Photo by E. J. Tarbuck)*

Cloud Varieties. In addition to the names given to the 10 basic cloud types, adjectives may also be used to describe variations of a particular cloud type. For example, the term *uncinus*, meaning "hook-shaped," is applied to streaks of cirrus clouds that are shaped like a comma resting on its side. Cirrus uncinus are often precursors of bad weather.

When stratus or cumulus clouds appear to be broken into smaller pieces, the adjective *fractus* may be used in their description. In addition, some clouds have rounded protuberances on their bottom surface, not unlike the udders of cows. When these structures are present, the term *mammatus* can be applied. This configuration is sometimes associated with stormy weather and cumulonimbus clouds.

Lens-shaped clouds are referred to as *lenticular*. They are common in areas that have rugged or mountainous

Figure 5–6 Cumulus clouds. These small, white, billowy clouds generally form on sunny days and, therefore, are often called "fair weather clouds." *(Photo by E. J. Tarbuck)*

Figure 5–7 Cumulonimbus clouds. These dense, billowy clouds have great vertical extent and can produce heavy precipitation. *(Photo by Tom Bean/DRK Photo)*

topography, where they are called *lenticular altocumulus* (Figure 5–8a). Although lenticular clouds can form whenever the airflow undulates sharply in the vertical, they most frequently form on the lee side of mountains. As air passes over mountainous terrain, a wave pattern develops, as shown in Figure 5–8b. Clouds form where the wavy flow causes air to ascend, whereas areas with descending air are cloud-free.

Figure 5–8 Lenticular clouds. (a) These lens-shaped clouds are relatively common in mountainous areas. *(Photo by Henry Lansford)* (b) This diagram depicts the formation of lenticular clouds in the turbulent flow that develops in the lee of a mountain range.

(a)

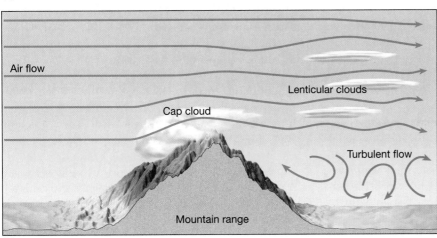

Air flow

Lenticular clouds

Cap cloud

Turbulent flow

Mountain range

(b)

Fog

Fog is generally considered an atmospheric hazard. When it is light, visibility is reduced to 2 or 3 kilometers (1 or 2 miles). When it is dense, visibility may be cut to a few dozen meters or less, making travel by any mode not only difficult but dangerous. Official weather stations only report fog when the visibility is reduced to 1 kilometer or less. Although arbitrary, this figure does permit a more objective criterion for comparing fog frequencies at different locations.

Fog is defined as *a cloud with its base at or very near the ground*. Physically, there is basically no difference between a fog and a cloud; their appearance and structure are the same. The essential difference is the method and place of formation. Clouds result when air rises and cools adiabatically. Fog results from cooling or by the addition of enough water vapor to cause saturation. Let us look at fogs, first those formed by cooling and then those formed by the addition of water vapor.

Fogs Formed by Cooling

When the temperature of a layer of air in contact with the ground falls below its dew point, condensation produces fog. Depending upon the prevailing conditions, the ground may become shrouded in radiation fog, advection fog, or upslope fog.

Radiation Fog. As the name implies, **radiation fog** results from radiation cooling of the ground and adjacent air. It is a nighttime phenomenon requiring clear skies and a fairly high relative humidity. Under these circumstances, the ground and the air immediately above it will cool rapidly. Because the relative humidity is high, just a small amount of cooling will lower the temperature to the dew point. If the air is calm, the fog may be patchy and less than a meter deep. For radiation fog to be more extensive vertically, a light breeze of 3 to 5 kilometers (2 to 3 miles) per hour is necessary. Then the light wind creates enough turbulence to carry the fog upward 10 to 30 meters (30 to 100 feet) without dispersing it.

Because the air containing the fog is relatively cold and dense, it drains downslope in hilly terrain. As a result, radiation fog is thickest in valleys, whereas the surrounding hills are clear (Figure 5–9). Normally, these fogs dissipate within 1 to 3 hours after sunrise. Often the fog is said to "lift." However, it does not rise. Instead, the Sun warms the ground, which, in turn, heats the lowest layer of air first. Consequently, the fog evaporates from the bottom up, giving the impression of lifting. The last vestiges of radiation fog may appear to be a low white cloud layer.

Advection Fog. When warm and moist air is blown over a cold surface, it becomes chilled by contact and, to a certain extent, by mixing with the cold air created by the cold surface below. If cooling is sufficient, the result will be a blanket of fog called **advection fog**. The term *advection* refers to air moving horizontally. Therefore, advection fogs are a consequence of air giving up heat to the surface below during horizontal movement. A classic example is the frequent advection fog around San Francisco's Golden Gate Bridge (Figure 5–10).

A certain amount of turbulence is needed for proper development of advection fog. Thus, winds between 10 and 30 kilometers (6 and 18 miles) per hour are usually associated with it. Not only does the turbulence facilitate cooling through a thicker layer of air, but it also carries the fog to greater heights. Unlike radiation fogs, advection fogs are often thick (300 to 600 meters deep) and persistent.

Examples of such fogs are common. The foggiest location in the United States, and perhaps in the world, is Cape Disappointment, Washington. The name is indeed appropriate because the station averages 2552 hours (106 days) of fog each year. The fog experienced at Cape Disappointment, as well as that at other West Coast locations during the summer and early autumn, is produced when warm, moist air from the Pacific Ocean moves over the cold California Current (Figure 5–10). It is then carried onshore by westerly winds or a local sea breeze.

Advection fog is also a common wintertime phenomenon in central and eastern North America. Here, warm, moist air from the Gulf of Mexico moves over cold and occasionally snow-covered surfaces to produce widespread foggy conditions. These fogs are frequently thick and produce hazardous driving conditions.

Upslope Fog. As its name implies, **upslope fog** is created when relatively humid air moves up a gradual sloping plain or, in some cases, up the steep slopes of a mountain. Because of the upward movement, air expands and cools adiabatically (this is the only type of fog that forms adiabatically). If the dew point is reached, an extensive layer of fog may form.

It is easy to visualize how upslope fog might form in mountainous terrain. However, in the United States, upslope fog also occurs in the Great Plains. Here, when humid Gulf air moves westward from the Mississippi River toward the Rocky Mountains, it gradually glides upslope. (Recall that Denver, Colorado, is called the "mile-high city" and the Gulf of Mexico is at sea level.) Air flowing "up" the Great Plains expands and cools adiabatically by as much as 12°C (22°F). The result can be an extensive upslope fog in the western plains.

Figure 5–9 Early morning radiation fog in downtown Peoria, Illinois, which is located in the Illinois River valley. (Photo by David Zalaznik/Peoria Journal Star)

Fogs Formed by Evaporation

When saturation occurs primarily because of the addition of water vapor, the resulting fogs are called *evaporation fogs*. Two types of evaporation fogs are recognized: steam fog and frontal (precipitation) fog.

Steam Fog. When cool air moves over warm water, enough moisture may evaporate from the water surface to saturate the air immediately above. As the rising water vapor meets the cold air, it condenses and rises with the air that is being warmed from below. Because the rising air looks like the "steam" that forms above a hot cup of coffee the phenomenon is called **steam fog** (Figure 5–11). It is a fairly common occurrence over lakes and rivers on clear crisp mornings in the fall when the waters are still relatively warm while the air is rather cold. Steam fog is often shallow, for as it rises, the water droplets evaporate as they mix with the unsaturated air above.

Steam fogs can be dense, however. During the winter, cold arctic air pours off the continents and ice shelves onto the comparatively warm open ocean. The temperature contrast between the warm ocean and cold air has been known to exceed 30°C (54°F). The result is an intense steam fog produced as the rising water vapor saturates a large volume of air. Because of its source and appearance, this type of steam fog is given the name *arctic sea smoke*.

Figure 5–10 Advection fog rolling into San Francisco Bay. (Photo by Ed Pritchard/Tony Stone Images)

Figure 5–11 Steam fog rising from upper St. Regis Lake, Adirondack Mountains, New York. *(Photo by Jim Brown/The Stock Market)*

Frontal Fog. When frontal wedging occurs, warm air is lifted over colder air. If the resulting clouds yield rain, and the cold air below is near the dew point, enough rain can evaporate to produce fog. A fog formed in this manner is called **frontal** or **precipitation fog**. The result is a more or less continuous zone of condensed water droplets reaching from the ground up through the clouds.

In summary, both steam fog and frontal fog result from the addition of moisture to a layer of air. As you saw, the air is usually cool or cold and already near saturation. Thus, only a relatively modest amount of evaporation is necessary to produce saturated conditions and fog.

The frequency of dense fog varies considerably from place to place. As might be expected, fog incidence is highest in coastal areas, especially where cold currents prevail, as along the Pacific and New England coasts. Relatively high frequencies are also found in the Great Lakes region and in the humid Appalachian Mountains of the East. In contrast, fogs are rare in the interior of the continent, especially in the arid and semiarid areas of the West.

Dew and Frost

Clouds and fog are the most conspicuous and meteorologically important forms of condensation. Dew and white frost must be considered minor by comparison. These common forms of condensation generally result from radiation cooling on clear, cool nights.

Dew is the condensation of water vapor on objects that have radiated sufficient heat to lower their temperature below the dew point of the surrounding air. Because different objects radiate heat at different rates, dew may form on some surfaces but not on others. An automobile, for example, may be covered with dew shortly after sunset, whereas the concrete driveway surrounding the car remains free of condensation throughout the night.

Dew is a common sight on lawns in the early morning. In fact, the grass will frequently have a coating of dew when nothing else does. Dew is more frequent on grass because the transpiration of water vapor by the blades raises the relative humidity to higher levels directly above the grass. Therefore, only modest radiation cooling may be necessary to bring about saturation and condensation.

Although dew is an unimportant source of moisture in humid areas, plant life in some arid regions depends on it for survival. In parts of Israel, for example, dew may supply as much as 55 millimeters (2.17 inches) of water annually. Further, this moisture is available mainly during the dry summer months when plants are experiencing the greatest stress.

Contrary to popular belief, white frost is not frozen dew. Rather, **white frost** (*hoar frost*) forms when the dew point of the air is below freezing. Thus, frost forms when water vapor changes directly from a gas into a solid (ice), without entering the liquid state. This process, called *deposition*, produces delicate patterns of ice crystals that frequently decorate windows in northern winters (see Figure 4–4).

How Precipitation Forms

Although all clouds contain water, why do some produce precipitation and others drift placidly overhead? This seemingly simple question perplexed meteorologists for many years. Before examining the processes that generate precipitation, we need to examine a couple of facts.

First, cloud droplets are very tiny, averaging under 20 micrometers (0.02 millimeter) in diameter (Figure 5–12). (One micrometer equals 0.001 millimeter.) For comparison, a human hair is about 75 micrometers in diameter. The small size of cloud droplets results mainly because condensation nuclei are usually very abundant and the available water is distributed among numerous droplets rather than concentrated into fewer large droplets.

Second, because of their small size, the rate at which cloud droplets fall is incredibly slow (see Box 5–2). An average cloud droplet falling from a cloud base at 1000 meters would require several hours to reach the ground. However, it would never complete its journey. This cloud droplet would evaporate before it fell a few meters from the cloud base into the unsaturated air below.

How large must a droplet grow in order to fall as precipitation? A typical raindrop has a diameter of about 2000 micrometers (2 millimeters) or 100 times that of the average cloud droplet having a diameter of 20 micrometers (0.02 millimeter). However, as shown in Box 5–2, the *volume* of a typical raindrop is a million times that of a cloud droplet. Thus, for precipitation to form, cloud droplets must grow in volume by roughly one million times. You might suspect that additional condensation creates drops large enough to survive the descent to the surface. However, clouds consist of many billions of tiny cloud droplets that all compete for the available water. Thus, condensation provides an inefficient means of raindrop formation.

For precipitation to form, millions of cloud droplets must somehow coalesce (join together) into drops large enough to sustain themselves during their descent. Two mechanisms that give rise to these "massive" drops are: the Bergeron process and the collision-coalescence process.

Precipitation from Cold Clouds: The Bergeron Process

You have probably watched a TV documentary in which mountain climbers brave intense cold and a ferocious snowstorm to scale an ice-covered peak. Although it is hard to imagine, very similar conditions exist in the upper portions of towering cumulonimbus clouds, even on sweltering summer days. (In fact, in the upper troposphere where commercial aircraft cruise, the temperature typically approaches –50°C (–58°F) or lower.) It turns out that the frigid conditions high in the troposphere provide an ideal environment to initiate precipitation. In fact, in the middle latitudes much of the rain that falls begins with the birth of snowflakes high in the cloud tops where temperatures are considerably below freezing. Obviously, in the winter, even low clouds are cold enough to trigger precipitation.

The process that generates much of the precipitation in the middle latitudes is named the **Bergeron process** for its discoverer, the highly respected Swedish meteorologist, Tor Bergeron (see Box 5–3). This mechanism relies on two interesting properties of water. First, *cloud droplets do not freeze at 0°C as expected.* In fact, pure water suspended in air does not freeze until it reaches a temperature of nearly –40°C (–40°F). Water in the liquid state below 0°C (32°F) is referred to as **supercooled**. Supercooled water will readily freeze if it impacts an object, which explains why airplanes collect ice when they pass through a liquid cloud made up of supercooled droplets. This also explains why the stuff we call *freezing rain* or *glaze* falls as a liquid but then turns to a sheet of ice when it strikes the pavement or a tree branch.

In addition, supercooled droplets will freeze on contact with solid particles that have a crystal form closely resembling that of ice (silver iodide is an example). These materials have been termed **freezing nuclei.** The need for freezing nuclei to initiate the freezing process is similar to the requirement for condensation nuclei in the process of condensation.

In contrast to condensation nuclei, however, freezing nuclei are sparse in the atmosphere and do not generally become active until the temperature reaches –10°C (14°F) or below. Thus, at temperatures between 0 and –10°C, clouds consist mainly of supercooled water droplets. Between –10 and –20°C, liquid droplets coexist with ice crystals, and below –20°C (–4°F), clouds are generally composed entirely of ice crystals—for example, high-altitude cirrus clouds.

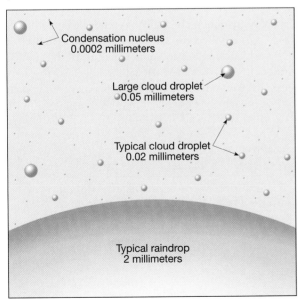

Condensation nucleus
0.0002 millimeters

Large cloud droplet
0.05 millimeters

Typical cloud droplet
0.02 millimeters

Typical raindrop
2 millimeters

Figure 5–12 Comparative diameters of particles involved in condensation and precipitation processes.

Box 5–2 Forces Acting on Cloud Droplets and Raindrops

Gregory J. Carbone*

Two opposing forces act on cloud droplets—*gravity* and *friction*. Gravitational force pulls a water droplet toward Earth's surface and is equal to the mass of the droplet times gravitational acceleration (gravitational force = droplet mass × 9.8 m/s²). Because 1 cubic centimeter (1 cm³) of volume equals 1 gram of mass, we can substitute volume for mass in the above equation and say that the gravitational force of a droplet increases with its volume. Assuming the droplet is a sphere, its volume is calculated as follows:

$$\text{Volume} = \tfrac{4}{3}\pi r^3$$

As a droplet falls it encounters air resistance, or frictional force. The magnitude of this force depends on the size of the drop's "bottom"—that is, the surface area resisting the fall (Figure 5–B). Again, assuming the droplet is spherical, frictional force will change with the area of a circle:

$$\text{Area} = \pi r^2$$

Frictional drag increases as a droplet accelerates, because a faster droplet encounters more air molecules.

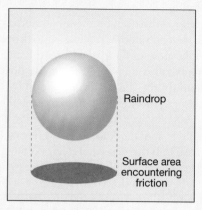

Figure 5–B Raindrops and friction.

Eventually, frictional and gravitational forces balance and the droplet no longer accelerates, but falls at a constant speed. This speed is referred to as the droplet's *terminal velocity*. Terminal velocity depends on size. Smaller droplets have a lower terminal velocity than do larger droplets because as droplet radius increases, gravitational force increases to the third power and frictional force increases to the second power. Consequently, larger droplets accelerate to higher speeds before reaching

This brings us to a second important property of water. *The saturation vapor pressure above ice crystals is somewhat lower than above supercooled liquid droplets.* This occurs because ice crystals are solid, which means that the individual water molecules are held together more tightly than those forming a liquid droplet. As a result, it is easier for water molecules to escape from the supercooled liquid droplets. This fact accounts for the higher saturation vapor pressure over supercooled liquid droplets compared with ice crystals.

Consequently, when air is saturated (100 percent relative humidity) with respect to liquid droplets, it is supersaturated with respect to ice crystals. Table 5–2, for example, shows that at –10°C (14°F), when the relative humidity is 100 percent with respect to water, the relative humidity with respect to ice is about 110 percent.

With these facts in mind, we can now explain how the Bergeron process produces precipitation. Visualize a cloud at a temperature of –10°C (14°F), where each ice

crystal is surrounded by many thousands of liquid droplets (Figure 5–13). Because the air was initially saturated (100 percent) with respect to liquid water, it will be supersaturated (over 100 percent) with respect to the newly formed ice crystals. As a result of this supersaturated condition, the ice crystals collect more water molecules than

Table 5–2 Relative humidity with respect to ice when relative humidity with respect to water is 100 percent

Temperature (°C)	Relative Humidity wth Respect to:	
	Water	Ice
0	100%	100%
–5	100%	105%
–10	100%	110%
–15	100%	115%
–20	100%	121%

Table 5–A Maximum fall distance before evaporation

Drop Diameter (µm)	Maximum Fall Distance (m)
2500	280,000
1000	42,000
100	150
10	0.033
1	0.0000033

terminal velocity. (Refer to Table 5–3 to find the terminal velocity of different sized droplets.)

The terminal velocity of individual droplets is important because it determines the probability that a droplet will reach Earth without evaporating or being carried in updrafts. In fact, it is possible to estimate the maximum fall distance before evaporation for droplets of various size. Table 5–A shows such values assuming a barometric pressure of 900 mb, temperature of 5°C, and relative humidity equal to 90 percent. Note that droplets with a radius less than 100 µm (micrometers) are unlikely to reach Earth's surface because cloud bases are typically higher than their maximum fall distance.

Because most cloud droplets are too small to reach the surface, precipitation requires processes that cause droplet growth. But condensation is a very slow growth process. Consider a typical cloud droplet of 10-µm radius. To grow to the size of a typical raindrop (1000

µm, or 100 times its original size), the cloud droplet must increase its volume by 1 million times:

$$
\begin{aligned}
\text{Volume}_{10\mu m} &= 4 \times \pi \times 10^3 \\
&= 12,566 \ \mu m^3 \\
\text{Volume}_{1000\mu m} &= 4 \times \pi \times 1000^3 \\
&= 12,566,370,610 \ \mu m^3 \\
12,566,370,610 \ \mu m/12,566 \ \mu m & \\
&= 1,000,000
\end{aligned}
$$

Clearly, condensation provides an inefficient means of cloud droplet growth, thus emphasizing the importance of other mechanisms such as the collision–coalescence and Bergeron process.

*Professor Carbone is a faculty member in the Department of Geography at the University of South Carolina.

they lose by sublimation. Thus, continued evaporation from the liquid drops provides a source of water vapor to feed the growth of ice crystals (Figure 5–13).

Because the level of supersaturation with respect to ice can be great, the growth of snow crystals is generally sufficiently rapid to generate crystals large enough to fall. During their descent, these crystals enlarge as they intercept cloud drops that freeze on them. Air movement will sometimes break up these delicate crystals, and the fragments will serve as freezing nuclei for other liquid droplets. A chain reaction develops and produces many snow crystals, which, by accretion, will form into larger masses called snowflakes. Large snowflakes may consist of 10 to 30 individual crystals.

In summary, the Bergeron process can produce precipitation throughout the year in the middle latitudes, provided at least the upper portions of clouds are cold enough to generate ice crystals. The type of precipitation (snow, sleet, rain or freezing rain) that reaches the ground depends on the temperature profile in the lower few kilometers of

the atmosphere. When the surface temperature is above 4°C (39°F), snowflakes usually melt before they reach the ground and continue their descent as rain. Even on a hot summer day, a heavy downpour may have begun as a snowstorm high in the clouds overhead.

Precipitation from Warm Clouds: The Collision–Coalescence Process

A few decades ago, meteorologists believed that the Bergeron process was responsible for the formation of most precipitation except for light drizzle. Later, it was discovered that copious rainfall is often associated with clouds located well below the freezing level (called *warm clouds*), especially in the tropics. Clearly a second mechanism also must trigger precipitation. Researchers discovered the **collision–coalescence process**.

Research has shown that clouds made entirely of liquid droplets must contain some droplets larger than 20

Box 5-3 Science and Serendipity*

Serendipity is defined by Nobel Laureate Irving Langmuir as "the art of profiting from unexpected occurrences." In other words, if you are observing something and the entirely unexpected happens, and if you see in this accident a new and meaningful discovery, then you have experienced serendipity. Most nonscientists, some scientists, and, alas, many teachers are not aware that many of the great discoveries in science were serendipitous.

An excellent example of serendipity in science occurred when Tor Bergeron, the great Swedish meteorologist, discovered the importance of ice crystals in the initiation of precipitation in supercooled clouds. Bergeron's discovery occurred when he spent several weeks at a health resort at an altitude of 430 meters (1400 feet) on a hill near Oslo. During his stay, Bergeron noted that this hill was often "fogged-in" by a layer of supercooled clouds. As he walked along a narrow road in the fir forest along the hillside, he noticed that the "fog" did not enter the "road tunnel" at temperatures below –5°C, but did enter it when the temperature was warmer than 0°C. (Profiles of the hill, trees, and fog for the two temperature regimes is shown in Figure 5–C.)

Bergeron immediately concluded that at temperatures below about –5°C the branches of the firs acted as freezing nuclei upon which some of the supercooled droplets crystallized. Once the ice crystals developed, they grew rapidly at the expense of the remaining water droplets (see Figure 5–C). The result was the growth of ice crystals

(rime) on the branches of the firs accompanied by a "clearing-off" between the trees and along the "road-tunnel."

From this experience, Bergeron realized that, if ice crystals somehow were to appear in the midst of a cloud of supercooled droplets, they would grow rapidly as water molecules diffused toward them from the evaporating cloud droplets. This rapid growth forms snow crystals that, depending on the air temperature beneath the cloud, fall to the ground as snow or rain. Bergeron had thus discovered one way that minuscule cloud droplets can grow large enough to fall as precipitation (see the discussion entitled "Precipitation From Cold Clouds: The Bergeron Process").

Serendipity influences the entire realm of science. Can we conclude that anyone who makes observations will necessarily make a major discovery? Not at all. A perceptive and inquiring mind is required, a mind that has been searching for order in a labyrinth of facts. As Langmuir said, the unexpected occurrence is not enough; you must know how to profit from it. Louis Pasteur observed that "In the field of observation, chance favors only the prepared mind." The discoverer of vitamin C, Nobel Laureat Albert Szent-Gyorgyi, remarked that discoveries are made by those who "see what everybody else has seen, and think what nobody else has thought." Serendipity is at the heart of science itself.

*Based on material prepared by Duncan C. Blanchard.

---

micrometers (0.02 millimeters) if precipitation is to form. These large droplets form when "giant" condensation nuclei are present, or when hygroscopic particles exist (such as sea salt). Recall that hygroscopic particles begin to remove water vapor from the air at relative humidities under 100 percent. Because the rate at which drops fall is size-dependent, these "giant" droplets fall most rapidly. (Table 5–3 summarizes drop size and falling velocities.)

As the larger droplets fall through a cloud, they collide with the smaller, slower droplets and coalesce (see Box 5–2). Becoming larger in the process, they fall even more rapidly (or, in an updraft, they rise more slowly) and increase their chances of collision and rate of growth (Figure 5–14a). After collecting the equivalent of a million or so cloud droplets, they are large enough to fall to the surface without evaporating.

Because of the huge number of collisions required for growth to raindrop size, clouds that have great vertical thickness and contain large cloud droplets have the best

Table 5–3 Fall velocity of water drops

Types	Diameter (millimeters)	Fall Velocity (km/hr)	Fall Velocity (miles/hr)
Small cloud droplets	0.01	0.01	0.006
Typical cloud droplets	0.02	0.04	0.03
Large cloud droplets	0.05	0.3	0.2
Drizzle drops	0.5	7	4
Typical rain drops	2.0	23	14
Large rain drops	5.0	33	20

Data from Smithsonian Meteorological Tables.

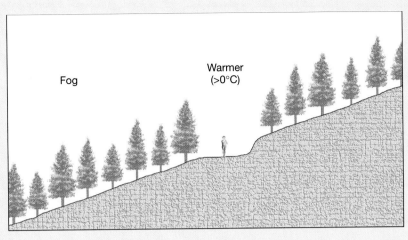

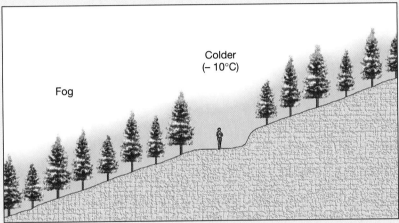

Figure 5–C Distribution of fog when the temperature is above freezing and when the temperature falls to –10°C.

chance of producing precipitation. Updrafts also aid this process because they allow the droplets to traverse the cloud repeatedly, colliding with more droplets.

As raindrops grow in size their fall velocity increases. This in turn increases the frictional resistance of the air, which causes the drop's "bottom" to flatten out (Figure 5–14b). As the drop approaches 4 millimeters in diameter it develops a depression as shown in Figure 5–14c. Raindrops can grow to a maximum of 5 millimeters when they fall at the rate of 33 kilometers (20 miles) per hour. At this size and speed, the water's surface tension, which holds the drop together, is surpassed by the frictional drag of the air. At this point the depression grows almost explosively, forming a donutlike ring that immediately breaks apart. The resulting breakup of a large raindrop produces numerous smaller drops that begin anew the task of sweeping up cloud droplets (Figure 5–14d).

The collision–coalescence process is not that simple, however. First, as the larger droplets descend, they produce an airstream around them similar to that produced by an automobile when driven rapidly down the highway. The airstream repels objects, especially small ones. If an automobile is driven at night and we use the bugs that fill the air on a summer evening as being like cloud droplets, it is easy to visualize how most cloud droplets, which are tiny, are swept aside. The larger the cloud droplet (or bug), the better chance it will have of colliding with the giant droplet (or car).

Next, collision does not guarantee coalescence. Experimentation has indicated that the presence of atmospheric electricity may be the key to what holds these droplets together once they collide. If a droplet with a negative charge should collide with a positively charged droplet, their electrical attraction may bind them together.

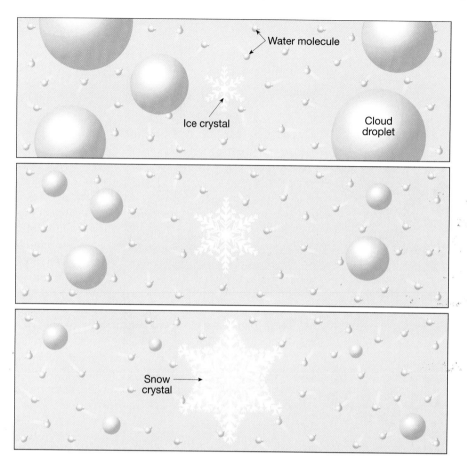

Figure 5–13 The Bergeron process. Ice crystals grow at the expense of cloud droplets until they are large enough to fall. The size of these particles has been greatly exaggerated.

From the preceding discussion, it should be apparent that the collision–coalescence mechanism is most efficient in environments where large cloud droplets are plentiful. It turns out that the air over the tropics, particularly the tropical oceans, is ideal. Here the air is very humid and relatively clean, so fewer condensation nuclei exit compared to the air over more populated regions. With fewer condensation nuclei to compete for available water vapor (which is plentiful), condensation is fast-paced and produces comparatively few large cloud droplets. Within developing cumulus clouds the largest drops quickly gather smaller droplets to generate the warm afternoon showers associated with tropical climates.

In the middle latitudes, the collision–coalescence process may contribute to the precipitation from a large cumulonimbus cloud by working in tandem with the Bergeron process—particularly during the hot, humid summer months. High in these towers the Bergeron process generates snow that melts as it passes below the freezing level. Melting generates relatively large drops with fast fall velocities. As these large drops descend they overtake and coalesce with the slower and smaller cloud droplets that comprise much of the lower regions of the cloud. The result can be a heavy downpour.

In summary, two mechanisms are known to generate precipitation: the *Bergeron process* and the *collision–coalescence process*. The Bergeron process is dominant in the middle latitudes where cold clouds (or cold cloud tops) are the rule. In the tropics, abundant water vapor and comparatively few condensation nuclei are the norm. This leads to the formation of fewer, larger drops with fast fall velocities that grow by collision and coalescence. No matter which process initiates precipitation, further growth in drop size is through collision–coalescence.

Forms of Precipitation

Because atmospheric conditions vary greatly both geographically and seasonally, several different forms of precipitation are possible (Figure 5–15). Rain and snow are the most common and familiar forms, but others listed in Table 5–4 are important as well. The occurrence of sleet, glaze, and hail is often associated with important weather events. Although limited in occurrence and sporadic in both time and space, these forms, especially glaze and hail, may on occasion cause considerable damage.

Rain

In meteorology, the term **rain** is restricted to drops of water that fall from a cloud and have a diameter of at least 0.5 millimeter. (This excludes drizzle and mist, which have

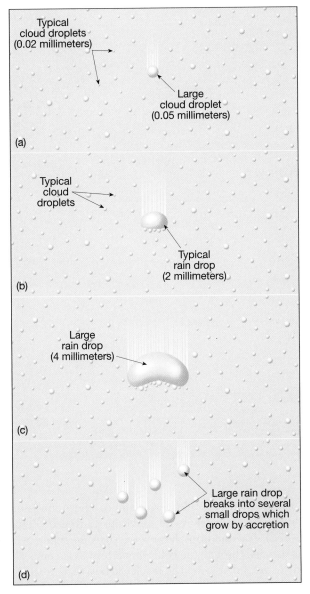

Figure 5–14 The collision–coalescence process. (a) Most cloud droplets are so small that the motion of the air keeps them suspended. Because large cloud droplets fall more rapidly than smaller droplets, they are able to sweep up the smaller ones in their path and grow. (b) As these drops increase in size, their fall velocity increases, resulting in increased air resistance, which causes the raindrop to flatten. (c) As the raindrop approaches 4 millimeters in size it develops a depression in the bottom. (d) Finally, when the diameter exceeds about 5 millimeters, the depression grows upward almost explosively, forming a donutlike ring of water that immediately breaks into smaller drops. (Note the drops are not drawn to scale—a typical raindrop has a volume equal to roughly 1 million cloud droplets.)

smaller droplets.) Most rain originates in either nimbostratus clouds or in towering cumulonimbus clouds that are capable of producing unusually heavy rainfalls known as *cloudbursts*. No matter what the rainfall intensity is, the size of raindrops rarely exceeds about 5 millimeters. Larger drops cannot survive because surface tension, which holds the drops together, is exceeded by the frictional drag of the air. Consequently, large raindrops regularly break apart into smaller ones.

Much of the world's rainfall begins as snow crystals or other solid forms such as hail or graupel, as shown in Figure 5–15a. Entering the warmer air below the cloud, these ice particles often melt and reach the ground as raindrops. In some parts of the world, particularly the subtropics, precipitation often forms in clouds that are warmer than 0°C (32°F). These rains frequently occur over the ocean where cloud condensation nuclei are not plentiful and those that do exist vary in size. Under such conditions, cloud droplets can grow rapidly by the collision–coalescence process to produce copious amounts of rain.

Fine, uniform drops of water having a diameter less than 0.5 millimeter are called *drizzle*. Drizzle and small raindrops generally are produced in stratus or nimbostratus clouds where precipitation may be continuous for several hours, or on rare occasions for days.

Precipitation containing the very smallest droplets able to reach the ground is called *mist*. Mist can be so fine that the tiny droplets appear to float and their impact is almost imperceptible.

As rain enters the unsaturated air below the cloud, it begins to evaporate. Depending on the humidity of the air and size of the drops, the rain may completely evaporate before reaching the ground. This phenomenon produces *virga*, which appear as streaks of precipitation falling from a cloud that extend only part of the way to Earth's surface (Figure 5–16).

Snow

Snow is precipitation in the form of ice crystals (snowflakes) or, more often, aggregates of ice crystals (Figure 5–15b). The size, shape, and concentration of snowflakes depend to a great extent on the temperature at which they form.

Recall that at very low temperatures, the moisture content of air is small. The result is the generation of very light and fluffy snow made up of individual six-sided ice crystals (Figure 5–17). This is the "powder" that downhill skiers talk so much about. By contrast, at temperatures warmer than about –5°C (23°F), the ice crystals join together into larger clumps consisting of tangled aggregates of crystals. Snowfalls consisting of these composite snowflakes are generally heavy and have a high moisture content, which makes them ideal

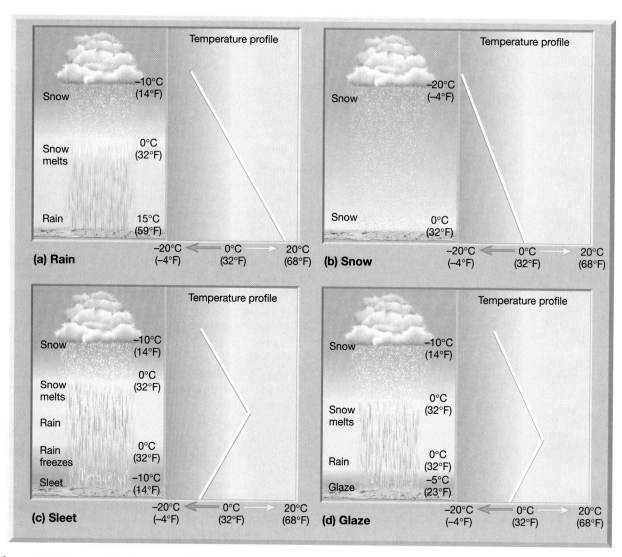

Figure 5–15 Four precipitation types and their temperature profiles.

for making snowballs. (For more on winter weather, see Box 5–4.)

Sleet and Glaze

Sleet is a wintertime phenomenon and refers to the fall of small particles of ice that are clear to translucent. Figures 5–15c and 5–18 shows how sleet is produced: An above-freezing air layer must overlie a subfreezing layer near the ground. When the raindrops, which are often melted snow, leave the warmer air and encounter the colder air below, they freeze and reach the ground as small pellets of ice roughly the size of the raindrops from which they formed.

On some occasions, when the vertical distribution of temperatures is similar to that associated with the formation of sleet, **freezing rain** or **glaze** results instead. In

such situations, the subfreezing air near the ground is not thick enough to allow the raindrops to freeze (Figure 5–15d). The raindrops, however, do become supercooled as they fall through the cold air and turn to ice on colliding with solid objects. The result can be a thick coating of ice having sufficient weight to break tree limbs, to down power lines, and to make walking and driving extremely hazardous (Figure 5–19).

In January 1998 an ice storm of historic proportions caused enormous damage in southeastern Canada. Here five days of freezing rain deposited a heavy layer of ice on all exposed surfaces from eastern Ontario to the Atlantic coast. The 8 centimeters (3 inches) of precipitation caused trees, power lines and high-voltage towers to collapse, leaving over a million households without power—many for nearly a month following the storm.

Table 5–4 Types of precipitation

Type	Approximate Size	State of Water	Description
Mist	0.005 to 0.05 mm	Liquid	Droplets large enough to be felt on the face when air is moving 1 meter/second. Associated with stratus clouds.
Drizzle	Less than 0.5 mm	Liquid	Small uniform drops that fall from stratus clouds, generally for several hours.
Rain	0.5 to 5 mm	Liquid	Generally produced by nimbostratus or cumulonimbus clouds. When heavy, size can be highly variable from one place to another.
Sleet	0.5 to 5 mm	Solid	Small, spherical to lumpy ice particles that form when raindrops freeze while falling through a layer of subfreezing air. Because the ice particles are small, any damage is generally minor. Sleet can make travel hazardous.
Glaze	Layers 1 mm to 2 cm thick	Solid	Produced when supercooled raindrops freeze on contact with solid objects. Glaze can form a thick coating of ice having sufficient weight to seriously damage trees and power lines.
Rime	Variable accumulations	Solid	Deposits usually consisting of ice feathers that point into the wind. These delicate frostlike accumulations form as supercooled cloud or fog droplets encounter objects and freeze on contact.
Snow	1 mm to 2 cm	Solid	The crystalline nature of snow allows it to assume many shapes, including six-sided crystals, plates, and needles. Produced in supercooled clouds where water vapor is deposited as ice crystals that remain frozen during their descent.
Hail	5 mm to 10 cm or larger	Solid	Precipitation in the form of hard, rounded pellets or irregular lumps of ice. Produced in large convective, cumulonimbus clouds, where frozen ice particles and supercooled water coexist.
Graupel	2 mm to 5 mm	Solid	Sometimes called "soft hail," graupel forms as rime collects on snow crystals to produce irregular masses of "soft" ice. Because these particles are softer than hailstones, they normally flatten out upon impact.

Figure 5–16 Virga, latin for "streak." In the arid west, rain frequently evaporates before reaching the ground. *(Photo by Pekka Parviainen/Science Photo Library/Photo Researchers, Inc.)*

At least 25 deaths were blamed on the storm, which caused damages in excess of $1 billion. Much of the damage was to the electrical grid, which one Canadian climatologist summed up this way: "What it took human beings a half-century to construct, took nature a matter of hours to knock down."

Hail

Hail is precipitation in the form of hard, rounded pellets or irregular lumps of ice. Large hailstones, when cut in half, often reveal nearly concentric shells of differing densities and degrees of opaqueness (Figure 5–20). The layers of ice accumulate as the hailstone travels up and down in a strong convective cloud.

Figure 5–17 All snow crystals are six-sided, but they come in an infinite variety of forms. *(Courtesy of NOAA, Seattle)*

reported. Many of these were probably composites of several stones frozen together.

The heaviest authenticated hailstone on record fell on Coffeyville, Kansas, September 3, 1970 (Figure 5–20). With a 14-centimeter (5.5-inch) diameter, this "giant" weighed 766 grams (1.67 pounds). It is estimated that this stone hit the ground faster than 160 kilometers (100 miles) per hour.

The destructive effects of large hailstones are well known, especially to farmers whose crops have been devastated in a few minutes and to people whose windows, roofs, and cars have been damaged (Figure 5–21). In the United States, hail damage each year can run into the hundreds of millions of dollars. One of the most severe hail storms to occur in the United States took place on March 25, 1992, in the Orlando, Florida, area where $60 million in damages were reported. It was the city's costliest natural disaster.

Hail is produced only in large cumulonimbus clouds where updrafts can sometimes reach speeds approaching 160 kilometers (100 miles) per hour, and where there is an abundant supply of supercooled water. Figure 5–22 shows the process. Hailstones begin as small embryonic ice pellets that grow by collecting supercooled droplets as they fall through the cloud. If they encounter a strong updraft, they may be carried upward again and begin the downward journey anew. Each trip through the supercooled portion of the cloud might be represented by an additional layer of ice. Hailstones can also form from a single descent through an updraft. Either way, the process continues until the hailstone encounters a downdraft or grows too heavy to remain suspended by the thunderstorm's updraft.

Most hailstones have diameters between 1 centimeter (pea size) and 5 centimeters (golf ball size), although some can be as big as an orange or even bigger. Occasionally, hailstones weighing a pound or more have been

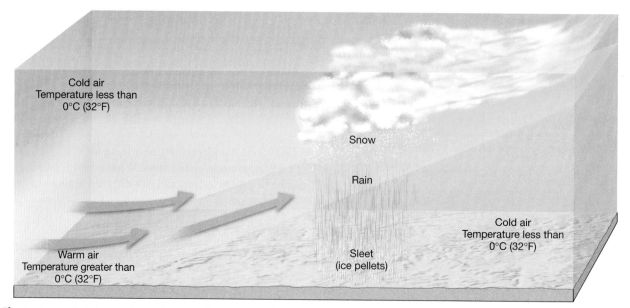

Figure 5–18 Sleet forms when rain passes through a cold layer of air and freezes into ice pellets. This occurs most often in the winter when warm air is forced over a layer of cold air.

Figure 5–20 Hail. A cross section of the Coffeyville hailstone. This largest recorded hailstone fell over Coffeyville, Kansas, in 1970 and weighed 0.75 kilogram (1.67 pounds). *(Photo courtesy of University Corporation for Atmospheric Research/National Science Foundation/National Center for Atmospheric Research)*

Figure 5–19 Glaze forms when supercooled raindrops freeze on contact with objects. *(Photo by Annie Griffiths Belt/DRK Photo)*

A single hailstone might contain several ice layers consisting of alternating zones of clear and milky ice. The milky layers are produced by the rapid freezing of small

supercooled water droplets that trap tiny air bubbles in the ice. By contrast, the clear ice is produced when larger droplets collide with the growing stone. Because larger droplets freeze more slowly than do smaller ones, the air escapes, leaving the ice clear.

Rime

Rime is a deposit of ice crystals formed by the freezing of supercooled fog or cloud droplets on objects whose surface temperature is below freezing. When rime forms on

Figure 5–21 Hail damage to a used car lot in Fort Worth, Texas, May 6, 1995. This storm, which packed high winds and hail the size of baseballs, killed at least nine people and injured more than 100 as it swept through the northern part of the state. *(Photo by Ron Heflin/AP Wide World Photos)*

Box 5–4 Atmospheric Hazard: Worst Winter Weather

Extremes, whether they be the tallest building or the record low temperature for a location, fascinate us. When it comes to weather, some places take pride in claiming to have the worst winters on record. In fact, Fraser, Colorado, and International Falls, Minnesota, have both proclaimed themselves the "ice box of the nation." Although Fraser recorded the low temperature for the 48 contiguous states 23 times in 1989, its neighbor, Gunnison, Colorado, recorded the low temperature 62 times, far more than any other location.

Such facts do not impress the residents of Hibbing, Minnesota, where the temperature dropped to –38°C (–37°F) during the first week of March 1989. But this is mild stuff, say the old-timers in Parshall, North Dakota, where the temperature fell to –51°C (–60°F) on February 15, 1936. Not to be left out, Browning, Montana, holds the record for the most dramatic 24-hour temperature drop. Here, the temperature plummeted 56°C (100°F), from a cool 7°C (44°F) to a frosty –49°C (–56°F) during a January evening in 1916.

Although impressive, the temperature extremes cited here represent only one aspect of winter weather. What about snowfall (Figure 5–D)? Cooke City holds the seasonal snowfall record for Montana with 1062 centimeters (418.1 inches) during the winter of 1977–1978. But what about cities like Sault Ste. Marie, Michigan, or Buffalo, New York? The winter snowfalls associated with the Great Lakes are legendary. Even larger snowfalls occur in many sparsely inhabited mountainous areas. The record for the largest snowfall from a single snowstorm in the United States goes to the Mt. Shasta Ski Bowl in California, where a storm that raged from February 13–19, 1959, dropped more than 480 centimeters (189 inches).

Try telling residents of the eastern United States that heavy snowfall by itself makes for the worst weather. A blizzard in March 1993 produced heavy snowfall along with hurricane-force winds and record low temperatures

that immobilized much of the region from Alabama to the Maritime Provinces of eastern Canada. This event quickly earned the well-deserved title of "storm of the century."

As we can see, determining which location has the worst winter weather depends on how you measure it. Most snowfall in a season? Longest cold spell? Coldest temperature? Most disruptive storm?

Here are the meanings of some common terms used by the National Weather Service for winter weather events.

Snow flurries Snow falling for short durations at intermittent periods and resulting in generally little or no accumulation.

Blowing snow Snow lifted from the surface by the wind and blown about to a degree that horizontal visibility is reduced.

Drifting snow Significant accumulations of falling or loose snow caused by strong wind.

Blizzard A winter storm characterized by winds of at least 56 kilometers (35 miles) per hour for at least three hours. The storm must also be accompanied by low temperatures and considerable falling and/or blowing snow that reduces visibility to a quarter of a mile or less.

Severe blizzard A storm with winds of at least 72 kilometers (45 miles) per hour, a great amount of falling or drifting snow, and temperatures –12°C (10°F) or lower.

Heavy snow warning A snowfall in which at least 4 inches in 12 hours or 6 inches in 24 hours is expected.

Freezing rain Rain falling in a liquid form through a shallow subfreezing layer of air near the ground. The rain (or drizzle) freezes on impact with the ground or other objects, resulting in a clear coating of ice known as *glaze*.

Sleet Also called *ice pellets*. Sleet is formed when raindrops or melted snowflakes freeze as they pass through a subfreezing layer of air near Earth's surface. Sleet does not stick to trees and wires, and it usually bounces when it hits the ground. An accumulation of sleet sometimes has the consistency of dry sand.

trees, it adorns them with its characteristic ice feathers, which can be spectacular to behold (Figure 5–23). In these situations, objects such as pine needles act as freezing nuclei, causing the supercooled droplets to freeze on contact. On occasions when the wind is blowing, only the windward surfaces of objects will accumulate the layer of rime.

Precipitation Measurement

The most common form of precipitation, rain, is probably the easiest to measure. Any open container that has a consistent cross section throughout can be a rain gauge (Figure 5–24a). In general practice, however, more sophisticated

Figure 5–D Rochester, New York was hit by blizzard conditions that left 24 inches of snow in 24 hours Thursday March 4, 1999.

Travelers' advisory Issued to inform the public of hazardous driving conditions caused by snow, sleet, freezing precipitation, fog, wind, or dust.

Cold wave A rapid fall of temperature in a 24-hour period, usually signifying the beginning of a spell of very cold weather.

Windchill A measure of apparent temperature that uses the effects of wind and temperature on the human body by translating the cooling power of wind to a temperature under calm conditions. It is an approximation only for the human body and has no meaning for cars, buildings, or other objects. (See Box 3–5 for more details.)

---

devices are used to measure small amounts of rainfall more accurately and to reduce loss from evaporation.

Standard Instruments

The **standard rain gauge** (Figure 5–24b) has a diameter of about 20 centimeters (8 inches) at the top. Once the water is caught, a funnel conducts the rain through a narrow opening into a cylindrical measuring tube that has a cross-sectional area only one-tenth as large as the receiver. Consequently, rainfall depth is magnified 10 times, which allows for accurate measurements to the nearest 0.025 centimeter (0.01 inch), and the narrow opening minimizes evaporation. When the amount of rain is less than 0.025 centimeter (0.01 inch), it is generally reported as being a **trace of precipitation**.

In addition to the standard rain gauge, several types of recording gauges are routinely used. These instruments not only record the amount of rain, but also its time of occurrence and intensity (amount per unit of time). Two of the most common gauges are the tipping-bucket gauge and the weighing gauge.

As can be seen in Figure 5–24c, the **tipping-bucket gauge** consists of two compartments, each one capable of holding 0.025 centimeter (0.01 inch) of rain, situated at the base of a funnel. When one "bucket" fills, it tips and

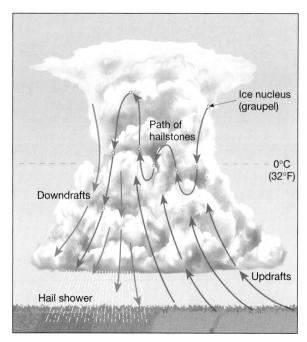

Figure 5–22 Growth of hailstones. Hailstones begin as small ice pellets that grow by adding supercooled water droplets as they move through a cloud. Strong updrafts may carry stones upward in several cycles, increasing the size of the hail by adding a new layer with each cycle. Eventually, the hailstones encounter a downdraft or grow too large to be supported by the updraft.

Figure 5–23 Rime consists of delicate ice crystals that form when supercooled fog or cloud droplets freeze on contact with objects. *(Photo by John Cancalosi/Stock Boston)*

empties its water. Meanwhile, the other "bucket" takes its place at the mouth of the funnel. Each time a compartment tips, an electrical circuit is closed and 0.025 centimeter (0.01 inch) of precipitation is automatically recorded on a graph.

The **weighing gauge**, as the name would indicate, works on a different principle. Precipitation is caught in a cylinder that rests on a spring balance. As the cylinder fills, the movement is transmitted to a pen that records the data.

Measuring Snowfall

When snow records are kept, two measurements are normally taken: depth and water equivalent. Usually, the depth of snow is measured with a calibrated stick. The actual measurement is not difficult, but choosing a representative spot can be. Even when winds are light or moderate, snow drifts freely. As a rule, it is best to take several measurements in an open place away from trees and obstructions and then average them. To obtain the water equivalent, samples may be melted and then weighed or measured as rain.

Sometimes large cylinders are used to collect snow. A major problem that hinders accurate measurement by

snow gauges is the wind. Snow will blow around the top of the cylinder instead of falling into it. Therefore, the amount caught by the gauge is generally less than the actual fall. As is often the practice with rain gauges, shields designed to break up wind eddies are placed around the snow gauge to ensure a more accurate catch.

The quantity of water in a given volume of snow is not constant. A general ratio of 10 units of snow to 1 unit of water is often used when exact information is not available. You may have heard TV weathercasters use this ratio, saying "Every 10 inches of snow equals 1 inch of rain." But the actual water content of snow may deviate widely from this figure. It may take as much as 30 centimeters of light and fluffy dry snow (30:1) or as little as 4 centimeters of wet snow (4:1) to produce 1 centimeter of water.

Measurement Errors

We tend to be unquestioning when we hear weather reports concerning the past day's events. However, unlike temperature and pressure, which tend to vary only slightly over a

1 inch
of rain

1 inch

(a) Simple rain gauge

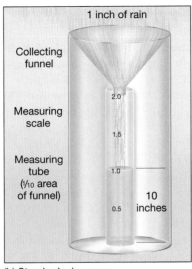

1 inch of rain

Collecting
funnel

Measuring
scale

2.0

1.5

Measuring
tube
(¹⁄₁₀ area
of funnel)

1.0

0.5

10
inches

(b) Standard rain gauge

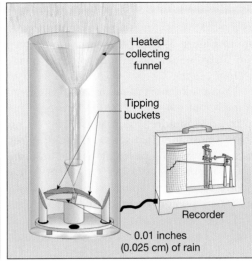

Heated
collecting
funnel

Tipping
buckets

Recorder

0.01 inches
(0.025 cm) of rain

(c) Tipping–bucket gauge

Figure 5–24 Precipitation measurement. (a) The simplest gauge is any container left in the rain. However, these homemade devices are hard to read precisely. (b) The standard rain gauge increases the height of water collected by a factor of 10, allowing for accurate rainfall measurement to the nearest 0.025 centimeter (0.01 inch). Because the cross-sectional area of the measuring tube is only one-tenth as large as the collector, rainfall is magnified 10 times. (c) The tipping-bucket rain gauge contains two "buckets" that hold the equivalent of 0.025 centimeter (0.01 inch) of precipitation. When one bucket fills, it tips and the other bucket takes its place. Each event is recorded as 0.01 inch of rainfall.

few dozen miles, precipitation on one side of the street can be significantly greater than on the other. Further, by its very nature precipitation is difficult to collect, and the techniques used have numerous potential sources for error.

When measuring rain, a certain amount goes unrecorded—some because it splashes out upon impact, and some "wets" the collection funnel, and never makes it into the cylinder. Snow in particular is often underestimated by rain gauges. In addition, in certain climates evaporation removes some of the precipitation before it is measured.

No matter which rain gauge is used, proper exposure is critical. Errors arise when gauges are located too close to buildings, trees, or other high objects that block obliquely falling rain. For best results the instrument should be twice as far away from such obstructions as the objects are high. Another cause for error is wind. It has been shown that as wind speed strengthens, turbulence increases, and it becomes more difficult to collect a representative quantity of rain. To offset this effect, a windscreen is often placed around the instrument so that rain falls into the gauge and is not carried across it (Figure 5–25).

Studies have shown that in the United States annual precipitation errors range between about 7 and 20 percent. Most often the precipitation is underestimated. In some high latitudes, percentage errors are thought to exceed 80 percent.

Figure 5–25 This standard rain gauge is fitted with metal slats that serve as a windscreen to minimize the undercatch that results because of windy conditions. *(Photo by Bobbé Christopherson)*

Precipitation Measurement by Weather Radar

Today's TV weathercasts show helpful maps like the one in Figure 5–26 to depict precipitation patterns. The instrument that produces these images is the *weather radar*.

The development of radar has given meteorologists an important tool to probe storm systems that may be up to a few hundred kilometers away. All radar units have a transmitter that sends out short pulses of radio waves. The specific wavelengths that are used depend on the objects the user wants to detect. When radar is used to monitor precipitation, wavelengths between 3 and 10 centimeters are employed.

These wavelengths can penetrate small cloud droplets, but are reflected by larger raindrops, ice crystals, or hailstones. The reflected signal, called an *echo*, is received and displayed on a TV monitor. Because the echo is "brighter" when the precipitation is more intense, modern radar is able to depict not only the regional extent of the precipitation but also the rate of rainfall. Figure 5–26 is a typical radar display in which colors show precipitation intensity. As you will see in Chapters 10 and 12, weather radar can also measure the rate and direction of storm movement.

Intentional Weather Modification

Intentional weather modification is deliberate human intervention to influence atmospheric processes that constitute the weather—that is, to alter the weather for human purposes. This desire to change the weather is nothing new. From earliest recorded times people have used prayer, wizardry, dances, and even black magic in attempts to alter the weather.

During the American Civil War, rainfall seemed to increase following some battles, leading to experiments in which cannons were fired into clouds to trigger more rain. These experiments and many others proved unsuccessful. However, by the nineteenth century, smudge pots, sprinklers, and wind machines were used successfully to fight frost.

Weather-modification strategies fall into three broad categories. The first employs energy to forcefully alter the weather. Examples are the use of intense heat sources or the mechanical mixing of air (such as by helicopters) to disperse fog at some airports.

The second category involves modifying land and water surfaces to change their natural interaction with the lower atmosphere. One often discussed but untried example is

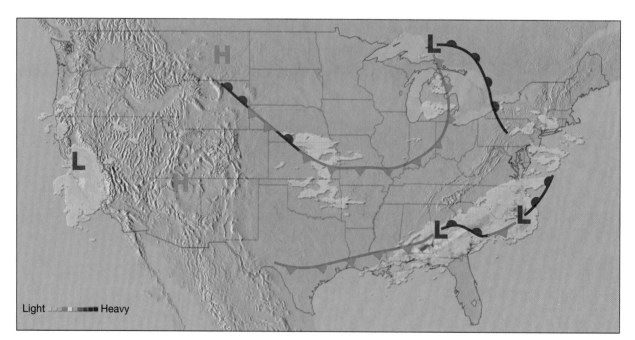

Light ⋯⋯▪▪▪▪ Heavy

Figure 5–26 Color weather radar display, commonly seen on The Weather Channel and local TV weathercasts. Colors indicate different intensities of precipitation. Note the area of heavy precipitation in southeastern Alabama (red color). This map is a composite that uses data acquired from several radar installations. Also, superimposed on this map is a satellite image showing cloud cover.

the blanketing of a land area with a dark substance. The additional solar energy absorbed by this dark surface would warm the layer of air near the surface and encourage the development of updrafts that might aid cloud formation.

The third category involves triggering, intensifying, or redirecting atmospheric processes. The seeding of clouds with agents such as dry ice (frozen carbon dioxide) and silver iodide to stimulate precipitation is the primary example. Because **cloud seeding** sometimes seems to show promising results and is a relatively inexpensive technique, it has been a primary focus of modern weather-modification technology.

Cloud Seeding

The first breakthrough in weather modification came in 1946 when Vincent J. Schaefer discovered that dry ice dropped into a supercooled cloud spurred the growth of ice crystals. Recall that once ice crystals form in a supercooled cloud, they grow larger at the expense of the remaining liquid cloud droplets and, on reaching a sufficient size, fall as precipitation.

Later it was learned that silver iodide could also be used for cloud seeding. The similarity in the structure of silver iodide crystals and ice crystals accounts for silver iodide's ability to initiate ice crystal growth. Thus, unlike dry ice, which simply chills the air, silver iodide crystals act as freezing nuclei. Because silver iodide can be more easily delivered to clouds from burners on the ground or from aircraft, it is a better alternative than dry ice (Figure 5–27).

If cloud seeding is to trigger precipitation, certain atmospheric conditions must exist. Clouds must be present, for seeding cannot create clouds. Also, a portion of the cloud must be supercooled—it must be made of liquid droplets below 0°C (32°F) in temperature.

Figure 5–27 Cloud seeding using silver iodide flares is one way that freezing nuclei are supplied to clouds. *(Courtesy of National Science Foundation/National Center for Atmospheric Research)*

One type of cloud seeding assumes that cold cumulus clouds are deficient in freezing nuclei and that adding them will stimulate precipitation. The object is to produce just enough ice crystals so they will grow large enough (via the Bergeron process) to fall as precipitation. Overseeding will simply produce billions of ice crystals that are too tiny to fall.

Seeding of winter clouds that form along mountain barriers (orographic clouds) has been tried repeatedly. These clouds are thought to be good candidates for seeding because only a small percentage of the water that condenses in cold orographic clouds actually falls as precipitation. The idea is to increase the winter snowpack, which melts and runs off during warmer months and is collected in reservoirs for irrigation and hydroelectric power generation.

Several experimental operations aimed at increasing winter precipitation have been attempted. These include the Sierra Cooperative Pilot Project and the Colorado Orographic Seeding Program. Mounting statistical evidence from these projects show that precipitation from supercooled orographic clouds can be increased by about 10 percent. Although the cause-and-effect relationship is not well documented, the potential to increase winter precipitation in mountainous terrains has been established.

In recent years, the seeding of warm convective clouds with hygroscopic (water-seeking) particles has received renewed attention. The interest in this technique arose when it was discovered that a pollution-belching paper mill near Nelspruit, South Africa, seemed to be triggering precipitation. Flying through clouds near the paper mill, research aircraft collected samples of the particulate matter emitted from the mill. It turned out that the mill was emitting tiny salt crystals (potassium chloride and sodium chloride), which rose with updrafts into the clouds. Because these salts attract moisture, they quickly form large cloud droplets, which grow into raindrops by the collision–coalescence process. Ongoing experiments being conducted over the arid landscape of northern Mexico are attempting to duplicate this process by seeding clouds using flares mounted on airplanes that spread hygroscopic salts. Although the results of these studies are promising, evidence that such seeding can increase rainfall over economically significant areas is not yet available.

In summary, researchers are confident that winter precipitation can be enhanced by seeding supercooled orographic clouds. Further, the use of large hygroscopic particles to seed warm convective clouds appears to hold promise—although the processes involved are poorly understood. However, because these studies have been limited in scope, the economic viability of cloud seeding remains uncertain. The major outcome of the last five

Figure 5–28 Effects produced by seeding a cloud deck with dry ice. Within 1 hour, a hole developed over the seeded area. *(Photo by E. J. Tarbuck)*

decades of cloud seeding experimentation has been the sobering realization that even the simplest weather events are exceedingly complex and not yet fully understood.

Fog and Cloud Dispersal

One of the most successful applications of cloud seeding involves spreading dry ice, particles of solid carbon dioxide at temperatures of $-78°C$ ($-108°F$), into layers of supercooled fog or stratus clouds to disperse them and thereby improve visibility. Airports, harbors, and foggy stretches of interstate highway are obvious candidates. Such applications trigger a transformation in cloud composition from supercooled water droplets to ice crystals. The ice crystals then settle out, leaving an opening in the cloud or fog (Figure 5–28). Initially, the ice crystals are too small to fall to the ground, but as the Bergeron process proceeds and they grow at the expense of the remaining water droplets, and as turbulent air disperses them, they can grow large enough to produce a snow shower. The snowfall is light, and visibility almost always improves. Often a large hole is opened in the stratus clouds or supercooled fog.

The U.S. Air Force has practiced this technology for many years at airbases, and commercial airlines have used this method at selected foggy airports in the western United States. The possibility of opening large holes in winter clouds to increase the solar radiation reaching the ground in northerly locations has been discussed, but little experimentation has followed.

Unfortunately, most fog does not consist of supercooled water droplets. The more common "warm fogs" are more expensive to combat because seeding will not diminish them. Successful attempts at dispersing warm fogs have involved mixing drier, warmer air from above into the fog, or heating the air. When the layer of fog is very shallow, helicopters have been used. By flying just above the fog, the helicopter creates a strong downdraft that forces drier air toward the surface where it mixes with the saturated foggy air.

At some airports where warm fogs are common, it has become more usual to heat the air and thus evaporate the fog. A sophisticated thermal fog-dissipation system, called Turboclair, was installed in 1970 at Orly Airport in Paris. It uses eight jet engines in underground chambers along the upwind edge of the runway. Although expensive to install, the system is capable of improving visibility for about 900 meters along the approach and landing zones.

Hail Suppression

Hail causes serious crop loss and property damage in many parts of the world (Figure 5–29). Consequently, hail-suppression efforts date from classical Greece to the present. A major attempt in Europe near the turn of the twentieth century actually involved firing cannons at thunderstorms (see Box 5–5).

To review, hail is believed to form by the collecting and freezing of supercooled droplets around a frozen nucleus. Updrafts of moist air in a cumulonimbus cloud are often

Box 5–5 The Hail Cannons of Europe

Some of history's more interesting efforts at weather modification have focused on suppressing hail. During the 1500s, village priests often rang church bells and called for prayers to shield nearby farms from hail. Near the turn of the twentieth century, serious attempts at hail suppression in Europe involved the firing of cannons into developing thunderstorms. It was thought that the formation of hailstones could be prevented by injecting smoke particles as condensation nuclei into the developing clouds.

These "hail cannons" were vertical muzzle-loading mortars that resembled huge megaphones. The one shown in Figure 5–E weighed 9000 kilograms (10 tons), was 9 meters (30 feet) tall, and pivoted in all directions. When fired, these cannons produced a loud whistling noise and a large smoke ring that rose to a height of about 300 meters.

During a two-year test of hail cannons in Austria, no hail was observed. At the same time, nearby provinces suffered severe hail damage. Believing from this that hail cannons were effective, these devices spread to other areas in Europe where crops of great value, such as vineyards, frequently experienced hail losses. Soon much of Europe acquired "cannon fever," and by 1899, over 2000 cannons were in use in Italy alone. However, the hail cannons proved to be ineffective, and the practice was abandoned during the next decade.

It is interesting to note, however, that by the 1960s, Russian scientists were experimenting with a similar

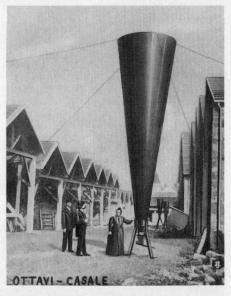

Figure 5–E Around 1900, cannons like this fired smoke particles into thunderstorms to prevent the formation of hail. However, hail shooting proved to be ineffective, and the practice was abandoned after a decade. *(Photo courtesy of J. Loreena Ivens, Illinois State Water Survey)*

approach to hail suppression. Their experiments employed rockets and artillery shells to carry freezing nuclei into the clouds. However, like earlier efforts at hail suppression, the Russian attempts were not much more effective than ringing church bells.

so strong that small hailstones may be recirculated in the upper part of the cloud, collecting additional layers of water that freeze and thus cause the stone to enlarge. Normally, only a small number of hail embryos exist, because freezing nuclei are relatively scarce; consequently, the embryos grow freely until they fall from the cloud.

Modern attempts at hail suppression have introduced silver iodide crystals into storm clouds to interrupt the formation and growth of hailstones. It is assumed that each of these crystals, acting as a freezing nucleus, attracts a portion of the cloud's water supply and thereby increases the competition for the available supercooled water droplets. With a diminished supply of supercooled water droplets, hailstones cannot grow large enough to be destructive.

Because of dramatic crop and property losses, hail-suppression technology has been the focus of many field and laboratory studies. The results, however, have been mixed at best. In Russia during the 1960s, scientists used rocket and artillery shells to carry freezing nuclei to the clouds, and they claimed to have extraordinary success.

In the United States during the 1970s, the federal government established the National Hail Research Experiment in northeastern Colorado. This experiment included a randomized seeding test to verify the Russian experience. An analysis of the data collected after three years revealed no statistically significant difference in the occurrence of hail between the seeded and nonseeded clouds, cutting short the planned five-year experiment.

Figure 5–29 Hail damage to a soybean field north of Sioux Falls, South Dakota. *(AP/Wide World Photos)*

Frost Prevention

Frost, the fruit grower's plight, occurs when the air temperature falls to 0°C (32°F) or below. Thus, frost conditions are dependent on temperature and need not be accompanied by deposits of ice crystals (called *white frost*), which only form if the air becomes saturated.

A frost hazard is produced by two conditions: when a cold air mass moves into a region or when sufficient radiation cooling occurs on a clear night. Frost associated with an invasion of cold air is characterized by low daytime temperatures and long periods of freezing conditions that inflict widespread crop damage. By contrast, frost induced by radiation cooling is strictly a nighttime phenomenon that tends to be confined to low-lying areas. Obviously, the latter phenomenon is much easier to combat.

Several methods of frost prevention are being used with varying success. They either conserve heat (reduce the heat lost at night) or add heat to warm the lowermost layer of air.

Heat-conservation methods include covering plants with insulating material, such as paper or cloth, or generating particles that, when suspended in air, reduce the rate of radiation cooling. *Smudge fires*, which produce dense black smoke, have been used to fill the air with carbon particles. Generally, this method has proven unsatisfactory. In addition to polluting the air, the carbon particles impede daytime warming by reducing the solar radiation that can reach the surface.

Warming methods employ water sprinklers, air-mixing techniques, and/or orchard heaters. Sprinklers (Figure 5–30a) add heat in two ways: first, from the warmth of the water and, more importantly, from the latent heat of fusion released when the water freezes. As long as an ice–water mixture remains on the plants, the latent heat released will keep the temperature from dropping below 0°C (32°F).

Air mixing works best when the air temperature 15 meters (50 feet) above the ground is at least 5°C (9°F) warmer than the surface temperature. By using a wind machine, the warmer air aloft is mixed with the colder surface air (Figure 5–30b).

Orchard heaters probably produce the most successful results (Figure 5–31). However, because as many as 30 to 40 heaters per acre are required, fuel cost can be significant.

Inadvertent Weather Modification: Urban-Induced Precipitation

Many studies show that cities influence the occurrence and amount of precipitation in their vicinities. After comparing urban and rural precipitation, these studies conclude that the amount of precipitation over a city is about 10 percent greater than over the nearby countryside. This work also demonstrates that even greater effects occur downwind from the city center. One comprehensive urban rain investigation in St. Louis, Missouri, clearly showed increases in precipitation at downwind locations that grew in magnitude over the years as industrial development expanded in the city. Further, if cities are really modifying precipitation, the effect should be greater on weekdays, when urban activities are most intense, than on weekends, when much of the activity ceases. This is indeed the case. The St. Louis study revealed a significantly greater frequency of rain per weekday than per weekend day in the affected downwind area. Figure 5–32 shows similar findings for Paris.

Several factors explain why an urban complex might be expected to increase precipitation.

(a)

(b)

Figure 5–30 Two common frost-prevention methods. (a) Sprinklers distribute water, which releases latent heat as it freezes on citrus. (b) Wind machines mix warmer air aloft with cooler surface air. *(Photos by (a) Bruce Borich/Silver Image Photo Agency, Inc. and (b) Christi Carter/Grant Heilman Photography, Inc.)*

air that rise because of their buoyancy. Further, recall that surface heating tends to steepen the environmental lapse rate, which enhances atmospheric *instability*.

3. Cities contain numerous tall buildings and other structures that tend to slow low-level airflow. The result is surface convergence, which enhances the upward flow of air. In addition, the rougher urban landscape impedes the progress of weather systems. Therefore, when rain-producing clouds are moving through an area they may linger over the city and increase its rainfall.

In summary, cities are thought to influence the occurrence and amount of precipitation by inadvertent cloud seeding, by enhancing the upward flow of air, and by impeding the progress of weather systems.

1. Because of their associated industries and extensive transportation systems, cities discharge large quantities of particulate matter into the atmosphere. It is believed that hygroscopic particles and freezing nuclei released into the atmosphere enhance precipitation by modifying cloud development.
2. Cities are warmer than the surrounding countryside (see Box 3–4 on the urban heat island). This differential heating can produce warm parcels of

Figure 5–31 Orchard heaters used to prevent frost damage to pear trees, Hood River, Oregon. *(Photo by Bruce Hands/The Image Works)*

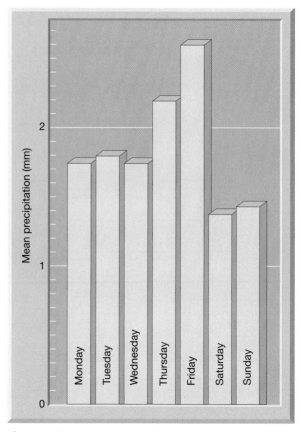

Figure 5–32 Average precipitation by day of the week in Paris for an 8-year period (1960–1967). Notice the gradual increase from Monday to Friday and the sharp drop in precipitation totals on Saturday and Sunday. The average for weekdays was 1.93 millimeters compared with only 1.47 on weekends, a difference of 24 percent. *(After I. Dettwiller, in Helmut E. Landsberg, "Inadvertent Atmospheric Modification," in Weather and Climate Modification, edited by Wilmot N. Hess, p. 755. New York: John Wiley & Sons, Inc., 1974)*

Chapter Summary

- *Condensation* occurs when water vapor changes to a liquid. For condensation to take place the air must be saturated and there must be a surface on which the vapor can condense. In the air above the ground, tiny *hygroscopic* (water-absorbent) particles known as *condensation nuclei* serve as the surfaces on which water vapor condenses.

- *Clouds*, visible aggregates of minute droplets of water or tiny crystals of ice, are one form of condensation. Clouds are classified on the basis of two criteria: form and height. The three basic cloud forms are *cirrus* (high, white, and thin), *cumulus* (globular, individual cloud masses), and *stratus* (sheets or layers). Cloud

heights can be either *high*, with bases above 6000 meters (20,000 feet), *middle*, from 2000 to 6000 meters, or *low*, below 2000 meters (6500 feet). Based on the two criteria, ten basic cloud types, including *cirrostratus*, *altocumulus*, and *stratocumulus*, are recognized.

- *Fog*, generally considered an atmospheric hazard, is a cloud with its base at or very near the ground. Fogs formed by cooling include *radiation fog* (from radiation cooling of the ground and adjacent air), *advection fog* (when warm and moist air flows over a cold surface), and *upslope fog* (created when air moves up a slope and cools adiabatically). Those formed by the addition of water vapor are *steam fog* (when water vapor evaporates from a warm water body and condenses in cool air above) and *frontal fog* (when warm air that is lifted over colder air generates precipitation that evaporates as it descends and saturates the air near the surface).

- *Dew* is the condensation of water vapor on objects that have radiated sufficient heat to lower their temperature below the dew point of the surrounding air. *White frost (hoar frost)* forms when the dew point of the air is below freezing.

- For precipitation to form, millions of cloud droplets must somehow coalesce into drops large enough to sustain themselves during their descent. The two mechanisms that have been proposed to explain this phenomenon are: the *Bergeron process*, which produces precipitation from cold clouds primarily in the middle latitudes, and the warm cloud process most associated with the tropics called the *collision-coalescence process*.

- The two most common and familiar forms of precipitation are *rain* (drops of water that fall from a cloud and have a diameter of at least 0.5 millimeter) and *snow* (precipitation in the form of ice crystals or, more often, aggregates of ice crystals). Other forms include *sleet* (falling small particles of ice that are clear to translucent), *glaze* (formed when supercooled raindrops turn to ice on colliding with solid objects), *hail* (hard, rounded pellets or irregular lumps of ice produced in large cumulonimbus clouds), and *rime* (a deposit of ice crystals formed by the freezing of supercooled fog or cloud droplets on objects whose surface temperature is below freezing).

- Rain, the most common form of precipitation, is probably the easiest to measure. The most common instruments used to measure rain are the *standard rain gauge*, which is read directly, and the *tipping bucket gauge* and *weighing gauge*, both of which record the amount and intensity of rain. The two most common measurements of snow are depth and water equivalent. Although the quantity of water in a given volume of snow is not constant, a general ratio of 10 units of snow to 1 unit of water is often used when exact information is not available.

- *Intentional weather modification* is deliberate human intervention to influence atmospheric processes that constitute the weather. Weather modification falls into three categories: (1) the use of energy to forcefully alter the weather, (2) modifying land and water surfaces to change their natural interaction with the lower atmosphere, and (3) triggering, intensifying, or redirecting atmospheric processes. The focus of intentional weather modification using modern weather technology is on *cloud seeding, fog and cloud dispersal, hail suppression*, and *frost prevention*.

Vocabulary Review

advection fog (p. 128)
Bergeron process (p. 131)
cirrus (p. 120)
clouds (p. 120)
cloud seeding (p. 147)
clouds of vertical development (p. 120)
collision–coalescence process (p. 133)
condensation nuclei (p. 119)
cumulus (p. 120)
dew (p. 130)
fog (p. 128)
freezing nuclei (p. 131)
freezing rain (p. 138)
frontal or precipitation fog (p. 130)
frost (p. 150) (prevention)
glaze (p. 138)
hail (p. 139)
high clouds (p. 120)

hygroscopic nuclei (p. 119)
intentional weather modification (p. 146)
low clouds (p. 120)
middle clouds (p. 120)
radiation fog (p. 128)
rain (p. 137)
rime (p. 141)
sleet (p. 138)
snow (p. 137)
standard rain gauge (p. 143)
steam fog (p. 129)
stratus (p. 120)
supercooled (p. 131)
tipping-bucket gauge (p. 143)
trace of precipitation (p. 143)
upslope fog (p. 128)
weighing gauge (p. 144)
white frost (p. 130)

Review Questions

1. Clouds are classified on the basis of two criteria. Name them.   *Height + form*
2. Why are high clouds always thin in comparison to low and middle clouds?   *Little water, ice crystals*
3. Which cloud types are associated with the following characteristics: thunder, halos, precipitation, hail, mackerel sky, lightning, and mares' tails?
4. What do layered clouds indicate about the stability of the air? What do clouds of vertical development indicate about the stability of air?
5. What is the importance of condensation nuclei?
6. Distinguish between clouds and fog.
7. List five types of fog and discuss how they form.
8. What actually happens when a radiation fog "lifts"?
9. Identify the fogs described in the following situations.
   a. You have stayed the night in a motel and decide to take an early morning swim. As you approach the heated swimming pool, you notice a fog over the water.
   b. You are located in the western Great Plains and the winds are from the east and fog is extensive.
   c. You are driving through hilly terrain during the early morning hours and experience fog in the valleys but clearing on the hills.
10. Why is there a relatively high frequency of dense fog along the Pacific Coast?
11. Describe the steps in the formation of precipitation according to the Bergeron process. Be sure to include (a) the importance of supercooled cloud droplets, (b) the role of freezing nuclei, and (c) the difference in saturation vapor pressure between liquid water and ice.
12. How does the collision–coalescence process differ from the Bergeron process?
13. If snow is falling from a cloud, which process produced it? Explain.
14. Describe sleet and glaze and the circumstances under which they form. Why does glaze result on some occasions and sleet on others?
15. How does hail form? What factors govern the ultimate size of hailstones?
16. Although an open container can serve as a rain gauge, what advantages does a standard rain gauge provide?
17. How do recording rain gauges work? Do they have advantages over a standard rain gauge?
18. Describe some of the factors that could lead to an inaccurate measurement of rain or snow.
19. Why are silver iodide crystals used to seed supercooled clouds?
20. If cloud seeding is to have a chance of success, certain atmospheric conditions must exist. Name them.
21. How do frost and white frost differ?
22. Describe how smudge fires, sprinkling, and air mixing are used in frost prevention.
23. List three factors that contribute to greater precipitation in and downwind of cities.

Problems

1. Suppose that the air temperature is 20°C (68°F) and the relative humidity is 50 percent at 6:00 P.M., and that during the evening the air temperature drops but its water vapor content does not change. If the air temperature drops 1°C (1.8°F) every two hours, will fog occur by sunrise (6:00 A.M.) the next morning? Explain your answer. (*Hint:* The data you need are found in Table 4–1.)
2. By using the same conditions as in Problem 1, will fog occur if the air temperature drops 1 degree every hour? If so, when will it first appear? What name is given to fog of this type that forms because of surface cooling during the night?
3. Assuming the air is still, how long would it take a large raindrop (5 mm) to reach the ground if it fell from a cloud base at 3000 meters? (See Table 5–3.) How long would a typical raindrop (2 mm) take if it fell from the same cloud? How about a drizzle drop (0.5 mm)?
4. Assuming the air is still, how long would it take for a typical cloud droplet (0.2 mm) to reach the ground if it fell from a cloud base at 1000 meters? It is very unlikely that a cloud droplet would ever reach the ground, even if the air were perfectly still. Can you explain why?
5. What is the volume of a small raindrop that is 1000 μm diameter (1 mm diameter)?
6. A large raindrop 5000 μm (5 mm) in diameter is 100 times the diameter of a large cloud droplet (50 μm in diameter). How many times greater is the volume of a large raindrop than a cloud droplet? (*Hint:* See Box 5–2.)

7. What is the area of the "bottom" of a falling drop of 1 mm diameter?

8. How many times greater is the "bottom" area of a falling 5-mm drop?

Atmospheric Science Online

The following are informative and interesting Internet sites that address topics related to those presented in the chapter:

Current U.S. Precipitation Radar Map (Intellicast):
- **http://www.intellicast.com/LocalWeather/World/UnitedStates/Radar/**

Meteorology: Online Guides (University of Illinois):
- **http://ww2010.atmos.uiuc.edu/(Gh)/guides/mtr/cld/prcp/home.rxml**

For direct links to these sites and others, chapter objectives and reviews, quiz questions, and topical investigations that utilize Web resources, visit *The Atmosphere, Eighth Edition* Home Page at:
- **http://www.prenhall.com/lutgens**

CHAPTER SIX

Air Pressure and Winds

Kite Festival, Long Beach, Washington. *(Photo by Jay Syverson/Stock Boston)*

Figure 6–1 Winds from Hurricane Irene, October 1999, Titusville, Florida. *(Photo by Marc Epstein, DRK Photo)*

Of the various elements of weather and climate, changes in air pressure are the least noticeable. In listening to a weather report, generally we are interested in moisture conditions (humidity and precipitation), temperature, and perhaps wind. It is the rare person, however, who wonders about air pressure. Although the hour-to-hour and day-to-day variations in air pressure are not perceptible to human beings, they are very important in producing changes in our weather. For example, it is variations in air pressure from place to place that generate winds that in turn can bring changes in temperature and humidity (Figure 6–1). Air pressure is one of the basic weather elements and a significant factor in weather forecasting. As we shall see, air pressure is closely tied to the other elements of weather (temperature, moisture, and wind) in a cause-and-effect relationship.

Understanding Air Pressure

In Chapter 1 we noted that **air pressure** is simply the pressure exerted by the weight of air above. Average air pressure at sea level is about 1 kilogram per square centimeter, or 14.7 pounds per square inch. This is roughly the same pressure that is produced by a column of water 10 meters (33 feet) in height. With some simple arithmetic you can calculate that the air pressure exerted on the top of a small (50 centimeter by 100 centimeter) school desk exceeds 5000 kilograms (11,000 pounds), or about the weight of a 50-passenger school bus. Why doesn't the desk collapse under the weight of the ocean of air above? Simply, air pressure is

exerted in all directions—down, up and sideways. Thus, the air pressure pushing down on the desk exactly balances the air pressure pushing up on the desk.

You might be able to visualize this phenomenon better if you imagine a tall aquarium that has the same dimensions as the desktop. When this aquarium is filled to a height of 10 meters (33 feet), the water pressure at the bottom equals 1 atmosphere (14.7 pounds per square inch). Now, imagine what will happen if this aquarium is placed on top of our student desk so that all the force is directed downward. Compare this to what results when the desk is placed inside the aquarium and allowed to sink to the bottom. In the latter situation the desk survives because the water pressure is exerted in all directions, not just downward as in our earlier example. The desk, like your body, is "built" to withstand the pressure of 1 atmosphere. It is important to note that although we do not generally notice the pressure exerted by the ocean of air around us, except when ascending or descending in an elevator or airplane, it is nonetheless substantial. The pressurized suits used by astronauts on space walks are designed to duplicate the atmospheric pressure experienced at Earth's surface. Without these protective suits to keep body fluids from boiling away, astronauts would perish in minutes.

The concept of air pressure can be better understood if we examine the behavior of gases. Gas molecules, unlike those of the liquid and solid phases, are not "bound" to one another but are freely moving about, filling all space available to them. When two gas molecules collide, which happens frequently under normal atmospheric conditions, they bounce off each other like very elastic balls. If a gas is confined to a container, this motion is restricted by its sides, much like the walls of a handball court redirect the motion of the handball. The continuous bombardment of gas molecules against the sides of the container exerts an outward push that we call air pressure. Although the atmosphere is without walls, it is confined from below by Earth's surface and effectively from above because the force of gravity prevents its escape. Here we define *air pressure* as the force exerted against a surface by the continuous collision of gas molecules.

Measuring Air Pressure

To measure atmospheric pressure, meteorologists use a unit of force from the science of physics called the **newton**.° At sea level the standard atmosphere exerts a force of 101,325 newtons per square meter. To simplify this large number, the U.S. National Weather Service adopted the **millibar** (mb), which equals 100 newtons per square meter. Thus,

°A newton is the force needed to accelerate a 1 kilogram mass 1 meter per second squared.

standard sea-level pressure is stated as 1013.25 millibars (Figure 6–2)†. The millibar has been the unit of measure on all U.S. weather maps since January 1940.

Although millibars are used almost exclusively by meteorologists, you might be better acquainted with the expression "inches of mercury," which is used by the media to describe atmospheric pressure. This expression dates from 1643 when Torricelli, a student of the famous Italian scientist Galileo, invented the **mercury barometer**. Torricelli correctly described the atmosphere as a vast ocean of air that exerts pressure on us and all things about us. To measure this force, he closed one end of a glass tube and filled it with mercury. He then inverted the tube into a dish of mercury (Figure 6–3). Torricelli found that the mercury flowed out of the tube until the weight of the mercury column was balanced by the pressure exerted on the surface of the mercury by the air above. In other words, the weight of the mercury in the column equaled the

†The standard unit of pressure in the SI system is the pascal, which is the name given to a newton per square meter (N/m^2). In this notation a standard atmosphere has a value of 101,325 pascals, or 101.325 kilopascals. If the National Weather Service officially converts to the metric system, it will probably adopt this unit.

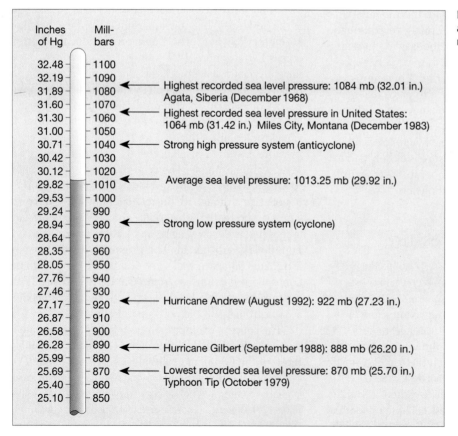

Figure 6–2 A comparison of atmospheric pressure in inches of mercury and millibars.

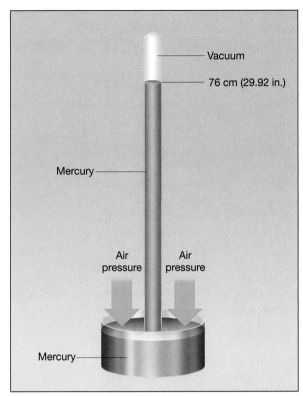

Figure 6–3 Simple mercury barometer. The weight of the column of mercury is balanced by the pressure exerted on the dish of mercury by the air above. If the pressure decreases, the column of mercury falls; if the pressure increases, the column rises.

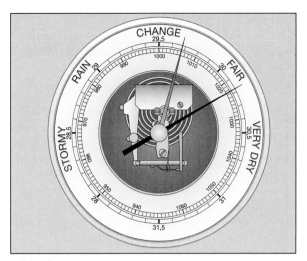

Figure 6–4 Aneroid barometer. The aneroid barometer has a partially evacuated chamber that changes shape, compressing as atmospheric pressure increases and expanding as pressure decreases.

weight of a similar diameter column of air that extended from the ground to the top of the atmosphere.

Torricelli noted that when air pressure increased, the mercury in the tube rose; conversely, when air pressure decreased, so did the height of the column of mercury. The length of the column of mercury, therefore, became the measure of the air pressure, or "inches of mercury." With some refinements the mercury barometer invented by Torricelli is still the standard pressure-measuring instrument used today. Standard atmospheric pressure at sea level equals 29.92 inches (760 millimeters) of mercury. In the United States the National Weather Service converts millibar values to inches of mercury for public and aviation use (Figure 6–3).

The need for a smaller and more portable instrument for measuring air pressure led to the development of the **aneroid barometer** (*aneroid* means without liquid). Instead of having a mercury column held up by air pressure, the aneroid barometer uses a partially evacuated metal chamber (Figure 6–4). The chamber, being very sensitive to variations in air pressure, changes shape, compressing as the pressure increases and expanding as the

pressure decreases. An advantage of the aneroid barometer is that it can easily be connected to a recording mechanism. The resulting instrument is a **barograph**, which provides a continuous record of pressure changes with the passage of time (Figure 6–5). Another important adaptation of the aneroid barometer is its use to indicate altitude for aircraft, mountain climbers, and map makers (see Box 6–1).

As we shall see later, meteorologists are most concerned with pressure differences that occur horizontally across an area. Thus, they obtain pressure readings from a number of stations. To compare pressure readings from

Figure 6–5 An aneroid barograph makes a continuous record of pressure changes. *(Photo courtesy of Qualimetrics, Inc., Sacramento, California.)*

Box 6–1 Air Pressure and Aviation

The cockpit of nearly every aircraft contains a *pressure altimeter*, an instrument that allows a pilot to determine the altitude of a plane. A pressure altimeter is essentially an aneroid barometer and, as such, responds to changes in air pressure. Recall that air pressure decreases with an increase in altitude and that the pressure distribution with height is well established. To make an altimeter, an aneroid is simply marked in meters instead of millibars. For example, we see in Table 6–1 that a pressure of 795 millibars "normally" occurs at a height of 2000 meters. Therefore, if such a pressure were experienced, the altimeter would indicate an altitude of 2000 meters.

Because of temperature variations and moving pressure systems, actual conditions are usually different from those represented as standard. Consequently, the altitude of an aircraft is seldom exactly that shown by its altimeter. When the barometric pressure aloft is lower than specified by the standard atmosphere, the plane will be flying lower than the height indicated by the altimeter. This could be especially dangerous if the pilot is flying a small plane through mountainous terrain with poor visibility. To correct for these situations, pilots make altimeter corrections before takeoffs and landings, and in some cases corrections are made en route.

Above 5.5 kilometers (18,000 feet), where commercial jets fly and pressure changes are more gradual, corrections cannot be made as precisely as at lower levels. Consequently, such aircraft have their altimeters set at the standard atmosphere and fly paths of constant pressure instead of constant altitude (Figure 6–A). Stated another way, when an aircraft flies at a constant altimeter setting, a pressure variation will result in a change in the plane's elevation. When pressure increases along a flight path, the plane will climb, and when pressure decreases, the plane will descend. There is little risk of midair collisions because all high-flying aircraft adjust their altitude in a similar manner.

Large commercial aircraft also use radio altimeters to measure heights above the terrain. The time required for a radio signal to reach the surface and return is used to accurately determine the height of the plane above the ground. This system is not without its drawbacks. Because a radio altimeter provides the elevation above the ground rather than above sea level, a knowledge of the underlying terrain is required.

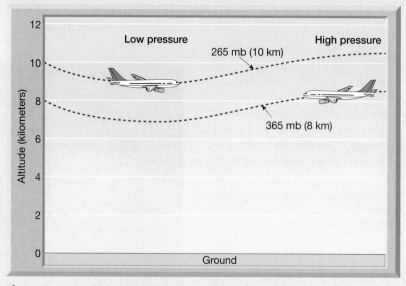

Figure 6–A Aircraft above 5.5 kilometers (18,000 feet) generally fly paths of constant pressure instead of constant altitude.

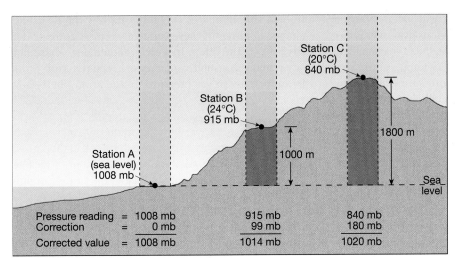

Figure 6–6 To compare atmospheric pressures, meteorologists first convert all pressure measurements to sea-level values. This is done by adding the pressure that would be exerted by an imaginary column of air (shown in red) to the station's pressure reading. (See Table D-2 in Appendix D.)

	Station A	Station B	Station C
Pressure reading =	1008 mb	915 mb	840 mb
Correction =	0 mb	99 mb	180 mb
Corrected value =	1008 mb	1014 mb	1020 mb

various weather stations, compensation must be made for the *elevation* of each station. This is done by converting all pressure measurements to sea-level equivalents (Figure 6–6). Doing so requires meteorologists to determine the pressure that would be exerted by an imaginary column of air equal in height to the elevation of the recording station and adding it to the station's pressure reading. Because temperature greatly affects the density, and hence the weight of this imaginary column, temperature also must be considered in the calculations. Thus, the corrected reading would give the pressure, at that time, as if it were taken at sea level under the same conditions.*

A comparison of pressure records taken over the globe reveals that horizontal variations in pressure are rather small. Extreme pressure readings are rarely greater than 30 millibars (1 inch of mercury) above average sea-level pressure or 60 millibars (2 inches) below average sea-level pressure. Occasionally, the barometric pressure measured in severe storms, such as hurricanes, is even lower (see Figure 6–2).

Factors Affecting Air Pressure

Recall that air pressure is the force exerted against a surface by the continuous collision of gas molecules. In still air two factors—namely, *temperature* and *density*—largely determine the pressure exerted by a column of air. The relationship that exists among these variables can be shown by the expression

$$pressure = temperature \times density \times constant.$$

---

*Appendix D explains how barometer corrections can be computed.

This simple relationship, called the **ideal gas law**, states that pressure is equal to temperature times density times a constant.** Although the constant is important when doing calculations, it does not affect the way these variables change relative to one another; therefore, we will not consider it in the following discussion.

What the gas law tells us is that a change in *any one* of these three variables (pressure, temperature, or density) will cause a predictable change in the other two. Unfortunately, since a change in one variable causes changes in *both* of the other two, the outcome is not always straightforward. For example, from the ideal gas law we see that an increase in temperature should result in increased air pressure. However, as we shall discuss later, an increase in temperature affects density in a way that causes a decrease in air pressure. Hence, it will be easier to consider the behavior of a gas if we keep one variable from changing when considering the behavior of the other two.

Let us first examine the effect of temperature on air pressure when density is kept constant. We do so by observing the behavior of a gas in a closed container. (In a closed container the volume and the density of a gas remains constant.) When the density is kept constant and the temperature of the air is raised, the speed of the gas molecules increases and, hence, so does their force. From this observation, we conclude that in a closed container an increase in temperature results in an increase in pressure.

Conversely, a decrease in temperature causes a decrease in pressure. The fact that air pressure is proportional to temperature is precisely why aerosol spray cans have a warning

---

**The relationship among pressure, temperature, and density described in this section is stated in the *ideal gas law*. A mathematical treatment of the ideal gas law is provided in Appendix E.

that cautions to keep them away from heat. Overheating of these containers can result in a dangerous explosion if the internal gas pressure exceeds the strength of the container.

Because air pressure increases with temperature, you might reason that on warm days air pressure will be high and that on cold days it will be low. Such, however, is not generally the case. Over the continents in the midlatitudes, for example, the highest pressures are recorded in winter, when temperatures are lowest. The reason is that air pressure is also proportional to *density* (the *number* or *total mass* of gas molecules in a volume) as well as temperature. On cold days the air molecules are more closely packed (greater density) than on warm days. Often the decrease in molecular motion associated with low temperatures is more than offset by the increased number of molecules exerting pressure. Thus, on a cold winter day, the air is relatively dense and surface pressures generally high. Conversely, when air is heated in the atmosphere, it expands (increases its volume) because of the increased speed of the gas molecules.

One might conclude from the preceding discussion that warm air is usually associated with low pressure. Not necessarily! Because the atmosphere is a very dynamic body, other factors that influence surface pressure come into play. As air moves around the globe, it sometimes piles up and increases the mass of air over an area, thus creating a region of higher pressure. The opposite also occurs when air aloft moves away from a region faster than it is being replaced—resulting in lower surface pressure.

In summary, air pressure is the force exerted against a surface by the continuous collisions of gas molecules. Both the temperature and the density of the gas determine the pressure a gas will exert. However, what appears to be a simple matter is complicated by the fact that a change in either of these variables (temperature or density) affects the other. In a closed container, where density remains constant, an increase in temperature results in an increase in gas pressure. In the atmosphere, however, a decrease in temperature results in an increase in density, which usually causes an increase in air pressure. Thus, because cold air masses are denser than warm air masses, they are generally associated with higher atmospheric pressure. In addition, the movement of air from one place to another plays a major role in determining the atmospheric pressure observed at any location.

The importance of atmospheric pressure to Earth's weather cannot be overemphasized. As you shall see shortly, differences in air pressure create global winds that become organized into the systems that "bring us our weather." Thus, much of the remainder of this book will consider the relationship between temperature (and thus density) and air pressure and the effect of air pressure on airflow, and vice versa.

Pressure Changes With Altitude

Let us consider the decrease in pressure with altitude mentioned in Chapter 1. The relationship between air pressure and density largely explains this observed decrease. To illustrate, imagine a cylinder fitted with a movable piston, as shown in Figure 6–7. If the temperature is kept constant and weights are added to the piston, the downward force exerted by gravity will squeeze (compress) the gas molecules together. The result is to increase the density of the gas and hence the number of gas molecules (per unit area) bombarding the cylinder wall as

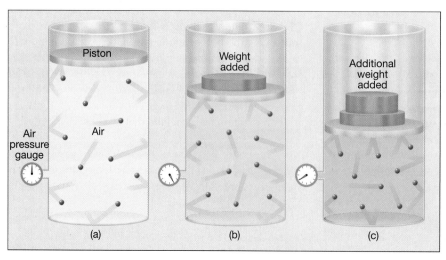

(a)   (b)   (c)

Figure 6–7 Schematic drawing showing the relationship between air pressure and density. Cylinder (a) of air fitted with movable piston. As more weight is added (b and c) to the piston, the increased number of molecules per unit volume (density) causes an increase in the pressure exerted on the walls of the cylinder and the gauge.

well as the bottom of the piston. Therefore, an increase in density results in an increase in pressure.

The piston will continue to compress the air molecules until the downward force is balanced by the ever-increasing gas pressure. If more weight is added to the piston, the air will compress further until the pressure of the gas once again balances the new weight of the piston.

Similarly, *the pressure at any given altitude in the atmosphere is equal to the weight of the air directly above that point.* Recall that at sea level a column of air weighs 14.7 pounds per square inch and therefore exerts that amount of pressure. As we ascend through the atmosphere, we find that the air becomes less dense because of the lesser amount (weight) of air above. As would be expected, there is a corresponding *decrease in pressure with an increase in altitude.*

The fact that density decreases with altitude is why the term "thin air" is normally associated with mountainous regions. Except for the Sherpas (indigenous peoples of Nepal), most of the climbers who have reached the summit of Mount Everest (elevation 8852 meters or 29,029 feet) and survived used supplementary oxygen for the final leg of the journey. Even with the aid of supplementary oxygen, most of these climbers experienced periods of disorientation because of an inadequate supply of oxygen to their brains. The decrease in pressure with altitude also affects the boiling temperature of water, which at sea level is 100°C (212°F). For example, in Denver, Colorado—the "Mile High City"—water boils at 95°C (203°F). Although water comes to a boil faster in Denver than in San Diego because of its lower boiling temperature, it takes longer to cook spaghetti in Denver.

Recall in Chapter 1 that the rate at which pressure decreases with altitude is not a constant. The rate of decrease is much greater near Earth's surface, where pressure is high, than aloft, where air pressure is low. The "normal" decrease in pressure experienced with increased altitude is provided by the standard atmosphere in Table 6–1. The **standard atmosphere** depicts the idealized vertical distribution of atmospheric pressure (as well as temperature and density), which is taken to represent average conditions in the real atmosphere. We see from Table 6–1 that atmospheric pressure is reduced by approximately one-half for each 5-kilometer increase in altitude. Therefore, at 5 kilometers the pressure is one-half its sea-level value; at 10 kilometers it is one-fourth; at 15 kilometers it is one-eighth; and so forth. Thus, at the altitude at which commercial jets fly (10 kilometers), the air exerts a pressure equal to only one-fourth that at sea level.

Table 6–1 U.S. standard atmosphere

Height (KM)	Pressure (MB)	Temperature (°C)
50.0	0.798	−2
40.0	2.87	−22
35.0	5.75	−36
30.0	11.97	−46
25.0	25.49	−51
20.0	55.29	−56
18.0	75.65	−56
16.0	103.5	−56
14.0	141.7	−56
12.0	194.0	−56
10.0	265.0	−50
9.0	308.0	−43
8.0	356.5	−37
7.0	411.0	−30
6.0	472.2	−24
5.0	540.4	−17
4.0	616.6	−11
3.5	657.8	−8
3.0	701.2	−4
2.5	746.9	−1
2.0	795.0	2
1.5	845.6	5
1.0	898.8	9
0.5	954.6	12
0	1013.2	15

Factors Affecting Wind

We discussed the upward movement of air and its importance in cloud formation. As important as vertical motion is, far more air is involved in horizontal movement, the phenomenon we call **wind**. Although we know that air will move vertically if it is warmer and thus more buoyant than surrounding air, what causes air to move horizontally?

Simply stated, *wind is the result of horizontal differences in air pressure.* Air flows from areas of higher pressure to areas of lower pressure. You may have experienced this condition when opening a vacuum-packed can of coffee. The noise you hear is caused by air rushing from the area of higher pressure outside the can to the lower pressure inside. Wind is nature's attempt to balance inequalities in air pressure. Because unequal heating of Earth's surface continually generates these pressure differences, solar radiation is the ultimate energy source for most wind.

If Earth did not rotate and if there were no friction, air would flow directly from areas of higher pressure to areas of lower pressure. Because both factors exist, however, wind is controlled by a combination of forces, including:

1. the pressure-gradient force
2. the Coriolis force
3. friction

Pressure-Gradient Force

To get anything to accelerate (change its velocity) requires an unbalanced force in one direction. The force that generates winds results from horizontal pressure differences. When air is subjected to greater pressure on one side than on another, the imbalance produces a force that is directed from the region of higher pressure toward the area of lower pressure. Thus, pressure differences cause the wind to blow, and the greater these differences, the greater the wind speed.

Variations in air pressure over Earth's surface are determined from barometric readings taken at hundreds of weather stations. These pressure data are shown on surface weather maps by means of isobars. **Isobars** are lines connecting places of equal air pressure

(*iso* = equal, *bar* = pressure), as shown in Figure 6–8. The *spacing* of the isobars indicates the amount of pressure change occurring over a given distance and is expressed as the **pressure gradient**. The mathematical expression of the pressure gradient force is provided in Box 6–2.

You might find it easier to visualize the concept of a pressure gradient if you think of it as being analogous to the slope of a hill. A steep pressure gradient, like a steep hill, causes greater acceleration of a parcel of air than does a weak pressure gradient (a gentle hill). Thus, the relationship between wind speed and the pressure gradient is straightforward. Closely spaced isobars indicate a steep pressure gradient and strong winds; widely spaced isobars indicate a weak pressure gradient and light winds. Figure 6–9 illustrates the relationship between the spacing of isobars and wind speed. Note also that the pressure-gradient force is always directed at *right angles* to the isobars.

Pressure differences observed on the daily weather map result from complex factors. However, the underlying cause of these differences is simply *unequal heating of Earth's land–sea surface*.

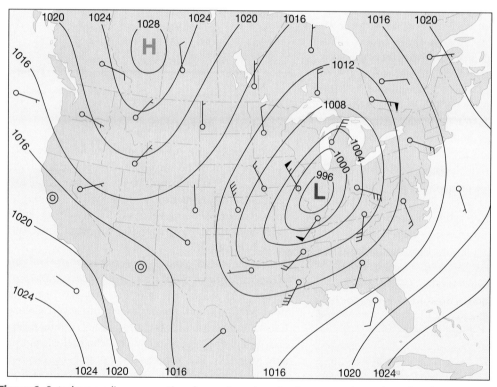

Figure 6–8 Isobars are lines connecting places of equal sea-level pressure. They are used to show the distribution of pressure on daily weather maps. Isobars are seldom straight, but usually form broad curves. Concentric rings of isobars indicate cells of high and low pressure. The "wind flags" indicate the expected airflow surrounding pressure cells and are plotted as "flying" with the wind (that is, the wind blows toward the station circle). Notice on this map that the isobars are more closely spaced and the wind speed is faster around the low-pressure center than around the high.

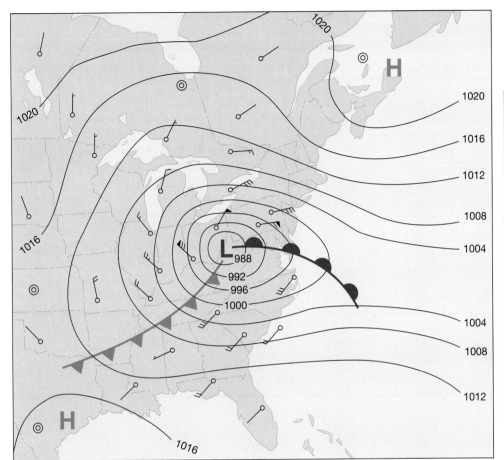

ff	Miles per hour
⊚	Calm
——	1–2
⌐	3–8
⌐	9–14
⌐	15–20
⌐⌐	21–25
⌐⌐	26–31
⌐⌐⌐	32–37
⌐⌐⌐	38–43
⌐⌐⌐⌐	44–49
⌐⌐⌐⌐⌐	50–54
▮	55–60
▮⌐	61–66
▮⌐	67–71
▮⌐⌐	72–77
▮⌐⌐⌐	78–83
▮⌐⌐⌐	84–89
▮▮	119–123

Figure 6–9 Pressure-gradient force. Closely spaced isobars indicate a strong pressure gradient and high wind speeds, whereas widely spaced isobars indicate a weak pressure gradient and low wind speeds.

Horizontal Pressure Gradients and Wind. To illustrate how temperature differences can generate a horizontal pressure gradient and thereby create winds, let us look at a common example, the *sea breeze*. Figure 6–10a shows a vertical cross-section of a coastal location just before sunrise. At this time we are assuming that temperatures and pressures do not vary horizontally at any level. This assumption is shown in Figure 6–10a by the horizontal pressure surfaces that indicate the equal pressure at equal heights. Because there is no horizontal variation in pressure (zero horizontal pressure gradient), there is no wind.

After sunrise, however, the unequal rates at which land and water heat will initiate pressure differences and, therefore, airflow (Figure 6–10b). Recall from Chapter 3 that surface temperatures over the ocean change only slightly on a daily basis. On the other hand, land surfaces and the air above can be substantially warmed during a single daylight period. As air over the land warms, it expands, causing a reduction in density. This in turn results in pressure surfaces that are bent upward, as shown in Figure 6–10b.

Although this warming does not by itself produce a surface pressure change, a given pressure surface aloft does become elevated over the land compared to over the ocean. The resultant pressure gradient aloft causes the air aloft to move from the land toward the ocean.

The mass transfer of air seaward creates a surface high-pressure area over the ocean, where the air is collecting, and a surface low over the land. The surface circulation that develops from this redistribution of air aloft is from the sea toward the land (sea-breeze), as shown in Figure 6–10c. Thus, a simple thermal circulation develops, with a seaward flow aloft and a landward flow at the surface. Note that *vertical* movement also is required to make the circulation complete. (We will consider the vertical pressure gradient next.)

An important relationship exists between pressure and temperature, as you saw in the preceding discussion. Temperature variations create pressure differences and hence wind. The greater these temperature differences, the stronger the horizontal pressure gradient and

Box 6–2   Pressure-Gradient Force

Gregory J. Carbone*

The magnitude of the pressure-gradient force is a function of the pressure difference between two points and air density. It can be expressed as

$$F_{PG} = \frac{1}{d} \times \frac{\Delta_p}{\Delta_n}; \text{ where}$$

F_{PG} = pressure-gradient force per unit mass
d = density of air
Δ_p = pressure difference between two points
Δ_n = distance between two points

Let us consider an example where the pressure 5 kilometers above Little Rock, Arkansas, is 540 millibars, and at 5 kilometers above St. Louis, Missouri, it is 530 millibars. The distance between the two cities is 450 kilometers, and the air density at 5 kilometers is 0.75 kilogram per cubic meter. In order to use the pressure-gradient equation, we must use compatible units. We must first convert pressure from millibars to pascals, another measure of pressure that has units of (kilograms × meters^{-1} × second^2).

In our example, the pressure difference above the two cities is 10 millibars, or 1000 pascals (1000 kg/m•s^2). Thus, we have

$$F_{PG} = \frac{1}{0.75} \times \frac{1000}{450,000} = 0.0029 \frac{m}{s^2}$$

Newton's second law states that force equals mass times acceleration ($F = m \times a$). In our example, we have considered pressure-gradient force *per unit mass*; therefore, our result is an acceleration ($F/m = a$). Because of the small units shown, pressure-gradient *acceleration* is often expressed as centimeters per second squared. In this example, we have 0.296 cm/s^2.

*Professor Carbone is a faculty member in the Department of Geography at the University of South Carolina.

resultant wind. Daily temperature differences and the pressure gradients so generated are generally confined to a zone that is only a few kilometers thick. On a global scale, however, variations in the amount of solar radiation received due to variations in Sun angle with latitude generate the much larger pressure systems that in turn produce the planetary atmospheric circulation. This is the topic of the next chapter.

If the pressure gradient force was the only force acting on the wind, air would flow directly from areas of higher pressures to areas of lower pressure and the lows would be quickly filled. Thus, the well-developed high and low pressure systems that dominate our daily weather maps could never sustain themselves. Instead, Earth's atmosphere would, at best, develop only very weak and short-lived pressure systems. Therefore, most locations would experience long periods of very still air that would occasionally be disrupted by gentle breezes.

Fortunately, for those of us who like more varied weather, this is not the case. Once the pressure-gradient force starts the air in motion, the Coriolis force, friction, and other forces come into play. Although these forces cannot generate wind (with the exception of gravity) they do *greatly modify airflow*. By doing so, these forces help sustain and even enhance the development of the Earth's pressure systems. We will consider some of these important *modifying* forces shortly.

Vertical Pressure Gradient. As you learned earlier, airflow is from areas of higher pressure to areas of lower pressure. Further, you are aware that air pressure is highest near Earth's surface and gets progressively lower as you move upward through the atmosphere. Combining these two ideas, you might wonder why air does not rapidly accelerate upward and escape into space. In fact, it would do just that if it were not for the force of gravity, which acts in the opposite direction to the upward directed, *vertical pressure gradient*. The important balance that is usually maintained between these two opposing forces is called **hydrostatic equilibrium**.

In general, the atmosphere is in or near hydrostatic balance. On those occasions when the gravitational force slightly exceeds the vertical pressure-gradient force, slow downward airflow results. This occurs, for example, within a dense Arctic air mass, where cold air is subsiding and spreading out at the surface. Most large-scale downward (and upward) motions are slow, averaging a few centimeters per second (less than 1 mile per day). However, some smaller-scale vertical motions are much more rapid. For example, strong updrafts and downdrafts associated with

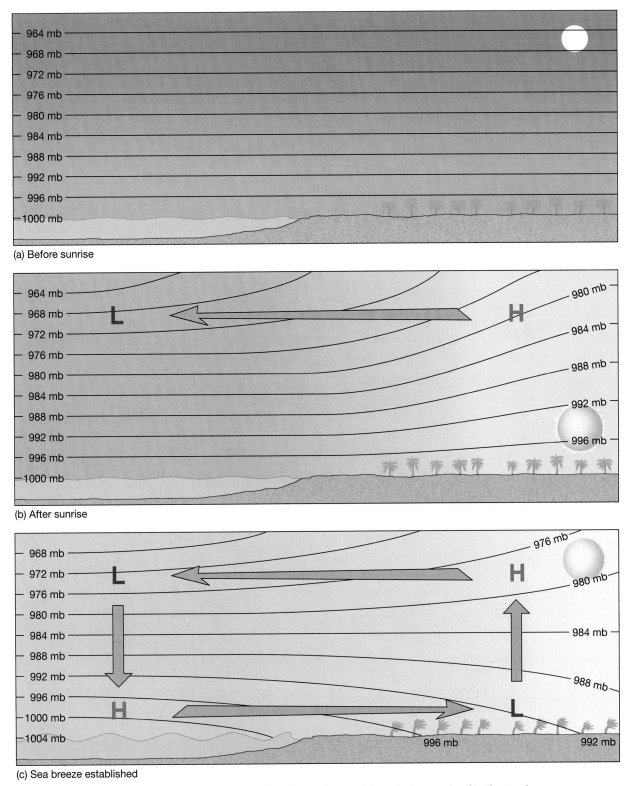

(a) Before sunrise

(b) After sunrise

(c) Sea breeze established

Figure 6–10 Cross-sectional view illustrating the formation of a sea breeze. (a) Just before sunrise (b) After sunrise (c) Sea breeze established.

severe thunderstorms can reach 100 kilometers (62 miles) per hour or more. Fortunately, these phenomena are sporadic and localized. Overall, the state of the atmosphere is in hydrostatic equilibrium, where the upward pressure-gradient force is balanced by the downward force of gravity.

In summary, the horizontal pressure gradient is the driving force of wind. It has both magnitude and direction. Its magnitude is determined from the spacing of isobars, and the direction of force is always from areas of higher pressure to areas of lower pressure and at right angles to the isobars. By contrast, the vertical pressure gradient is usually in, or near, balance with gravity. Thus, upward and downward flow in the atmosphere is comparatively slow (with the exception of localized updrafts and downdrafts).

Coriolis Force

The weather map in Figure 6–8 shows the typical air movements associated with surface high- and low-pressure systems. As expected, the air moves out of the regions of higher pressure and into the regions of lower pressure. However, the wind does not cross the isobars at right angles, as the pressure-gradient force directs. This deviation is the result of Earth's rotation and has been named the **Coriolis force** after the French scientist, Gaspard Gustave Coriolis, who first expressed its magnitude quantitatively.

All free-moving objects, including wind, are deflected to the *right* of their path of motion in the Northern Hemisphere and the *left* in the Southern Hemisphere. The reason for this deflection can be illustrated by imagining the path of a rocket launched from the North Pole toward a target on the equator (Figure 6–11). If the rocket took an hour to reach its target, Earth would have rotated 15° to the east during its flight. To someone standing on Earth, it would look as if the rocket veered off its path and hit Earth 15° west of its target. The true path of the rocket was straight and would appear so to someone out in space looking down at Earth. It was Earth turning under the rocket that gave it its apparent deflection. Note that the rocket was deflected to the right of its path of motion because of the counterclockwise rotation of the Northern Hemisphere. Clockwise rotation produces a similar deflection in the Southern Hemisphere, but to the left of the path of motion.

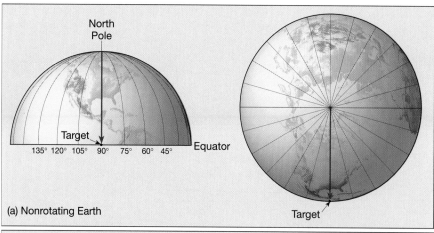

(a) Nonrotating Earth

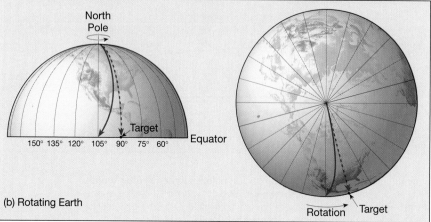

(b) Rotating Earth

Figure 6–11 The Coriolis force illustrated using a 1-hour flight of a rocket traveling from the North Pole to a location on the equator. (a) On a nonrotating Earth, the rocket would travel straight to its target. (b) However, Earth rotates 15° each hour. Thus, although the rocket travels in a straight line, when we plot the path of the rocket on Earth's surface, it follows a curved path that veers to the right of the target.

Although it is usually easy for people to visualize the Coriolis deflection when the motion is from north to south, as in our rocket example, it is not so easy to see how a west-to-east flow would be deflected. Figure 6–12 illustrates this situation using winds blowing eastward at four different latitudes (0°, 20°, 40°, and 60°). Notice that after a few hours the winds along the 20th, 40th, and 60th parallels appear to be veering off course. However, when viewed from space, it is apparent that these winds have maintained their original direction. It is the North American continent changing its orientation as Earth turns on its axis that produces the deflection we observe.

We can also see in Figure 6–12 that the amount of deflection is greater at 60° latitude than at 40° latitude, which is greater than at 20°. Furthermore, there is no deflection observed for the airflow along the equator. We conclude, therefore, that the magnitude of the deflecting force (Coriolis force) is dependent on latitude; it is strongest at the poles, and it weakens equatorward, where it eventually becomes nonexistent. We can also see that the amount of Coriolis deflection increases with wind speed. This results because faster winds cover a greater distance than do slower winds in the same time period. A mathematical treatment of the Coriolis force is provided in Box 6–3.

It is of interest to point out that any "free-moving" object will experience a similar deflection. This fact was dramatically discovered by our navy in World War II. During target practice long-range guns on battleships continually missed their targets by as much as several hundred yards until ballistic corrections were made for the changing position of a seemingly stationary target. Over a short distance, however, the Coriolis force is relatively small. Nevertheless, in the middle latitudes this deflecting force is great enough to potentially affect the outcome of a baseball game. A ball hit a horizontal distance of 100 meters (330 feet) in 4 seconds down the right field line will be deflected 1.5 centimeters (more than 1/2 inch) to the right by the Coriolis force. This could be just enough to turn a potential home run into a foul ball!

In summary, on a rotating Earth the Coriolis force acts to change the direction of a moving body to the right in the Northern Hemisphere and to the left in the Southern Hemisphere. This deflecting force (1) is always directed at right angles to the direction of airflow; (2) affects only wind direction, not wind speed; (3) is affected by wind speed (the stronger the wind, the greater the deflecting force); and (4) is strongest at the poles and weakens equatorward, becoming nonexistent at the equator.

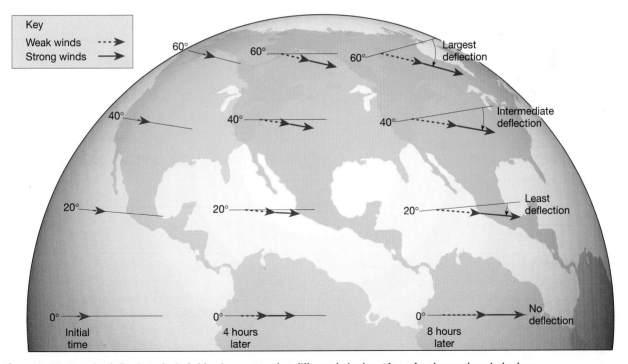

Figure 6–12 Coriolis deflection of winds blowing eastward at different latitudes. After a few hours, the winds along the 20th, 40th, and 60th parallels appear to veer off course. This deflection (which does not occur at the equator) is caused by Earth's rotation, which changes the orientation of the surface over which the winds are moving.

Box 6–3 Coriolis Force as a Function of Wind Speed and Latitude

Gregory J. Carbone*

Figure 6–12 shows how wind speed and latitude conspire to affect the Coriolis force. Consider a west wind at four different latitudes (0°, 20°, 40°, and 60°). After several hours Earth's rotation has changed the orientation of latitude and longitude of all locations except the equator such that the wind appears to be deflected to the right. The degree of deflection for a given wind speed increases with latitude because the orientation of latitude and longitude lines changes more at higher latitudes. The degree of deflection of a given latitude increases with wind speed because greater distances are covered in the period of time considered.

We can show mathematically the importance of latitude and wind speed on Coriolis force:

$$F_{co} = 2v \, \Omega \, sin \, \phi,$$

where F_{co} = Coriolis force *per unit mass of air*
v = wind speed
Ω = Earth's rate of rotation or angular velocity (which is 7.29×10^{-5} radians per second)
ϕ = latitude

(Note that $sin \, \phi$ is a trigonometric function equal to zero for an angle of 0° (equator) and 1 when $\phi = 90°$ (poles).)

As an example, the Coriolis force per unit mass that must be considered for a 10-meter-per-second (m/s) wind at 40° is calculated as:

$$F_{co} = 2\Omega \, sin \, \phi v$$
$$F_{co} = 2\Omega \, sin \, 40 \times 10 \text{ m/s}$$
$$F_{co} = 2(7.29 \times 10^{-5} \text{ s}^{-1}) \, 0.64 \, (10/ms)$$
$$F_{co} = 0.00094 \text{ meter per second squared}$$
$$= 0.094 \text{ cm s}^{-2}$$

The result (0.094 cm s^{-2}) is expressed as an acceleration because we are considering force per unit mass and force = mass × acceleration.

Using this equation, one could calculate the Coriolis force for any latitude or wind speed. Consider Table 6–A, which shows the Coriolis force per unit mass for three specific wind speeds at various latitudes. All values are expressed in centimeters per second squared (cm s^{-2}). Because pressure gradient and Coriolis force approximately balance under geostrophic conditions, we can see from our table that the pressure-gradient force (per unit mass) of 0.296 cm s^{-2} illustrated in Box 6–2 would produce relatively high winds.

Table 6–A Coriolis force for three wind speeds at various latitudes

Wind Speed		Latitude (ϕ)			
		0°	20°	40°	60°
(M/S)	(KPH)	Coriolis Force (CM/S²)			
5	18	0	0.025	0.047	0.063
10	36	0	0.050	0.094	0.126
25	90	0	0.125	0.235	0.316

*Professor Carbone is a faculty member in the Department of Geography at the University of South Carolina.

Friction

Earlier we stated that the pressure-gradient force is the primary driving force of the wind. As an unbalanced force, it causes air to accelerate from regions of higher pressure to regions of lower pressure. Thus, wind speeds should continually increase (accelerate) for as long as this imbalance exists. But we know from personal experience that winds do not become faster indefinitely. Some other force, or forces, must oppose the pressure-gradient force to moderate airflow. From our everyday experience we know that friction acts to slow a moving object. Although friction significantly influences airflow near Earth's surface, its effect is negligible above a height of a few kilometers. For this reason, we will divide our discussion. First, we will examine the flow aloft, where the effect of friction is small. Then we will analyze surface winds, where friction significantly influences airflow.

Winds Aloft and Geostrophic Flow

This section will deal only with airflow above a few kilometers, where the effects of friction are small enough to disregard.

Aloft, the Coriolis force is responsible for balancing the pressure-gradient force and thereby directing airflow. Figure 6–13 shows how a balance is reached between these opposing forces. For illustration only, we assume a nonmoving parcel of air at the starting point in Figure 6–13. (Remember that air is rarely stationary in the atmosphere.) Because our parcel of air has no motion, the Coriolis force exerts no influence; only the pressure-gradient force can act at the starting point. Under the influence of the pressure-gradient force, which is always directed perpendicularly to the isobars, the parcel begins to accelerate directly toward the area of low pressure. As soon as the flow begins, the Coriolis force comes into play and causes a deflection to the right for winds in the Northern Hemisphere. As the parcel continues to accelerate, the Coriolis force intensifies (recall that the magnitude of the Coriolis force is proportional to wind speed). Thus, the increased speed results in further deflection.

Eventually the wind turns so that it is flowing parallel to the isobars. When this occurs, the pressure-gradient force is balanced by the opposing Coriolis force, as shown in Figure 6–13. As long as these forces remain balanced, the resulting wind will continue to flow parallel to the isobars at a constant speed. Stated another way, the wind can be considered to be coasting (not accelerating or decelerating) along a pathway defined by the isobars.

Under these idealized conditions, when the Coriolis force is exactly equal and opposite to the pressure-gradient force, the airflow is said to be in *geostrophic balance*. The winds generated by this balance are called **geostrophic winds** (geostrophic means "turned by Earth"). Geostrophic winds flow in a straight path, parallel to the isobars, with velocities proportional to the pressure-gradient force. A steep pressure gradient creates strong winds, and a weak pressure gradient produces light winds.

It is important to note that the geostrophic wind is an idealized model that only approximates the actual behavior of airflow aloft. In the real atmosphere, winds are never purely geostrophic. Nonetheless, the geostrophic model offers a useful approximation of the actual winds aloft. By measuring the pressure field (orientation and

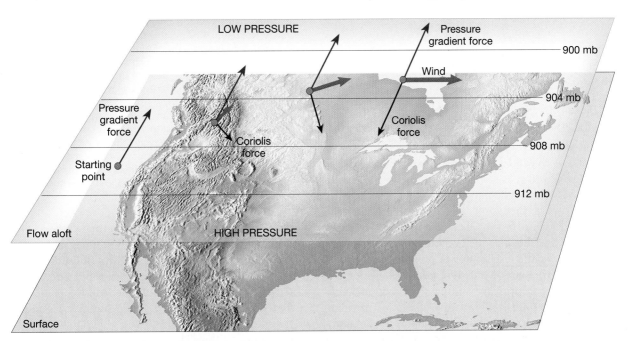

Figure 6–13 The geostrophic wind. The only force acting on a stationary parcel of air is the pressure-gradient force. Once the air begins to accelerate, the Coriolis force deflects it to the right in the Northern Hemisphere. Greater wind speeds result in a stronger Coriolis force (deflection) until the flow is parallel to the isobars. At this point the pressure-gradient force and Coriolis force are in balance and the flow is called a *geostrophic wind*. It is important to note that in the "real" atmosphere, airflow is continually adjusting for variations in the pressure field. As a result, the adjustment to geostrophic equilibrium is much more irregular than shown.

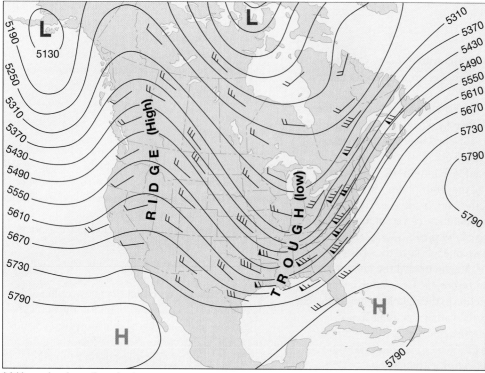

(a) Upper-level weather chart

ff	Miles per hour
◎	Calm
	1–2
	3–8
	9–14
	15–20
	21–25
	26–31
	32–37
	38–43
	44–49
	50–54
	55–60
	61–66
	67–71
	72–77
	78–83
	84–89
	119–123

(b) Representation of upper-level chart

Figure 6–14 Upper-air weather chart. (a) This simplified weather chart shows the direction and speed of the upper-air winds. Note from the flags that the airflow is almost parallel to the contours. Like most upper-air charts, this one shows variations in the height (in meters) at which a selected pressure (500 millibars) is found, instead of showing variations in pressure at a fixed height like surface maps. Do not let this confuse you, because there is a simple relationship between height contours and pressure. Places experiencing 500-millibar pressure at higher altitudes (toward the south) are experiencing higher pressures than places where the height contours indicate lower altitudes. Thus, *higher-elevation* contours indicate *higher* pressures, and *lower-elevation* contours indicate *lower* pressures. (b) Representation of the 500-millibar surface shown in the upper-air weather chart above.

spacing of isobars) that exists aloft, meteorologists can determine both wind direction and speed (Figure 6–14).

The idealized geostrophic flow predicts winds parallel to the isobars with speeds that depend on isobaric spacing. That is, the closer the isobars, the higher the wind speed (Figure 6–14). Of equal importance to meteorologists is that the same method can be used in reverse. In other words, the distribution of pressure aloft can be determined from airflow measurements. The interrelationship between pressure and wind greatly enhances the reliability of upper-air weather charts by providing checks and balances. In addition, it minimizes the number of direct observations required to describe adequately the conditions aloft, where accurate data are most expensive and difficult to obtain.

As we have seen, wind direction is directly linked to the prevailing pressure pattern. Therefore, if we know the wind direction, we can, in a crude sort of way, also establish the pressure distribution. This rather straightforward

relationship between wind direction and pressure distribution was first formulated by the Dutch meteorologist Buys Ballott in 1857. Essentially **Buys Ballott's law** states: In the Northern Hemisphere *if you stand with your back to the wind, low pressure will be found to your left and high pressure to your right.* In the Southern Hemisphere, the situation is reversed.

Although Buys Ballott's law holds for airflow aloft, it must be used with caution when applied to surface winds. At the surface, friction and topography interfere with the idealized circulation. At the surface, if you stand with your back to the wind, then turn clockwise about 30°, low pressure will be to your left and high pressure to your right.

In summary, winds above a few kilometers can be considered geostrophic—that is, they flow in a straight path parallel to the isobars at speeds that can be calculated from the pressure gradient. The major discrepancy from true geostrophic winds involves the flow along highly curved paths, a topic considered next.

Curved Flow and the Gradient Wind

Even a casual glance at a weather map shows that the isobars are not generally straight; instead, they make broad sweeping curves (see Figure 6–8). Occasionally the isobars connect to form roughly circular cells of either high or low pressure. Thus, unlike geostrophic winds that flow in a straight path, winds around cells of high or low pressure follow curved paths in order to parallel the isobars. Winds of this nature, which blow at a constant speed parallel to curved isobars, are called **gradient winds**.

Let us examine how the pressure-gradient force and Coriolis force combine to produce gradient winds. Figure 6–15a shows the gradient flow around a center of low pressure. As soon as the flow begins, the Coriolis force causes the air to be deflected. In the Northern Hemisphere, where the Coriolis force deflects the flow to the

right, the resulting wind blows counterclockwise about a low (Figure 6–15a). Conversely, around a high-pressure cell, the outward-directed pressure-gradient force is opposed by the inward-directed Coriolis force, and a clockwise flow results (Figure 6–15b).

Because the Coriolis force deflects the winds to the left in the Southern Hemisphere, the flow is reversed there—clockwise around low-pressure centers and counterclockwise around high-pressure centers.

It is common practice to call all centers of low pressure **cyclones** and the flow around them cyclonic. **Cyclonic flow** has the same direction of rotation as Earth: counterclockwise in the Northern Hemisphere and clockwise in the Southern Hemisphere. Centers of high pressure are frequently called **anticyclones** and exhibit **anticyclonic flow** (opposite that of Earth's rotation). Whenever isobars curve to form elongated regions of low and high pressure, these areas are called **troughs** and **ridges**, respectively (Figure 6–14). The flow about a trough is cyclonic; the flow around a ridge is anticyclonic.

Now let us consider the forces that produce the gradient flow associated with cyclonic and anticyclonic circulations. Wherever the flow is curved, a force has deflected the air (changed its direction), even when no change in speed results. This is a consequence of Newton's first law of motion, which states that a moving object will

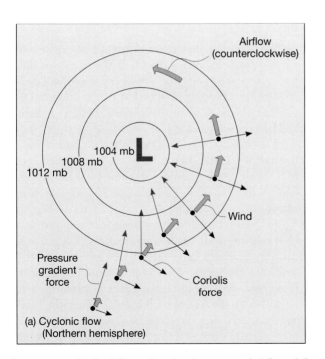

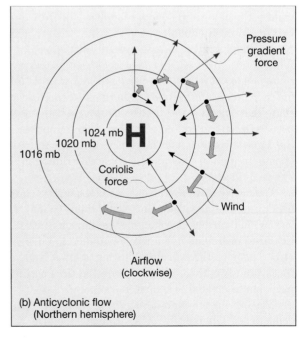

Figure 6–15 Idealized illustration showing expected airflow aloft around low- and high-pressure centers. It is important to note that in the "real" atmosphere, airflow is continually adjusting for variations in the pressure field. As a result, the adjustment to gradient balance is much more irregular than shown.

continue to move in a straight line unless acted upon by an unbalanced force. You have undoubtedly experienced the effect of Newton's law when the automobile in which you were riding made a sharp turn and your body tried to continue moving straight ahead (see Box 6–4).

Referring back to Figure 6–15a, we see that in a low-pressure center, the inward-directed pressure-gradient force is opposed by the outward-directed Coriolis force. But to keep the path curved (parallel to the isobars), the inward pull of the pressure-gradient force must be strong enough to balance the Coriolis force as well as to turn (accelerate) the air inward. The inward turning of the

air is called *centripetal acceleration*. Stated another way, the pressure-gradient force must exceed the Coriolis force to overcome the air's tendency to continue moving in a straight line.[*]

The opposite situation exists in anticyclonic flows, where the inward-directed Coriolis force must balance the pressure-gradient force as well as provide the inward acceleration needed to turn the air. Notice in Figure 6–15 that the pressure-gradient and Coriolis force are not balanced

[*]The tendency of a particle to move in a straight line when rotated creates an imaginary outward force called *centrifugal force*.

Box 6–4 Newton's Laws of Motion

Because air is composed of atoms and molecules, its motion is governed by the same natural laws that apply to all matter. Simply put, when a force is applied to air, it will be displaced from its original position. Depending on the direction from which the force is applied, the air may move horizontally to produce winds, or in some situations vertically to generate convective flow. To better understand the forces that produce global winds, it is helpful to become familiar with Newton's first two laws of motion.

Newton's first law of motion states that *an object at rest will remain at rest, and an object in motion will continue moving at a uniform speed and in a straight line unless a force is exerted upon it*. In simple terms this law states that objects at rest tend to stay at rest and objects in motion tend to continue moving at the same rate in the same direction. The tendency of things to resist change in motion (including a change in direction) is known as *inertia*.

You have experienced Newton's first law if you have ever pushed a stalled auto along flat terrain. To start the automobile moving (accelerating) requires a force sufficient to overcome its inertia (resistance to change). However, once this vehicle is moving, a force equal to that of the frictional force between the tires and the pavement is enough to keep it moving.

Moving objects often deviate from straight paths or come to rest, whereas objects at rest begin to move. The changes in motion we observe in daily life are the result of one or more applied forces.

Newton's second law of motion describes the relationship between the forces that are exerted on objects and the observed accelerations that result. Newton's second law states that *the acceleration of an object is directly proportional to the net force acting on that body and inversely proportional to the mass of the body*. The first part of Newton's second law means that the acceleration of an object changes as the intensity of the applied force changes.

We define acceleration as the rate of change in velocity. Because velocity describes both the speed and direction of a moving body, the velocity of something can be changed by changing its speed or its direction or both. Further, the term *acceleration* refers both to decreases and increases in velocity.

For example, we know that when we push down on the gas pedal of an automobile, we experience a positive acceleration (increase in velocity). On the other hand, using the brakes retards acceleration (decreases velocity).

In the atmosphere, three forces are responsible for changing the state of motion of winds. These are the pressure-gradient force, the Coriolis force, and friction. From the preceding discussion, it should be clear that the relative strengths of these forces will determine to a large degree the role of each in establishing the flow of air. Further, these forces can be directed in such a way as to increase the speed of airflow, decrease the speed of airflow, or, in many instances, just change the direction of airflow.

(the arrow lengths are different), as they are in geostrophic flow. This imbalance provides the change in direction (centripetal acceleration) that generates curved flow.

Despite the importance of centripetal acceleration in establishing curved flow aloft, near the surface, friction comes into play and greatly overshadows this much weaker force. Consequently, except for rapidly rotating storms such as tornadoes and hurricanes, the effect of centripetal acceleration is negligible and therefore is not considered in the discussion of surface circulation.

Surface Winds

Friction as a factor affecting wind is important only within the first few kilometers of Earth's surface. We know that friction acts to slow the movement of air. By slowing air movement, friction also reduces the Coriolis force, which is proportional to wind speed. Because the pressure-gradient force is not affected by wind speed, it wins the tug of war against the Coriolis force and wind direction changes (Figure 6–16). The result is the movement of air at an angle across the isobars, toward the area of lower pressure.

The roughness of the terrain determines the angle at which the air will flow across the isobars as well as influence the speed at which it will move. Over the relatively smooth ocean surface, where friction is low, air moves at an angle of 10° to 20° to the isobars and at speeds roughly two-thirds

of geostrophic flow. Over rugged terrain, where friction is high, the angle can be as great as 45° from the isobars, with wind speeds reduced by as much as 50 percent.

Near the surface, friction plays a major role in redistributing air within the atmosphere by changing the direction of airflow. This is especially noticeable when considering the motion around surface cyclones and anticyclones, two of the most common features on surface weather maps.

We have learned that above the friction layer in the Northern Hemisphere, winds blow counterclockwise around a cyclone and clockwise around an anticyclone, with winds nearly parallel to the isobars. When we add the effect of friction, we notice that the airflow crosses the isobars at varying angles, depending on the roughness of the terrain, but always from higher to lower pressure. In a cyclone, in which pressure decreases inward, friction causes a net flow *toward* its center (Figure 6–17). In an anticyclone just the opposite is true: The pressure decreases outward, and friction causes a net flow *away* from the center. Therefore, the resultant winds blow into and counterclockwise about a surface cyclone (Figure 6–18), and outward and clockwise about a surface anticyclone. Of course, in the Southern Hemisphere the Coriolis force deflects the winds to the left and reverses the direction of flow.

In whatever hemisphere, however, friction causes a net inflow (**convergence**) around a cyclone and a net outflow (**divergence**) around an anticyclone. This very important relationship between cyclonic flow and convergence and anticyclonic flow and divergence will be considered again.

How Winds Generate Vertical Air Motion

So far we have discussed wind without regard to how airflow in one region might affect airflow elsewhere. As one researcher put it, a butterfly flapping its wings in South America can generate a tornado in the United States. Although this is an exaggeration, it does illustrate how airflow in one region might cause a change in weather at some later time and at a different location.

Of particular importance is the question of how horizontal airflow (winds) relates to vertical flow. Although vertical transport is small compared to horizontal motion, it is very important as a weather maker. Rising air is associated with cloudy conditions and precipitation, whereas subsidence produces adiabatic heating and clearing conditions. In this section we will discern how the movement of air (dynamic effect) can itself create pressure change and hence generate winds. In doing so, we will examine the interrelationship between horizontal and vertical flow and its effect on the weather.

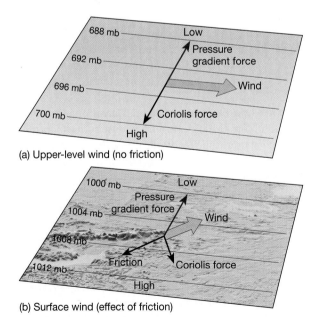

(a) Upper-level wind (no friction)

(b) Surface wind (effect of friction)

Figure 6–16 Comparison between upper-level winds and surface winds showing the effects of friction on airflow. Friction slows surface wind speed, which weakens the Coriolis force, causing the winds to cross the isobars.

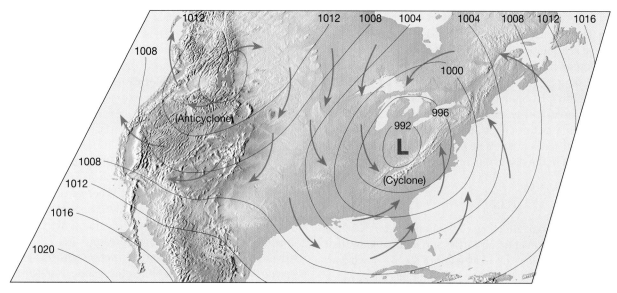

Figure 6–17 Cyclonic and anticyclonic winds in the Northern Hemisphere. Arrows show the winds blowing inward and counterclockwise around a low, and outward and clockwise around a high.

Figure 6–18 Color infrared satellite image of Hurricane Fran over the Caribbean Sea, September, 1996 shows the storm's cyclonic flow. By examining the cloud pattern, you can see the inward and counterclockwise circulation. *(Image by NASA/Science Photo Library/Photo Researchers, Inc.)*

Vertical Airflow Associated with Cyclones and Anticyclones

Let us first consider the situation around a surface low-pressure system (cyclone) in which the air is spiraling inward (see Figure 6–17). Here the net inward transport of air causes a shrinking of the area occupied by the air mass, a process called *horizontal convergence* (Figure 6–19). Whenever air converges horizontally, it must pile up—that is, it must increase in height to allow for the decreased area it now occupies. This process generates a "taller" and therefore heavier air column, yet a surface low can exist only as long as the column of air above remains light. We seem to have encountered a paradox—low-pressure centers cause a net accumulation of air, which increases their pressure. Consequently, a surface cyclone should quickly eradicate itself in a manner not unlike what happens to the vacuum in a coffee can when it is opened.

You can see that for a surface low to exist for very long, compensation must occur aloft. For example, surface convergence could be maintained if *divergence* (spreading out) aloft occurred at a rate equal to the inflow below. Figure 6–19 diagrammatically shows the relationship between surface convergence (inflow) and the divergence aloft (outflow) that is needed to maintain a low-pressure center.

Divergence aloft may even exceed surface convergence, thereby resulting in intensified surface inflow and accelerated vertical motion. Thus, divergence aloft can intensify storm centers as well as maintain them. On the

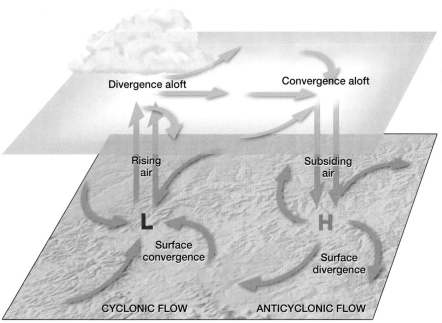

Figure 6–19 Airflow associated with surface cyclones and anticyclones. A low, or cyclone, has converging surface winds and rising air causing cloudy conditions. A high, or anticyclone, has diverging surface winds and descending air, which leads to clear skies and fair weather.

other hand, inadequate divergence aloft permits surface flow to "fill" and weaken the accompanying cyclone.

Note that surface convergence about a cyclone causes a *net upward movement*. The rate of this vertical movement is slow, generally less than 1 kilometer per day. (Recall that updrafts in thunderstorms sometimes exceed 100 kilometers per hour.) Nevertheless, because rising air often results in cloud formation and precipitation, the passage of a low-pressure center generally is related to unstable conditions and stormy weather.

As often as not, it is divergence aloft that creates a surface low. Spreading out aloft initiates upflow in the atmosphere directly below, eventually working its way to the surface, where inflow is encouraged.

Like their cyclonic counterparts, anticyclones must also be maintained from above. Outflow near the surface is accompanied by convergence aloft and general subsidence of the air column (Figure 6–19). Because descending air is compressed and warmed, cloud formation and precipitation are unlikely in an anticyclone. Thus, fair weather can usually be expected with the approach of a high-pressure system.

For reasons that should now be obvious, it has been common practice to write the word "stormy" at the low end of household barometers and "fair" at the high end (see Figure 6–5). By noting the pressure trend—rising, falling, or steady—we have a good indication of forthcoming weather. Such a determination, called the **pressure tendency** or **barometric tendency**, is useful in

short-range weather prediction. The generalizations relating cyclones and anticyclones to weather conditions are stated nicely in this verse ("glass" refers to the barometer):

> When the glass falls low,
> Prepare for a blow;
> When it rises high,
> Let all your kites fly.

In conclusion, you should now be better able to understand why local television weather broadcasters emphasize the positions and projected paths of cyclones and anticyclones. The "villain" on these weather programs is always the cyclone, which produces "bad" weather in any season. Lows move in roughly a west-to-east direction across the United States and require from a few days to more than a week for the journey. Because their paths can be erratic, accurate prediction of their migration is difficult and yet essential for short-range forecasting. Meteorologists must also determine if the flow aloft will intensify an embryo storm or act to suppress its development.

Factors That Promote Vertical Airflow

Because of the close tie between vertical motion in the atmosphere and our daily weather, we will consider some other factors that contribute to surface convergence (uplifting) and surface divergence (subsidence).

Friction can cause convergence and divergence in several ways. When air moves from the relatively smooth ocean surface onto land, for instance, the increased friction causes an abrupt drop in wind speed. This reduction of wind speed downstream results in a pile-up of air upstream. Thus, converging winds and ascending air accompany flow off the ocean. This effect contributes to the cloudy conditions over land often associated with a sea breeze in a humid region like Florida. Conversely, when air moves from land onto the ocean, general divergence and subsidence accompany the seaward flow of air because of lower friction and increasing wind speed over the water. (It should be noted that if cool air moves over a comparatively warm water body, heating from below tends to destabilize the air.)

Mountains also hinder the flow of air and cause divergence and convergence. As air passes over a mountain range, it is compressed vertically, which produces horizontal spreading (divergence) aloft. On reaching the lee side of the mountain, the air experiences vertical expansion, which causes horizontal convergence. This effect greatly influences the weather in the United States east of the Rocky Mountains, as we shall examine later. When air flows equatorward, where the Coriolis force is weakened, divergence and subsidence prevail; during poleward migration, convergence and slow uplift are favored.

As a result of the close tie between surface conditions and those aloft, great emphasis has been placed on understanding total atmospheric circulation, especially in the midlatitudes. Once we have examined the workings of global atmospheric circulation in the next chapter, we will again consider the close tie between horizontal airflow and vertical motion in light of this information.

Wind Measurement

Two basic wind measurements—direction and speed—are important to the weather observer. Winds are always labeled by the direction *from* which they blow. A north wind blows from the north toward the south; an east wind blows from the east toward the west. One instrument that is commonly used to determine wind direction is the **wind vane** (Figure 6–20). This instrument, which is a common sight on many buildings, always points into the wind. Sometimes the wind direction is shown on a dial that is connected to the wind vane. The dial indicates the direction of the wind either by points of the compass—that is, N, NE, E, SE, and so on—or by a scale of 0 to 360°. On the latter scale 0° (or 360°) is north, 90° is east, 180° is south, and 270° is west.

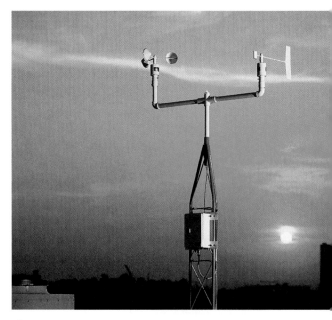

Figure 6–20 Wind vane and cup anemometer. The wind vane shows wind direction and the anemometer measures wind speed. *(Photo by Belfort Instrument Company)*

When the wind consistently blows more often from one direction than from any other, it is called a **prevailing wind**. You may be familiar with the prevailing westerlies that dominate the circulation in the midlatitudes. In the United States, for example, these winds consistently move the "weather" from west to east across the continent. Embedded within this general eastward flow are cells of high and low pressure with their characteristic clockwise and counterclockwise flow. As a result, the winds associated with the westerlies, as measured at the surface, often vary considerably from day to day and from place to place. By contrast, the direction of airflow associated with the belt of trade winds is much more consistent, as can be seen in Figure 6–21.

A *wind rose* provides a method of representing prevailing winds by indicating the percentage of time the wind blows from various directions (Figure 6–21). The length of the lines on the wind rose indicates the percentage of time the wind blew from that direction. Knowledge of the wind patterns for a particular area can be useful. For example, during the construction of an airport, the runways are aligned with the prevailing wind to assist in takeoffs and landings. Furthermore, prevailing winds greatly affect the weather and climate of a region. North–south trending mountain ranges, such as the Cascade Range of the Pacific Northwest, for example, causes the ascent of the prevailing westerlies. Thus, the windward (west) slopes of these ranges are rainy, whereas the leeward (east) sides are dry.

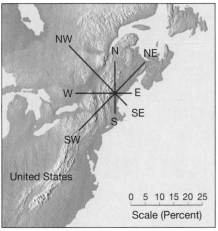

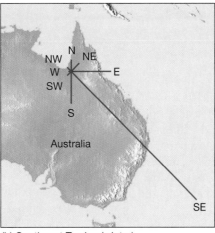

(a) Westerlies (winter)   (b) Southeast Trades (winter)

Figure 6–21 Wind roses showing the percentage of time airflow is from various directions. (a) Wind frequency for the winter in the eastern United States. (b) Wind frequency for the winter in northern Australia. Note the reliability of the southeast trades in Australia as compared to the westerlies in the eastern United States. *(Data from G. T. Trewartha)*

Wind speed is often measured with a **cup anemometer** (Figure 6–20). The wind speed is read from a dial much like the speedometer of an automobile. Sometimes an **aerovane** is used instead of a wind vane and cup anemometer. As can be seen in Figure 6–22, this instrument resembles a wind vane with a propeller at one end. The fin keeps the propeller facing into the wind, allowing the blades to rotate at a rate that is proportional to the wind speed. This instrument is commonly attached to a recorder to keep a continuous record of wind speed and direction. Places where winds are steady and speeds are relatively high are potential sites for tapping wind energy (see Box 6–5).

At small airstrips *wind socks* are frequently used. They consist of a cone-shaped bag that is open at both ends and free to change position with shifts in wind direction. The degree to which the sock is inflated is an indication of the strength of the wind.

If you are aware of the locations and intensities of low-pressure centers and high-pressure centers, you can predict the changes in wind speed and direction that will be experienced as the pressure centers move past. Changes in wind direction often bring changes in temperature and moisture conditions; therefore, the ability to predict the winds can be useful. In the Midwest, for example, a north wind may bring cool, dry air from Canada, whereas a south wind may bring warm, humid air from the Gulf of Mexico. Sir Francis Bacon summed it up nicely when he wrote, "Every wind has its weather."

Figure 6–22 An aerovane. *(Photo by Warren Faidley/Weatherstock)*

Box 6–5 Wind Energy: An Alternative with Potential

Approximately 0.25 percent (one-quarter of 1 percent) of the solar energy that reaches the lower atmosphere is transformed into wind. Although it is just a minuscule percentage, the absolute amount of energy is enormous. According to one estimate, if the winds of North and South Dakota could be harnessed, they would provide 80 percent of the electrical energy used in the United States.

Wind has been used for centuries as an almost free and nonpolluting source of energy. Sailing ships and wind-powered grist mills represent two of the early ways that this renewable resource was harnessed. Further, as rural America was settled, there was a strong reliance on wind power to pump water and later to generate small amounts of electricity.

Following the "energy crisis" that was brought about by an oil embargo in the 1970s, interest in wind power as well as other alternative forms of energy increased dramatically. In 1980 the federal government initiated a program to develop wind-power systems. A project sponsored by the U.S. Department of Energy involved setting up experimental wind farms in mountain passes known to have strong persistent winds. One of these facilities, located at Altamont Pass near San Francisco, now employs more than 7000 wind turbines. In 1995 wind supplied about 1 percent of California's electricity (Figure 6–B).

In September 1998, an array of 143 20-story-high wind turbines went on line near Lake Benton in southwestern Minnesota. The 107-megawatt project is the largest single wind farm in the world. When the wind is blowing, the Lake Benton farm can power 40,000 homes. When the project is completed in 2002, the facility's output will be 425 megawatts, nearly four times the 1998 output.

As technology has improved, efficiency has increased and the costs of wind-generated electricity have become more competitive. Between 1983 and 1999 technological advances cut the cost of wind power by about 85 percent. Some estimate that in the next 50 to 60 years wind power could meet between 5 and 10 percent of the country's demand for electrical energy. One area for the expansion of wind energy will likely be islands and other regions far from electrical grids that must import fuel for generating power.

Although the future for wind power appears promising, it is not without difficulties. In addition to technical advances that must continue to be made, noise pollution and the costs of large tracts of land in populated areas present significant obstacles to development. In the Netherlands, where windmills have a long history, proposals to locate wind turbines atop the country's dike system have been met with strong opposition. Part of the problem is that the Netherlands, like much of

Chapter Summary

- *Air pressure* is the pressure exerted by the weight of air above. Average air pressure at sea level is about 1 kilogram per square centimeter, or 14.7 pounds per square inch. Another way to define air pressure is that it is the force exerted against a surface by the continuous collision of gas molecules.

- The *newton* is the unit of force used by meteorologists to measure atmospheric pressure. A *millibar* (mb) equals 100 newtons per square meter. Standard sea-level pressure is 1013.25 millibars. Two instruments used to measure atmospheric pressure are the *mercury barometer*, where the height of a mercury column provides a measure of air pressure (standard

atmospheric pressure at sea level equals 29.92 inches or 760 millimeters), and the *aneroid barometer*, which uses a partially evacuated metal chamber that changes shape as air pressure changes.

- In still air, the two factors that largely determine the amount of pressure that a particular gas brings to bear are temperature and density. The relationship that exists among these variables, called the *ideal gas law*, can be shown by the expression: pressure = temperature × density × constant. A change in any one of the variables will cause a predictable change in the others.

- The pressure at any given altitude is equal to the weight of the air above that point. Furthermore, the

Figure 6–B These wind turbines are operating near Palm Springs, California. *(Photo by John Mead/Science Photo Library/Photo Researchers, Inc.)*

Europe, is densely populated and, hence, has few remote locations.

Another significant limitation of wind energy is that it is intermittent. If wind were to constitute a high proportion of the total energy supply, any shortfall could cause severe economic disruptions. Better

means of storage would allow wind to contribute a significantly higher percentage of our energy needs. One proposal would use wind energy to produce hydrogen by electrolysis of water. This combustible gas would then be distributed and stored in a manner similar to natural gas.

rate at which pressure decreases with an increase in altitude is much greater near Earth's surface. The "normal" decrease in pressure experienced with increased altitude is provided by the *standard atmosphere*, which depicts the idealized vertical distribution of atmospheric pressure.

- *Wind* is the result of horizontal differences in air pressure. If Earth did not rotate and there were no friction, air would flow directly from areas of higher pressure to areas of lower pressure. However, because both factors exist, wind is controlled by a combination of (1) the *pressure-gradient force*, (2) the *Coriolis force*, and (3) *friction*. The pressure-gradient force is the primary driving force of wind that results from pressure differences that occur over a given distance, as

depicted by the spacing of *isobars*, lines drawn on maps that connect places of equal air pressure. The spacing of isobars indicates the amount of pressure change occurring over a given distance, expressed as the *pressure gradient*. Closely spaced isobars indicate a steep pressure gradient and strong winds; widely spaced isobars indicate a weak pressure gradient and light winds. There is also an upward directed, vertical pressure gradient which is usually balanced by gravity in what is referred to as *hydrostatic equilibrium*. On those occasions when the gravitational force slightly exceeds the vertical pressure gradient force, slow downward airflow results. The Coriolis force produces a deviation in the path of wind due to Earth's rotation (to the right in the Northern Hemisphere and to the left in the

Southern Hemisphere). The amount of deflection is greatest at the poles and decreases to zero at the equator. The amount of Coriolis deflection also increases with wind speed. Friction, which significantly influences airflow near Earth's surface, is negligible above a height of a few kilometers.

- Above a height of a few kilometers, the effect of friction on airflow is small enough to disregard. Here, as the wind speed increases, the deflection caused by the Coriolis force also increases. Winds in which the Coriolis force is equal to and opposite the pressure gradient force are called *geostrophic winds*. Geostrophic winds flow in a straight path, parallel to the isobars, with velocities proportional to the pressure-gradient force.

- Winds that blow at a constant speed parallel to curved isobars are termed *gradient winds*. In centers of low pressure, called *cyclones*, the circulation of air, referred to as *cyclonic flow*, is counterclockwise in the Northern Hemisphere and clockwise in the Southern Hemisphere. Centers of high pressure, called *anticyclones*, exhibit *anticyclonic flow* which is clockwise in the Northern Hemisphere and counterclockwise in the Southern Hemisphere. Whenever isobars curve to form elongated regions of low and high pressure, these areas are called *troughs* and *ridges*, respectively.

- Near the surface, friction plays a major role in redistributing air within the atmosphere by changing the direction of airflow. The result is a movement of air at an angle across the isobars, toward the area of lower pressure. Therefore, the resultant winds blow into and counterclockwise about a Northern Hemisphere surface cyclone. In a Northern Hemisphere surface anticyclone, winds blow outward and clockwise. Regardless of the hemisphere, friction causes a net inflow (*convergence*) around a cyclone and a net outflow (*divergence*) around an anticyclone.

- A surface low-pressure system with its associated horizontal convergence is maintained or intensified by divergence (spreading out) aloft. Inadequate divergence aloft will weaken the accompanying cyclone. Because surface convergence about a cyclone accompanied by divergence aloft causes a net upward movement of air, the passage of a low pressure center is often associated with stormy weather. By contrast, fair weather can usually be expected with the approach of a high-pressure system. As the result of the general weather patterns usually associated with cyclones and anticyclones, the *pressure tendency* or *barometric tendency* (the nature of the change of the barometer over the past several hours) is useful in short-range weather prediction.

- Two basic wind measurements—direction and speed—are important to the weather observer. Wind direction is commonly determined using a *wind vane*. When the wind consistently blows more often from one direction than from any other, it is called a *prevailing wind*. Wind speed is often measured with a *cup anemometer*.

Vocabulary Review

aerovane (p. 179)
air pressure (p. 157)
aneroid barometer (p. 159)
anticyclone (p. 173)
anticyclonic flow (p. 173)
barograph (p. 159)
barometric tendency (p. 177)
Buys Ballott's law (p. 172)
convergence (p. 175)
Coriolis force (p. 168)
cup anemometer (p. 179)
cyclone (p. 173)
cyclonic flow (p. 173)
divergence (p. 175)
geostrophic wind (p. 171)

gradient wind (p. 173)
hydrostatic equilibrium (p. 166)
ideal gas law (p. 161)
isobar (p. 164)
mercury barometer (p. 158)
millibar (p. 158)
newton (p. 158)
pressure gradient (p. 164)
pressure tendency (p. 177)
prevailing wind (p. 178)
ridge (p. 173)
standard atmosphere (p. 163)
trough (p. 173)
wind (p. 163)
wind vane (p. 178)

Review Questions

1. What is standard sea-level pressure in millibars? In inches of mercury? In pounds per square inch?

2. Describe the operating principles of the mercury barometer and the aneroid barometer. List two advantages of the aneroid barometer.

3. When density remains constant and the temperature is raised, how will the pressure of a gas change?

4. When gases in the atmosphere are heated, air pressure normally decreases. Based on your answer to the first question, explain this apparent paradox.

5. Explain why air pressure decreases with an increase in altitude.

6. What force is responsible for *generating* wind?

7. Write a generalization relating the spacing of isobars to the speed of wind.

8. Temperature variations create pressure differences, which in turn produce winds. On a small scale, the sea breeze illustrates this principle nicely. Describe how a sea breeze forms.

9. Although vertical pressure differences may be great, such variations do not generate strong vertical currents. Explain.

10. Briefly describe how the Coriolis force modifies the movement of air.

11. Which two factors influence the magnitude of the Coriolis force?

12. Explain the formation of a geostrophic wind.

13. Unlike winds aloft, which blow nearly parallel to the isobars, surface winds generally cross the isobars. Explain what causes this difference.

14. Prepare a diagram (isobars and wind arrows) showing the winds associated with surface cyclones and anticyclones in both the Northern and Southern Hemispheres.

15. For surface low pressure to exist for an extended period, what condition must exist aloft?

16. What are the general weather conditions to be expected when the pressure tendency is rising? When the pressure tendency is falling?

17. Converging winds and ascending air are often associated with the flow of air from the oceans onto land. Conversely, divergence and subsidence often accompany the flow of air from land to sea. What causes this convergence over land and divergence over the ocean?

18. A southwest wind blows from the _____ (direction) toward the _____ (direction).

19. The wind direction is 315°. From what compass direction is the wind blowing?

Problems

1. Calculate the magnitude of the pressure-gradient force (per unit mass) between two cities 500 kilometers apart if pressure at the respective cities is 1010 mb and 1017 mb?

2. Calculate the magnitude of Coriolis force acting on air moving at:

   a. 36 kilometers per hour (10 meters per second) at 35° latitude.

   b. 36 kilometers per hour (10 meters per second) at 65° latitude.

   c. 54 kilometers per hour (15 meters per second) at 35° latitude.

Atmospheric Science Online

The following are informative and interesting Internet sites that address topics related to those presented in the chapter:

Current U.S. Pressures and Winds (Penn State University):
• **http://www.ems.psu.edu/wx/usstats/pressstats.html**

Forces and Winds: Online Meteorology Guide (University of Illinois):
• **http://ww2010.atmos.uiuc.edu/(Gh)/guides/mtr/fw/home.rxml**

For direct links to these sites and others, chapter objectives and reviews, quiz questions, and topical investigations that utlize Web resources, visit *The Atmosphere, Eighth Edition* Home Page at:
• **http://www.prenhall.com/lutgens**

Circulation of the Atmosphere

Windsurfers at San Simeon, California. *(Photo by George D. Lepp/Photo Researchers, Inc.)*

The main goal of this chapter is to gain an understanding of Earth's highly integrated wind system. Global atmospheric circulation can be thought of as a series of deep rivers of air that encircle the planet. Embedded in the main currents are vortices of various sizes including hurricanes, tornadoes, and midlatitude cyclones. Like eddies in a stream, these rotating wind systems develop and die out with somewhat predictable regularity. In general, the smallest eddies, such as dust devils, last only a few minutes, whereas larger and more complex systems such as hurricanes may survive for several days.

Recall from Chapter 6 that winds are generated by pressure differences that arise because of unequal heating of Earth's surface. Global winds are generated because the tropics receive more solar radiation than Earth's polar regions. Thus, Earth's winds blow in an unending attempt to balance inequalities in surface temperatures. Because the zone of maximum solar heating migrates with the seasons—moving northward during the Northern Hemisphere summer and southward as winter approaches—the wind patterns that make up the general circulation also migrate latitudinally.

Although we will focus on the global circulation, local wind systems will also be considered. The chapter concludes with a discussion of global precipitation patterns. As you will see, the global distribution of precipitation is closely linked to the patterns of atmospheric pressure, and thus, the global wind system.

Scales of Atmospheric Motion

Those who live in the United States are familiar with the term "westerlies" to describe the winds that predominately blow from west to east. But all of us have experienced winds from the south and north and even directly from the east. You may even recall being in a storm when shifts in wind direction and speed came in such rapid succession that it was impossible to determine the wind's direction. With such variations, how can we describe our winds as westerly? The answer lies in our attempt to simplify descriptions of the atmospheric circulation by sorting out events according to *size*. On the scale of a weather map, for instance, where observing stations are spaced about 150 kilometers apart, small whirlwinds that carry dust skyward are far too small to show up. Instead, weather maps reveal larger-scale wind patterns, such as those associated with traveling cyclones and anticyclones.

Not only do we separate winds according to the size of the system, but equal consideration is also given to the time frame in which they occur. In general, large weather patterns have longer life spans than do their smaller counterparts. For example, dust devils usually last a few minutes and rarely occur for more than an hour. By contrast, midlatitude cyclones typically take a few days to cross the United States and occasionally dominate the weather for a week or longer. The time and size scales we will use for atmospheric motions are provided in Table 7–1.

Large- and Small-Scale Circulation

The wind systems shown in Figure 7–1 illustrate the three major categories of atmospheric circulation: macroscale, mesoscale and microscale. *Macroscale* circulation includes large planetary-scale flow, such as the trade winds that blow consistently for weeks or longer as well as smaller features like hurricanes (Figure 7–1a). *Mesoscale* circulation is associated with tornadoes, thunderstorms, and numerous *local winds* that generally last from minutes to hours. Finally, *microscale* events have life spans of from a few seconds to minutes and include dust devils, gusts, and general atmospheric turbulence.

Macroscale Winds. The largest wind patterns, called **macroscale winds**, are exemplified by the westerlies and trade winds that carried sailing vessels back and forth across the Atlantic during the opening of the New World. These *planetary-scale* flow patterns extend around the entire globe and can remain essentially unchanged for weeks at a time.

Table 7–1 Time and space scales for atmospheric motions			
Scale	**Time Scale**	**Distance Scale**	**Examples**
Macroscale			
Planetary	Weeks or longer	1000–40,000 km	Westerlies and trade winds
Synoptic	Days to weeks	100–5000 km	Midlatitude cyclones, anticyclones, and hurricanes
Mesoscale	Minutes to hours	1–100 km	Thunderstorms, tornadoes, and land–sea breeze
Microscale	Seconds to minutes	<1 km	Turbulence, dust devils, and gusts

+ all local winds

(a)

Figure 7–1 Three scales of atmospheric motion. (a) Satellite image of Hurricane Nora, an example of macroscale circulation. (b) Tornadoes exemplfy mesoscale wind systems. (c) Gusts illustrate microscale winds. *(Photos by (a) NASA/Science Photo Library/Photo Researchers, Inc., (b) A. and J. Verkaik/The Stock Market, (c) E.J. Tarbuck)*

(b)

(c)

A somewhat smaller macroscale circulation is called *synoptic scale*, or *weather-map scale*. Two well-known synoptic scale systems are the individual traveling cyclones and anticyclones that appear on weather maps as areas of low and high pressure, respectively. These weather producers are found in the middle latitudes where they move from west-to-east as part of the larger westerly flow. Furthermore, these rotating systems usually persist for days or occasionally weeks and have a horizontal dimension of hundreds to thousands of kilometers.

Recall that the direction of surface flow around a cyclone is toward the center with a general upward component. Anticyclones, in contrast, are areas of subsidence associated with outward flow near the surface. The average rate of vertical motion within these systems is slow, typically less than 1 kilometer per day.

Somewhat smaller macroscale systems are the tropical cyclones and hurricanes that develop in late summer and early fall over the warm tropical oceans. Airflow in these systems is inward and upward as in the larger midlatitude cyclones. However, the rate of horizontal flow associated with hurricanes is usually more rapid than that of their more poleward cousins.

Mesoscale Winds. **Mesoscale winds** generally last for several minutes and may exist for hours. These middle-size phenomena are usually less than 100 kilometers (62 miles) across. Further, some mesoscale winds—for example, thunderstorms and tornadoes—also have a strong vertical component (Figure 7–1b). It is important to remember that thunderstorms and tornadoes are always imbedded within, and thus move as part of, the larger macroscale circulation. Further, much of the vertical air movement within a midlatitude cyclone, or hurricane, is provided by thunderstorms, where updrafts in excess of 100 kilometers (62 miles) per hour have been measured. Land and sea breezes, as well as mountain and valley winds, also fall into this category and will be discussed in the next section along with other mesoscale winds.

Microscale Winds. The smallest scale of air motion is referred to as **microscale circulation**. These small, often chaotic winds normally last for seconds or at most minutes. Examples include simple gusts, which hurl debris into the air (Figure 7–1c) and small, well-developed vortices such as dust devils (see Box 7–1).

Structure of Wind Patterns

Although it is common practice to divide atmospheric motions according to scale, remember that global winds are a composite of all scales of motion—much like a meandering river that contains large eddies composed of smaller eddies containing still smaller eddies. For example, let us examine the flow associated with hurricanes that form over the North Atlantic. When we view one of these tropical cyclones on a satellite image, the storm appears as a large whirling cloud migrating slowly across the ocean (see Figure 7–1a). From this perspective, which is at the weather-map (synoptic) scale, the general counterclockwise rotation of the storm can be easily seen.

When we average the winds of hurricanes, we find that they often have a net motion from east to west, thereby indicating that these larger eddies are embedded in a still larger flow (planetary scale) that is moving westward across the tropical portion of the North Atlantic.

If we examine a hurricane more closely by flying an airplane through it, some of the small-scale aspects of the storm become noticeable. As the plane approaches the outer edge of the system, it becomes evident that the large rotating cloud that we saw in the satellite images is made of many individual cumulonimbus towers (thunderstorms).

Each of these mesoscale phenomena lasts for only a few hours and must be continually replaced by new ones if the hurricane is to persist. As we fly into these storms, we quickly realize that the individual clouds are made up of even smaller-scale turbulences. The small thermals of rising air that occur in these clouds make for a rather rough trip.

Thus, a typical hurricane exhibits several scales of motion, including many mesoscale thunderstorms, which, in turn, consist of numerous microscale turbulences. Furthermore, the counterclockwise circulation of the hurricane (weather-map scale) is imbedded in the global winds (planetary scale) that flow from east to west in the tropical North Atlantic.

Local Winds

Before examining the large macroscale circulation for Earth, let us turn to some mesoscale winds (time frame of minutes to hours and size of 1 to 100 kilometers—Table 7–1). Remember that all winds have the same cause: pressure differences that arise because of temperature differences caused by unequal heating of Earth's surface. Local winds are medium-scale winds produced by a locally generated pressure gradient.

Although many winds are given local names, some are actually part of the global wind system. The "norther" of Texas, for instance, is a cold southward flow produced by the circulation around anticyclones that invade the United States from Canada in the winter. Because these winds are not locally generated, they cannot be considered true local winds. Others, like those about to be described, are truly mesoscale and are caused either by topographic effects or variations in local surface composition.

Recall that winds are named for the direction *from which they blow*. This holds true for local winds. Thus, a sea breeze originates over water and blows toward the land, whereas a valley breeze blows upslope away from its source.

Land and Sea Breezes

The daily temperature contrast between the land and the sea, and the pressure pattern that generates a sea breeze, was discussed in the preceding chapter (see Figure 6–10). Recall that land is heated more intensely during daylight hours than is an adjacent body of water. As a result, the air above the land surface heats and expands, creating an area of low pressure. A **sea breeze** then develops, as cooler air over the water moves onto the land (Figure 7–2a). At night, the reverse may take place; the land cools more rapidly than the sea and a **land breeze** develops (Figure 7–2b).

Box 7–1 Dust Devils

A common phenomenon in arid regions of the world is the whirling vortex called the *dust devil* (Figure 7–A). Although they resemble tornadoes, dust devils are generally much smaller and less intense than their destructive cousins. Most dust devils are only a few meters in diameter and reach heights no greater than about 100 meters (300 feet). Further, these whirlwinds are usually short-lived microscale phenomena. Most form and die out within minutes. In rare instances, dust devils have lasted for hours.

Unlike tornadoes, which are associated with convective clouds, dust devils form on days when clear skies dominate. Further, these whirlwinds form from the ground upward, exactly opposite of tornadoes. Because surface heating is critical to their formation, dust devils occur most frequently in the afternoon when surface temperatures are highest.

Recall that when the air near the surface is considerably warmer than the air a few dozen meters overhead, the layer of air near Earth's surface becomes unstable. In this situation, warm surface air begins to rise, causing air near the ground to be drawn into the developing whirlwind. The rotating winds that are associated with dust devils are produced by the same phenomenon that causes ice skaters to spin faster as they pull their arms closer to their body. (For an expanded discussion of this process, see Box 11–1, Chapter 11.) As the inwardly spiraling air rises, it carries sand, dust, and other loose debris dozens of meters into the air. It is this material that makes a dust devil visible. Occasionally, dust devils form above vegetated surfaces.

Figure 7–A Dust devil. Although these whirling vortices resemble tornadoes, they have different origins and are much smaller and less intense. *(Courtesy of St. Meyers/Okapia/Photo Researchers, Inc.)*

Under these conditions, the vortices may go undetected unless they interact with objects at the surface.

Most dust devils are small and short-lived; consequently, they are not generally destructive. Occasionally, however, these whirlwinds grow to be 100 meters or more in diameter and over a kilometer high. With wind speeds that may reach 100 kilometers (60 miles) per hour, large dust devils can do considerable damage. Fortunately, such occurrences are few and far between.

The sea breeze has a significant moderating influence on coastal temperatures. Shortly after the breeze begins, air temperatures over the land may drop by as much as 5° to 10°C. However, the cooling effect of these breezes generally reaches a maximum of only about 100 kilometers (60 miles) inland in the tropics and often only half that distance in the middle latitudes. These cool sea breezes generally begin shortly before noon and reach their greatest intensity, about 10 to 20 kilometers per hour, in the mid-afternoon.

Smaller-scale sea breezes can also develop along the shores of large lakes. People who live in a city near the

Great Lakes, such as Chicago, recognize the "lake effect," especially in the summer. Residents are reminded daily by reports of the cooler temperatures near the lake compared to warmer outlying areas. In many places, sea breezes also affect the amount of cloud cover and rainfall. The peninsula of Florida, for example, experiences a summer precipitation maximum caused partly by the convergence of sea breezes from both the Atlantic and Gulf coasts.

The intensity and extent of land and sea breezes depend on the location and the time of year. Tropical areas where intense solar heating is continuous throughout the

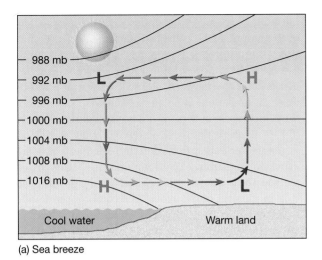

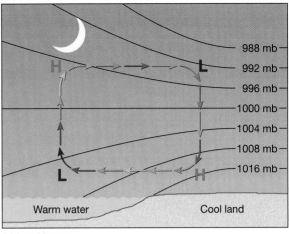

(a) Sea breeze

(b) Land breeze

Figure 7–2 Illustration of a sea breeze and a land breeze. (a) During the daylight hours the air above the land heats and expands, creating an area of lower pressure. Cooler and denser air over the water moves onto the land, generating a sea breeze. (b) At night the land cools more rapidly than the sea, generating an offshore flow called a *land breeze.*

year experience more frequent and stronger sea breezes than do midlatitude locations. The most intense sea breezes develop along tropical coastlines adjacent to cool ocean currents. In the middle latitudes, sea breezes are most common during the warmest months, but the counterpart, the land breeze, is often missing, for the land does not always cool below the ocean temperature. In the higher middle latitudes, the frequent migration of high- and low-pressure systems dominates the circulation, so land and sea breezes are less noticeable.

Mountain and Valley Breezes

A daily wind similar to land and sea breezes occurs in many mountainous regions. During the day, air along mountain slopes is heated more intensely than air at the same elevation over the valley floor (Figure 7–3a). This warmer air glides up along the mountain slope and generates a **valley breeze**. The occurrence of these daytime upslope breezes can often be identified by the isolated cumulus clouds that develop over adjacent mountain peaks (Figure 7–4). This also causes the late afternoon

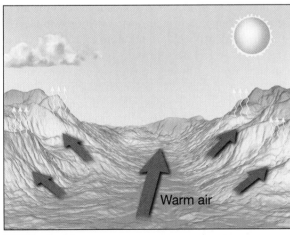

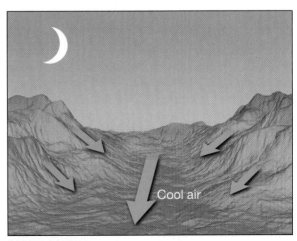

(a) Valley breeze

(b) Mountain breeze

Figure 7–3 Valley and mountain breezes. (a) Heating during the daylight hours warms the air along the mountain slopes. This warm air rises, generating a valley breeze. (b) After sunset, cooling of the air near the mountain can result in cool air drainage into the valley, producing the mountain breeze.

Figure 7–4 The occurrence of a daytime upslope (valley) breeze is identified by cloud development on mountain peaks, sometimes progressing to a mid-afternoon thunderstorm. *(Photo by James E. Patterson/James Patterson Collection)*

thundershowers so common on warm summer days in the mountains. After sunset, the pattern is reversed. Rapid radiation heat loss along the mountain slopes cools the air, which drains into the valley below and causes a **mountain breeze** (Figure 7–3b). Similar cool air drainage can occur in regions that have little slope. The result is that the coldest pockets of air are usually found in the lowest spots, a phenomenon you likely have experienced while walking in hilly terrain. Consequently, low areas are the first to experience radiation fog and are the most likely spots for frost damage to crops.

Like many other winds, mountain and valley breezes have seasonal preferences. Although valley breezes are most common during the warm season when solar heating is most intense, mountain breezes tend to be more frequent during the cold season.

Chinook (Foehn) Winds

Warm, dry winds sometimes move down the east slopes of the Rockies, where they are called **chinooks**, and the Alps, where they are called **foehns**. Such winds are often created when a pressure system on the leeward side of the mountains, such as a cyclone, pulls air over these imposing barriers. As the air descends the leeward slopes of the mountains, it is heated adiabatically (by compression). Because condensation may have occurred as the air ascended the windward side, releasing latent heat, the air descending the leeward side will be warmer and drier than at a similar elevation on the windward side. Although the temperature of these winds is generally less than 10°C (50°F), which is not particularly high, they usually occur in the winter and spring when the affected area may be experiencing subfreezing temperatures. Thus, by comparison, these dry, warm winds often bring drastic change. Within minutes of the arrival of a chinook, the temperature may climb 20°C (36°F). When the ground has a snow cover, these winds melt it in short order. The Native American word *chinook* means "snoweater."

The chinook is viewed by some as beneficial to ranchers east of the Rockies, for it keeps their grasslands clear of snow during much of the winter, but this benefit is offset by the loss of moisture that the snow would bequeath to the land if it remained until the spring melt.

Another chinooklike wind that occurs in the United States is the **Santa Ana**. Found in southern California, these hot, desiccating winds greatly increase the threat of fire in this already dry area (see Box 7–2).

Katabatic (Fall) Winds

In the winter, areas adjacent to highlands may experience a local wind called a **katabatic wind** or **fall wind**. These winds originate when cold air, situated over a highland area such as the ice sheets of Greenland or Antarctica, is set in motion. Under the influence of gravity, the cold air cascades over the rim of a highland like a waterfall. Although the air is heated adiabatically, as are chinooks, the initial temperatures are so low that the wind arrives in the lowlands still colder and more dense than the air it displaces. In fact, this air *must* be colder than the air it invades, for it is the air's greater density that causes it to descend. As this frigid air descends, it occasionally is channeled into narrow valleys, where it acquires velocities capable of great destruction.

A few of the better-known katabatic winds have local names. Most famous is the **mistral**, which blows from the French Alps toward the Mediterranean Sea. Another is the **bora**, which originates in the mountains of Yugoslavia and blows to the Adriatic Sea.

Country Breeze

One mesoscale wind, called a **country breeze**, is associated with large urban areas. As the name implies, this circulation pattern is characterized by a light wind blowing

Box 7–2 Atmospheric Hazard: Santa Ana Winds and Wildfires

People living in southern California are well aware of chinook-type winds called the *Santa Anas*. These hot, dry, dust-bearing winds invade California most often in autumn when they bring temperatures that often approach 32°C (90°F) and may occasionally exceed 38°C (100°F). When a strong anticyclone is centered over the Great Basin, the clockwise flow directs desert air from Arizona and Nevada westward toward the Pacific. The wind gains speed as it is funneled through the canyons of the Coast Ranges, in particular the Santa Ana Canyon, from which the winds derive their name. Compressional heating of this already warm, dry air as it descends mountain slopes further accentuates the already parched conditions. Vegetation, seared by the summer heat, is dried even further by these desiccating winds.

In late October 1993, Santa Anas began blowing toward the coast of southern California. They had sustained speeds of 64 kilometers (40 miles) per hour with gusts up to 126 kilometers (78 miles) per hour. With low humidities and nearly eight months without rain, the region was ripe for fires. Just a spark could set off a major blaze (Figure 7–B). Once started, these fires could move almost as fast as the ferocious Santa Ana winds sweeping through the canyons.

During the first week, about a dozen major wildfires broke out from Ventura County, just north of Los Angeles, to the Mexican border. A few of these fires, including the Thousand Oaks and Laguna Beach fires, were probably set by arsonists. At week's end, Santa Ana–fanned fires had charred 720 structures, 152,000 acres, and caused over $500 million in damages. Remarkably, no one was killed.

Firefighters received a lucky break over the weekend when the Santa Ana winds were replaced by light, cool sea breezes. However, the respite was short-lived and the Santa Anas returned the following week. When an arsonist lit a small fire in the hills near Calabasas, California, it quickly became a roaring wildfire that scorched 18,000 acres and destroyed more than 350 homes in the exclusive Malibu area. Unfortunately, this time people perished.

Strong Santa Ana winds, coupled with dry summers, have produced wildfires in southern California for a millennia. These fires are nature's way of burning out chaparral thicket and sage scrub to prepare the land for new growth. The problem began when people started to crowd into the fire-prone area between Santa Barbara and San Diego. Residents have compounded the danger by landscaping their yards with highly flammable eucalyptus and pine trees. Furthermore, successful fire prevention has allowed the buildup of even more flammable material. Clearly, wildfires will remain a major threat in these areas into the foreseeable future.

Figure 7–B Firestorms rage near exclusive homes in the hills around Malibu, California, fall, 1993. *(Photo by Louis D. Bernstein/AP/Wide World Photos)*

into the city from the surrounding countryside. The country breeze is best developed on relatively clear, calm nights. Under these conditions cities, because they contain massive buildings composed of rocklike materials, tend to retain the heat accumulated during the day more than the less built up outlying areas (see Box 3–4 on the urban heat island). The result is that the warm, less dense air over the city rises, which in turn initiates the country-to-city flow.

One investigation in Toronto showed that heat accumulated within this city created a rural/city pressure difference that was sufficient to cause an inward and counterclockwise circulation centered on the downtown area. One of the unfortunate consequences of the country breeze is that pollutants emitted near the urban perimeter tend to drift in and concentrate near the city's center.

Global Circulation

Our knowledge of global winds comes from two sources: the patterns of pressure and winds observed worldwide, and theoretical studies of fluid motion. We first consider the classical model of global circulation that was developed largely from average worldwide pressure distribution. We then add to this idealized circulation more recently discovered aspects of the atmosphere's complex motions.

Single-Cell Circulation Model

One of the first contributions to the classical model of global circulation came from George Hadley in 1735. Hadley was well aware that solar energy drives the winds. He proposed that the large temperature contrast between the poles and the equator creates one large *convection cell* in each hemisphere (Northern and Southern), as shown in Figure 7–5.

In Hadley's model, intensely heated equatorial air rises until it reaches the tropopause, where it begins to spread toward the poles. Eventually, this upper-level flow would reach the poles, where cooling would cause it to sink and spread out at the surface as equatorward-moving winds. As this cold polar air approached the equator, it would be

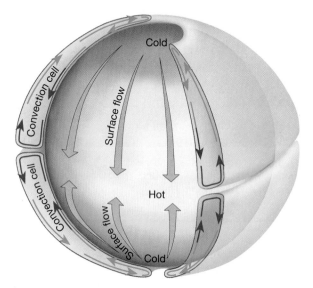

Figure 7–5 Global circulation on a nonrotating Earth. A simple convection system is produced by unequal heating of the atmosphere on a nonrotating Earth.

reheated and rise again. Thus, the circulation proposed by Hadley has upper-level air flowing poleward and surface air moving equatorward.

Although correct in principle, Hadley's model does not take into account the fact that Earth rotates on its axis. (Hadley's model would better approximate the circulation of a *nonrotating* planet.) As Earth's pressure and wind patterns became better known it was clear that the single-cell model (in each hemisphere) could not create the global circulation that was actually observed. Consequently, the Hadley model was replaced by a model that better fit observations.

Three-Cell Circulation Model

In the 1920s, a three-cell circulation model (for each hemisphere) was proposed. Although this model has been modified to fit upper-air observations, it remains a useful way to examine global circulation. Figure 7–6 illustrates the idealized three-cell model and the surface winds that result.

In the zones between the equator and roughly 30° latitude north and south, the circulation closely resembles the model used by Hadley for the whole Earth. Consequently, the name **Hadley cell** is generally applied. Near the equator, the warm rising air that releases latent heat during the formation of cumulus towers is believed to provide the energy to drive the Hadley cells. The clouds also provide the rainfall that maintains the lush vegetation of the rain forests of southeast Asia, equatorial Africa, and South America's Amazon Basin.

As the flow aloft in Hadley cells moves poleward, it begins to subside in a zone between 20° and 35° latitude. Two factors contribute to this general subsidence: (1) As upper-level flow moves away from the stormy equatorial region, where the release of latent heat of condensation keeps the air warm and buoyant, radiation cooling increases the density of the air. Satellites that monitor radiation emitted in the upper troposphere record considerable outward-emitted radiation over the tropics. (2) Because the Coriolis force becomes stronger with increasing distance from the equator, the poleward-moving upper air is deflected into a nearly west-to-east flow by the time it reaches 25° latitude. Thus, a restricted poleward flow of air ensues. Stated another way, the Coriolis force causes a general pileup of air (convergence) aloft. As a result, general subsidence occurs in the zones located between 20° and 35° latitude.

This subsiding air is relatively dry, because it has released its moisture near the equator. In addition, the effect of adiabatic heating during descent further reduces the relative humidity of the air. Consequently, this zone of

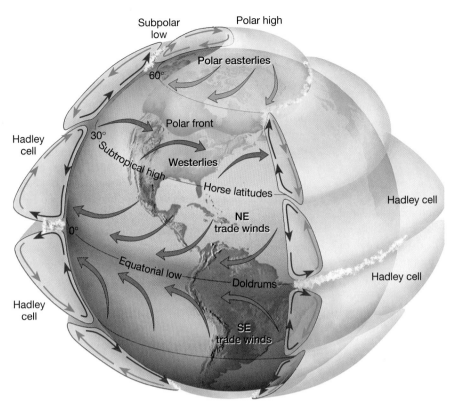

Figure 7–6 Idealized global circulation proposed for the three-cell circulation model of a rotating Earth.

subsidence is the site of the world's subtropical deserts. The Sahara Desert of North Africa and the Great Australian Desert are located in these regions of sinking air.

Because winds are generally weak and variable near the center of this zone of descending air, this region is popularly called the **horse latitudes**. The name is believed to have been coined by Spanish sailors, who, while crossing the Atlantic, were sometimes becalmed in these waters and reportedly were forced to throw horses overboard when they could no longer water or feed them.

From the center of the horse latitudes, the surface flow splits into a poleward branch and an equatorward branch. The equatorward flow is deflected by the Coriolis force to form the reliable **trade winds**, so called because they enabled early sailing ships to trade between continents. In the Northern Hemisphere, the trades blow from the northeast, where they provided the sail power for exploration of the New World in the sixteenth and seventeenth centuries. In the Southern Hemisphere, the trades are from the southeast. The trade winds from both hemispheres meet near the equator in a region that has a weak pressure gradient. This region is called the **doldrums**. Here light winds and humid conditions provide the monotonous weather that is the basis for the expression "down in the doldrums."

In the three-cell model, the circulation between 30° and 60° latitude (north and south) is more complicated than that within the Hadley cells (Figure 7–6). The net surface flow is poleward and, because of the Coriolis force, the winds have a strong westerly component. These **prevailing westerlies** were known to Benjamin Franklin, perhaps the first American weather forecaster, who noted that storms migrated from west to east across the colonies. Franklin also observed that the westerlies were much more sporadic and, therefore, less reliable than the trade winds for sail power. We now know that it is the migration of cyclones and anticyclones across the midlatitudes that disrupts the general westerly flow at the surface. Because of the importance of the midlatitude circulation in producing our daily weather, we will consider the westerlies in more detail in a later section.

Relatively little is known about the circulation in high (polar) latitudes. It is generally understood that subsidence near the poles produces a surface flow that moves equatorward and is deflected into the **polar easterlies** of both hemispheres. As these cold polar winds move equatorward, they eventually encounter the warmer westerly flow of the midlatitudes. The region where the flow of warm air clashes with cold air has been named the **polar front**. The significance of this region will be considered later.

Observed Distribution of Pressure and Winds

As you might expect, Earth's global wind patterns are associated with a distinct distribution of surface air pressure. To simplify this discussion, we will first examine the idealized pressure distribution that would be expected if Earth's surface were uniform, that is, composed of all sea or all smooth land.

Idealized Zonal Pressure Belts

If Earth had a uniform surface, two latitudinally oriented belts of high and two of low pressure would exist (Figure 7–7a). Near the equator, the warm rising branch of the Hadley cells is associated with the pressure zone known as the **equatorial low**. This region of ascending moist, hot air is marked by abundant precipitation. Because it is the region where the trade winds converge, it is also referred to as the **intertropical convergence zone (ITCZ)**. In Figure 7–8, the ITCZ is visible as a band of clouds near the equator.

In the belts about 20° to 35° on either side of the equator, where the westerlies and trade winds originate and go their separate ways, are the pressure zones known as the **subtropical highs**. These zones of high pressure are caused mainly by the Coriolis deflection, which restricts the poleward movement of the upper-level

branch of the Hadley cells. As a result, a high-level pileup of air occurs around 20° to 35° latitude. Here a subsiding air column and diverging winds at the surface result in warm and clear weather. Recall that many large deserts lie near 30° latitude, within this zone of sinking air. Generally the rate at which air accumulates in the upper troposphere exceeds the rate at which the air descends and spreads out at the surface. Thus, the subtropical highs exist throughout most of the year and are regarded as *semipermanent* features of the general circulation.

Another low-pressure region is situated at about 50° to 60° latitude, in a position corresponding to the polar front. Here the polar easterlies and westerlies clash to form a convergent zone known as the **subpolar low**. As you will see later, this zone is responsible for much of the stormy weather in the middle latitudes, particularly in the winter.

Finally, near Earth's poles are the **polar highs**, from which the polar easterlies originate (Figure 7–7a). The high-pressure centers that develop over the cold polar areas are generated by entirely different processes than those that create the subtropical highs. Recall that the high-pressure zones in the subtropics result because the rate at which air piles up aloft exceeds the rate at which it spreads out at the surface. Stated another way, more air accumulates near 30° latitude than leaves these air columns. By contrast, the polar highs exhibit high surface pressure mainly because of surface cooling. Because air near the poles is cold and dense, it exerts a higher than average pressure.

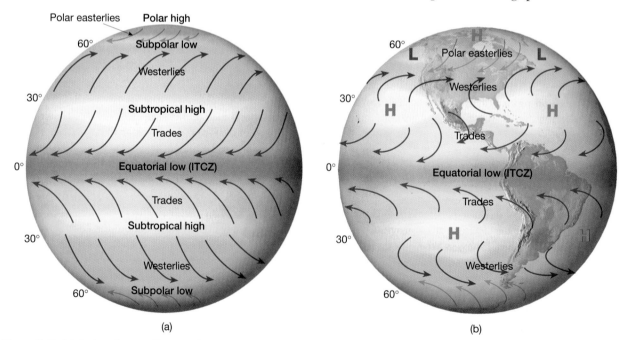

Figure 7–7 (a) An imaginary uniform Earth with idealized zonal (continuous) pressure belts. (b) The real Earth with disruptions of the zonal pattern caused by large landmasses. These disruptions break up pressure zones into semipermanent high- and low-pressure cells.

Figure 7–8 The Intertropical Convergence Zone (ITCZ) is seen as the broad band of clouds across central Africa and extending across the equatorial ocean. *(Courtesy of NASA/Goddard Space Flight Center)*

Semipermanent Pressure Systems: The Real World

Up to this point, we have considered the global pressure systems as if they were continuous belts around Earth. However, because Earth's surface is not uniform, the only true zonal distribution of pressure exists along the subpolar low in the Southern Hemisphere, where the ocean is continuous. To a lesser extent, the equatorial low is also continuous. At other latitudes, particularly in the Northern Hemisphere, where there is a higher proportion of land compared to ocean, the zonal pattern is replaced by semipermanent cells of high and low pressure.

The idealized pattern of pressure and winds for the "real" Earth is illustrated in Figure 7–7b. Although representative, the pattern shown is always in a state of flux because of seasonal temperature changes, which serve to either strengthen or weaken these pressure cells. In addition, the latitudinal position of these pressure systems moves either poleward or equatorward along with the seasonal migration of the zone of maximum solar heating. This is particularly true of the low-pressure belt associated with the intertropical convergence zone. The position of this thermally produced belt of low pressure is highly dependent on solar heating. As a consequence of these factors, Earth's pressure patterns vary in strength or location during the course of the year. A view of the average global pressure patterns and resulting winds for the months of January and July are shown in Figure 7–9. Notice on these pressure maps that, for the most part, the observed pressure regimes are cellular (or elongated) instead of zonal. The most prominent features on both maps are the subtropical highs. These systems are centered between 20° and 35° latitude over all the larger subtropical oceans.

When we compare Figures 7–9a (January) and 7–9b (July), we see that some pressure cells are more or less year-round features, like the subtropical highs. Others, however, are seasonal, such as the low over the southwestern United States in the summer, which appears on only the July map. Relatively little pressure variation occurs from midsummer to midwinter in the Southern Hemisphere, a fact we attribute to the dominance of water in that hemisphere. Numerous departures from the idealized zonal pattern are evident in the Northern Hemisphere. The main cause of these variations is the seasonal temperature fluctuations experienced over the landmasses, especially those in the middle and higher latitudes.

January Pressure and Wind Patterns. On the January pressure map shown in Figure 7–9a, note that a very strong high-pressure center, called the **Siberian high**, is positioned over the frozen landscape of northern Asia. A

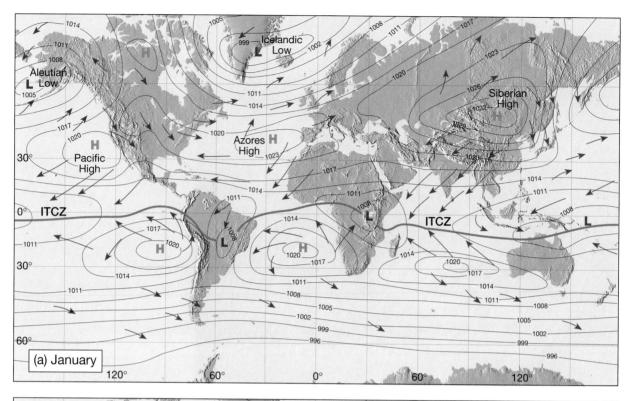

(a) January

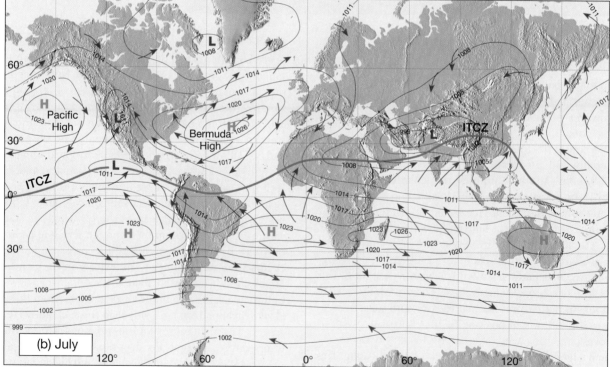

(b) July

Figure 7–9 Average surface pressure and associated global circulation for (a) January and (b) July.

weaker polar high is located over the chilled North American continent. These cold anticyclones consist of very dense air that accounts for the weight of these air columns. In fact, the highest sea-level pressure ever measured, 1,084 millibars (32.01 inches of mercury) was recorded in December 1968 at Agata, Siberia.

The polar highs are prominent features of the winter circulation over the northern continents. Subsidence within these air columns results in clear skies and divergent surface flow. The resulting winds are called *polar easterlies*.

As the Arctic highs strengthen over the continents, a weakening is observed in the subtropical anticyclones situated over the oceans. Further, the average position of the subtropical highs tends to be closer to the eastern margin of the oceans in January than in July. Notice in Figure 7–9a that the center of the subtropical high located in the North Atlantic (sometimes called the **Azores high**) is positioned close to the northwest coast of Africa.

Also shown on the January map but absent in July are two intense semipermanent low-pressure centers (Figure 7–9). Named the **Aleutian** and **Icelandic lows**, these cyclonic cells are situated over the North Pacific and North Atlantic, respectively. They are not stationary cells, but rather the composite of numerous cyclonic storms that traverse these regions. In other words, so many cyclones are present that these regions of the globe are almost always experiencing low pressure, hence the term *semipermanent*. Remember that cyclones are traveling low-pressure centers with low-level convergence and an upward flow. As a result, the areas affected by the Aleutian and Icelandic lows experience cloudy conditions and abundant winter precipitation.

The cyclones that form the Aleutian low are produced as frigid air, directed by the Siberian high, flows off the continent of Asia and overruns comparatively warm air over the Pacific. The strong temperature contrast creates a steep pressure gradient that becomes organized into a counterclockwise rotating storm cell. Just how a cyclone of this type is generated from the clash of two different air masses will be considered in Chapter 9. Nevertheless, with the large number of cyclonic storms that form over the North Pacific and travel eastward, it should be no surprise that the coastal areas of southern Alaska receive abundant precipitation. This fact is exemplified by the climate data for Sitka, Alaska, a coastal town that receives 215 centimeters (85 inches) of precipitation each year, over five times that received in Churchill, Manitoba, Canada. Although both towns are situated at roughly the same latitude, Churchill is located in the continental interior far removed from the influence of the Aleutian low.

July Pressure and Wind Patterns. The pressure pattern over the Northern Hemisphere changes dramatically with the onset of summer as increased amounts of radiation strike the northern landmasses (Figure 7–9b). High surface temperatures over the continents generate lows that replace the winter highs. These thermal lows consist of warm ascending air that induces inward directed surface flow. The strongest of these low-pressure centers develops over southern Asia. A weaker thermal low is also found in the southwestern United States.

Notice in Figure 7–9 that, during the summer months, the subtropical highs in the Northern Hemisphere migrate westward and become more intense than during the winter months. These strong high-pressure centers dominate the summer circulation over the oceans and pump warm moist air onto the continents that lie to the west of these highs. This results in an increase in precipitation over parts of eastern North America and Southeast Asia.

During the peak of the summer season, the subtropical high found in the North Atlantic is positioned near the island of Bermuda, hence the name **Bermuda high**. (Bermuda is located about 1500 kilometers, or 900 miles, east of the South Carolina coast.) Recall that in the Northern Hemisphere winter the Bermuda high is located near Africa and goes by the alias *Azores high* (Figure 7–9).

Monsoons

The greatest *seasonal* change in Earth's global circulation is the monsoon. Contrary to popular belief, **monsoon** does not mean "rainy season"; rather, it refers to a wind system that exhibits a pronounced seasonal reversal in direction. In general, winter is associated with winds that blow predominantly off the continents and that produce a dry winter monsoon. By contrast, in summer, warm moisture-laden air blows from the sea toward the land. Thus, the summer monsoon, which is usually associated with abundant precipitation, is the source of the misconception.

The Asian Monsoon

The best-known and most pronounced monsoon circulation is found in southern and southeastern Asia. Like all winds, the Asian monsoon is driven by pressure differences that are generated by unequal heating of Earth's surface. As summer approaches, the temperatures in India and surrounding Southeast Asia soar. For example, summertime temperatures at New Delhi, India, often exceed 40°C (104°F). This intense solar heating generates a low-pressure area over southern Asia. Recall that thermal lows form because intense surface heating causes expansion of the overlying air column. This in turn generates outflow aloft that encourages inward flow at the surface. With the development of the low-pressure center over India, moisture-laden air from the Indian Ocean flows landward, thereby generating a pattern of precipitation typical of the summer monsoon.

One of the world's rainiest regions is found on the slopes of the Himalayas where orographic lifting of moist air from the India Ocean produces copious precipitation. Cherrapunji, India, once recorded an annual rainfall of 25 meters (82.5 feet), most of which fell during the four months of the summer monsoon (Figure 7–10b).

As winter approaches, long nights and low sun angles result in the accumulation of frigid air over the vast landscape of northern Russia. This generates a cold anticyclone called the *Siberian high*, which begins to dominate the winter circulation over Asia. The subsiding dry air of the Siberian high produces surface flow that moves across southern Asia producing predominantly offshore winds (Figure 7–10b). By the time this flow reaches India, it has warmed considerably but remains extremely dry. For example, Bombay, India, receives less than 1 percent of its annual precipitation in the winter. The remainder comes in the summer, with the vast majority falling from June through September.

The Asian monsoon is complex and is strongly influenced by the seasonal change in the amount of solar heating received by the vast Asian continent. However, another factor, related to the annual migration of the Sun, also contributes to the pronounced monsoon circulation of southeastern Asia. As shown in Figure 7–10, the Asian monsoon is associated with a larger than average seasonal migration of the intertropical convergence zone (ITCZ). With the onset of summer, the ITCZ moves northward over the continent and is accompanied by peak rainfall. The opposite occurs in the Asian winter as the ITCZ moves south of the equator. (Recall the ITCZ is a belt of low pressure and rising air that drives the Hadley cells.)

The migration of the ITCZ is accompanied by a dramatic change in pressure. The strong high-pressure system and subsidence that dominates the winter flow is replaced by low pressure and convergence during the summer months. This significant shift in pressure is thought to aid the northward movement of the ITCZ.

Of major importance are the Himalaya Mountains and the huge Tibetan Plateau, which has an *average* elevation that is higher than the highest peaks in the Colorado Rockies. During the winter months, these topographic barriers contribute to the extreme temperature difference that exists between the cold continental interior and

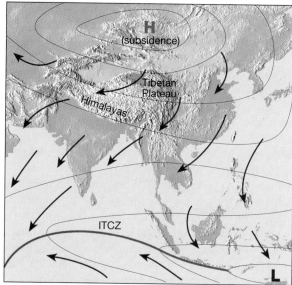

(a) Winter monsoon

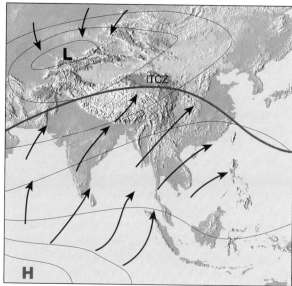

(b) Summer monsoon

Figure 7–10 Asia's monsoon circulation occurs in conjunction with the seasonal shift of the intertropical convergence zone (ITCZ). (a) In January, a strong high pressure develops over Asia and cool, dry continental air generates the dry winter monsoon. (b) With the onset of summer, the ITCZ migrates northward and draws warm, moist air onto the continent.

the milder coastal areas. This temperature contrast produces a strong jet stream that becomes anchored over these highlands. The role that this jet stream plays in the dry winter monsoon is not known with certainty. Nevertheless, in the summer, as the temperature differential over the continent diminishes, the jet stream breaks down.

Clearly, these topographic barriers and the resulting upper air flow play a major role in the migration of the ITCZ.

Nearly half the world's population inhabits regions affected by monsoonal circulation. Further, many of these people depend on subsistence agriculture for their livelihood. Therefore, the timely arrival of the monsoon rains often means the difference between adequate nutrition and widespread malnutrition.

The North American Monsoon

Many regions of the globe experience seasonal wind shifts like those associated with the Asian monsoon. Although none are as dramatic as the Asian monsoon, these smaller features are important elements of the global circulation and affect most of Earth's landmasses.

For example, a relatively small seasonal wind shift influences a portion of North America. Sometimes called the *North American monsoon*, this circulation pattern produces a dry spring followed by a comparatively rainy summer that impacts large areas of the southwestern United States and northwestern Mexico.° This is illustrated by

°This event is also called the *Arizona monsoon* and the *Southwest monsoon* because it has been extensively studied in this part of the United States.

Figure 7–12 High summer temperatures over the southwestern United States create a thermal low that draws moisture from the Gulf of California and the Gulf of Mexico. This summer monsoon produces an increase in precipitation, which often comes in the form of thunderstorms, over the southwestern United States and northwestern Mexico.

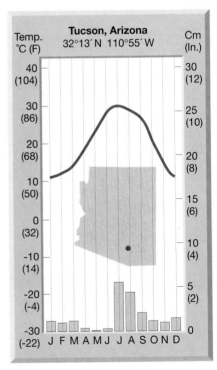

Figure 7–11 Climate diagram for Tucson, Arizona, showing a summer precipitation maximum produced by monsoon circulation that draws moist air in from the Gulf of California and to a lesser extent from the Gulf of Mexico.

Tucson, Arizona, which typically receives almost sixty times more precipitation in July than in May. As shown in Figure 7–11, these summer rains typically last into September when drier conditions reestablish themselves.

Summer daytime temperatures in the American Southwest, particularly in the low deserts, can be extremely hot. This intense surface heating generates a low-pressure center over Arizona that draws in warm, moist air from the Gulf of California (Figure 7–12). The Gulf of Mexico is also thought to be a source of some of the moisture responsible for the summer precipitation. The supply of atmospheric moisture from nearby marine sources, coupled with the convergence and upward flow of the thermal low, is conducive to generating the precipitation this region experiences during the hottest months. Although often associated with the state of Arizona, this monsoon is actually strongest in northwestern Mexico and is quite pronounced in New Mexico.

The Westerlies

Prior to World War II, upper-air observations were scarce. Since then, aircraft and radiosondes have provided a great deal of data about the upper troposphere. Among the most important discoveries was that airflow aloft in the middle latitudes has a strong west-to-east component, thus the name *westerlies*.

Why Westerlies?

Let us consider the reason for the predominance of westerly flow aloft. Recall that winds are created and maintained by pressure differences that are a result of temperature differences. In the case of the westerlies, it is the temperature contrast between the poles and equator that drives these winds. Figure 7–13 illustrates the pressure distribution with height over the cold polar region as compared to the much warmer tropics. Because cold air is more dense (compact) than warm air, air pressure decreases more rapidly in a column of cold air than in a column of warm air. The pressure surfaces (planes) in Figure 7–13 represent a grossly simplified view of the pressure distribution we would expect to observe from pole to equator.

Over the equator, where temperatures are higher, air pressure decreases more gradually than over the cold polar regions. Consequently, at the same altitude above Earth's surface, higher pressure exists over the tropics and lower pressure is the norm above the poles. Thus, the resulting pressure gradient is directed from the equator (area of higher pressure) toward the poles (area of lower pressure).

Once the air from the tropics begins to advance poleward in response to this pressure gradient force (red arrow in Figure 7–13), the Coriolis force comes into play to change the direction of airflow. Recall that in the Northern Hemisphere, the Coriolis force causes winds to be deflected to the right. Eventually, a balance is reached between the poleward-directed pressure-gradient force and the Coriolis force to generate a wind with a strong west-to-east component (Figure 7–13). Recall that such winds are called *geostrophic winds*. Because the equator-to-pole temperature gradient shown in Figure 7–13 is typical over the globe, a westerly flow aloft should be expected, and on most occasions it is observed.

It can also be shown that the pressure gradient increases with altitude; as a result, so should wind speeds. This increase in wind speed continues only to the tropopause, where it starts to decrease upward into the stratosphere.

Jet Streams

Imbedded within the westerly flow aloft are narrow ribbons of high-speed winds that meander for thousands of kilometers (Figure 7–14a). These fast streams of air once were considered analogous to jets of water and thus were named **jet streams**. The best-known occurs in the middle latitudes at elevations between 7500 and 12,000 meters (25,000 and 40,000 feet) and is appropriately named the **midlatitude jet stream**. These high-speed air currents have widths that vary from less than 100 kilo-

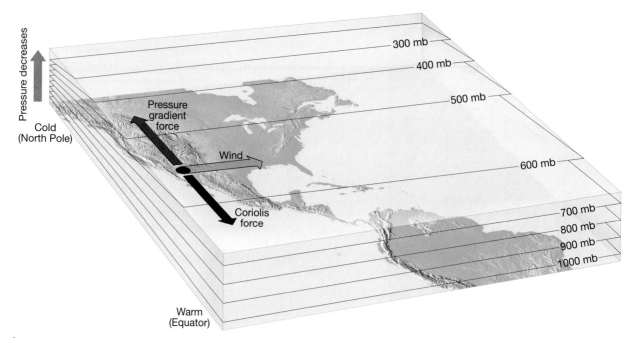

Figure 7–13 Idealized pressure gradient that develops aloft because of density differences between cold polar air and warm tropical air. Notice that the poleward-directed pressure-gradient force is balanced by an equatorward-directed Coriolis force. The result is a prevailing flow from west to east, which is called the *westerlies*.

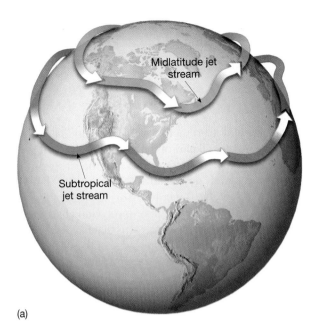

(a)

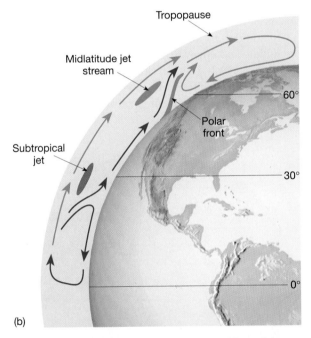

(b)

Figure 7–14 Jet streams. (a) Approximate positions of the midlatitude and subtropical jet streams. Note that these fast-moving currents are generally not continuous around the entire globe. (b) A cross-sectional view of the midlatitude and subtropical jets.

meters (60 miles) to over 500 kilometers (300 miles) and are generally a few kilometers thick. Wind speeds are frequently in excess of 200 kilometers (120 miles) per hour, but rarely exceed 400 kilometers (240 miles) per hour.

Although jet streams had been predicted earlier, their existence was first dramatically illustrated during World War II. American bombers heading westward toward Japanese-occupied islands occasionally made little headway. On abandoning their missions, the planes on their return flight experienced westerly tail winds that sometimes exceeded 300 kilometers per hour. Even today, commercial aircraft use these strong tail winds to increase their speed when making eastward flights around the globe. On westward flights, of course, these fast currents of air are avoided.

Origin of the Midlatitude Jet Stream

What is the origin of these distinctive, energetic winds that exist within the slower, general westerly flow? The key is that large temperature contrasts at the surface produce steep pressure gradients aloft, and hence faster upper air winds. In winter, it is not unusual to have a warm balmy day in southern Florida and near-freezing temperatures in Georgia, only a few hundred kilometers to the north. Such large wintertime temperature contrasts lead us to expect faster westerly flow at that time of year. Observations substantiate these expectations. In general, the fastest upper air winds are located above regions of the globe having large temperature contrasts across very narrow zones. Stated another way, jet streams are located in regions of the atmosphere where large horizontal temperature differences occur over short distances.

These large temperature contrasts occur along linear zones called *fronts*. The midlatitude jet stream occurs along a major frontal zone called the *polar front* (Figure 7–14b). (Because the midlatitude jet stream occurs in association with the polar front it is also known as the *polar jet stream*.) Recall that the polar front is situated between the cool winds of the polar easterlies and the relatively warm westerlies. Instead of flowing nearly straight west-to-east, the midlatitude jet stream usually has a meandering path. Occasionally, it flows almost due north–south. Sometimes it splits into two jets that may, or may not, rejoin. Like the polar front, this jet is not continuous around the globe.

Occasionally, the midlatitude jet exceeds 500 kilometers (300 miles) per hour. On the average, however, it travels at 125 kilometers (75 miles) per hour in the winter and roughly half that speed in the summer (Figure 7–15). This seasonal difference is due to the much stronger temperature gradient that exists in the middle latitudes during the winter.

Because the location of the midlatitude jet roughly coincides with that of the polar front, its latitudinal position migrates with the seasons. Thus, like the zone of maximum

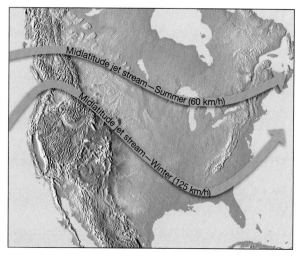

Figure 7–15 The position and speed of the midlatitude jet stream changes with the seasons, migrating freely between about 30° and 70° latitude. Shown are flow patterns that are common for summer and winter.

solar heating, the jet moves northward during summer and southward in winter. During the cold winter months, the midlatitude jet stream may extend as far south as 30° north latitude (Figure 7–15). With the coming of spring, the zone of maximum solar heating and therefore the jet begins a gradual northward migration. By midsummer, its average position is about 50° north latitude, but it may penetrate much farther poleward.

As the midlatitude jet shifts northward, there is a corresponding change in the region where outbreaks of severe thunderstorms and tornadoes occur. For example, in February, most tornadoes occur in the states bordering the Gulf of Mexico. By midsummer, the center of this activity shifts to the northern plains and Great Lakes states. As you can see, the midlatitude jet stream plays a very important role in the weather of the midlatitudes. In addition to supplying energy to the circulation of surface storms, it also directs their paths of movement. Consequently, determining changes in the location and flow pattern of the midlatitude jet is an important part of modern weather forecasting.

Subtropical Jet Stream

Other jet streams are known to exist, but none have been studied in as much detail as the midlatitude jet stream. A semipermanent jet exists over the subtropics and as such is called the **subtropical jet** (see Figure 7–14b). The subtropical jet is a wintertime phenomenon. It is absent in summer because of the lack of a strong temperature contrast in the subtropics. Somewhat slower than the midlatitude jet, this west-to-east flowing current is centered

at 25° latitude at an altitude of about 13 kilometers (8 miles). On occasion, the subtropical jet stream will migrate poleward and merge with the midlatitude jet stream.

Waves in the Westerlies

It is important to remember that the midlatitude jet stream is an integral part of the westerlies. It is not a dramatic anomaly like a hurricane. In fact, the jet stream can be more accurately described as the fast core of the overall westerly flow, like the fastest moving portion of a river (Figure 7–16). Studies of upper-level wind charts reveal that the westerlies follow wavy paths that have rather long wavelengths. Much of our knowledge of these large-scale motions is attributed to C. G. Rossby, who first explained the nature of these waves. The longest wave patterns (called *Rossby waves*) have wavelengths of 4000 to 6000 kilometers, so that three to six waves will fit around the globe (Figure 7–16). Although the air flows eastward along this wavy path, these long waves tend to remain stationary or to move slowly.

In addition to Rossby waves, shorter waves occur in the middle and upper troposphere. These shorter waves are often associated with cyclones at the surface and, like these storms, they travel from west to east around the globe at rates of up to 15° of longitude per day.

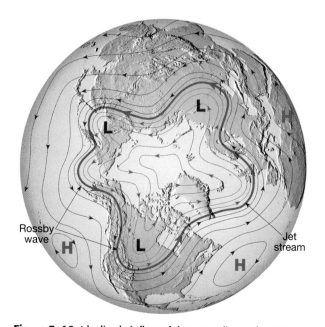

Figure 7–16 Idealized air flow of the westerlies at the 500-millibar level. The five long-wavelength undulations, called *Rossby waves*, compose this flow. The jet stream is the fast core of this wavy flow.

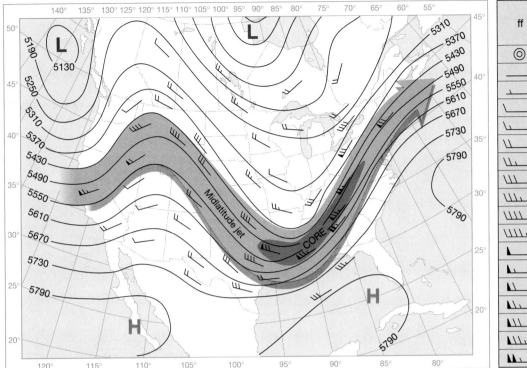

Figure 7–17 Simplified 500-millibar height-contour chart for January. The position of the jet-stream core is shown in dark red.

ff	Miles per hour
◎	Calm
	1–2
	3–8
	9–14
	15–20
	21–25
	26–31
	32–37
	38–43
	44–49
	50–54
	55–60
	61–66
	67–71
	72–77
	78–83
	84–89
	119–123

Although much remains to be learned about the wavy flow of the westerlies, its basic features are understood with some certainty. Among the most obvious characteristics of the flow aloft are its seasonal changes. The higher wind speeds in the cool season are depicted on upper-air charts by more closely spaced contour lines, as shown in Figure 7–17. The summer-to-winter change in wind speed is a consequence of the seasonal contrasts of the temperature gradients. The steep temperature gradient across the middle latitudes in winter corresponds to a stronger flow aloft.

In addition to seasonal changes in the strength of its flow, the position of the midlatitude jet stream also shifts from summer to winter. With the approach of winter, the jet migrates equatorward. As summer nears, it moves back toward the poles (Figure 7–15). By midwinter, the jet core may penetrate as far south as central Florida. Because the paths of cyclonic systems are guided by the flow aloft, we can expect the southern tier of states to encounter most of their severe weather in the winter. During hot summer months, the storm track is across the northern states, and some cyclones never leave Canada. In addition to seasonal changes in storm tracks, the number of cyclones generated also varies seasonally. The largest number form in the cooler months when temperature contrasts are most pronounced.

As you might expect, a close relationship exists between the location of the midlatitude jet stream and conditions near the surface, particularly temperatures. When the midlatitude jet is situated equatorward of your location, the weather will usually be colder and stormier than normal. Conversely, when the midlatitude jet moves poleward of your location, warmer and drier conditions are likely to prevail. Further, when the position of the jet stream remains fixed for extended periods, weather extremes can result. Thus, depending on the position of the jet stream, the weather could be hotter, colder, drier, or wetter than normal. We shall return later to this important relationship between the jet stream and middle latitude weather.

Westerlies and Earth's Heat Budget

Now let us return to the wind's function of maintaining Earth's heat budget by transporting heat from the equator toward the poles. In Chapter 2, we showed that the equator receives more solar radiation than it radiates back into space, whereas the poles experience the reverse situation. Thus, the equator has excess heat, whereas the poles experience a deficit. Although the flow near the equator is somewhat *meridional* (north to south), at most other latitudes, the flow is *zonal* (west to east). The reason for the zonal flow, as we have seen, is the Coriolis force. The question we now consider is: *How can wind with a west-to-east flow transfer heat from south to north?*

In addition to its seasonal migrations, the midlatitude jet can change positions on shorter time scales as well. There may be periods of a week or more when the flow is nearly west to east, as shown in Figure 7–18a. When this condition prevails, relatively mild temperatures occur and few disturbances are experienced in the region south of the jet stream. Then without warning, the flow aloft begins to meander and produces large-amplitude waves and a general north-to-south flow (Figure 7–18b and c). Such a change allows cold air to advance southward. Because this influx intensifies the temperature gradient,

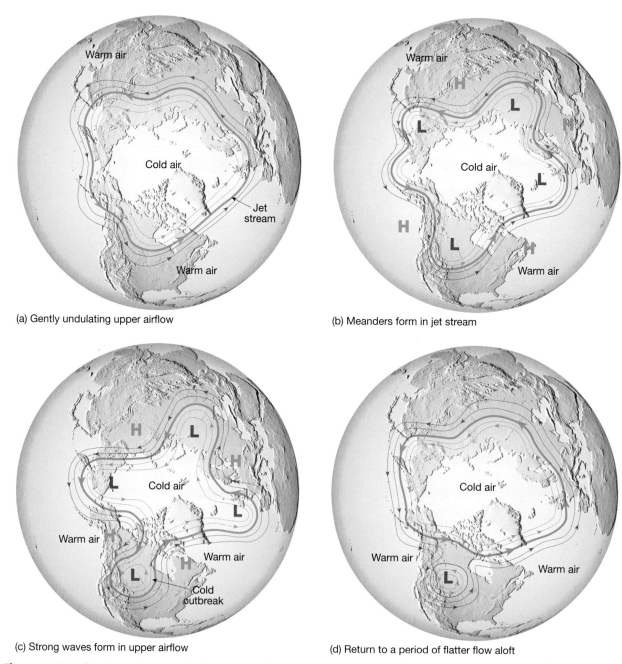

(a) Gently undulating upper airflow

(b) Meanders form in jet stream

(c) Strong waves form in upper airflow

(d) Return to a period of flatter flow aloft

Figure 7–18 Cyclic changes that occur in the upper-level airflow of the westerlies. The flow, which has the jet stream as its axis, starts out nearly straight and then develops meanders and cyclonic activity that dominates the weather.

the flow aloft is also strengthened. Recall that strong temperature contrasts create steep pressure gradients that organize into rotating cyclonic systems. The jet stream supports these lows by providing divergence aloft that enhances the inward surface flow of these systems.

During these periods, cyclonic activity dominates the weather. For a week or more, cyclonic storms redistribute large quantities of heat across the middle latitudes by moving cold air equatorward and warm air poleward. This redistribution eventually results in a weakened temperature gradient and a return to a flatter flow aloft and less intense weather at the surface (Figure 7–18d). Cycles such as these, which consist of alternating periods of calm and stormy weather, can last from one to six weeks.

Laboratory experiments using rotating fluids to simulate Earth's circulation support the existence of waves and eddies that carry on the task of heat transfer in the middle latitudes. In these studies, called *dishpan experiments*, a large circular pan is heated around the outer edge to represent the equator and the center is cooled to duplicate a pole. Colored particles are added so that the flow can be easily observed and photographed. When the pan is heated,

but not rotated, a simple convection cell forms to redistribute the heat. This cell is similar to the Hadley cell considered earlier. When the pan is rotated, however, the simple circulation breaks down and the flow develops a wavy pattern with eddies embedded between meanders, as seen in Figure 7–19. These experiments indicate that changing the rate of rotation and varying the temperature gradient largely determine the flow pattern produced. Such studies have added greatly to our understanding of global circulation.

In summary, we now understand that the wavy flow aloft is largely responsible for producing surface weather patterns. During periods when the flow aloft is relatively flat (small-amplitude waves), little cyclonic activity is generated at the surface. Conversely, when the flow exhibits large-amplitude waves having short wavelengths, vigorous cyclonic storms are created. This important relationship between the flow aloft and cyclonic storms is considered in more detail in Chapter 9.

Global Winds and Ocean Currents

Where the atmosphere and ocean are in contact, energy is passed from moving air to the water through friction. As a consequence, the drag exerted by winds blowing steadily across the ocean causes the surface layer of water to move. Thus, because winds are the primary driving force of **ocean currents**, a relationship exists between the oceanic circulation and the general atmospheric circulation. A comparison of Figures 7–20 and 7–9 illustrates this. A further clue to the influence of winds on ocean circulation is provided by the currents in the northern Indian Ocean, where there are seasonal wind shifts known as the *summer* and *winter monsoons*. When the winds change directions, the surface currents also reverse direction.

North and south of the equator are two westward-moving currents, the North and South Equatorial currents, which derive their energy principally from the trade winds that blow from the northeast and southeast, respectively, toward the equator. These equatorial currents can be thought of as the backbone of the system of ocean currents. Because of the Coriolis force, these currents are deflected poleward to form clockwise spirals in the Northern Hemisphere and counterclockwise spirals in the Southern Hemisphere. These nearly circular ocean currents are found in each of the major ocean basins centered around the subtropical high-pressure systems (Figure 7–20).

In the North Atlantic, the equatorial current is deflected northward through the Caribbean, where it becomes the *Gulf Stream*. As the Gulf Stream moves along the eastern coast of the United States, it is strengthened by

Figure 7–19 Photograph obtained from a *dishpan experiment*. The pan is heated around the edge to represent the equator, and the center is cooled to duplicate the poles. When the pan is rotated, a wavy flow develops with eddies embedded between meanders, as shown. This flow pattern closely parallels that produced in the "real" atmosphere, where the meanders represent the wavy flow of the westerlies and the eddies are cyclonic and anticyclonic systems embedded within this larger circulation. *(Courtesy of D. H. Fultz, University of Chicago Hydrodynamics Laboratory)*

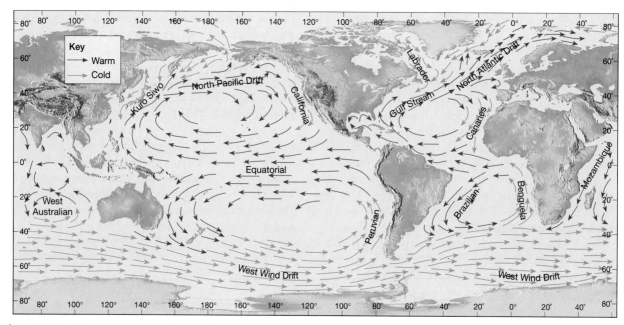

Figure 7–20 Major ocean currents. Poleward-moving currents are warm, and equatorward-moving currents are cold.

the prevailing westerly winds and is deflected to the east (to the right) between 35° north and 45° north latitude (Figure 7–21). As it continues northeastward beyond the Grand Banks, it gradually widens and decreases speed until it becomes a vast, slowly moving current known as the North Atlantic Drift. As the North Atlantic Drift approaches Western Europe, it splits, with part moving northward past Great Britain and Norway. The other part is deflected southward as the cool Canaries current. As the Canaries current moves south, it eventually merges into the North Equatorial current.

The Importance of Ocean Currents

In addition to being significant considerations in ocean navigation, currents have an important effect on climate. The moderating effect of poleward-moving warm ocean currents is well known. The North Atlantic Drift, an extension of the warm Gulf Stream, keeps Great Britain and much of northwestern Europe warmer than one would expect for their latitudes.

In addition to influencing temperatures of adjacent land areas, cold currents have other climatic influences. For example, where tropical deserts exist along the west coasts of continents, cold ocean currents have a dramatic impact. The principal west-coast deserts are the Atacama in Peru and Chile, and the Namib in southern Africa. The aridity along these coasts is intensified because the lower air is chilled by cold offshore waters. When this occurs,

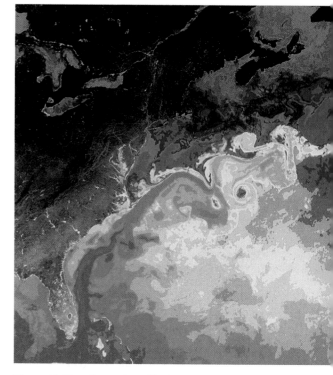

Figure 7–21 Enhanced satellite image of the complex flow of the Gulf Stream. Reds and yellows indicate warmer waters. Note that the Gulf Stream flows northward along the Florida coast (lower left) and then diagonally toward the upper right of the image. *(Courtesy of O. Brown, R. Evans, and M. Carle/Rosenstiel School of Marine and Atmospheric Science)*

the air becomes very stable and resists the upward movements necessary to create precipitation-producing clouds. In addition, the presence of cold currents causes temperatures to approach and often reach the dew point. As a result, these areas are characterized by high relative humidities and much fog. Thus, not all tropical deserts are hot with low humidities and clear skies. Rather, the presence of cold currents transforms some tropical deserts into relatively cool, damp places that are often shrouded in fog.

Ocean currents also play a major role in maintaining Earth's heat balance. They accomplish this task by transferring heat from the tropics, where there is an excess of heat, to the polar regions, where a deficit exists. Ocean water movements account for about a quarter of this total heat transport and winds the remaining three-quarters.

Ocean Currents and Upwelling

In addition to producing surface currents, winds can also cause vertical water movements. **Upwelling**, the rising of cold water from deeper layers to replace warmer surface water, is a common wind-induced vertical movement. It is most characteristic along the eastern shores of the oceans, most notably along the coasts of California, Peru, and West Africa.

Upwelling occurs in these areas when winds blow toward the equator parallel to the coast. Because of the Coriolis force, the surface-water movement is directed away from the shore. As the surface layer moves away from the coast, it is replaced by water that "upwells" from below the surface. This slow upward flow from depths of 50 to 300 meters (165 to 1000 feet) brings water that is cooler than the original surface water and creates a characteristic zone of lower temperatures near the shore.

For swimmers who are accustomed to the waters along the mid-Atlantic shore of the United States, a dip in the Pacific off the coast of central California can be a chilling surprise. In August, when temperatures in the Atlantic are 21°C (70°F) or higher, central California's surf is only about 15°C (60°F). Coastal upwelling also brings to the ocean surface greater concentrations of dissolved nutrients, such as nitrates and phosphates. These nutrient-enriched waters from below promote the growth of plankton, which in turn supports extensive populations of fish.

El Niño and La Niña

As can be seen in Figure 7–22a, the cold Peruvian current flows equatorward along the coast of Ecuador and Peru. This flow encourages upwelling of cold nutrient-filled waters that serve as the primary food source for millions of fish, particularly anchovies. Near the end of each year, however, a warm current that flows southward along the coasts of Ecuador and Peru replaces the cold Peruvian current. During the nineteenth century the local residents named this warm countercurrent El Niño ("the child") after the Christ child because it usually appeared during the Christmas season. Normally, these warm countercurrents last for at most a few weeks when they again give way to the cold Peruvian flow. However, at irregular intervals of three to seven years, these countercurrents become unusually strong and replace normally cold offshore waters with warm equatorial waters (Figure 7–22b). Today, scientists use the term **El Niño** for these episodes of ocean warming that affect the eastern tropical Pacific.

The onset of El Niño is marked by abnormal weather patterns that drastically affect the economies of Ecuador and Peru. As shown in Figure 7–22b, these unusually strong undercurrents amass large quantities of warm water that block the upwelling of colder, nutrient-filled water. As a result, the anchovies starve, devastating the fishing industry. At the same time, some inland areas that are normally arid receive an abnormal amount of rain. Here, pastures and cotton fields have yields far above the average. These climatic fluctuations have been known for years, but they were originally considered local phenomena. Today, we know that El Niño is part of the global circulation and affects the weather at great distances from Peru and Ecuador.

Two of the strongest El Niño events on record occurred between 1982–83 and 1997–98, and were responsible for weather extremes of a variety of types in many parts of the world. During the 1982–83 El Niño event, heavy rains and flooding plagued normally dry portions of Ecuador and Peru. Some locations that usually receive only 10 to 13 centimeters of rain each year had as much as 350 centimeters of precipitation. At the same time, severe drought beset Australia, Indonesia, and the Philippines. Huge crop losses, property damage, and much human suffering were recorded.

The 1997–98 El Niño brought ferocious storms that struck the California coast, causing unprecedented beach erosion, landslides, and floods. In the southern United States, heavy rains also brought floods to Texas and the Gulf states. The same energized jet stream that produced storms in the South, upon reaching the Atlantic, sheared off the northern portions of hurricanes, destroying the storms. It was one of the quietest Atlantic hurricane seasons in years.

Major El Niño events, such as the one in 1997 and 1998, are intimately related to the large-scale atmospheric circulation. Each time an El Niño occurs, the barometric pressure drops over large portions of the

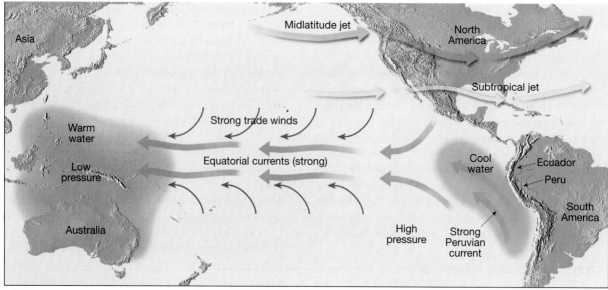

(a) Normal conditions

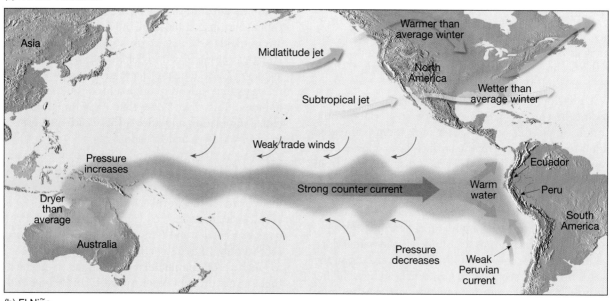

(b) El Niño

Figure 7–22 The relationship between the Southern Oscillation and El Niño is illustrated on these simplified maps. (a) Normally, the trade winds and strong equatorial currents flow toward the west. At the same time, the strong Peruvian current causes upwelling of cold water along the west coast of South America. (b) When the Southern Oscillation occurs, the pressure over the eastern and western Pacific flip-flops. This causes the trade winds to diminish, leading to an eastward movement of warm water along the equator. As a result, the surface waters of the central and eastern Pacific warm, with far-reaching consequences to weather patterns.

southeastern Pacific, whereas in the western Pacific, near Indonesia and northern Australia, the pressure rises (Figure 7–23). Then, as a major El Niño event comes to an end, the pressure difference between these two regions swings back in the opposite direction. This seesaw pattern of atmospheric pressure between the eastern and western Pacific is called the **Southern Oscillation**. It is an inseparable part of the El Niño warmings that occur in the central and eastern Pacific every three to seven years. Therefore, this phenomenon is often termed El Niño/Southern Oscillation, or ENSO for short.

Winds in the lower atmosphere are the link between the pressure change associated with the Southern Oscillation and the extensive ocean warming associated with El Niño (see Box 7–3). During a typical year, the trade winds converge near the equator and flow westward toward Indonesia (Figure 7–23a). This steady westward flow creates a warm surface current that moves from east to west along the equator. The result is a "piling up" of a thick layer of warm surface water that produces higher sea levels (by 30 centimeters) in the western Pacific. Meanwhile, the eastern Pacific is characterized by a strong Peruvian current, upwelling of cold water, and lower sea levels.

Then when the Southern Oscillation occurs, the normal situation just described changes dramatically. Barometric pressure rises in the Indonesian region, causing the pressure gradient along the equator to weaken or even to reverse. As a consequence, the once-steady trade winds diminish and may even change direction. This reversal creates a major change in the equatorial current system, with warm water flowing eastward (Figure 7–23b). With time, water temperatures in the central and eastern Pacific increase and sea level in the region rises. This eastward shift of the warmest surface water marks the onset

of El Niño and sets up changes in atmospheric circulation that affect areas far outside the tropical Pacific.

When an El Niño began in the summer of 1997, forecasters predicted that the pool of warm water over the Pacific would displace the paths of both the subtropical and midlatitude jet streams, which steer weather systems across North America (Figure 7–22). As predicted, the subtropical jet brought rain to the Gulf Coast, where Tampa, Florida, received more than three times its normal winter precipitation. Furthermore, the midlatitude jet pumped warm air far north into the continent. As a result, winter temperatures west of the Rockies were significantly above normal.

The effects of El Niño are somewhat variable depending in part on the temperatures and size of the warm pools. Nevertheless, some locales appear to be affected more consistently. In particular, during most El Niños, warmer-than-normal winters occur in the northern United States and Canada. In addition, normally arid portions of Peru and Ecuador, as well as the eastern United States, experience wet conditions. By contrast, drought conditions are generally observed in Indonesia, Australia, and the Philippines. One major benefit of the circulation asso-

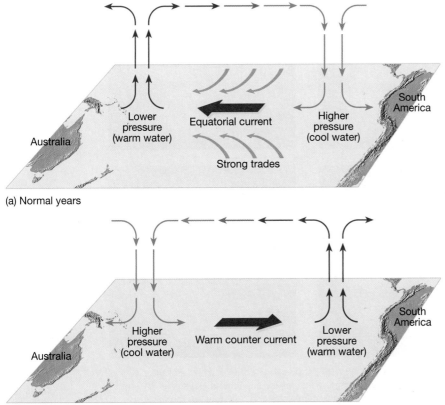

(a) Normal years

(b) El Niño years

Figure 7–23 Simplified illustration of the see-saw pattern of atmospheric pressure between the eastern and western Pacific, called the *Southern Oscillation*. (a) During average years, high pressure over the eastern Pacific causes surface winds and warm equatorial waters to flow westward. The result is a pileup of warm water in the western Pacific, which promotes the lowering of pressure. (b) An El Niño event begins as surface pressure increases in the western Pacific and decreases in the eastern Pacific. This air pressure reversal weakens, or may even reverse the trade winds, and results in an eastward movement of the warm waters that had accumulated in the western Pacific.

Box 7–3 Tracking El Niño From Space

The images in Figure 7–C show the progression of the 1997–98 El Niño as derived from the TOPEX/Poseidon satellite.* This satellite uses radar altimeters to bounce radar signals off the ocean surface to obtain precise measurements of the distance between the satellite and the sea surface. This information is combined with high-precision orbital data from the Global Positioning System (GPS) of satellites to produce maps of sea-surface height. Such maps show the topography of the sea surface. Elevated topography ("hills") indicates warmer-than-average water, whereas areas of low topography ("valleys") indicate cooler-than-normal water. With this detailed information, scientists can determine the speed and direction of surface ocean currents.

The colors represented in these images show sea-level height relative to average. Remember, "hills" are warm and "valleys" are cool. The white and red areas indicate places of higher-than-normal sea-surface heights ("hills"). In the white areas, the sea surface is between 14 and 32 centimeters above normal; in the red areas, sea level is elevated by about 10 centimeters. Green areas indicate average conditions, whereas purple shows zones that are at least 18 centimeters below average sea level.

The images depict the progression of the large warm water mass from west to east across the equatorial Pacific. At its peak in November 1997, the surface area covered by the warm water mass was about one and a half times the size of the contiguous 48 states. The added amount of warm water in the eastern Pacific with a temperature between 21° and 30°C was about 30 times the combined volume of the water in all of the U.S. Great Lakes.

*This box is based on material published by NASA's Goddard Space Flight Center.

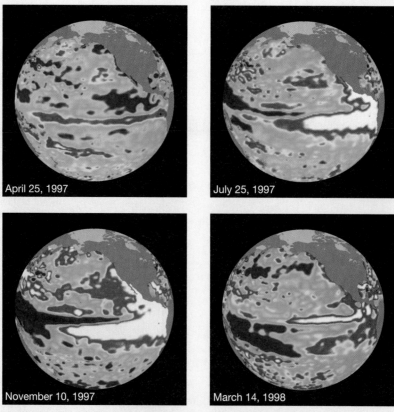

Figure 7–C The image for April 25, 1997, shows the Pacific on the eve of the 1997–98 El Niño event. By examining the images for July 25 and November 10, you can see the dramatic buildup of warm water (white and red areas) in the eastern Pacific and the enlarging region of cool water (purple) in the western Pacific. By the time of the March 14, 1998, image, the area of warm water in the eastern Pacific was *much* smaller. *(Courtesy of NASA Goddard Space Flight Center).*

ciated with El Niño is a suppression of the number of Atlantic hurricanes.

The opposite of El Niño is an atmospheric phenomenon known as **La Niña**. Once thought to be the normal conditions that occur between two El Niño events, meteorologists now consider La Niña an important atmospheric phenomenon in its own right. Researchers have come to recognize that when surface temperatures in the eastern Pacific are *colder than average*, a La Niña event is triggered that has a distinctive set of weather patterns. A typical La Niña winter blows colder than normal air over the Pacific Northwest and the northern Great Plains while warming much of the rest of the United States. Further, greater precipitation is expected in the Northwest. During the La Niña winter of 1998–99, a world record snowfall for one season occurred in Washington State. Another La Niña impact is greater hurricane activity. A recent study concluded that the cost of hurricane damages in the United States is 20 times greater in La Niña years as compared to El Niño years.

In summary, the effects of El Niño and La Niña on world climate are widespread and variable. There is no place on Earth where the weather is indifferent to air and ocean conditions in the tropical Pacific. Events associated with El Niño and La Niña are now understood to have a significant influence on the state of weather and climate almost everywhere.

Global Distribution of Precipitation

A casual glance at Figure 7–24a reveals a complex pattern for the distribution of precipitation. Although the map appears to be complicated, the general features of the pattern can be explained using our knowledge of global winds and pressure systems. In general, regions influenced by high pressure, with its associated subsidence and divergent winds, experience dry conditions. Conversely, regions under the influence of low pressure and its converging winds and ascending air receive ample precipitation. However, if the wind-pressure regimes were the only control of precipitation, the pattern shown in Figure 7–24a would be much simpler.

The inherent nature of the air is also important in determining precipitation potential. Because cold air has a low capacity for moisture compared with warm air, we would expect a latitudinal variation in precipitation, with low latitudes receiving the greatest amounts of precipitation and high latitudes receiving the least. Figure 7–24a indeed reveals heavy precipitation in equatorial regions and meager moisture near the landmasses poleward of 60° latitude. A noticeably arid region, however, is also found in the subtropics. This situation can be explained by examining the global wind-pressure regimes.

In addition to latitudinal variations in precipitation, the distribution of land and water complicates the precipitation pattern. Large landmasses in the middle latitudes commonly experience decreased precipitation toward their interiors. For example, central North America and central Eurasia receive considerably less precipitation than do coastal regions at the same latitude. Furthermore, the effects of mountain barriers alter the idealized precipitation regimes we would expect solely from the global wind systems. Windward mountain slopes receive abundant precipitation, whereas leeward slopes and adjacent lowlands are usually deficient in moisture.

Zonal Distribution of Precipitation

Let us first examine the zonal distribution of precipitation that we would expect on a uniform Earth, and then add the variations caused by land and water influences. Recall from our earlier discussion that on a uniform Earth, four major pressure zones emerge in each hemisphere (Figure 7–7a). These zones include the equatorial low (ITCZ), the subtropical high, the subpolar low, and the polar high. Also, remember that these pressure belts show a marked seasonal shift toward the summer hemisphere.

The idealized precipitation regimes expected from these pressure systems are shown in Figure 7–25, in which we can see that the equatorial regime is centered over the equatorial low throughout most of the year. In this region, where the trade winds converge (ITCZ), heavy precipitation is experienced in all seasons. Poleward of the equatorial low in each hemisphere lie the belts of subtropical high pressure. In these regions, subsidence contributes to the dry conditions found there throughout the year. Between the wet equatorial regime and the dry subtropical regime lies a zone that is influenced by both pressure systems. Because the pressure systems migrate seasonally with the Sun, these transitional regions receive most of their precipitation in the summer when they are under the influence of the ITCZ. They experience a dry season in the winter when the subtropical high moves equatorward.

The midlatitudes receive most of their precipitation from traveling cyclonic storms (Figure 7–26). This region is the site of the polar front, the convergent zone between cold polar air and the warmer westerlies. It is along the polar front that cyclones are frequently generated. Because the position of the polar front migrates freely between approximately 30° and 70° latitude, most midlatitude areas receive ample precipitation. But the mean position of this zone also moves north and south with the Sun, so that a narrow belt between 30° and 40° latitude experiences a marked seasonal fluctuation in precipitation.

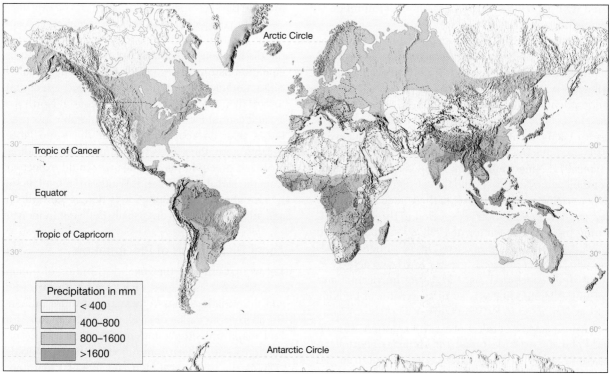

(a) Annual

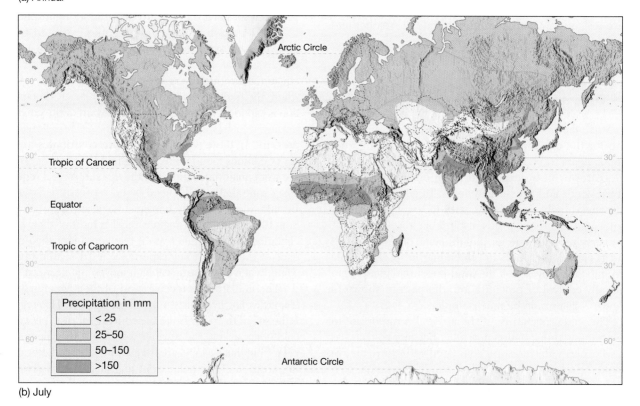

(b) July

Figure 7–24 Global distribution of precipitation: (a) annual, (b) July, and (c) January.

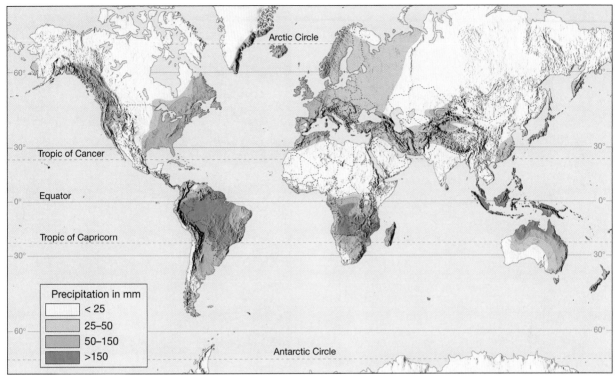

(c) January

In winter, this zone receives precipitation from numerous cyclones as the position of the polar front moves equatorward. During the summer, however, this region is dominated by subsidence associated with the dry subtropical high. Compare the July and January precipitation data for the west coast of North America by referring to Figure 7–24b and c.

The polar regions are dominated by cold air that holds little moisture. Throughout the year, these regions experience only meager precipitation. Even in the summer, when temperatures rise, these areas of ice and snow are dominated by high pressure that blocks the movement of the few cyclones that do travel poleward.

Distribution of Precipitation Over the Continents

The zonal pattern outlined in the previous section roughly approximates general global precipitation. Abundant precipitation occurs in the equatorial and midlatitude regions, whereas substantial portions of the subtropical and polar realms are relatively dry. Yet numerous exceptions to this idealized zonal pattern are obvious in Figure 7–24a.

For example, several arid areas are found in the midlatitudes. The desert region of southern South America,

known as Patagonia, is one example. Midlatitude deserts such as Patagonia exist mostly on the leeward (rain shadow) side of a mountain barrier or in the interior of a continent, cut off from a source of moisture. Most other departures from the idealized zonal scheme result from the distribution of continents and oceans.

The most notable anomaly in the zonal distribution of precipitation occurs in the subtropics. Here we find not only many of the world's great deserts, but also regions of abundant rainfall (Figure 7–24a). This pattern results because the subtropical high-pressure centers that dominate the circulation in these latitudes have different characteristics on their eastern and western flanks (Figure 7–27). Subsidence is most pronounced on the eastern side of these oceanic highs, and a strong temperature inversion is encountered very near the surface and results in stable atmospheric conditions. The upwelling of cold water along the west coasts of the adjacent continents cools the air from below and adds to the stability on the eastern sides of these highs.

Because these anticyclones tend to crowd the eastern side of an ocean, particularly in the winter, we find that the western sides of the continents adjacent to these subtropical highs are arid (Figure 7–27). Centered at approximately 25° north or south latitude on the western side of their respective continents, we find the Sahara Desert of

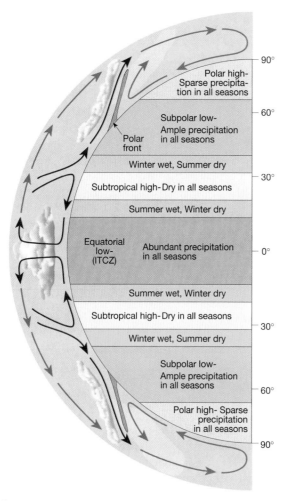

Figure 7–25 Zonal precipitation patterns.

Figure 7–26 Satellite image of Earth's cloud cover. The comma-shaped cloud pattern situated over the eastern United States is a well-developed cyclone. These traveling storm systems produce most of the precipitation in the middle latitudes. *(Courtesy of NASA Headquarters)*

North Africa, the Namib of southwest Africa, the Atacama of South America, the deserts of the Baja Peninsula of North America, and the Great Desert of Australia.

On the western side of these highs, however, subsidence is less pronounced, and convergence with associated rising

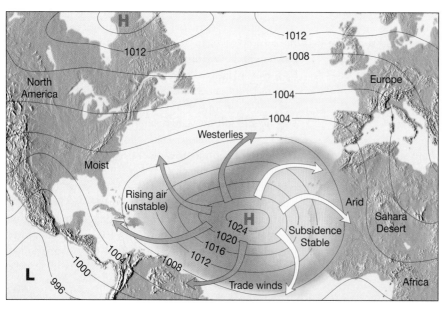

Figure 7–27 Characteristics of subtropical high-pressure systems. Subsidence on the east side of these systems produces stable conditions and aridity. Convergence and uplifting on the western flank promote uplifting and instability.

air appears to be more prevalent. In addition, as this air travels over a large expanse of warm water, it acquires moisture through evaporation that acts to enhance its instability. Consequently, the eastern regions of subtropical continents generally receive ample precipitation throughout the year. Southern Florida is a good example.

Precipitation Regimes on a Hypothetical Continent

When we consider the influence of land and water on the distribution of precipitation, the pattern illustrated in Figure 7–28 emerges. This figure is a highly idealized precipitation scheme for a hypothetical continent located in the Northern Hemisphere. The "continent" is shaped as it is to approximate the percentage of land found at various latitudes.

Notice that this hypothetical landmass has been divided into seven zones. Each zone represents a different precipitation regime. Stated another way, all locations within the same zone (precipitation regime) experience roughly the same precipitation pattern. As we shall see, the odd shape and size of these zones reflect the way precipitation is normally distributed across landmasses located in the Northern Hemisphere. The cities shown are representatives of the precipitation pattern of each location. With this in mind, we will now examine each precipitation regime (numbered 1 through 7) to see the influences of land and water on the distribution of precipitation.

First, compare the precipitation patterns for the nine cities. Notice that the precipitation regimes for west coast locations correspond to the zonal pressure and precipitation pattern shown in Figure 7–25: Zone 6 equates to the equatorial low (wet all year); Zone 4 matches the subtropical high (dry all year), and so on.

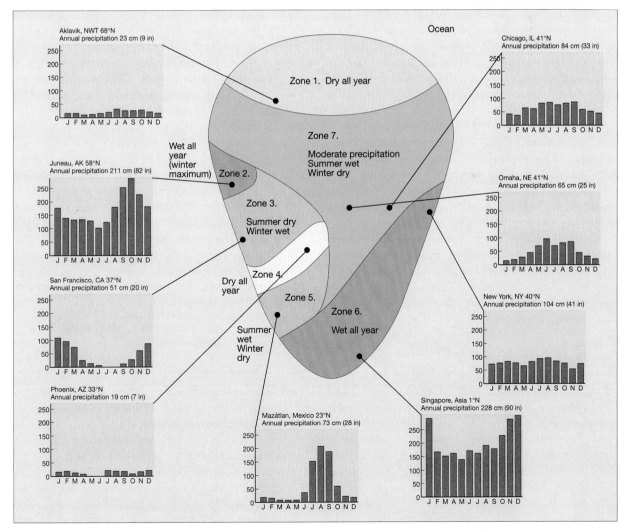

Figure 7–28 Idealized precipitation regimes for a hypothetical landmass in the Northern Hemisphere.

For illustration, Aklavik, Canada, which is strongly influenced by the polar high, has scant moisture, whereas Singapore, near the equator, has plentiful rain every month. Also note that precipitation varies seasonally with both latitude and each city's location with respect to the ocean.

For example, precipitation graphs for San Francisco and Mazatlan, Mexico, illustrate the marked seasonal variations found in zones 3 and 5. Recall that it is the seasonal migration of the pressure systems that causes these fluctuations. Notice that zone 2 shows a precipitation maximum in the fall and early winter, as illustrated by data for Juneau, Alaska. This pattern occurs because cyclonic storms over the North Pacific are more prevalent in the coolest half of the year. Further, the average path of these storms moves equatorward during the winter, impacting locations as far south as southern California.

The eastern half of the continent shows a marked contrast to the zonal precipitation pattern on the western side. Only the dry polar regime (Zone 1) is similar in size and position on both the eastern and western seaboards. The most noticeable departure from the zonal pattern is the absence of the arid subtropical region from the east coast. As noted earlier, this difference is caused by the behavior of subtropical anticyclones located over the oceans. Recall that southern Florida, which is located on the east coast of the continent, receives ample precipitation year round, whereas the Baja Peninsula on the west coast is arid. Moving inland from the east coast, observe how precipitation decreases (compare total precipitation westward for New York, Chicago, and Omaha). This decrease, however, does not hold true for mountainous regions; even relatively small ranges like the Appalachians are able to extract a greater share of precipitation.

Another variation is seen with latitude. As we move poleward along the eastern half of our hypothetical continent, note the decrease in precipitation (compare Singapore, New York, and Aklavik). This is the decrease we would expect with cooler air temperatures and corresponding lower moisture capacities. We do not, however, find a similar decrease in precipitation in the higher middle latitudes along the west coast. Because this region lies in the westerly wind belt, the west coasts of midlatitude regions, such as Alaska, Canada, and Norway, receive abundant moisture from storms that originate over the adjacent oceans. In fact, some of the rainiest regions on Earth are found in such settings. In contrast, the east coasts of midlatitude locations experience temperature and moisture regimes more typical of continental locations, particularly in the winter when the air flow is off the land.

Consider next the continental interior, which shows a somewhat different precipitation pattern, especially in the middle latitudes. Because cyclones are more common in the winter, we would expect midlatitude locations also to experience maximum precipitation in the winter. The precipitation regimes of the interior, however, are affected to a degree by a type of monsoon circulation. Cold winter temperatures and dominance of flow off the land in the winter make it the drier season (Omaha and Chicago). Although cyclonic storms frequent the continental interior in the winter, they do not become precipitation producers until they have drawn in abundant moist air from the Gulf of Mexico. Thus, most of the winter precipitation falls in the eastern one-third of the United States. The lack of abundant winter precipitation in the Plains states, compared with the East, can be seen in the data for Omaha and New York City.

In the summer, the inflow of warm, moist air from the ocean is aided by the thermal low that develops over the land and causes a general increase in precipitation over the midcontinent. The dominance of summer monsoon-type precipitation is evident when we compare the patterns for India in summer and winter by using Figure 7–24b and c. In the United States, places in the Southeast begin their rainy season in early spring, whereas more northerly and westerly locations do not experience their peak precipitation until summer or early fall.

Chapter Summary

- The largest planetary-scale wind patterns, called *macroscale winds*, include the westerlies and trade winds. A somewhat smaller macroscale circulation is called *synoptic scale*, or *weather-map scale*. *Mesoscale winds*, such as thunderstorms, tornadoes, and land and sea breezes, influence smaller areas and often exhibit intense vertical flow. The smallest scale of air motion is the *microscale*. Examples of these very local, often chaotic winds include gusts and dust devils.

- All winds have the same cause: pressure differences that arise because of temperature differences that are caused by unequal heating of Earth's surface. In addition to *land* and *sea breezes* brought about by the daily temperature contrast between land and water, other mesoscale winds include *mountain and valley breezes*, *chinook (foehn)* winds, *katabatic (fall)* winds, and *country breezes*. Mountain and valley breezes develop as air along mountain slopes is heated more intensely than air at the same elevation over the valley floor. Chinooks are warm, dry winds that sometimes move down the east slopes of the Rockies. In the Alps, winds similar to chinooks are called foehns. Katabatic (fall) winds originate when cold air, situated over a highland area such as the ice sheets of Greenland and Antarctica, is set in motion under the influence of gravity. Country breezes are associated with large urban areas where the circulation pattern is characterized by a light wind blowing into the city from the surrounding countryside.

- A simplified view of global circulation is a three-cell circulation model for each hemisphere. Because the circulation patterns between the equator and roughly 30° latitude north and south closely resemble a single-cell model developed by George Hadley in 1735, the name *Hadley cell* is generally applied. According to the three-cell circulation model, in each hemisphere, atmospheric circulation cells are located between the equator and 30° latitude, 30° and 60° latitude, and 60° latitude and the pole. The areas of general subsidence in the zone between 20° and 35° are called the *horse latitudes*. In each hemisphere, the equatorward flow from the horse latitudes forms the reliable *trade winds*. Convergence of the trade winds from both hemispheres near the equator produces a region of light winds called the *doldrums*. The circulation between 30° and 60° latitude (north and south) results in the *prevailing westerlies*. Air that moves equatorward from the poles produces the *polar easterlies* of both hemispheres. The area where the cold polar easterlies clash with the warm westerly flow of the midlatitudes is referred to as the *polar front*, an important meteorological region.

- If Earth's surface were uniform, two latitudinally oriented belts of high and two of low pressure would exist. Beginning at the equator, the four belts would be the (1) *equatorial low*, also referred to as the *intertropical convergence zone* (ITCZ) because it is the region where the trade winds converge, (2) *subtropical high*, at about 20° to 35° on either side of the equator, (3) *subpolar low*, situated at about 50° to 60° latitude, and (4) *polar high*, near Earth's poles.

- In reality, the only true zonal pattern of pressure exists along the subpolar low in the Southern Hemisphere. At other latitudes, particularly in the Northern Hemisphere, where there is a higher proportion of land compared to ocean, the zonal pattern is replaced by semipermanent cells of high and low pressure. January pressure and wind patterns show a very strong high-pressure center, called the *Siberian High*, positioned over the frozen landscape of northern Asia. Also evi-

dent in January, but absent in July, are two intense semipermanent low-pressure centers, the *Aleutian* and *Icelandic lows*, situated over the North Pacific and North Atlantic, respectively. With the onset of summer the pressure pattern over the Northern Hemisphere changes dramatically and high temperatures over the continents generate lows that replace the winter highs. During the peak summer season, the subtropical high found in the North Atlantic (called the *Azores high* in winter) is positioned near the island of Bermuda and is called the *Bermuda high*.

- The greatest seasonal change in Earth's global circulation is the development of *monsoons*, wind systems that exhibit a pronounced seasonal reversal in direction. The best-known and most pronounced monsoonal circulation, the *Asian monsoon* found in southern and southeastern Asia, is a complex seasonal change that is strongly influenced by the amount of solar heating received by the vast Asian continent. The *North American monsoon* (also called the *Arizona monsoon* and *Southwest monsoon*), a relatively small seasonal wind shift, produces a dry spring followed by a comparatively rainy summer that impacts large areas of southwestern United States and northwestern Mexico.

- The temperature contrast between the poles and equator drives the westerly winds (*westerlies*) located in the middle latitudes. Imbedded within the westerly flow aloft are narrow ribbons of high-speed winds, called *jet streams*, that meander for thousands of kilometers. The key to the origin of midlatitude jet streams is found in great temperature contrasts at the surface. In the region between 30° and 70°, the *midlatitude jet stream* occurs in association with the *polar front*.

- Studies of upper-level wind charts reveal that the westerlies follow wavy paths that have rather long wavelengths. The longest wave patterns are called *Rossby waves*. During periods when the flow aloft is relatively flat, little cyclonic activity is generated at the surface. Conversely, when the flow exhibits large-amplitude waves having short wavelengths, vigorous cyclonic storms are created. In addition to seasonal changes in the strength of its flow, the position of the midlatitude jet also shifts from summer to winter.

- Because winds are the primary driving force of ocean currents, a relationship exists between the oceanic circulation and the general atmospheric circulation. In general, in response to the circulation associated with the subtropical highs, ocean currents form clockwise spirals in the Northern Hemisphere and counterclockwise spirals in the Southern Hemisphere. In addition to influencing temperatures of adjacent land areas, cold ocean currents also transform some subtropical deserts into relatively cool, damp places that are often shrouded in fog. Furthermore, ocean currents play a major role in maintaining Earth's heat balance. In addition to producing surface currents, winds may also cause vertical water movements, such as the *upwelling* of cold water from deeper layers to replace warmer surface-water.

- *El Niño* refers to episodes of ocean warming caused by a warm counter-current flowing southward along the coasts of Ecuador and Peru that replaces the cold Peruvian current. The events are part of the global circulation and related to a seesaw pattern of atmospheric pressure between the eastern and western Pacific called the *Southern Oscillation*. El Niño events influence weather at great distances from Peru and Ecuador. Two of the strongest El Niño events (1982–83 and 1997–98) were responsible for a variety of weather extremes in many parts of the world. When surface temperatures in the eastern Pacific are colder than average a *La Niña* event is triggered. A typical La Niña winter blows colder than normal air over the Pacific Northwest and the northern Great Plains while warming much of the rest of the United States.

- The general features of the global distribution of precipitation can be explained by global winds and pressure systems. In general, regions influenced by high pressure, with its associated subsidence and divergent winds, experience dry conditions. On the other hand, regions under the influence of low pressure and its converging winds and ascending air receive ample precipitation.

- On a uniform Earth, throughout most of the year, heavy precipitation would occur in the equatorial region, the midlatitudes would receive most of their precipitation from traveling cyclonic storms, and polar regions would be dominated by cold air that holds little moisture. The most notable anomaly to this zonal distribution of precipitation occurs in the subtropics, where many of the world's great deserts are located. In these regions, pronounced subsidence on the eastern side of subtropical high-pressure centers results in stable atmospheric conditions. Upwelling of cold water along the west coasts of the adjacent continents further adds to the stable and dry conditions. On the other hand, the east coast of the adjacent continent receives abundant precipitation year round due to the convergence and rising air associated with the western side of the oceanic high.

Vocabulary Review

Aleutian low (p. 197)
Azores high (p. 197)
bora (p. 190)
Bermuda high (p. 197)
chinook (p. 190)
country breeze (p. 190)
doldrums (p. 193)
El Niño (p. 207)
equatorial low (p. 194)
foehn (p. 190)
Hadley cell (p. 192)
horse latitudes (p. 193)
Icelandic low (p. 197)
intertropical convergence zone (ITCZ) (p. 194)
jet stream (p. 200)
katabatic or fall wind (p. 190)
land breeze (p. 187)
La Niña (p. 211)
macroscale winds (p. 185)
mesoscale winds (p. 187)

microscale circulation (p. 187)
midlatitude jet stream (p. 200)
mistral (p. 190)
monsoon (p. 197)
mountain breeze (p. 190)
ocean currents (p. 205)
polar easterlies (p. 193)
polar front (p. 193)
polar high (p. 194)
prevailing westerlies (p. 193)
Santa Ana (p. 190)
sea breeze (p. 187)
Siberian high (p. 195)
Southern Oscillation (p. 208)
subpolar low (p. 194)
subtropical high (p. 194)
subtropical jet stream (p. 202)
trade winds (p. 193)
upwelling (p. 207)
valley breeze (p. 189)

Review Questions

1. Distinguish among macroscale, mesoscale, and microscale winds. Give an example of each.

2. If you were to view a weather map of the entire world for any single day of the year, would the global pattern of winds likely be visible? Explain your answer.

3. Why is the well-known Texas "norther" not a true local (mesoscale) wind?

4. The most intense sea breezes develop along tropical coasts adjacent to cool ocean currents. Explain.

5. What are katabatic (fall) winds? Name two examples.

6. Explain how cities create their own local winds.

7. Briefly describe the idealized global circulation proposed by George Hadley. What are the shortcomings of the Hadley model?

8. Which factors cause air to subside between 20° and 35° latitude?

9. If Earth rotated more rapidly, the Coriolis force would be stronger. How would a faster rate of rotation affect the location (latitude) of Earth's belts of subtropical high pressure?

10. Referring to the idealized three-cell model of atmospheric circulation, most of the United States is situated in which belt of prevailing winds?

11. Briefly explain each of the following statements that relate to the global distribution of surface pressure:
    a. The only true zonal distribution of pressure exists in the region of the subpolar low in the Southern Hemisphere.
    b. The subtropical highs are stronger in the North Atlantic in July than in January.
    c. The subpolar low in the Northern Hemisphere is represented by individual cyclonic storms that are more common in the winter.
    d. A strong high-pressure cell develops in the winter over northern Asia.

12. Explain the cause of the Asian monsoon. In which season (summer or winter) does it rain?

13. What area of North America experiences a pronounced monsoon circulation?

14. Why is the flow aloft predominantly westerly?

15. At what time of year should we expect the fastest westerly flow? Explain.

16. Describe the situation in which jet streams were first encountered.

17. Describe the manner in which pressure distribution is shown on upper-air charts. How are high- and low-pressure areas depicted on these charts?

18. What were the *dishpan experiments* and what have we learned from them?

19. Describe how a major El Niño event in the tropical Pacific might affect the weather in other parts of the globe.

20. How is La Niña different from El Niño?

21. What factors, other than global wind and pressure systems, exert an influence on the world distribution of precipitation?

Atmospheric Science Online

The following are informative and interesting Internet sites that address topics related to those presented in the chapter:

El Niño Theme Page (NOAA):
- **http://www.pmel.noaa.gov/toga-tao/el-nino/nino-home.html**

Jet Stream Analyses and Forecasts (California Regional Weather Server):
- **http://squall.sfsu.edu/crws/jetstream.html**

For direct links to these sites and others, chapter objectives and reviews, quiz questions, and topical investigations that utilize Web resources, visit *The Atmosphere, Eighth Edition* Home Page at:
- **http://www.prenhall.com/lutgens**

Air Masses

A remarkable six-day lake-effect snowstorm in November 1996 dropped 175 centimeters (nearly 69 inches) of snow on Chardon, Ohio, setting a new state record. Such snows are associated with the movement of continental polar air masses across the Great Lakes. *(Photo by Tony Dejak/AP/Wide World Photos)*

Figure 8–1 When an air mass that forms over a desert in summer moves out from its region of origin, it will bring hot, dry conditions to other regions. *(Photo by Warren Faidley/Weatherstock)*

Most people living in the middle latitudes have experienced hot, "sticky" summer heat waves and frigid winter cold waves. In the first case, after several days of sultry weather, the spell may come to an abrupt end that is marked by thundershowers and followed by a few days of relatively cool relief. In the second case, thick stratus clouds and snow may replace the clear skies that had prevailed and temperatures may climb to values that seem mild compared with what preceded them. In both examples, what was experienced was a period of generally uniform weather conditions followed by a relatively short period of change and the subsequent reestablishment of a new set of weather conditions that remained for perhaps several days before changing again.

What Is an Air Mass?

The weather patterns just described are the result of the movements of large bodies of air, called air masses. An **air mass**, as the term implies, is an immense body of air, usually 1600 kilometers (1000 miles) or more across and perhaps several kilometers thick, which is characterized by homogeneous physical properties (in particular, temperature and moisture content) at any given altitude. When this air moves out of its region of origin, it will carry these temperatures and moisture conditions elsewhere, eventually affecting a large portion of a continent (Figure 8–1).

An excellent example of the influence of an air mass is illustrated in Figure 8–2. Here a cold, dry mass from

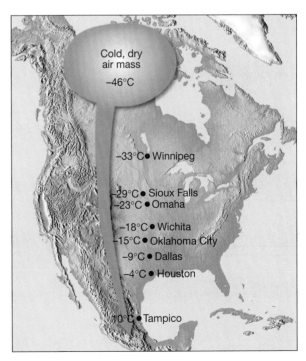

Figure 8–2 As this frigid Canadian air mass moved southward, it brought some of the coldest weather of the winter. *(After Tom L. McKnight, Physical Geography, 5th ed. Upper Saddle River, NJ: Prentice Hall, 1996, p. 174)*

northern Canada moves southward. With a beginning temperature of –46°C (–51°F), the air mass warms 13°C (24°F), to –33°C (–27°F), by the time it reaches Winnipeg. It continues to warm as it moves southward through the Great Plains and into Mexico. Throughout its southward journey, the air mass becomes warmer. But, it also brings some of the coldest weather of the winter to the places in its path. Thus, the air mass is modified, but it also modifies the weather in the areas over which it moves.

The horizontal uniformity of an air mass is not complete because it may extend through 20 degrees or more of latitude and cover hundreds of thousands to millions of square kilometers. Consequently, small differences in temperature and humidity from one point to another at the same level are to be expected. Still, the differences observed within an air mass are small in comparison to the rapid rates of change experienced across air mass boundaries.

Because it may take several days for an air mass to traverse an area, the region under its influence will probably experience generally constant weather conditions,

a situation called **air-mass weather**. Certainly, some day-to-day variations may exist, but the events will be very unlike those in an adjacent air mass.

The air mass concept is an important one because it is closely related to the study of atmospheric disturbances. Most disturbances in the middle latitudes originate along the boundary zones that separate different air masses.

Source Regions

Where do air masses form? What factors determine the nature and degree of uniformity of an air mass? These two basic questions are closely related, for the site where an air mass forms vitally affects the properties that characterize it.

Areas in which air masses originate are called **source regions**. Because the atmosphere is heated chiefly from below and gains its moisture by evaporation from Earth's surface, the nature of the source region largely determines the initial characteristics of an air mass. An ideal source region must meet two essential criteria. First, it must be

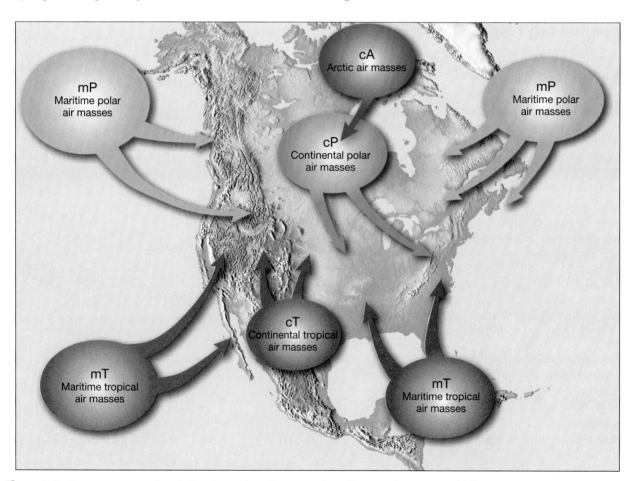

Figure 8–3 Air-mass source regions for North America. *(Courtesy of Ward's Natural Science Establishment, Inc., Rochester, N.Y.)*

an extensive and physically uniform area. A region having highly irregular topography or one that has a surface consisting of both water and land is not satisfactory.

The second criterion is that the area be characterized by a general stagnation of atmospheric circulation so that air will stay over the region long enough to come to some measure of equilibrium with the surface. In general, it means regions dominated by stationary or slow-moving anticyclones with their extensive areas of calms or light winds.

Regions under the influence of cyclones are not likely to produce air masses because such systems are characterized by converging surface winds. The winds in lows are constantly bringing air with unlike temperature and humidity properties into the area. Because the time involved is not long enough to eliminate these differences, steep temperature gradients result, and air-mass formation cannot take place.

Figure 8–3 shows the source regions that produce the air masses that most often influence North America. The waters of the Gulf of Mexico and Caribbean Sea and similar regions in the Pacific west of Mexico yield warm air masses, as does the land area that encompasses the southwestern United States and northern Mexico. In contrast, the North Pacific, the North Atlantic, and the snow- and ice-covered areas comprising northern North America and the adjacent Arctic Ocean are major source regions for cold air masses.

Notice that major source regions are not found in the middle latitudes, but instead are confined to subtropical and polar locations. The fact that the middle latitudes are the site where cold and warm air masses clash, often because the converging winds of a traveling cyclone draw them together, means that this zone lacks the conditions necessary to be a source region. Almost every air mass found in the middle latitudes originated elsewhere.

Classifying Air Masses

The classification of an air mass depends on the latitude of the source region and the nature of the surface in the area of origin—ocean or continent. The latitude of the source region indicates the temperature conditions within the air mass, and the nature of the surface below strongly influences the moisture content of the air.

Air masses are identified by two-letter codes. With reference to latitude (temperature), air masses are placed into one of three categories: **polar (P), arctic (A)**, or **tropical (T)**. The differences between polar and arctic are usually small and simply serve to indicate the degree of coldness of the respective air masses.

The lowercase letter m (for **maritime**) or the lowercase letter c (for **continental**) is used to designate the nature of the surface in the source region and hence the humidity characteristics of the air mass. Because maritime air masses form over oceans, they have a relatively high water-vapor content compared to continental air masses that originate over landmasses (Figure 8–4).

Figure 8–4 Air masses acquire their properties of temperature and moisture from extensive and physically uniform areas called *source regions*. When an air mass forms over the ocean, it is called *maritime*. Such air masses are relatively humid in contrast to continental air masses, which are dry. *(Photo by Arnulf Husmo/Tony Stone Images)*

When this classification scheme is applied, the following air masses can be identified:

cA	continental	arctic
cP	continental	polar
cT	continental	tropical
mT	maritime	tropical
mP	maritime	polar

Notice that the list does not include mA (maritime arctic). These air masses are not listed because they seldom, if ever, form. Although arctic air masses form over the Arctic Ocean, this water body is ice-covered throughout the year. Consequently, the air masses that originate here consistently have the moisture characteristics associated with a continental source region.

Air-Mass Modification

After an air mass forms, it normally migrates from the area where it acquired its distinctive properties to a region with different surface characteristics. Once the air mass moves from its source region, it not only modifies the weather of the area it is traversing, but it is also gradually modified by the surface over which it is moving. Warming or cooling from below, the addition or loss of moisture, and vertical movements all act to bring about changes in an air mass. The amount of modification can be relatively small or, as the following example illustrates, the changes can be profound enough to alter completely the original identity of the air mass.

When cA or cP air moves over the ocean in winter, it undergoes considerable change. Evaporation from the water surface rapidly transfers large quantities of moisture to the once-dry continental air. Furthermore, because the underlying water is warmer than the air above, the air is also heated from below. This factor leads to instability and vertically ascending currents that rapidly transport heat and moisture to higher levels. In a matter of days, cold, dry, and stable continental air is transformed into an unstable mP air mass.

When an air mass is colder than the surface over which it is passing, as in the preceding example, the lowercase letter k is added after the air-mass symbol. If, however, an air mass is warmer than the underlying surface, the lowercase letter w is added. It should be remembered that the k or w suffix does not mean that the air mass itself is cold or warm. It means only that the air is *relatively* cold or warm in comparison with the underlying surface over which it is traveling. For example, an mT air mass from the Gulf of Mexico is usually classified as mTk as it

moves over the southeastern states in summer. Although the air mass is warm, it is still cooler than the highly heated landmass over which it is passing.

The k or w designation gives an indication of the stability of an air mass and hence the weather that might be expected. An air mass that is colder than the surface is obviously going to be warmed in its lower layers. This fact causes greater instability that favors the ascent of the heated lower air and creates the possibility of cloud formation and precipitation. Indeed, a k air mass is often characterized by cumulus clouds, and should precipitation occur, it will be of the shower or thunderstorm variety. Also, visibility is generally good (except in rain) because of the stirring and overturning of the air.

Conversely, when an air mass is warmer than the surface over which it is moving, its lower layers are chilled. A surface inversion that increases the stability of the air mass often develops. This condition does not favor the ascent of air, and so it opposes cloud formation and precipitation. Any clouds that do form will be stratus, and precipitation, if any, will be light to moderate. Moreover, because of the lack of vertical movements, smoke and dust often become concentrated in the lower layers of the air mass and cause poor visibility. During certain times of the year, fogs, especially the advection type, may also be common in some regions.

In addition to modifications resulting from temperature differences between an air mass and the surface below, vertical movements induced by cyclones and anticyclones or topography can also affect the stability of an air mass. Such modifications are often called *mechanical* or *dynamic* and are usually independent of the changes caused by surface cooling or heating. For example, significant modification can result when an air mass is drawn into a low. Here convergence and lifting dominate and the air mass is rendered more unstable. Conversely, the subsidence associated with anticyclones acts to stabilize an air mass. Similar alterations in stability occur when an air mass is lifted over highlands or descends the leeward side of a mountain barrier. In the first case, the air's stability is reduced; in the second case, the air becomes more stable.

Properties of North American Air Masses

Air masses frequently pass over us, which means that the day-to-day weather we experience often depends on the temperature, stability, and moisture content of these large bodies of air. In this section, we briefly examine the properties of the principal North American air masses. In addition, Table 8–1 serves as a summary.

Table 8–1 Weather characteristics of North American air masses

Air Mass	Source Region	Temperature and Moisture Characteristics in Source Region	Stability in Source Region	Associated Weather
cA	Arctic basin and Greenland ice cap	Bitterly cold and very dry in winter	Stable	Cold waves in winter
cP	Interior Canada and Alaska	Very cold and dry in winter	Stable entire year	a. Cold waves in winter b. Modified to cPk in winter over Great Lakes bringing "lake-effect" snow to leeward shores
mP	North Pacific	Mild (cool) and humid entire year	Unstable in winter Stable in summer	a. Low clouds and showers in winter b. Heavy orographic precipitation on windward side of western mountains in winter c. Low stratus and fog along coast in summer; modified to cP inland
mP	Northwestern Atlantic	Cold and humid in winter Cool and humid in summer	Unstable in winter Stable in summer	a. Occasional "nor'easter" in winter b. Occasional periods of clear, cool weather in summer
cT	Northern interior Mexico and southwestern U.S. (summer only)	Hot and dry	Unstable	a. Hot, dry, and cloudless, rarely influencing areas outside source region b. Occasional drought to southern Great Plains
mT	Gulf of Mexico, Caribbean Sea, western Atlantic	Warm and humid entire year	Unstable entire year	a. In winter it usually becomes mTw moving northward and brings occasional widespread precipitation or advection fog b. In summer, hot and humid conditions, frequent cumulus development and showers or thunderstorms
mT	Subtropical Pacific	Warm and humid entire year	Stable entire year	a. In winter it brings fog, drizzle, and occasional moderate precipitation to N.W. Mexico and S.W. United States b. In summer this air mass occasionally reaches the western United States and is a source of moisture for infrequent convectional thunderstorms.

Continental Polar (cP) and Continental Arctic (cA) Air Masses

Continental polar and continental arctic air masses are, as their classification implies, cold and dry. Continental polar air originates over the snow-covered interior regions of Canada and Alaska, poleward of the 50th parallel. Continental arctic air forms farther north over the Arctic Basin and the Greenland ice cap (Figure 8–3).

Continental arctic air is distinguished from cP air by its generally lower temperatures, although at times the differences may be slight. In fact, some meteorologists do not differentiate between cP and cA.

During the winter, both air masses are bitterly cold and very dry. Because the winter nights are long, Earth radiates heat that is, for the most part, not replenished by incoming solar energy. Therefore, the surface reaches very low temperatures and the air near the ground is gradually chilled to heights of 1 kilometer (0.6 mile) or more. The result is a strong and persistent temperature inversion with the coldest temperatures near the ground. Marked stability is,

Box 8–1 The Siberian Express

The surface weather map for December 22, 1989, shows a high-pressure center covering the eastern two-thirds of the United States and a substantial portion of Canada (Figure 8–A). As is usually the case in winter, a large anticyclone such as this is associated with a huge mass of dense and bitterly cold arctic air. After such an air mass forms over the frozen expanses near the Arctic Circle, the winds aloft sometimes direct it toward the south and east. When an outbreak takes place, it is popularly called the "Siberian Express" by the news media.

November 1989 was unusually mild for late autumn. In fact, across the United States, over 200 daily high-temperature records were set. December, however, was different. East of the Rockies, the month's weather was dominated by two arctic outbreaks. The second brought record-breaking cold.

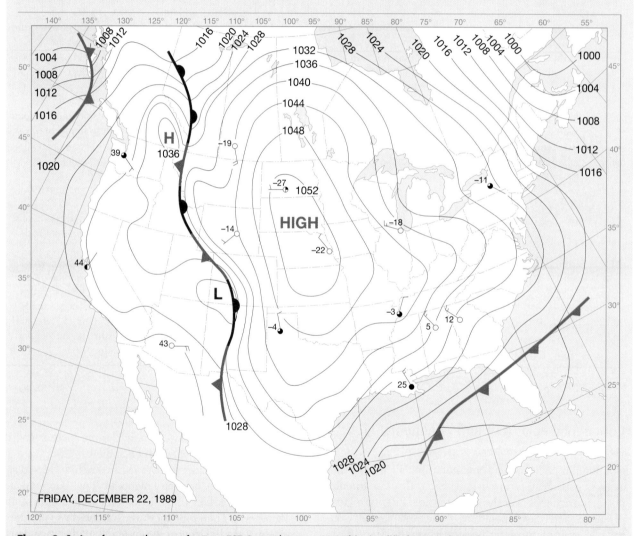

Figure 8–A A surface weather map for 7 A.M. EST, December 22, 1989. This simplified National Weather Service (NWS) map shows an intense winter cold wave caused by an outbreak of frigid continental arctic air. This event brought subfreezing temperatures as far south as the Gulf of Mexico. Temperatures on NWS maps are in degrees Fahrenheit.

Between December 21–25, as the frigid dome of high pressure advanced southward and eastward, more than 370 record low temperatures were reported. On December 21, Havre, Montana, had an overnight low of −42.2°C (−44°F), breaking a record set in 1884. Meanwhile, Topeka's −32.2°C (−26°F) was that city's lowest temperature for any date since record keeping started 102 years earlier.

The three days that followed saw the arctic air migrate toward the south and east. By December 24, Tallahassee, Florida, had a low temperature of −10°C (14°F). In the center of the state, the daily minimum at Orlando was −5.6°C (22°F). It was actually warmer in North Dakota on Christmas Eve than in central and northern Florida!

As we would expect, utility companies in many states reported record demand. When the arctic air advanced into Texas and Florida, agriculture was especially hard hit. Some Florida citrus growers lost 40 percent of their crop and many vegetable crops were wiped out completely (Figure 8–B).

After Christmas, the circulation pattern that brought this record-breaking Siberian Express from the arctic deep into the United States changed. As a result, temperatures for much of the country during January and

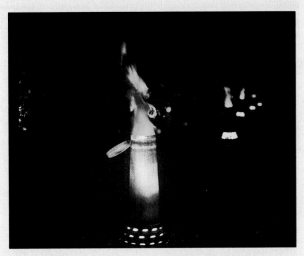

Figure 8–B When an arctic air mass invades the citrus regions of Florida and Texas, even modern freeze controls may not be able to prevent significant losses. *(Photo courtesy of Florida Citrus Commission)*

February 1990 were well above normal. In fact, January 1990 was the second warmest January in 96 years. Thus, despite frigid December temperatures, the winter of 1989–1990 "averaged out" to be a relatively warm one.

---

therefore, the rule. Because the air is very cold and the surface below is frozen, the mixing ratio of these air masses is necessarily low, ranging from perhaps 0.1 gram per kilogram in cA up to 1.5 grams per kilogram in some cP air.

As wintertime cP or cA air moves outward from its source region, it carries its cold, dry weather to the United States, normally entering between the Great Lakes and the Rockies. Because there are no major barriers between the high-latitude source regions and the Gulf of Mexico, cP and cA air masses can sweep rapidly and with relative ease far southward into the United States. The winter cold waves experienced in much of the central and eastern United States are closely associated with such polar outbreaks. One such cold wave is described in Box 8–1. Usually, the last freeze in spring and the first in autumn can be correlated with outbreaks of polar or arctic air.

Because cA air is present principally in the winter, only cP air has any influence on our summer weather, and this effect is considerably reduced when compared with winter. During summer months, the properties of the source region for cP air are very different from those during winter. Instead of being chilled by the ground, the air is warmed from below as the long days and higher Sun angle warm the

snow-free land surface. Although summer cP air is warmer and has a higher moisture content than its wintertime counterpart, the air is still cool and relatively dry compared with areas farther south. Summer heat waves in the northern portions of the eastern and central United States are often ended by the southward advance of cP air, which for a day or two brings cooling relief and bright, pleasant weather.

Lake-Effect Snow: Cold Air Over Warm Water

Continental polar air masses are not, as a rule, associated with heavy precipitation. Yet every late autumn and winter a unique and interesting weather phenomenon takes place along the downwind shores of the Great Lakes.° Periodically, brief, heavy snow showers issue from dark clouds that move onshore from the lakes (see Box 8–2). Seldom do these storms move more than about 80 kilometers (50 miles) inland from the shore before the snows come to an end. These highly localized storms, occurring along the leeward shores of the Great Lakes, create what are known as **lake-effect snows**.

°Actually the Great Lakes are just the best known example. Other large lakes can also experience this phenomenon.

Box 8–2 Atmospheric Hazard: An Extraordinary Lake-Effect Snowstorm

Northeastern Ohio is part of the Lake Erie snowbelt, a zone that extends eastward into northwestern Pennsylvania and western New York (see Figure 8–5). Here snowfall is enhanced when cold winds blow from the west or northwest across the relatively warm unfrozen waters of Lake Erie. Average annual snowfall in northeastern Ohio is between 200 and 280 centimeters (80 to 110 inches) and increases to 450 centimeters (175 inches) in western New York.

Although residents here are accustomed to abundant snows, even they were surprised in November 1996 by an especially early and strong storm. During a six-day span, from November 9 to 14, northeastern Ohio experienced record-breaking lake-effect snows (see chapter-opening photo). Rather than raking fall leaves, people were shoveling sidewalks and clearing snow from overloaded roofs.

Persistent snow squalls (narrow bands of heavy snow) frequently produced accumulation rates of 5 centimeters (2 inches) per hour during the six-day siege. The snow was the result of especially deep and prolonged lower atmospheric instability created by the movement of cold air across a relatively warm Lake Erie. The surface temperature of the lake was 12°C (54°F), several degrees above normal. The air temperature at a height of 1.5 kilometers (nearly 5000 feet) was –5°C (23°F).

The northwest flow of cold air across Lake Erie and the 17°C lapse rate generated lake-effect rain and snow almost immediately.* During some periods loud thunder accompanied the snow squalls!

Data from the Cleveland National Weather Service Snow Spotters Network show 100 to 125 centimeters (39 to 49 inches) of snow through the core of the Ohio snowbelt (Figure 8–C). The deepest accumulation was measured near Chardon, Ohio. Here the six-day total of 175 centimeters (68.9 inches) far exceeded the previous Ohio record of 107 centimeters (42 inches) set in 1901. In addition, the November 1996 snowfall total of 194.8 centimeters (76.7 inches) at the same site set a new

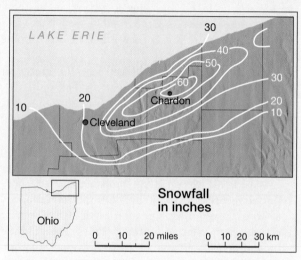

Figure 8–C Snowfall totals (in inches) for northeastern Ohio for the November 9–14, 1996, lake-effect storm. The deepest accumulation, nearly 69 inches (175 centimeters) occurred near Chardon. *(After National Weather Service)*

monthly snowfall record for Ohio. The previous one-month record had been 176.5 centimeters (69.5 inches).

The impact of the storm was considerable. Ohio's governor declared a state of emergency on November 12 and National Guard troops were dispatched to assist in snow removal and aid in the rescue of snowbound residents. There was widespread damage to trees and shrubs across northeastern Ohio because the snow was particularly wet and dense and readily clung to objects. Press reports indicated that about 168,000 homes were without electricity, some for several days. Roofs of numerous buildings collapsed under the excessive snow loads. Although residents of the northeast Ohio snowbelt are winter storm veterans, the six-day storm in November 1996 will be long remembered as an extraordinary event.

*Thomas W. Schmidlin and James Kasarik, "A Record Ohio Snowfall during 9–14 November 1996," *Bulletin of the American Meteorological Society*, Vol. 80, No. 6, June 1999, p. 1109.

Lake-effect storms account for a high percentage of the snowfall in many areas adjacent to the lakes. The strips of land that are most frequently affected, called *snowbelts*, are shown in Figure 8–5. When snowfall statistics for large cities were examined for a recent 10-year span, the two snowiest metropolitan areas in the United States turned out to be Buffalo and Rochester in upstate New York. Buffalo, on the eastern shore of Lake Erie, had more than 2700 centimeters (nearly 90 feet) of snow during the period, and Rochester, on the south side of Lake Ontario, recorded

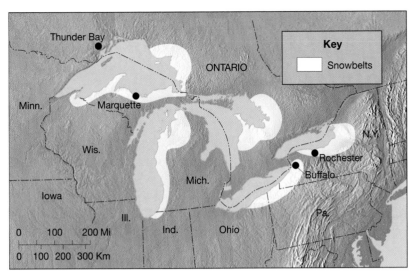

Figure 8–5 The snowbelts of the Great Lakes region are the zones that most frequently experience lake-effect snowstorms.

2635 centimeters. A comparison of average snowfall totals at Thunder Bay, Ontario, on the north shore of Lake Superior, and Marquette, Michigan, along the southern shore, provides another excellent example. Because Marquette is situated on the leeward shore of the lake, it receives substantial lake-effect snow and therefore has a much higher snowfall total than does Thunder Bay (Table 8–2).

What causes lake-effect snow? The answer is closely linked to the differential heating of water and land (Chapter 3) and to the concept of atmospheric instability. During the summer months, bodies of water, including the Great Lakes, absorb huge quantities of energy from the Sun and from the warm air that passes over them. Although these water bodies do not reach particularly high temperatures, they nevertheless represent huge reservoirs of heat. The surrounding land, in contrast, cannot store heat nearly as effectively. Consequently, during autumn and winter, the temperature of the land drops quickly, whereas water bodies lose their heat more gradually and cool slowly.

From late November through late January, the contrasts in average temperatures between water and land range from about 8°C in the southern Great Lakes to 17°C farther north. However, the temperature differences can be much greater (perhaps 25°C) when a very cold cP or cA air mass pushes southward across the lakes. When such a dramatic temperature contrast exists, the lakes interact with the air to produce major lake-effect storms. Figure 8–6 depicts the movement of a cP air mass across one of the Great Lakes. During its journey, the air acquires large quantities of heat and moisture from the relatively warm lake surface. By the time it reaches the opposite shore, this cPk air is humid and unstable and heavy snow showers are likely.

Maritime Polar (mP) Air Masses

Maritime polar air masses form over oceans at high latitudes. As the classification would indicate, mP air is cool to cold and humid, but compared with cP and cA air masses in winter, mP air is relatively mild because of the higher temperatures of the ocean surface as contrasted to the colder continents.

Two regions are important sources for mP air that influences North America: the North Pacific and the northwestern Atlantic from Newfoundland to Cape Cod (Figure 8–3). Because of the general west-to-east circulation in the middle latitudes, mP air masses from the North Pacific source region have a more profound influence on North American weather than do mP air masses generated in the northwestern Atlantic. Whereas air masses that form in the Atlantic generally move eastward toward Europe, mP air from the North Pacific has a strong influence on the weather along the western coast of North America, especially in the winter.

Table 8–2 Monthly snowfall at Thunder Bay, Ontario, and Marquette, Michigan

Thunder Bay, Ontario			
October	November	December	January
3.0 cm	14.9 cm	19.0 cm	22.6 cm
(1.2 in.)	(5.8 in.)	(7.4 in.)	(8.8 in.)
Marquette, Michigan			
October	November	December	January
5.3 cm	37.6 cm	56.4 cm	53.1 cm
(2.1 in.)	(14.7 in.)	(22 in.)	(20.7 in.)

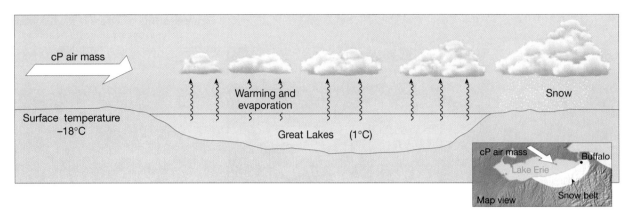

Figure 8–6 As continental polar air crosses the Great Lakes in winter, it acquires moisture and is made less stable because of warming from below. A "lake-effect" snow shower on the lee side of the lakes is often the consequence of this air-mass modification.

Pacific mP Air Masses. During the winter, mP air masses from the Pacific usually begin as cP air in Siberia (Figure 8–7). Although air rarely stagnates over this area, the source region is extensive enough to allow the air moving across it to acquire its characteristic properties. As the air advances eastward over the relatively warm water, active evaporation and heating occur in the lower levels. Consequently, what was once a very cold, dry, and stable air mass is changed into one that is mild and humid near the surface and relatively unstable. As this mP air arrives at the western coast of North America, it is often accompanied by low clouds and shower activity. When the mP air advances inland against the western mountains, orographic uplift produces heavy rain or snow on the windward slopes of the mountains (see Figure 15–12).

Maritime Polar Air from the North Atlantic. As is the case for mP air from the Pacific, air masses forming in the

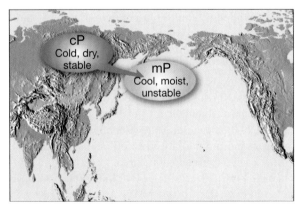

Figure 8–7 During winter, maritime polar (mP) air masses in the North Pacific usually begin as continental polar (cP) air masses in Siberia. The cP air is modified to mP as it slowly crosses the ocean.

northwestern Atlantic source region were originally cP air masses that moved from the continent and were transformed over the ocean. However, unlike air masses from the North Pacific, mP air from the Atlantic only occasionally affects the weather of North America. Nevertheless, this air mass does have an effect when the northeastern United States is on the northern or northwestern edge of a passing low-pressure center. On these occasions, cyclonic winds draw mP air into the region. Its influence is generally confined to the area east of the Appalachians and north of Cape Hatteras, North Carolina. The weather associated with an invasion of mP air from the Atlantic is known locally as a **nor'easter**. Strong northeast winds, freezing or near freezing temperatures, high relative humidity, and the possibility of precipitation make this weather phenomenon a most unwelcome event.

Summer brings a change in the characteristics of mP air masses from the North Pacific. During the warm season, the ocean is cooler than the surrounding continents. In addition, the Pacific anticyclone lies off the western coast of the United States (see Figure 7–9). Consequently, there is almost continuous southward flow of air having moderate temperatures. Although the air near the surface may often be conditionally unstable, the presence of the Pacific high means that there is subsidence and stability aloft. Thus, low stratus clouds and summer fogs characterize much of the western coast. Once summer mP air from the Pacific moves inland, it is heated at the surface over the hot and dry interior. The heating and resulting turbulence act to reduce the relative humidity in the lower layers, and the clouds dissipate.

Although mP air masses from the Atlantic may produce an occasional unwelcome nor'easter during the winter, summertime incursions of this air mass often bring pleasant weather. Like the Pacific source region, the northwestern Atlantic is dominated by high pressure during the summer (see Figure 7–7). Thus, the upper air is stable because of subsidence and the lower air is essentially stable because of the chilling effect of the cold water. As the circulation on the southern side of the anticyclone carries this stable and relatively dry mP air into New England, and occasionally as far south as Virginia, the region enjoys clear, cool weather and good visibility.

Maritime Tropical (mT) Air Masses

Maritime tropical air masses affecting North America most often originate over the warm waters of the Gulf of Mexico, the Caribbean Sea, or the adjacent western Atlantic Ocean (Figure 8–3). The tropical Pacific is also a source region for mT air. However, the land area affected by this latter source is small compared with the size of the region influenced by air masses produced in the Gulf of Mexico and adjacent waters.

As expected, mT air masses are warm to hot and they are humid. In addition, they are often unstable. It is through invasions of mT air masses that the subtropics export much heat and moisture to the less endowed areas to the north. Consequently, these air masses are important to the weather whenever present because they are capable of contributing significant precipitation.

North Atlantic mT Air. Maritime tropical air masses from the Gulf–Caribbean–Atlantic source region greatly affect the weather of the United States east of the Rocky Mountains. Although the source region is dominated by the North Atlantic subtropical high, the air masses produced are not stable but are neutral or unstable because the source region is located on the weak western edge of the anticyclone where pronounced subsidence is absent.

During winter, when cP air dominates the central and eastern United States, mT air only occasionally enters this part of the country. When an invasion does occur, the lower portions of the air mass are chilled and stabilized as it moves northward. Its classification is changed to mTw. As a result, the formation of convective showers is unlikely. Widespread precipitation does occur, however, when a northward-moving mT air mass is pulled into a traveling cyclone and forced to ascend. In fact, much of the wintertime precipitation over the eastern and central states results when mT air from the Gulf of Mexico is lifted along fronts in traveling cyclones.

Another weather phenomenon associated with a northward-moving wintertime mT air mass is advection fog. Dense fogs can develop as the warm, humid air is chilled as it moves over the cold land surface.

During the summer, mT air masses from the Gulf, Caribbean, and adjacent Atlantic affect a much wider area of North America and are present for a greater percentage of the time than during the winter. As a result, they exert a strong and often dominating influence over the summer weather of the United States east of the Rocky Mountains. This influence is due to the general sea-to-land (monsoonal) airflow over the eastern portion of North America during the warm months, which brings more frequent incursions of mT air that penetrate much deeper into the continent than during the winter months. Consequently, these air masses are largely responsible for the hot and humid conditions that prevail over the eastern and central United States.

Initially, summertime mT air from the Gulf is unstable. As it moves inland over the warmer land, it becomes an mTk air mass as daytime heating of the surface layers further increases the air's instability. Because the relative humidity is high, only modest lifting is necessary to bring about active convection, cumulus development, and thunderstorm or shower activity (Figure 8–8). This is, indeed, a common warm-weather phenomenon associated with mT air.

Figure 8–8 As mT air from the Gulf of Mexico moves over the heated land in the summer, cumulus development and afternoon showers frequently result. *(Photo by Rod Planck/Photo Researchers, Inc.)*

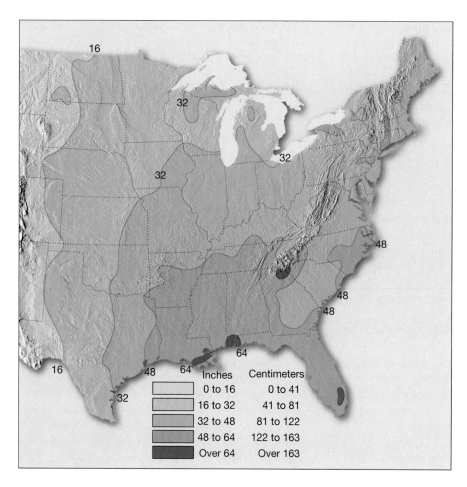

Figure 8–9 Average annual precipitation for the eastern two-thirds of the United States. Note the general decrease in yearly precipitation totals with increasing distance from the Gulf of Mexico, the source region for mT air masses. Isohyets are labeled in inches. *(Courtesy of Environmental Data Service, NOAA)*

Inches	Centimeters
0 to 16	0 to 41
16 to 32	41 to 81
32 to 48	81 to 122
48 to 64	122 to 163
Over 64	Over 163

It should be also noted here that air masses from the Gulf–Caribbean–Atlantic region are the primary source of much, if not most, of the precipitation received in the eastern two-thirds of the United States. Pacific air masses contribute little to the water supply east of the Rockies because the western mountains effectively "drain" the moisture from the air by numerous episodes of orographic uplift.

Figure 8–9, which shows the distribution of average annual precipitation for the eastern two-thirds of the United States by using **isohyets** (lines connecting places having equal rainfall), illustrates this situation nicely. The pattern of isohyets shows the greatest rainfall in the Gulf region and a decrease in precipitation with increasing distance from the mT source region.

North Pacific mT Air. Compared to mT air from the Gulf of Mexico, mT air masses from the Pacific source region have much less of an impact on North American weather. In winter, only northwestern Mexico and the extreme southwestern United States are influenced by air from the tropical Pacific. Because the source region lies along the eastern side of the Pacific anticyclone, subsidence

aloft produces upper-level stability. When the air mass moves northward, cooling at the surface also causes the lower layers to become more stable, often resulting in fog or drizzle. If lifted along a front or forced over mountains, moderate precipitation results.

For many years, the summertime influence of air masses from the tropical Pacific source region on the weather of the southwestern United States and northern Mexico was believed to be minimal. It was thought that the moisture for the infrequent summer thunderstorms that occur in the region came from occasional westward thrusts of mT air from the Gulf of Mexico. However, the Gulf of Mexico is no longer believed to be the primary supplier of moisture for the area west of the Continental Divide. Rather, it has been demonstrated that the tropical North Pacific west of central Mexico is a more important source of moisture for this area.

In summer, the mT air moves northward from its Pacific source region up the Gulf of California and into the interior of the western United States. This movement, which is confined largely to July and August, is essentially monsoonal in character. That is, the inflow of moist air is a response to the thermally produced low over the heated landmass. The July–August rainfall maximum for Tucson is a response to this incursion of Pacific mT air (Figure 8–10).

Continental Tropical (cT) Air Masses

North America narrows as it extends southward through Mexico; therefore, the continent has no extensive source region for continental tropical air masses. Only in summer do northern interior Mexico and adjacent parts of the arid southwestern United States produce hot, dry cT air. Because of the intense daytime heating at the surface, both a steep environmental lapse rate and turbulence extending to considerable heights are found. Nevertheless, although the air is unstable, it generally remains nearly cloudless because of extremely low humidity. Consequently, the prevailing weather is hot, with an almost complete lack of rainfall. Large daily temperature ranges are the rule. Although cT air masses are usually confined to the source region, occasionally they move into the southern Great Plains. If the cT air persists for long, drought may occur.

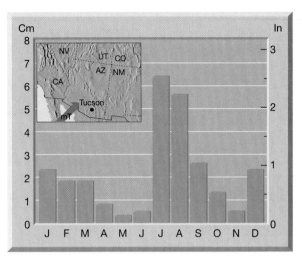

Figure 8–10 Monthly precipitation data for Tucson, Arizona. The July–August rainfall maximum in the desert Southwest results from the monsoonal flow of Pacific mT air into the region.

Chapter Summary

- In the middle latitudes, many weather events are associated with the movements of air masses. An *air mass* is a large body of air, usually 1600 kilometers (1000 miles) or more across and perhaps several kilometers thick, that is characterized by homogeneous physical properties (in particular temperature and moisture content) at any given altitude. A region under the influence of an air mass will probably experience generally constant weather conditions, a situation referred to as *air mass weather*.

- Areas in which air masses originate are called *source regions*. An ideal source region must meet two criteria. First, it must be an extensive and physically uniform area. The second criterion is that the area is characterized by a general stagnation of atmospheric circulation so that air will stay over the region long enough to come to some measure of equilibrium with the surface.

- The classification of an air mass depends on (1) the latitude of the source region, and (2) the nature of the surface in the area of origin—ocean or continent. Air masses are identified by two-letter codes. With reference to latitude (temperature), air masses are placed into one of three categories: *polar* (P), *arctic* (A), or *tropical* (T). A lowercase letter (m, for *maritime*, or c for *continental*) is placed in front of the uppercase letter to designate the nature of the surface in the source region and therefore the humidity characteristics of the air mass. Using this classification scheme, the following air masses are identified: cA, cP, cT, mT, and mP.

- Once an air mass moves from its source region, it not only modifies the weather of the area it is traversing, but it is also gradually modified by the surface over which it is moving. Changes to the stability of an air mass can result from temperature differences between an air mass and the surface and/or vertical movements induced by cyclones, anticyclones or topography.

- The day-to-day weather we experience depends on the temperature, stability, and moisture content of the air mass we are experiencing. *Continental polar* (cP) and *continental arctic* (cA) air masses are, as their classification implies, cold and dry. Although cP air masses are not, as a rule, associated with heavy precipitation, those that cross the Great Lakes during late autumn and winter sometimes bring *lake-effect snows* to the leeward shores. *Maritime polar* air masses (mP) form over oceans at high latitudes and are cool to cold and humid. The stormy winter weather associated with an invasion of mP air from the Atlantic into an area east of the Appalachians and north of Cape Hatteras is known as a *nor'easter*. *Maritime tropical* (mT) air masses affecting North America most often originate over the warm water of the Gulf of Mexico, the Caribbean Sea, or the adjacent western Atlantic Ocean. As expected, mT air masses are warm to hot and they are humid. During winter, when cP air dominates the central and eastern United States, mT air only occasionally enters this part of the country. However, during the summer, mT air masses from the Gulf, Caribbean, and adjacent Atlantic are more common and cover a much wider area of the continent. The mT

air masses from the Gulf-Caribbean-Atlantic source region are also the source of much, if not most, of the precipitation received in the eastern two-thirds of the United States. *Isohyets*, lines of equal rainfall drawn on a map, show that the greatest rainfall occurs in the Gulf region and decreases with increasing distance from the mT source region. Hot and dry *continental tropical* (cT) air is produced only in the summer in northern interior Mexico and adjacent parts of the arid southwestern United States.

Vocabulary Review

air mass (p. 221)
air-mass weather (p. 222)
arctic (A) air mass (p. 223)
continental (c) air mass (p. 223)
isohyet (p. 232)
lake-effect snow (p. 227)

maritime (m) air mass (p. 223)
nor'easter (p. 230)
polar (P) air mass (p. 223)
source region (p. 222)
tropical (T) air mass (p. 223)

Review Questions

1. Define the terms *air mass* and *air-mass weather*.
2. What two criteria must be met for an area to be an air-mass source region?
3. Why are regions that have a cyclonic circulation generally not conducive to air-mass formation?
4. On what bases are air masses classified? Compare the temperature and moisture characteristics of the following air masses: cP, mP, mT, and cT.
5. Why is mA left out of the air-mass classification scheme?
6. What do the lowercase letters k and w indicate about an air mass? List the general weather conditions associated with k and w air masses.
7. During winter, polar air masses are cold. Which should be coldest, a wintertime mP air mass or a wintertime cP air mass? Explain your choice.
8. How might vertical movements induced by a pressure system or topography act to modify an air mass?
9. What two air masses are most important to the weather of the United States east of the Rocky Mountains? Explain your choice.

10. What air mass influences the weather of the Pacific Coast more than any other?
11. Why do cA and cP air masses often sweep so far south into the United States?
12. Describe the modifications that occur as a cP air mass moves across one of the Great Lakes in the winter.
13. Why do mP air masses from the North Atlantic source region seldom affect the eastern United States?
14. What air mass and source region provide the greatest amount of moisture to the eastern and central United States?
15. For each statement below, indicate which air mass is most likely involved and from what source region it came:
    a. summer drought in the southern Great Plains
    b. wintertime advection fog in the Midwest
    c. heavy winter precipitation in the western mountains
    d. summertime convectional showers in the Midwest and East
    e. a nor'easter

Problems

1. Figure 8–11 shows the distribution of air temperatures (top number) and dew point temperatures (lower number) for a December morning. Two well-developed air masses are influencing North America

at this time. The air masses are separated by a broad zone that is not affected by either air mass. Draw lines on the map that show the boundaries of each air mass. Label each air mass with the proper classification.

2. The Great Lakes are not the only water bodies associated with lake-effect snow. For example, large lakes in Canada also experience this phenomenon. Shown here are snowfall data (in centimeters) for two Canadian stations that receive significant lake-effect snow. Fort Resolution is on the southeastern shore of Great Slave Lake and Norway House is at the northern end of Lake Winnipeg. During what month does each station have its snowfall maximum? Suggest an explanation as to why the maximum occurs when it does.

	Sept.	Oct.	Nov.	Dec.	Jan.
Fort Resolution	2.5	13.7	36.6	19.6	15.4
Norway House	3.8	6.9	29.5	20.1	16.3

3. Albuquerque, New Mexico, is situated in the desert Southwest. Its annual precipitation is just 21.2 centimeters (8.3 inches). Month-by-month data (in centimeters) are as follows:

J	F	M	A	M	J	J	A	S	O	N	D
1.0	1.0	1.3	1.3	1.3	1.3	3.3	3.8	2.3	2.3	1.0	1.3

What are the two rainiest months? The pattern here is similar to other southwestern cities, including Tuscon, Arizona. Briefly explain why the rainiest months occur when they do.

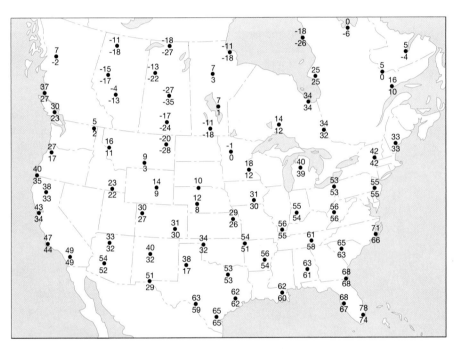

Figure 8–11 Map to accompany problem 1.

Atmospheric Science Online

The following are informative and interesting Internet sites that address topics related to those presented in the chapter:

Meteorology: Online Guides (University of Illinois):
- **http://ww2010.atmos.uiuc.edu/(Gh)/guides /mtr/home.rxml**

Lake-effect Weather Page (National Weather Service)
- **http://www.nws.noaa.gov/er/buf/lakeffect /newintro.html**

For direct links to these sites and others, chapter objectives and reviews, quiz questions, and topical investigations that utilize Web resources, visit *The Atmosphere, Eighth Edition* Home Page at:
- **http://www.prenhall.com/lutgens**

CHAPTER NINE

Weather Patterns

Satellite image of a midlatitude cyclone over the British Isles.
(Courtesy of European Space Agency/Science Photo Library/Photo Researchers, Inc.)

Figure 9–1 A New York City pedestrian struggles in the wind during the 1993 "storm of the century." *(Courtesy of AP/Wide World Photos)*

The winter of 1992–1993 came to a stormy conclusion in eastern North America during a weekend in mid-March. The daffodils were already out across the South and people were thinking about spring when the Blizzard of '93 struck on March 13 and 14. The huge storm brought record-low temperatures, record-low barometric-pressure readings, and record-high snowfalls to an area that stretched from Alabama to the Maritime Provinces of eastern Canada (Table 9–1). The monster storm with its driving winds and heavy snow combined the attributes of both hurricane and blizzard as it moved up the spine of the Appalachians, lashing and burying a huge swatch of territory (Figure 9–1). Although the atmospheric pressure at the storm's center was lower than the pressures at the centers of some hurricanes and the winds were frequently as strong as those in a hurricane, this was definitely not a tropical storm but a classic winter cyclone (Figure 9–2).

The event quickly earned the title "Storm of the Century." The name was well deserved. Although many notable storms occurred during the twentieth century, none could match the extreme dimensions of this event. A famous nineteenth-century storm known as the "Blizzard of '88" was equally severe but affected a much smaller area.

Despite excellent forecasts that permitted timely warnings, at least 270 people died as a result of the storm, more than three times the combined death tolls of hurricanes Hugo and Andrew. In Florida, severe weather that included 27 tornadoes killed 44 people. At one time or

Table 9–1 A look at the "Storm of the Century"

City	Snowfall Totals (CM)	(IN.)
Mt. LeConte, Tenn.	140	56
Mt. Mitchell, N.C.	125	50
Syracuse, N.Y.	108	43
Lincoln, N.H.	88	35
Mountain City, Ga.	60	24
Chattanooga, Tenn.	50	20
Asheville, N.C.	48	19

City	Record Low Temperatures in the South (°C)	(°F)
Mt. LeConte, Tenn.	−23	−10
Waynesville, N.C.	−20	−4
Birmingham, Ala.	−17	2
Knoxville, Tenn.	−14	6
Atlanta, Ga.	−8	18
Daytona Beach, Fla.	−0.6	31

City	Strong Wind Gusts (KPH)	(MPH)
Mt. Washington, N.H.	232	144
Flattop Mountain, N.C.	163	101
South Marsh Island, La.	148	92
Myrtle Beach, S.C.	145	90
Fire Island, N.Y.	143	89
Boston, Mass.	130	81

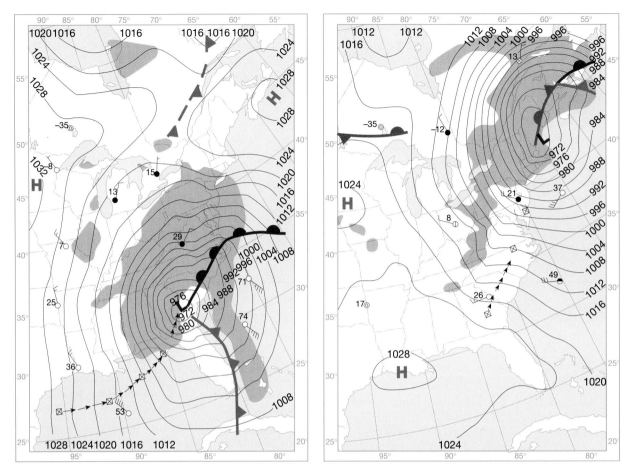

Figure 9–2 These simplified weather maps show the positions of an enormous cyclonic storm on March 13 and 14, 1993. After the storm formed in the Gulf of Mexico, it moved northeastward, spawning tornadoes in Florida and dumping huge quantities of snow from Alabama to Canada's Maritime Provinces. The closely spaced isobars depict the steep pressure gradient that generated hurricane-force winds in many places. The strong winds caused extreme drifting of snow and created storm waves in the Atlantic that pounded the coast, causing much beach erosion and damage to shoreline homes.

another, every major airport on the U.S. East Coast was closed, as were most highways. Hundreds of hikers were stranded in the mountains by the record snowfalls, and thousands of others were stranded away from home in a variety of settings. At one point, 3 million people were without electricity. As the blizzard buried the region, hundreds of roofs collapsed under the wet, heavy snow. The National Weather Service office in Asheville, North Carolina, reported an extraordinary snow-to-water ratio of 4.2 to 1. A general ratio of 10 units of snow to 1 unit of water is common. Meanwhile, storm-generated waves pounded the coast. About 200 homes along the Outer Banks of North Carolina were damaged, with many left uninhabitable, and on Long Island, 18 houses succumbed to the surf.

In the aftermath of the storm, temperatures plunged to record low levels as a frigid continental arctic air mass

spilled into the region. On Sunday, March 14, some 70 cities in the East and South experienced record low temperatures. The next day, 75 more record low daily minimums were established. Freezing temperatures reached as far south as Orlando, Florida, and Birmingham, Alabama, set a new March record of –17°C (2°F). The "Storm of the Century" was truly an awesome event that inhabitants of eastern North America will not soon forget.

Polar-Front Theory

In previous chapters, we examined the basic elements of weather as well as the dynamics of atmospheric motions. We are now ready to apply our knowledge of these diverse phenomena to an understanding of day-to-day weather patterns in the middle latitudes. For our purposes, *middle latitudes* refers to the region between southern Florida and Alaska, essentially the area of the westerlies. The primary weather producer here is the **midlatitude** or **middle-latitude cyclone**°. These are the same phenomena that a TV weather reporter calls a *low-pressure system* or simple a *low*.

Midlatitude cyclones are low-pressure systems with diameters often exceeding 1000 kilometers (600 miles) that travel from west to east across the planet (Figure 9–3). Lasting from a few days to more than a week, these weather systems have a counterclockwise circulation pattern with a flow inward toward their centers. Most midlatitude cyclones have a cold front and a warm front extending from the central area of low pressure. Surface convergence and upward flow initiate cloud development that frequently produces precipitation.

As early as the 1800s, it was known that cyclones were the bearers of precipitation and severe weather. Thus, the barometer was established as the main tool in "forecasting" day-to-day weather changes. However, this early method of weather prediction largely ignored the role of air-mass interactions in the formation of these weather systems. Consequently, it was not possible to determine the conditions under which cyclone development was favorable.

The first encompassing model to consider the development and intensification of a midlatitude cyclone was constructed by a group of Norwegian scientists during World War I. The Norwegians were then cut off from weather reports, especially those from the Atlantic. To counter this deficiency, a closely spaced network of weather stations was established throughout the country. Using this network, several Norwegian-trained meteorologists made great advances in broadening our understanding of the weather and in particular our understanding of the middle-latitude cyclone. Included in this early group were J. Bjerknes, V. Bjerknes, H. Solberg, and T. Bergeron. In 1918, J. Bjerknes published his hypothesis of cyclone formation in an article entitled "On the Structure of Moving Cyclones." These insights, which became known as the **polar-front theory**, provide us with a useful model of how a midlatitude cyclone develops. Although some changes to the original model were necessary because of subsequent discoveries, the main tenets of this theory remain an integral part of modern meteorological thought.

In the polar-front theory, middle-latitude cyclones develop in conjunction with the polar front. Recall that the polar front separates cold polar air from warm subtropical air (Chapter 7). During cool months, the polar front is generally well defined and forms a nearly continuous band around Earth that can be recognized on upper-air charts. At the surface, this frontal zone is often broken into distinct segments. These frontal segments are separated by regions of more gradual temperature change. It is along frontal zones where cold, equatorward-moving air collides with warm, poleward-moving air that most middle-latitude cyclones form.

Because midlatitude cyclones develop in conjunction with fronts, we will first consider the nature of fronts and the weather associated with their movement and then apply what we learn to the model of cyclone development (Figure 9–3).

Fronts

Fronts are boundary surfaces that separate air masses of different densities. One air mass is usually warmer and often contains more moisture than the other. However, fronts can form between any two contrasting air masses. When the vast sizes of air masses are considered, these 15- to 200-kilometer (9- to 120-mile)-wide bands of discontinuity are relatively narrow. On the scale of a weather map, they are normally thin enough to represent as a broad line.

Above the ground, the frontal surface slopes at a low angle so that warmer air overlies cooler air, as shown in Figure 9–4. In the ideal case, the air masses on both sides of the front would move in the same direction and at the same speed. Under this condition, the front would act simply as a barrier that travels along with the air masses. Generally, however, the distribution of pressure across a front is such that one air mass moves faster relative to the frontal boundary than does the other. Thus, one air mass actively advances into another and "clashes" with it. The Norwegian meteorologists visualized these zones of airmass interactions as analogous to battle lines and tagged them "fronts," like battle fronts. It is along these zones of conflict that midlatitude cyclones develop and produce much of the precipitation and severe weather in the belt of the westerlies.

As one air mass moves into another, limited mixing occurs along the frontal surface, but for the most part, the air masses retain their identity as one is displaced upward over the other. No matter which air mass is advancing, it is always the warmer, less dense air that is forced aloft, whereas the cooler, denser air acts as a wedge on which lifting takes place. The term **overrunning** is generally applied to warm air gliding up along a cold air mass. We now shall look at five types of fronts—*warm, cold, stationary, occluded,* and *drylines*.

°Midlatitude cyclones go by a number of different names, including wave cyclones, frontal cyclones, extratropical cyclones, low-pressure systems, and simply lows.

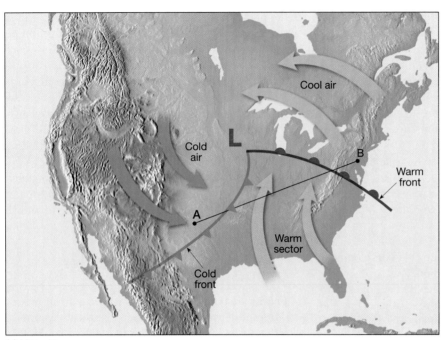

(a) Map view.

Figure 9–3 Idealized structure of a midlatitude cyclone. (a) Map view showing positions of cold front and warm front. (b) Three-dimensional view of the fronts along a line from point A to point B.

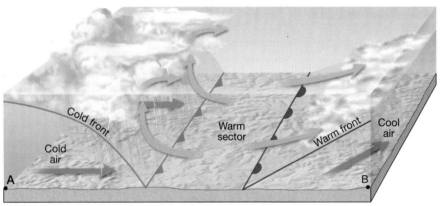

(b) Three-dimensional view from points A to B.

Warm Fronts

When the surface (ground) position of a front moves so that warm air occupies territory formerly covered by cooler air, it is called a **warm front** (Figure 9–4). On a weather map, the surface position of a warm front is shown by a red line with red semicircles protruding into the cooler air. East of the Rockies, maritime tropical (mT) air often enters the United States from the Gulf of Mexico and overruns receding cool air. As the colder air retreats, friction with the ground greatly slows the advance of the surface position of the front compared to its position aloft. Stated another way, less dense, warm air has a hard time displacing heavier, cold air. Consequently, the boundary separating these air masses acquires a very gradual slope. The

slope of a warm front (height compared to horizontal distance) averages only about 1:200. This means that if you traveled 200 kilometers (120 miles) ahead of the surface location of a warm front, the frontal surface would be only 1 kilometer (0.6 miles) overhead.

As warm air ascends the retreating wedge of cold air, it expands and cools adiabatically, causing moisture in the ascending air to condense into clouds that often produce precipitation. The cloud sequence in Figure 9–4 typically precedes a warm front. The first sign of the approaching warm front is cirrus clouds. These high clouds form where the overrunning warm air has ascended high up the wedge of cold air, 1000 kilometers (600 miles) or more ahead of the surface front. Another indication of an approaching

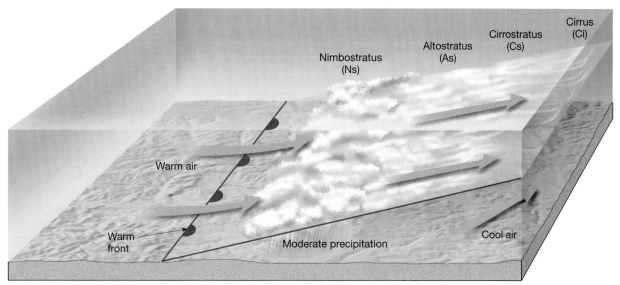

Figure 9–4 Warm front produced as warm air glides up over a cold air mass.

warm front is provided by aircraft contrails. On a clear day, when these condensation trails persist for several hours, you can be fairly certain that comparatively warm, moist air is ascending overhead.

As the front nears, cirrus clouds grade into cirrostratus that gradually blend into denser sheets of altostratus. About 300 kilometers (180 miles) ahead of the front, thicker stratus and nimbostratus clouds appear and precipitation commences.

Because of their slow rate of advance and very low slope, warm fronts usually produce light-to-moderate precipitation over a large area and for an extended period (Figure 9–5). Warm fronts, however, are occasionally associated with cumulonimbus clouds and thunderstorms. This occurs when the overrunning air is unstable and the temperatures on opposite sides of the front contrast sharply. When such conditions exist, cirrus clouds are generally followed by cirrocumulus clouds. These clouds create the familiar "mackerel sky" (resembling fish scales—see Figure 5–3c) that warned sailors of an impending storm:

> Mackerel scales and mares' tails
> Make lofty ships carry low sails.

At the other extreme, a warm front associated with a dry air mass could pass unnoticed at the surface.

Figure 9–5 Light rain associated with a warm front in late fall. *(Photo by Miro Vintoniv/Stock Boston)*

As you can see from Figure 9–4, the precipitation associated with a warm front occurs ahead of the surface position of the front. Some of the rain that falls through the cool air below the clouds evaporates. As a result, the air directly beneath the cloud base often becomes saturated and a stratus cloud deck develops. These clouds occasionally grow rapidly downward, which can cause problems for pilots of small aircraft that require visual landings. One minute pilots may have adequate visibility and the next be in a cloud mass (frontal fog) that has the landing strip "socked in." Occasionally during the winter, a relatively warm air mass is forced over a body of subfreezing air. When this occurs, it can create hazardous driving conditions. Raindrops become supercooled as they fall through the subfreezing air. Upon colliding with the road surface they flash-freeze to produce the icy layer called *glaze*.

When a warm front passes, temperatures gradually rise. As you would expect, the increase is most apparent when a large contrast exists between adjacent air masses. Moreover, a wind shift from the east to the southwest is generally noticeable. (The reason for this shift will become evident later.) The moisture content and stability of the encroaching warm air mass largely determine the time period required for clear skies to return. During the summer, cumulus and occasionally cumulonimbus clouds are embedded in the warm unstable air mass that follows the front. These clouds can produce precipitation, which can be heavy, but is usually scattered and of short duration.

Cold Fronts

When cold, continental polar air actively advances into a region occupied by warmer air, the zone of discontinuity is called a **cold front** (Figure 9–6). As with warm fronts, friction slows the surface position of a cold front compared to its position aloft. Thus, the cold front steepens as it moves. On the average, cold fronts are about twice as steep as warm fronts, having a slope of perhaps 1:100. In addition, cold fronts advance at speeds around 35 to 50 kilometers (20 to 35 miles) per hour compared to 25 to 35 kilometers (15 to 20 miles) per hour for warm fronts. These two differences, steepness of slope and rate of movement, largely account for the more violent nature of cold-front weather, compared to the weather generally accompanying a warm front.

The arrival of a cold front is sometimes preceded by altocumulus clouds. As the front approaches, generally from the west or northwest, towering clouds can often be seen in the distance. Near the front, a dark band of ominous clouds foretells the ensuing weather. The forceful lifting of warm, moist air along a cold front is often so rapid that the released latent heat increases the air's buoyancy. The heavy downpours and vigorous wind gusts associated with mature cumulonimbus clouds frequently result. Because a cold front produces roughly the same amount of lifting as a warm front, but over a shorter distance, the precipitation intensity is greater but of shorter duration (Figure 9–7).

In addition, a marked temperature drop and wind shift from the southwest to the northwest usually accompany frontal passage. The reason for this shift will be explained later in this chapter. The sharp temperature contrast and sometimes violent weather along cold fronts are symbolized on a weather map by a blue line with blue triangular points extending into the warm air mass (Figure 9–6).

Most often the weather behind a cold front is dominated by subsiding air within a continental polar air mass. Thus, the drop in temperature is accompanied by clearing

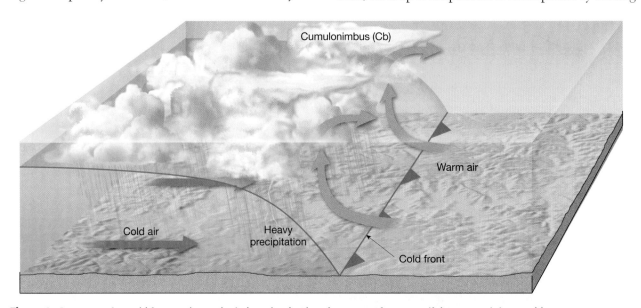

Figure 9–6 Fast-moving cold front and cumulonimbus clouds. Thunderstorms often occur if the warm air is unstable.

Figure 9–7 Thunderstorm development along a cold front over the Great Plains. *(Photo by Eddie Soloway/Tony Stone Images)*

that begins soon after the front passes. Although subsidence causes adiabatic heating aloft, the effect on surface temperatures is minor. In winter, the long cloudless nights that often follow the passage of a cold front allow for abundant radiation cooling that reduces surface temperatures.

When a cold front moves over a relatively warm surface, radiation emitted from Earth can heat the lower atmosphere enough to produce shallow convection. This, in turn, may generate low cumulus or stratocumulus clouds behind the front. However, subsidence aloft renders these air masses quite stable. Any clouds that form will not develop great vertical thickness and will seldom produce precipitation. One exception is lake-effect snow (discussed in Chapter 8) where the cold air behind the front acquires both heat and moisture when it traverses a comparatively warm body of water.

Stationary Fronts

Occasionally, the air flow on both sides of a front is neither toward the cold air mass nor toward the warm air mass, but almost parallel to the line of the front. Consequently, the surface position of the front does not move, or moves very slowly. This condition is called a **stationary front**. On a weather map, stationary fronts are shown with blue triangular points on one side of the line and red semicircles on the other. Because overrunning usually occurs along stationary fronts, gentle to moderate precipitation is likely. Stationary fronts may remain over an area for several days, in which case flooding is possible.

Occluded Fronts

The fourth type of front is the **occluded front**. Here a rapidly moving cold front overtakes a warm front, as shown in Figure 9–8a. As the cold air wedges the warm front upward, a new front forms between the advancing cold air and the air over which the warm front is gliding (Figure 9–8b). The weather of an occluded front is

generally complex. Most precipitation is associated with the warm air being forced aloft (Figure 9–8c). When conditions are suitable, however, the newly formed front is capable of initiating precipitation of its own.

As you might expect, there are cold-type occluded fronts and warm-type occluded fronts. In the occluded front shown in Figure 9–9a, the air behind the cold front is colder than the cool air it is overtaking. This is the most common type of occluded front east of the Rockies and is called a **cold-type occluded front**.

It is also possible for the air behind the advancing cold front to be warmer than the cold air it is overtaking. These **warm-type occluded fronts** (Figure 9–9b) frequently occur along the Pacific Coast, where milder maritime polar air invades more frigid polar air that had its origin over the continent.

Notice in Figure 9–9 that in the warm-type occluded front, the warm air aloft (and hence the precipitation) often precedes the arrival of the surface front. This situation is reversed for the cold-type occluded front, where the front aloft (and its associated precipitation) lags behind the surface front. Also note that cold-type occluded fronts frequently resemble cold fronts in the type of weather generated.

Because of the complex nature of occluded fronts, they are often drawn on weather maps as either warm or cold fronts, depending on what kind of air is the aggressor. In those cases when they are drawn as an occluded front, a purple line with alternating purple triangles and semicircles pointing in the direction of movement is employed.

Drylines

Classifying fronts based only on the temperature differences across the frontal boundary can be misleading. Humidity also influences the density of air, with humid air being less dense than dry air, all other factors being equal. In the summer it is not unusual for a southeastward moving air mass that originated over the northern

Map view

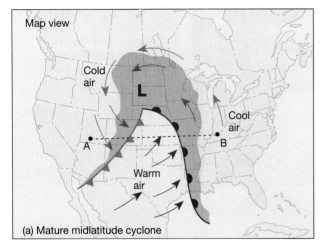

Cross sectional view

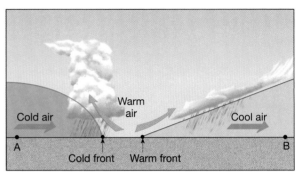

(a) Mature midlatitude cyclone

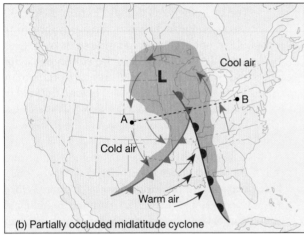

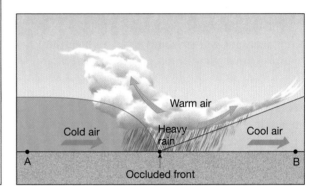

(b) Partially occluded midlatitude cyclone

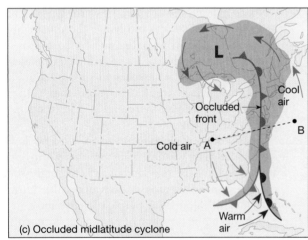

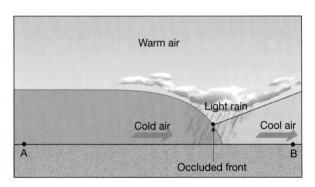

(c) Occluded midlatitude cyclone

Figure 9–8 Stages in the formation of an occluded front and its relationship to a developing midlatitude cyclone. After the warm air has been forced aloft, the system begins to dissipate. The shaded areas indicate regions where precipitation is most likely to occur.

Great Plains to displace warm, humid air over the lower Mississippi Valley. The front that develops is usually labeled a cold front, even though the advancing air may not be any colder than the air it displaces. Simply, the drier air is denser and forcefully lifts the moist air in its path, just like a cold front. The passage of this type of

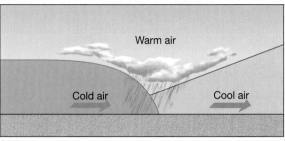

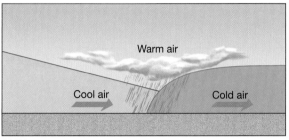

Figure 9–9 Occluded fronts of (a) the cold type and (b) the warm type.

frontal boundary is noticed as a sharp drop in humidity, without an appreciable drop in temperature.

A related type of frontal boundary, called a **dryline**, develops over the southern Great Plains. This occurs when dry, continental tropical (cT) air originating over the American Southwest meets moist, maritime tropical (mT) air from the Gulf of Mexico. Basically a spring and summer phenomenon, drylines most often generate a band of severe thunderstorms along a line extending from Texas to Nebraska that moves eastward across the Great Plains. A dryline is easily identified by comparing the dew point temperatures of the cT air west of the front with the dew points of the mT air mass to the east (see Figure 10–12, p. 280).

In summary, because cold fronts are the aggressor, the lifting associated along a cold front tends to be more concentrated and forceful than that along a warm front. Hence, the precipitation associated with a cold front usually occurs along a narrower zone and is heavier than that associated with a warm front. Further, most severe weather occurs along cold fronts or drylines in which dry continental air aggressively lifts warm maritime air.

Life Cycle of a Midlatitude Cyclone

The polar-front theory, which describes the development and intensification of a midlatitude cyclone, was created primarily from near-surface observations. As more data became available from the middle and upper troposphere, some modifications were necessary. Yet, this model is still a useful working tool for interpreting the weather. It helps us to visualize our dynamic atmosphere as it generates our day-to-day weather. If you keep this model in mind as you observe changes in the weather, the changes will no longer come as a surprise. Furthermore, you will begin to see some order in what had appeared to be disorder, and you might even correctly forecast the impending weather!

Formation: The Clash of Two Air Masses

According to the polar-front model, cyclones form along fronts and proceed through a generally predictable life cycle. This cycle can last a few days to over a week, depending on whether conditions for development are favorable. Figure 9–10 shows six stages in the life of a typical midlatitude cyclone. In part (a), the stage is set for **cyclogenesis** (cyclone formation). Here two air masses of different densities (temperatures) are moving roughly parallel to the front, but in opposite directions. (In the classic polar-front model, this would be continental polar air associated with the polar easterlies on the north side of the front and maritime tropical air driven by the westerlies on the south side of the front.)

Under suitable conditions, the frontal surface that separates these two contrasting air masses will take on a wave shape that is usually several hundred kilometers long (Figure 9–10b). These waves are analogous to the waves produced on water by moving air, except that the scale is different. Some waves tend to dampen, or die out, whereas others grow in amplitude. Those storms that intensify or "deepen" develop waves that change in shape over time, much like a gentle ocean swell does as it moves into shallow water and becomes a tall, breaking wave (see Figure 9–10c).

Development of Cyclonic Flow

As the wave develops, warm air advances poleward invading the area formerly occupied by colder air, while cold air moves equatorward. This change in the direction of the surface flow is accompanied by a readjustment in the pressure pattern that results in nearly circular isobars, with the low pressure centered at the apex of the wave. The resulting flow is a counterclockwise cyclonic circulation that can be seen clearly on the weather map shown in Figure 9–11. Once the cyclonic circulation develops, we would expect general convergence to result in vertical lifting, especially where warm air is overrunning colder air. You can see in Figure 9–11 that the air in the warm sector (over the southern states) is flowing northeastward toward

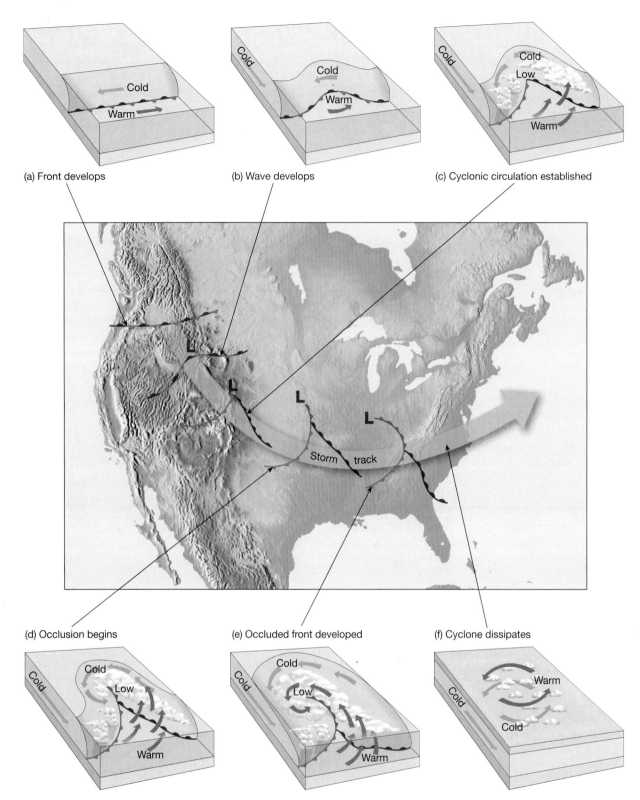

(a) Front develops

(b) Wave develops

(c) Cyclonic circulation established

(d) Occlusion begins

(e) Occluded front developed

(f) Cyclone dissipates

Figure 9-10 Stages in the life cycle of a middle-latitude cyclone, as proposed by J. Bjerknes.

colder air that is moving toward the northwest. Because the warm air is moving in a direction perpendicular to this front, we can conclude that the warm air is invading a region formerly occupied by cold air. Therefore, this must be a warm front. Similar reasoning indicates that, to the left (west) of the cyclonic disturbance, cold air from the northwest is displacing the air of the warm sector and generating a cold front.

Occlusion: The Beginning of the End

Usually, the position of the cold front advances faster than the warm front and begins to close (lift) the warm front, as shown in Figure 9–10c,d. This process, known as **occlusion**, forms an *occluded front*, which grows in length

as it displaces the warm sector aloft. As occlusion begins, the storm often intensifies. Pressure at the storm's center falls, and wind speeds increase. In the winter, heavy snowfalls and blizzard-like conditions are possible during this phase of the storm's evolution.

As more of the sloping discontinuity (front) is forced aloft, the pressure gradient weakens. In a day or two, the entire warm sector is displaced and cold air surrounds the cyclone at low levels (Figure 9–10f). Thus, the horizontal temperature (density) difference that existed between the two contracting air masses has been eliminated. At this point, the cyclone has exhausted its source of energy. Friction slows the surface flow, and the once highly organized counterclockwise flow ceases to exist.

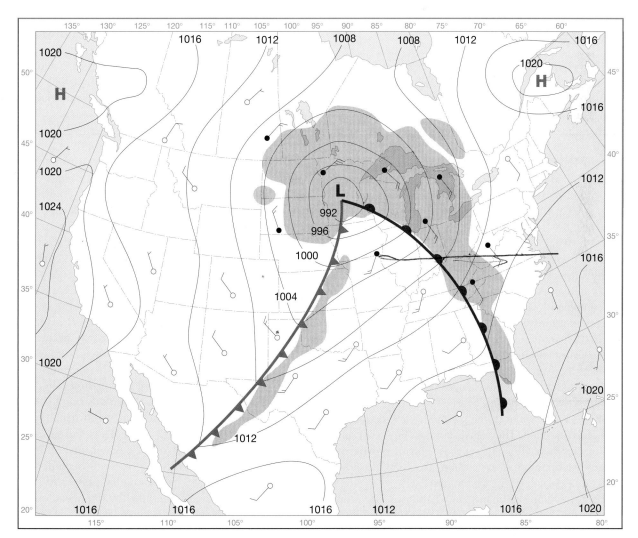

Figure 9–11 Simplified weather map showing the circulation of a middle-latitude cyclone. The gray areas indicate regions of probable precipitation.

A simple analogy may help you visualize what is happening to the cold and warm air masses in the preceding discussion. Imagine a large water trough that has a vertical divider separating the tank into two equal parts. Half of the tank is filled with hot water containing red dye, and the other half is filled with blue-colored ice water. Now, imagine what happens when the divider is removed. The cold, dense water will flow under the less dense warm water, displacing it upward. This rush of water will come to a halt as soon as all of the warm water is displaced toward the top of the container. In a similar manner, a middle-latitude cyclone dies once all of the warm air is displaced aloft and the horizontal discontinuity between the air masses no longer exists.

Idealized Weather of a Midlatitude Cyclone

As stated earlier, the polar-front model is a useful tool for examining weather patterns of the middle latitudes. Figure 9–12 illustrates a mature midlatitude cyclone; note the distribution of clouds and thus the regions of possible precipitation. Compare this map to the satellite image of a cyclone in Figure 9–13. It is easy to see why we often refer to the cloud pattern of a cyclone as having a "comma" shape.

Guided by the westerlies aloft, cyclones generally move eastward across the United States. Therefore, we can expect the first signs of a cyclone's arrival to appear in the western sky. In the region of the Mississippi Valley, however, cyclones often begin a more northeasterly trajectory and occasionally move directly northward. Typically, a midlatitude cyclone requires two to four days to pass completely over a region. During that period, abrupt changes in atmospheric conditions may occur, particularly in the winter and spring when the greatest temperature contrasts occur across the middle latitudes.

Using Figure 9–12 as a guide, let us examine these weather producers and the changes you could expect if a midlatitude cyclone passed over your area. To facilitate our discussion, profiles of the storm are provided above and below the map. They correspond to lines A–E and F–G on the map. (Remember, these storms move from west to east; therefore, the right side of the cyclone shown in Figure 9–12 will be the first to pass by.)

First, imagine the change in weather as you move from right to left along profile A–E (bottom of Figure 9–12). At point A, the sighting of high cirrus clouds is the first sign of the approaching cyclone. These high clouds can precede the surface front by 1000 kilometers (600 miles) or more, and they are normally accompanied by falling pressure. As the warm front advances, a lowering and thickening of the cloud deck is noticed. Within 12 to 24 hours after the first sighting of cirrus clouds, light precipitation usually commences (point B). As the front nears, the rate of precipitation increases, a rise in temperature is noticed, and winds begin to change from an easterly to a southerly flow.

With the passage of the warm front, the area is under the influence of the maritime tropical air mass of the warm sector (point C). Generally, the region affected by this part of the cyclone experiences relatively warm temperatures, southerly winds, and clear skies, although fair-weather cumulus or altocumulus clouds are not uncommon.

The rather pleasant weather of the warm sector passes quickly in the spring and is replaced by gusty winds and precipitation generated along the cold front. The approach of a rapidly advancing cold front is marked by a wall of rolling black clouds (point D). Severe weather accompanied by heavy precipitation and occasionally hail or a tornado is common at this time of year.

The passage of the cold front is easily detected by a wind shift. The warm flow from the south or southwest is replaced by cold winds from the west to northwest, resulting in a pronounced drop in temperature. Also, rising pressure hints of the subsiding cool, dry air behind the cold front. Once the front passes, the skies clear quickly as cooler air invades the region (point E). A day or two of almost cloudless deep blue skies are often experienced, unless another cyclone is edging into the region.

A very different set of weather conditions prevails in the portion of the cyclone that contains the occluded front along profile F–G (top of Figure 9–12). Here temperatures remain cool during the passage of the storm. The first hints of the approaching low-pressure center are a continual drop in air pressure and increasingly overcast conditions. This section of the cyclone most often generates snow or icing storms during the coldest months. Moreover, the occluded front usually moves more slowly than the other fronts. Thus, the entire wishbone-shaped frontal structure shown in Figure 9–12 rotates counterclockwise so that the occluded front appears to "bend over backward." This effect adds to the misery of the region influenced by the occluded front, for it remains over the area longer than the other fronts (see Box 9–1).

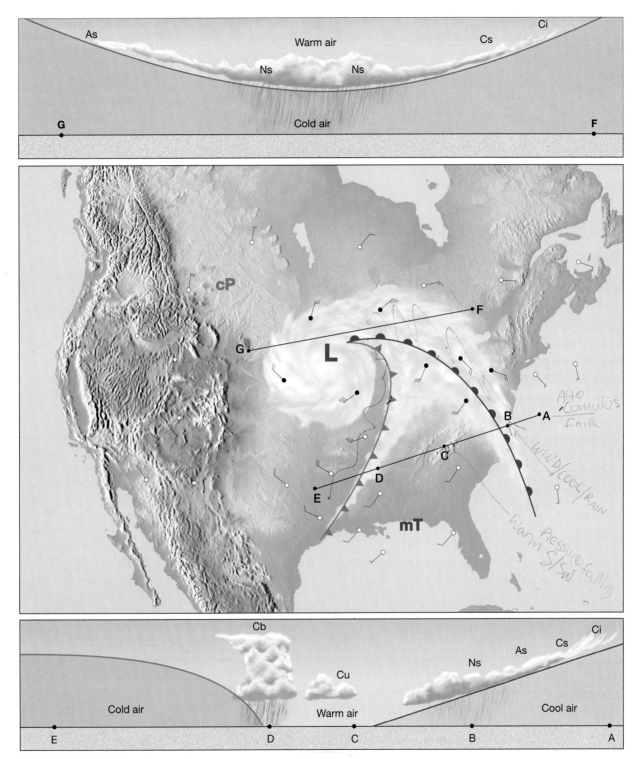

Figure 9–12 Cloud patterns typically associated with a mature middle-latitude cyclone. The middle section is a map view. Note the cross-section lines (F–G, A–E). Above the map is a vertical cross section along line F–G. Below the map is a section along A–E. For cloud abbreviations, refer to Figures 9–4 and 9–6.

Figure 9–13 Satellite view of a mature midlatitude cyclone over the eastern one-third of the United States. It is easy to see why we often refer to the cloud pattern of a cyclone as having a "comma" shape. (Courtesy of John Jensenius/National Weather Service)

Cyclogenesis

The polar-front model shows that cyclogenesis (cyclone formation) occurs where a frontal surface is distorted into a wave-shaped discontinuity. Several surface factors are thought to produce this wave in a frontal zone. Topographic irregularities (such as mountains), temperature contrasts (as between sea and land), or ocean-current influences can disrupt the general zonal flow sufficiently to produce a wave along a front. But more often than not, the formation of a cyclone is initiated by the flow aloft in the vicinity of the midlatitude jet stream.

When the earliest studies of cyclones were made, little data were available on airflow in the middle and upper troposphere. Since then, a close relationship has been established between surface disturbances and the flow aloft. Whenever the winds aloft exhibit a relatively straight zonal flow—that is, from west to east—little cyclonic activity occurs at the surface. However, when the upper air begins to meander widely from north-to-south, forming high-amplitude waves of alternating troughs (lows) and ridges (highs), cyclonic activity intensifies (Figure 9–14). Moreover, when surface cyclones form, almost invariably they are centered below the jet-stream axis and downwind from an upper-level trough (Figure 9–14).

Cyclonic and Anticyclonic Circulation

Before discussing how cyclones are generated by the flow aloft, let us review the nature of cyclonic and anticyclonic winds. Recall that airflow about a surface low is inward, a fact that leads to mass convergence (coming together). Because accumulation of air is accompanied by a corresponding increase in surface pressure we might expect a surface low-pressure center to "fill" rapidly and be eliminated, just as the vacuum in a coffee can is quickly equalized when we open it. However, cyclones often exist for a week or longer. For this to happen, surface convergence must be offset by outflow aloft (Figure 9–15). As long as divergence (spreading out) aloft is equal to or greater than the surface inflow, the low pressure can be sustained.

Because cyclones are bearers of stormy weather, they have received far more attention than their counterparts, anticyclones. Yet the close relationship between them makes it difficult to totally separate a discussion of these two pressure systems. The surface air that feeds a cyclone, for example, generally originates as air flowing out of an anticyclone (Figure 9–15). Consequently, cyclones and anticyclones are typically found adjacent to one another. Like the cyclone, an anticyclone depends on the flow aloft to maintain its circulation. In the anticyclone, divergence at the surface is balanced by convergence aloft and general subsidence of the air column (Figure 9–15).

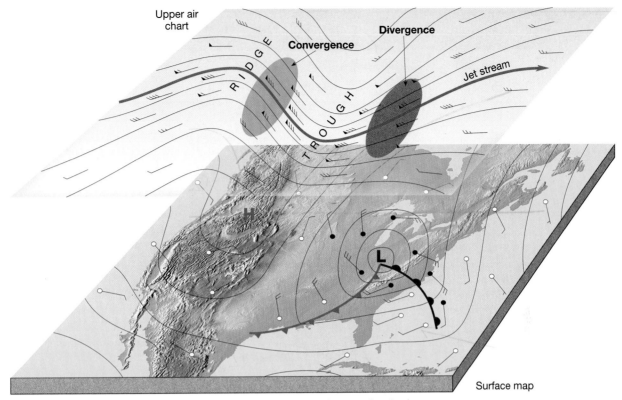

Figure 9–14 Relationship of the wavy flow pattern aloft to surface cyclones and anticyclones.

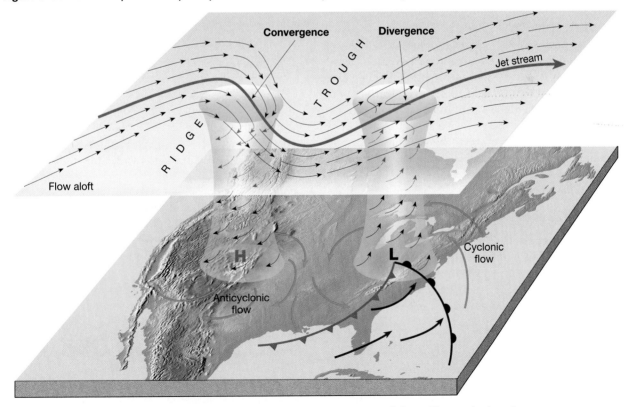

Figure 9–15 Idealized depiction of the support that divergence and convergence aloft provide to cyclonic and anticyclonic circulation at the surface.

Box 9–1 Winds as a Forecasting Tool

Every wind has its weather.

Francis Bacon

People living in the middle latitudes know well that during the winter, northerly winds can chill you to the bone. Conversely, a sudden change to a southerly flow can bring welcome relief from these frigid conditions. If you are more observant, you might have noticed that when the winds switch from a southerly flow to a more easterly direction, foul weather often follows. By contrast, a change in wind direction from the southwest to the northwest is usually accompanied by clearing conditions. Just how are the winds and the forthcoming weather related?

Modern weather forecasts require the processing capabilities of high-speed computers as well as the expertise of people with considerable professional training. Nevertheless, some reasonable insights into the impending weather can be gained through careful observation. The two most significant elements for this purpose are barometric pressure and wind direction. Recall that anticyclones (high-pressure cells) are associated with clear skies, and that cyclones (low-pressure cells) frequently bring clouds and precipitation. Thus, by noting whether the barometer is rising, falling, or steady, we have some indication of the forthcoming weather. For example, rising pressure indicates the approach of a high-pressure system and generally clearing conditions.

The use of winds in weather forecasting is also quite straightforward. Because cyclones are the "'villains" in the weather game, we are most concerned with the circulation around these storm centers. In particular, changes in wind direction that occur with the passage of warm and cold fronts are useful in predicting the impending weather. Notice in Figure 9–12, that with the passage of both the warm and cold fronts, the wind arrows change positions in a clockwise manner. Specifically, with the passage of the cold front, the wind shifts from southwest to northwest. From nautical terminology, the word *veering* is applied to this clockwise wind shift. Because clearing conditions normally occur with the passage of either front, veering winds are indicators that the weather will improve.

In contrast, the area in the northern portion of the cyclone will experience winds that shift in a counterclockwise direction, as can be seen in Figure 9–12. Winds that shift in this manner are said to be *backing*. With the approach of a midlatitude cyclone, backing winds indicate cool temperatures and continued foul weather.

A summary of the relationship among barometer readings, winds, and the impending weather is provided in Table 9–A. Although this information is applicable in a

---

Airflow aloft plays an important role in maintaining cyclonic and anticyclonic circulation. In fact, more often than not, these rotating surface-wind systems are actually generated by upper-level flow.

Divergence and Convergence Aloft

Because divergence aloft is so important to cyclogenesis, we need a basic understanding of its role. Divergence aloft does not involve the outward flow in all directions as occurs about a surface anticyclone. Instead, the winds aloft flow generally from west to east along sweeping curves. How does zonal flow aloft cause upper-level divergence?

One mechanism responsible for divergence aloft is a phenomenon known as **speed divergence**. It has been known for some time that wind speeds can change dramatically in the vicinity of the jet stream. On entering a zone of high wind speed, air accelerates and stretches out (divergence). In contrast, when air enters a zone of slower wind speed, an air pileup (convergence) results. Analogous situations occur every day on a toll highway. When exiting a toll booth and entering the zone of maximum speed, we find automobiles diverging (increasing the number of car lengths between them). As automobiles slow to pay the toll, they experience convergence (coming together).

In addition to speed divergence, several other factors contribute to divergence (or convergence) aloft. These include *directional divergence*, which is horizontal spreading of an air stream, and *vorticity*, which is the amount of rotation exhibited by a mass of moving air. Vorticity can either enhance or inhibit divergence aloft.

The combined effect of the phenomena that influence flow aloft is that an area of upper-air divergence and surface

Table 9–A Wind, barometric pressure, and impending weather

Changes in Wind Direction	Barometric Pressure	Pressure Tendency	Impending Weather
Any direction	1023 mb and above (30.20 in.)	Steady or rising	Continued fair with no temperature change
SW to NW	1013 mb and below (29.92 in.)	Rising rapidly	Clearing within 12 to 24 hours and colder
S to SW	1013 mb and below (29.2 in.)	Rising slowly	Clearing within a few hours and fair for several days
SE to SW	1013 mb and below (29.92 in.)	Steady or slowly falling	Clearing and warmer, followed by possible precipitation
E to NE	1019 mb and above (30.10 in.)	Falling slowly	In summer, with light wind, rain may not fall for several days; in winter, rain within 24 hours
E to NE	1019 mb and above (10.10 in.)	Falling rapidly	In summer, rain probable within 12 to 24 hours; in winter, rain or snow with strong winds likely
SE to NE	1013 mb and below (29.92 in.)	Falling slowly	Rain will continue for 1 to 2 days
SE to NE	1013 mb and below (29.92 in.)	Falling rapidly	Stormy conditions followed within 36 hours by clearing and, in winter, colder temperatures

Source: Adapted from the National Weather Service

very general way to much of the United States, local influences must be taken into account. For example, a rising barometer and a change in wind direction, from southwest to northwest, is usually associated with the passage of a cold front and indicates that clearing conditions should follow. However, in the winter, residents of the southeast shore of one of the Great Lakes may not be so lucky. As cold, dry northwest winds cross large expanses of open water, they acquire heat and moisture from the relatively warm lake surface. By the time this air reaches the leeward shore, it is often humid and unstable enough to produce heavy lake-effect snow (see Box 8–2).

cyclonic circulation generally develop downstream from an upper-level trough, as illustrated in Figure 9–15. Consequently, in the United States, surface cyclones generally form east of an upper-level trough. As long as divergence aloft exceeds convergence at ground level, surface pressures will fall and the cyclonic storm will intensify. Eventually, however, the inward flow of air at the surface will exceed the outward flow aloft and the low will "fill" and die.

Conversely, the zone in the jet stream that experiences convergence and anticyclonic rotation is located downstream from a ridge (Figure 9–15). The accumulation of air in this region of the jet stream leads to subsidence and increased surface pressure. Hence, this is a favorable site for the development of a surface anticyclone.

Because of the significant role that the upper-level flow has on cyclogenesis, it should be evident that any attempt at weather prediction must take into account the airflow aloft. This is why television weather reporters frequently illustrate the flow within the jet stream.

In summary, the flow aloft contributes to the formation and intensification of surface low- and high-pressure systems. Areas of upper-level convergence and divergence are located in the vicinity of jet stream, where dramatic changes in wind speeds cause air either to pileup (converge) or spread-out (diverge). Upper-level convergence is favored downstream (east) of a ridge, whereas divergence occurs downstream of an upper-level trough. Below regions of upper-level convergence are areas of high pressure (anticyclones), whereas upper-level divergence supports the formation and development of cyclonic systems (lows).

Traveling Cyclones

Waves in the westerlies are important not only as a cause of cyclonic development. The flow aloft is also important in determining how rapidly these pressure systems advance and the direction they follow. Compared with the general flow aloft at the 500-millibar level, cyclones generally travel at somewhat less than half that speed. They normally advance at a rate from 25 to 50 kilometers (15 to 30 miles) per hour so that distances of roughly 600 to 1200 kilometers (400 to 800 miles) are traversed each day. The faster speeds occur during the coldest months when temperature gradients are greatest.

One of the most exacting tasks in weather forecasting is predicting the paths of cyclonic storms. We have already seen that the flow aloft tends to steer developing pressure systems. Let us examine an example of this steering effect by seeing how changes in the upper-level flow correspond to changes in the path taken by a cyclone.

Figure 9–16a illustrates the changing position of a middle latitude cyclone over a four-day period. Notice in Figure 9–16b that on March 21, the 500-millibar contours are relatively flat. Also notice that, for the following two days, the cyclone moves in a rather straight southeasterly direction. By March 23, the 500-millibar contours make a sharp bend northward on the eastern side of a trough situated

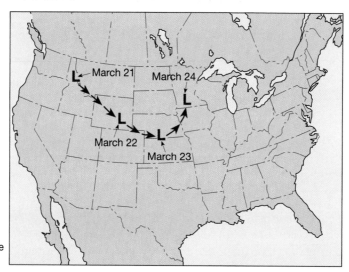

(a) Movement of cyclone from March 21-24

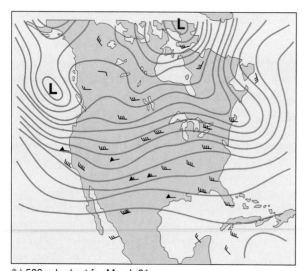

(b) 500-mb chart for March 21

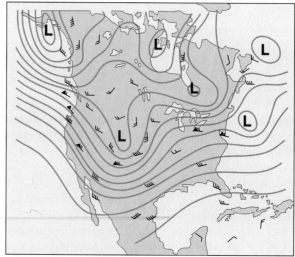

(c) 500-mb chart for March 23

Figure 9–16 Steering of midlatitude cyclones. (a) Notice that the cyclone (low) moved almost in a straight southeastward direction on March 21 and March 22. On the morning of March 23, it abruptly turned northward. This change in direction corresponds to the change from (b) rather straight contours on the upper-air chart for March 21 to (c) curved contours on the chart for March 23.

over Wyoming (Figure 9–16c). On the next day, the path of the cyclone makes a similar northward migration.

Although this is an oversimplified example, it illustrates the steering effect of upper-level flow. Here we have examined the influence of upper airflow on cyclonic movement after the fact. To make useful predictions of future positions of cyclones, accurate appraisals of changes in the westerly flow aloft are required. For this reason, predicting the behavior of the wavy flow in the jet stream is an important part of modern weather forecasting.

Patterns of Movement

The preceding example illustrates the tendency of midlatitude cyclones that form east of the Rockies to first migrate in roughly an easterly direction and then travel a more northeastward path. The cyclones that exit the Eastern Seaboard frequently cross the North Atlantic in the vicinity of the semipermanent low-pressure system (Icelandic low) that occurs at roughly 60° north latitude.

Cyclones that influence western North America originate over the Pacific Ocean. Most of these systems move northeastward across the Pacific toward the Gulf of Alaska, where they merge with the Aleutian low. During the winter

months, these storms develop farther south in the Pacific and often reach the coast of the contiguous 48 states, occasionally traveling as far south as southern California. These cyclonic systems provide the short winter rainy season that affects California and the West Coast in general.

Most Pacific storms do not cross the Rockies intact, but may redevelop on the lee (eastern) side of these mountains. A favorite site for redevelopment is Colorado, but other common sites of formation exist as far south as Texas and as far north as Alberta. The cyclones that form in Canada tend to move southward toward the Great Lakes and then turn northeastward and move out into the Atlantic. Cyclones that redevelop over the Great Plains generally migrate eastward until they reach the central United States, where a northeastward or even northward trajectory is followed. Furthermore, these storms usually occur as members of a "family" of cyclones in various stages of development, as shown in Figure 9–17. Many of these cyclones traverse the Great Lakes region, making this the most storm-ridden portion of the country.

Not all cyclones that affect the United States originate in the Pacific. Some form over the Great Plains and are associated with an influx of maritime tropical air from the Gulf of Mexico. Another area where cyclogenesis

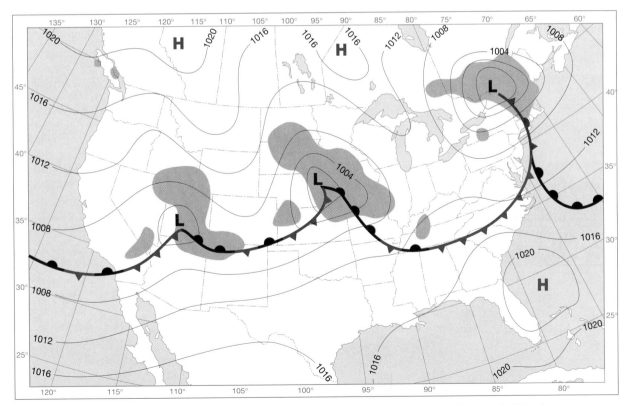

Figure 9–17 A family of midlatitude cyclones in various stages of development migrating eastward across the United States and Canada.

occurs is east of the southern Appalachians. These storms tend to move northward with the Gulf Stream and eventually merge with the Icelandic low.

In contrast to cyclones that tend to migrate toward the northeast, anticyclones more frequently move southeastward. However, some anticyclones are embedded between members of a cyclone family and thus travel northeastward with the group. After the last member of a cyclone family passes through the United States, the cold anticyclone situated behind the cold front breaks out and moves southward. This wintertime phenomenon produces the occasional cold waves that are experienced deep into the southern states.

Anticyclonic Weather and Blocking Highs

Owing to the gradual subsidence within them, anticyclones generally produce clear skies and calm conditions. Because these high-pressure systems are not associated with stormy weather, both their development and movement have not been studied as extensively as that of mid-latitude cyclones. This does not imply, however, that anticyclones always bring desirable weather. Large anticyclones often develop over the Arctic during the winter. These cold high-pressure centers are known to migrate as far south as the Gulf Coast where they can impact the weather over as much as two-thirds of the United States

(Figure 9–18). This dense frigid air often brings record-breaking cold temperatures (Figure 9–19).

Approximately one to three times each winter, and occasionally during other seasons, large anticyclones form and persist over the middle latitudes for nearly two weeks and sometimes longer than a month. These large highs deflect the nearly zonal west-to-east flow and send it poleward. Thus, they are sometimes called *blocking highs*. Once in place, these stagnant anticyclones block the eastward migration of cyclones. As a result, one section of the nation is kept dry for a week or more while another region remains continually under the influence of cyclonic storms. Such a situation prevailed during the summer of 1993 when a strong high-pressure system became anchored over the southeastern United States and caused migrating storms to stall over the Midwest. The result was the most devastating flooding on record for the central and upper Mississippi Valley (see Box 9–2). At the same time, the Southeast experienced severe drought.

Large stagnant anticylcones can also contribute to air pollution episodes. The subsidence within an anticyclone can produce a temperature inversion that acts like a lid to trap pollutants. (For more on this see Chapter 13.) Further, the light winds associated with the center of an anticyclone do little to disperse polluted air. Both Los Angeles and Mexico City experience air pollution episodes when strong, stagnant high-pressure systems dominate their circulation for extended periods.

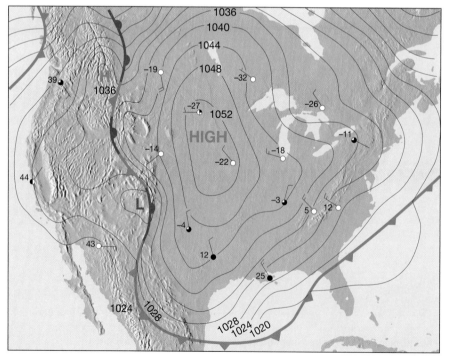

Figure 9–18 A cold anticyclone associated with an outbreak of frigid arctic air impacts the eastern two-thirds of North America. Temperatures are shown in degrees Fahrenheit.

Figure 9–19 Cold outbreak of arctic air invades New England, bringing subzero temperatures and mostly clear skies. *(Photo by Seth Resnick/Stock Boston)*

Case Study of a Midlatitude Cyclone

To give you a picture of the weather one might expect from a strong late-winter cyclonic storm we are going to look at an actual event. Our sample storm was one of three cyclones to migrate across the United States during the latter half of March. This cyclone reached the U.S. West Coast on March 21 a few hundred kilometers northwest of Seattle, Washington. Like many Pacific storms, this one rejuvenated over the western United States and moved eastward into the Plains states. By the morning of March 23, it was centered over the Kansas–Nebraska border (Figure 9–20). At this time, the central pressure had reached 985 millibars, and its well-developed cyclonic circulation exhibited a warm front and a cold front.

During the next 24 hours, the forward motion of the storm's center became sluggish. It curved slowly northward through Iowa, and the pressure deepened to 982 millibars (Figure 9–21). Although the storm center advanced slowly, the associated fronts moved vigorously toward the east and somewhat northward. The northern sector of the cold front overtook the warm front and generated an occluded front, which by the morning of March 24 was oriented nearly east-west (Figure 9–21).

This period in the storm's history marked one of the worst blizzards ever to hit the north-central states. While the winter storm was brewing in the north, the cold front marched from northwestern Texas (on March 23) to the Atlantic Ocean (March 25). During its two-day trek to the

ocean, this violent cold front generated numerous severe thunderstorms and 19 tornadoes.

By March 25, the low pressure had diminished in intensity (1000 millibars) and had split into two centers (Figure 9–22). Although the remnant low from this system, which was situated over the Great Lakes, generated some snow for the remainder of March 25, by the following day it had completely dissipated.

Violent Spring Weather

Now that you have read an overview of this storm, let us revisit this cyclone's passage in detail using the weather charts for March 23 through March 25 (Figures 9–20 to 9–22). The weather map for March 23 depicts a classic developing cyclone. The warm sector of this system, as exemplified by Fort Worth, Texas, is under the influence of a warm, humid air mass having a temperature of 70°F and a dew-point temperature of 64°F. Notice in the warm sector that winds are from the south and are overrunning cooler air situated north of the warm front. In contrast, the air behind the cold front is 20° to 40°F cooler than the air of the warm sector and is flowing from the northwest, as depicted by the data for Roswell, New Mexico.

Prior to March 23, this system had generated some precipitation in the Northwest and as far south as California. On the morning of March 23, little activity was occurring along the fronts. As the day progressed, however, the storm intensified and changed dramatically. The map for March 24 illustrates the highly developed cyclone that evolved during the next 24 hours. The extent and

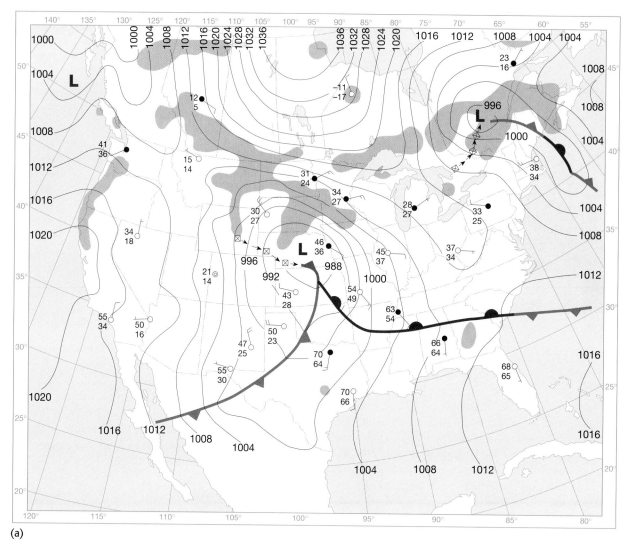

(a)

(b)

Figure 9–20 (a) Surface weather map for March 23. (b) Satellite image showing the cloud patterns for March 23. *(Courtesy of NOAA/Seattle)*

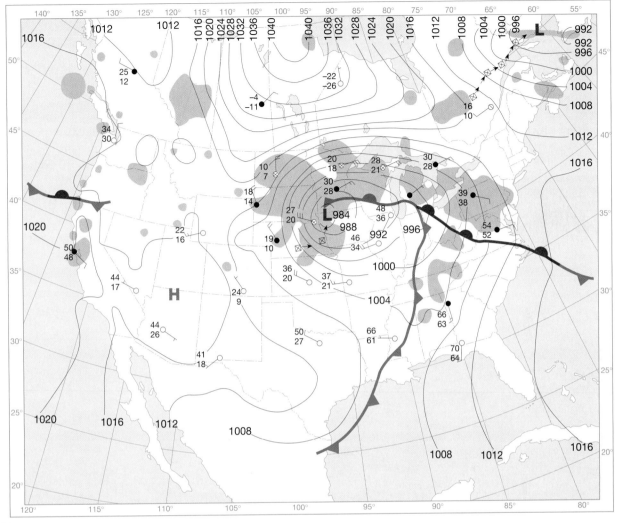

(a)

(b)

Figure 9–21 (a) Surface weather map for March 24. (b) Satellite image showing the cloud patterns on that day. *(Courtesy of NOAA/Seattle)*

Box 9–2 Atmospheric Hazard: The Great Flood of 1993

by Steven Hilberg*

Unprecedented rainfall produced the wettest spring and early summer of the century for the Upper Mississippi River basin (upstream of Quincy, Illinois), according to rainfall data gathered by the Midwestern Regional Climate Center at the Illinois State Water Survey. Portions of the basin received up to twice the normal rainfall (Figure 9–A).

The magnitude of rainfall over such a vast area of the Upper Midwest resulted in flooding of extraordinary and catastrophic proportions on the Mississippi and many of its tributaries, affecting large portions of Illinois, Iowa, Kansas, Minnesota, Missouri, Nebraska, South Dakota, and Wisconsin (Figure 9–B).

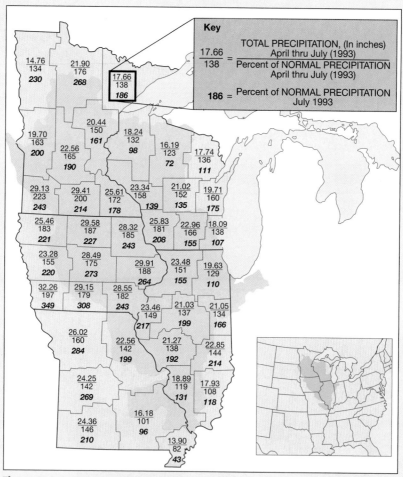

Figure 9–A Precipitation data for the upper Mississippi River basin between April 1 and July 31, 1993. *(Data from the Illinois State Water Survey)*

spacing of the isobars indicate a strong system that affected the circulation of the entire eastern two-thirds of the United States. A glance at the winds reveals a robust counterclockwise flow converging on the low.

The activity in the cold sector of the storm just north of the occluded front produced one of the worst March blizzards ever to affect the northcentral United States. In the Duluth–Superior area of Minnesota and Wisconsin,

Soils throughout the Midwest were already saturated from ample rainfall in summer and fall 1992, and soils remained moist as winter began. This pattern continued into spring and summer 1993.

Compared to the long-term average, rainfall in the Upper Mississippi River basin during April and May was 40 percent higher than average, and June rainfall was double the average. As the deluge continued through July, much of the basin received rainfall between two and three times the norm.

A stationary weather pattern over the United States was responsible for the Midwest's persistent, drenching rains. Most at the showers and thunderstorms developed in the boundary area between cooler air over the Northern Plains and warm, very humid air over the South. This front oscillated north and south over the Midwest during much of June and July. Meanwhile, a strong high-pressure system (the "Bermuda High") became anchored over the southeastern United States, blocking the progression of weather systems through the eastern half of the nation.

Some of the individual station reports were nothing short of astounding. In northwestern Missouri, Skidmore reported 25.35 inches of rain in July, and Worth County reported 30.30 inches through July 25. Normal July rainfall for this area is about 4 inches, while the average annual rainfall is 35 inches. Alton (northwestern Iowa) had a July rainfall total of 20.41 inches, and Leon (southcentral Iowa) reported 20.68 inches.

July rainfall in Illinois was also much above normal in most areas, but maximum amounts were in the 10-to-15-inch range: 10.65 inches at the Quincy Memorial Bridge, 11.45 inches at Monmouth, 11.83 inches at Galesburg, and 13.88 inches at Flora. Rainfall totals such as these leave no doubt that the record-breaking flooding on the Mississippi and other rivers in the Midwest was directly related to the exceptional rainfall during the spring and early summer.

*Steven Hilberg is Meteorologist and Director, Office of Extension Services and Operations, Illinois State Water Survey.

Figure 9–B Water rushes through a break in an artificial levee in Monroe County, Illinois. During the record-breaking 1993 Midwest floods, many artificial levees could not withstand the force of the floodwaters. Sections of many weakened structures were overtopped or simply collapsed. *(Photo by James A. Finley/AP/Wide World Photos)*

winds of up to 81 miles per hour were measured. Unofficial estimates of wind speeds in excess of 100 miles per hour were made on the aerial bridge connecting these cities. Winds blew 12 inches of snow into 10-to-15-foot

drifts, and some roads were closed for three days (Figure 9–23). One large supper club in Superior was destroyed by fire because drifts prevented firefighting equipment from reaching the blaze.

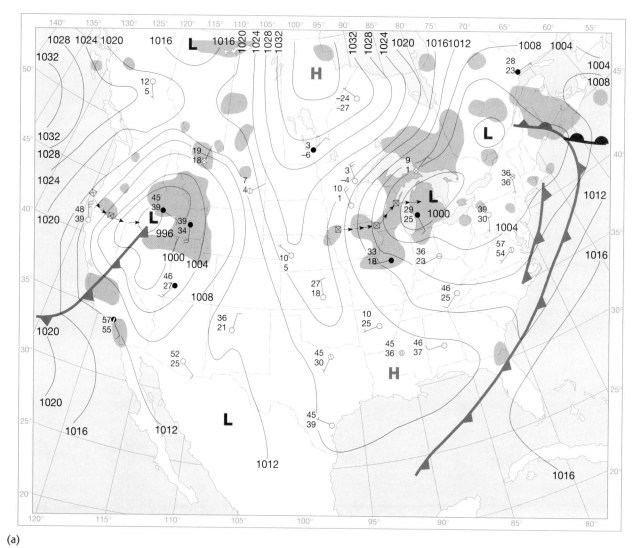

(a)

(b)

Figure 9–22 (a) Surface weather map for March 25. (b) Satellite image showing the cloud patterns on that day. *(Courtesy of NOAA/Seattle)*

Figure 9–23 Paralyzing blizzard strikes the northcentral United States. *(Photo by Mike McCleary/Bismarck Tribune)*

At the other extreme was the weather produced by the cold front as it closed in on the warm, humid air flowing into the warm sector. By the late afternoon of March 23, the cold front had generated a hailstorm in parts of eastern Texas. As the cold front moved eastward, it affected all of the southeastern United States except southern Florida. Throughout this region, numerous thunderstorms were spawned. Although high winds, hail, and lightning triggered extensive damage, the 19 tornadoes generated by the storm caused even greater death and destruction.

The path of the front can be easily traced from the reports of storm damage. By the evening of March 23, hail and wind damage were reported as far east as Mississippi and Tennessee. Early on the morning of March 24, golf-ball-sized hail was reported in downtown Selma, Alabama. About 6:30 A.M. that day the "Governor's Tornado" struck Atlanta, Georgia. Here the storm displayed its worst temper. Damage was estimated at over $50 million, three lives were lost, and 152 people were injured. The 12-mile path of the "Governor's Tornado" cut through an affluent residential area of Atlanta that included the governor's mansion (hence the name). (The official report on this tornado notes than no mobile homes lay in the tornado's path. Why was this fact worth noting?) The final damage along the cold front was reported at 4:00 A.M. on March 25 in northeastern Florida. Here hail and a small tornado caused minor damage. Thus,

a day and a half and some 1200 kilometers (about 750 miles) after the cold front became active in Texas, it left the United States to vent its energy harmlessly over the Atlantic Ocean.

By the morning of March 24, you can see that cold polar air had penetrated deep into the United States behind the cold front (Figure 9–21). Fort Worth, Texas, which just the day before was in the warm sector, now experienced cool northwest winds. Subfreezing temperatures had moved as far south as northern Oklahoma. Notice, however, that by March 25, Fort Worth was again experiencing a southerly flow. We can conclude that this was a result of the decaying cyclone that no longer dominated the circulation in the region. We can also safely assume that a warming trend was experienced in Fort Worth over the next day or so. Also notice on the map for March 25 that a high was situated over southwestern Mississippi. The clear skies and weak surface winds associated with the center of a subsiding air mass are well illustrated here.

You may have already noticed another cyclone moving in from the Pacific on March 25 as our storm exited to the east. This storm developed in a similar manner, but was centered somewhat farther north. As you might guess, another blizzard hammered the northern Plains states and a few tornadoes spun through Texas, Arkansas, and Kentucky, while precipitation dominated the weather in the central and eastern United States.

Weather in Peoria

Having examined the general weather associated with the passage of this cyclone from March 23 through March 25, let us now look at the weather experienced at a single location during this time. We have selected Peoria, a city in central Illinois located just north of the Springfield station shown on the weather charts.

Before you read the description of Peoria's weather, try to answer the following questions by using Table 9–2, which provides weather observations at three-hour intervals during this time period. Also use the three weather charts (Figures 9–20 through 9–22) and recall the general wind and temperature changes expected with the passage of fronts.

1. What type of clouds were probably present in Peoria during the early morning hours of March 23?
2. At approximately what time did the warm front pass through Peoria?
3. List two lines of evidence indicating that a warm front did pass through Peoria.
4. How did the wind and temperature changes during the early morning hours of March 23 indicate the approach of a warm front?
5. Explain the slight temperature increases experienced between 6:00 P.M. and 9:00 P.M. on March 23.
6. By what time had the cold front passed through Peoria?

Table 9–2 Weather data for Peoria, Illinois, March 23 to 25

	Temperature (°F)	Wind Direction	Cloud Coverage (Tenths)	Visibility (Miles)	Weather and Precipitation
March 23					
00:00	43	ENE	5	15	
3:00 A.M.	43	ENE	8	15	
6:00 A.M.	42	E	5	12	
9:00 A.M.	50	ESE	10	10	
12:00 P.M.	61	SE	10	12	
3:00 P.M.	64	SE	10	10	Thunderstorm with rain showers
6:00 P.M.	64	SE	10	6	Haze
9:00 P.M.	65	S	10	10	Thunderstorm with rain
March 24					
00:00	57	WSW	10	15	
3:00 A.M.	47	SW	2	15	
6:00 A.M.	42	SW	6	15	
9:00 A.M.	39	SW	8	15	
12:00 P.M.	37	SSW	10	15	Snow showers
3:00 P.M.	33	SW	10	10	Snow showers
6:00 P.M.	30	WSW	10	10	Snow showers
9:00 P.M.	26	WSW	10	6	Snow showers
March 25					
00:00	26	WSW	10	15	
3:00 A.M.	25	WSW	10	8	Snow shower
6:00 A.M.	24	WSW	10	1	Snow
9:00 A.M.	26	W	10	3	Snow
12:00 P.M.	25	W	10	12	Snow
3:00 P.M.	27	WNW	10	12	Snow
6:00 P.M.	25	NW	9	12	
9:00 P.M.	24	NW	4	15	

7. List some changes that indicate the passage of the cold front.

8. The cold front had already gone through Peoria by noon on March 24, so how do you account for the snow shower that occurred during the next 24 hours?

9. Did the thunderstorms in Peoria occur with the passage of the warm front, the cold front, or the occluded front?

10. Basing your answer on the apparent clearing skies late on March 25, would you expect the low temperature on March 26 to be lower or higher than on March 25? Explain.

Now, read the following discussion to obtain a more complete description of the weather you just reviewed.

We begin our weather observations in Peoria just after midnight on March 23 (Table 9–2). The sky contains cirrus and cirrocumulus clouds, and cool winds from the ENE dominate. As the early morning hours pass, we observe a slight wind shift toward the southeast and a very small drop in temperature. Three hours after sunrise, altocumulus clouds are replaced by stratus and nimbostratus clouds that darken the sky, yet the warm front passes without incident. The 20°F increase in temperature and the wind shift from an easterly to a southeasterly flow mark the passage of the warm front as the warm sector moves over Peoria. The pleasant 60°F temperatures are welcome in Peoria, which enjoys a day 14°F above normal.

But the mild weather is short-lived because the part of the cyclone that passed Peoria was near the apex of the storm, where the cold front is generally close behind the warm front. By afternoon, the cold front generates numerous thunderstorms from cumulonimbus clouds embedded in the warm sector just ahead of the cold front.

Strong winds, half-inch hail, and a tornado cause local damage. The temperature remains unseasonably warm during the thunderstorm activity.

Early on the morning of March 24, the passage of the cold front is marked by a wind shift, rapidly clearing skies, and a temperature drop of nearly 20°F, all in less than four hours. Throughout March 24, southwesterly winds bring cold air around the back (west) side of the intensifying storm (see Figure 9–22). Although the surface fronts have passed, this intense cyclone, with its occluded front aloft, is generating snowfall over a wide area. This is not unusual behavior. Occluding cyclones often slow their movement as this one did. For about 24 hours, snow flurries dominate Peoria's weather picture.

By noon on March 25, the storm has lost its punch and the pressure begins to climb. The skies start clearing and the winds become northwesterly as the once-tightly wound storm weakens. The mean temperature on March 25 is 16°F below normal, and the ground is covered with snow.

This example demonstrates the effect of a spring cyclone on the weather of a midlatitude location. Within just three days, Peoria's temperatures changed from unseasonably warm to unseasonably cold. Thunderstorms with hail were followed by snow showers. You can see how the north–south temperature gradient, which is most pronounced in the spring, generates these intense storms. Recall that it is the role of these storms to transfer heat from the tropics poleward. Because of Earth's rotation, however, this latitudinal heat exchange is complex, for the Coriolis force gives the winds a zonal (west-to-east) orientation. If Earth rotated more slowly, or not at all, a more leisurely north-south flow would exist that might reduce the temperature gradient. Thus, the tropics would be cooler and the poles warmer, and the midlatitudes would not be as stormy.

Chapter Summary

- The primary weather producer in the middle latitudes (for our purposes, the region between southern Florida and Alaska, essentially the area of the westerlies) is the *middle-latitude* or *midlatitude cyclone*. Midlatitude cyclones are large low pressure systems with diameters often exceeding 1000 kilometers (600 miles) that generally travel from west to east. They last a few days to more than a week, have a counterclockwise circulation pattern with a flow inward toward their centers, and have a cold front and frequently a warm front extending from the central area of low pressure. In the *polar front theory*, midlatitude cyclones develop in conjunction with the *polar front*.

- *Fronts* are boundary surfaces that separate air masses of different densities, one usually warmer and more moist than the other. As one air mass moves into another, the warmer, less dense air mass is forced aloft in a process called *overrunning*. The five types of fronts are (1) *warm front*, which occurs when the surface (ground) position of a front moves so that warm air occupies territory formerly covered by cooler air, (2) *cold front*, where cold continental polar air actively advances into a region occupied by warmer air, (3) *stationary front*, which occurs when the air flow on both sides of a front is neither toward the cold air mass nor toward the warm air mass, (4) *occluded front*, which develops when an active cold front overtakes a warm front and wedges the warm front upward, and (5) a *dryline*, a boundary between denser dry, air and less dense humid air often associated with severe thunderstorms during the spring and summer. The two types of occluded fronts are the *cold-type occluded front*, where the air behind the cold front is colder than the cool air it is overtaking, and the *warm-type occluded front*, where the air behind the advancing cold front is warmer than the cold air it overtakes.

- According to the polar front model, midlatitude cyclones form along fronts and proceed through a generally predictable life cycle. Along the polar front, where two air masses of different densities are moving parallel to the front and in opposite directions, *cyclogenesis* (cyclone formation) occurs and the frontal surface takes on a wave shape that is usually several hundred kilometers long. Once a wave forms, warm air advances poleward invading the area formerly occupied by colder air. This change in the direction of the surface flow causes a readjustment in the pressure pattern that results in almost circular isobars, with the low pressure centered at the apex of the wave. Usually, the position of the cold front advances faster than the warm front and gradually closes the warm sector and lifts the warm front. This process, known as *occlusion*, creates an occluded front. Eventually all the warm sector is forced aloft and cold air surrounds the cyclone at low levels. At this point, the cyclone has exhausted its source of energy and the once highly organized counterclockwise flow ceases to exist.

- Guided by the westerlies aloft, cyclones generally move eastward across the United States. As an idealized midlatitude cyclone moves over a region, the passage of a warm front places the area under the influence of a maritime tropical air mass and its generally warm temperatures, southerly winds, and clear skies. The passage of a cold front is easily detected by a wind shift, the replacement of south or southwesterly winds with winds from the west or northwest. There is also a pronounced drop in temperature. A passing occluded front is often associated with cool, overcast conditions, and snow or glaze during the cool months.

- Airflow aloft (divergence and convergence) plays an important role in maintaining cyclonic and anticyclonic circulation. In a cyclone, divergence aloft does not involve the outward flow of air in all directions. Instead, the winds flow generally from west to east, along sweeping curves. Also, at high altitudes, speed variations within the jet stream cause air to converge in areas where the velocity slows, and to diverge where air is accelerating. In addition to *speed divergence*, *directional divergence* (the horizontal spreading of an air stream) and *vorticity* (the amount of rotation exhibited by a mass of moving air) also contribute to divergence (or convergence) aloft.

- During the colder months, when temperature gradients are steepest, cyclonic storms advance at their fastest rate. Furthermore, the westerly airflow aloft tends to steer these developing pressure systems in a general west-to-east direction. Cyclones that influence western North America originate over the Pacific Ocean. Although most Pacific storms do not cross the Rockies intact, many redevelop on the lee (eastern) side of these mountains. Some cyclones that affect the United States form over the Great Plains and are associated with an influx of maritime tropical air from the Gulf of Mexico. Another area where cyclogenesis occurs is east of the southern Appalachians. These cyclones tend to migrate toward the northeast, impacting the eastern seaboard.

- Due to the gradual subsidence within them, anticyclones generally produce clear skies and calm conditions. One to three times each winter, large highs, called *blocking highs*, persist over the middle latitudes and deflect the nearly zonal west-to-east flow poleward. These stagnant anticyclones block the eastward migration of cyclones, keeping one section of the nation dry for a week or more while another region experiences one cyclonic storm after another. Also due to subsidence, large stagnant anticyclones can produce a temperature inversion that contributes to air pollution episodes.

- In the spring, Earth's pronounced north-south temperature gradient can generate intense cyclonic storms. At a midlatitude location, as a spring cyclone with its associated fronts passes, temperatures can change quickly from unseasonably warm to unseasonably cold, and thunderstorms with hail can be followed by snow showers.

Vocabulary Review

cold front (p. 242)
cold-type occluded front (p. 243)
cyclogenesis (p. 245)

dryline (p. 245)
front (p. 239)
middle-latitude (midlatitude) cyclone (p. 239)

occluded front (p. 243)
occlusion (p. 247)
overrunning (p. 239)
polar-front theory (p. 239)

speed divergence (p. 252)
stationary front (p. 243)
warm front (p. 240)
warm-type occluded front (p. 243)

Review Questions

1. How did the early Norwegian meteorologists describe fronts?

2. If you were located 400 kilometers ahead of the surface position of a typical warm front, how high would the frontal surface be above you?

3. Compare the weather of a warm front with that of a cold front.

4. Why is cold-front weather usually more severe than warm-front weather? *Cold fronts have cumulonimbus*

5. Explain the basis for the following weather proverb:

   Rain long foretold, long last;
   Short notice, soon past.

6. How does a stationary front produce precipitation when its position does not change, or changes very slowly?

7. Distinguish between cold-type and warm-type occluded fronts.

8. Describe the initial stage in the formation of a midlatitude cyclone. *Cold air meets warm air*

9. Midlatitude cyclones are sometimes called *wave cyclones*. Why do you think this is so?

10. Although the formation of an occluded front often represents a period of increased intensity for a midlatitude cyclone, it also marks the beginning of the end of the system. Explain why such is the case.

11. For each of the weather elements listed here, describe the changes that an individual experiences when a middle latitude cyclone passes with its center *north* of the observer. (Hint: See Figures 9–11 and 9–12)
    a. wind direction
    b. pressure tendency
    c. cloud type
    d. cloud cover
    e. precipitation
    f. temperature

12. Describe the weather conditions that an observer would experience if the center of a midlatitude cyclone passed to the south.

13. Distinguish between veering and backing winds (see Box 9–1).

14. Briefly explain how the flow aloft maintains cyclones at the surface.

15. What is speed divergence? Speed convergence?

16. Given an upper air chart, where do forecasters usually look to find favorable sites for cyclogenesis? Where do anticyclones usually form in relation to the upper-level flow?

17. What are two possible ways a blocking high might influence the weather?

18. Briefly describe the various weather phenomena that could be associated with a strong springtime cyclonic storm traveling across the United States.

Atmospheric Science Online

The following are informative and interesting Internet sites that address topics related to those presented in the chapter:

Fronts (Online Meteorology Guides, University of Illinois):

- **http://ww2010.atmos.uiuc.edu/(Gh)/guides/mtr/af/frnts/home.rxml**

Midlatitude Cyclones (Online Meteorology Guides, University of Illinois):

- **http://ww2010.atmos.uiuc.edu/(Gh)/guides/mtr/cyc/home.rxml**

United States Surface Weather Map (Intellicast):

- **http://www.intellicast.com/LocalWeather/World/UnitedStates/SurfaceAnalysis/**

For direct links to these sites and others, chapter objectives and reviews, quiz questions, and topical investigations that utilize Web resources, visit *The Atmosphere, Eighth Edition* Home Page at:

- **http://www.prenhall.com/lutgens**

Thunderstorms and Tornadoes

Tornado destroying buildings in Pampa, Texas. *(Photo by Alan R. Moller/Tony Stone Images)*

Figure 10–1 A time-exposure of cloud-to-ground lightning. *(Photo by Tom Ives/The Stock Market)*

The subject of this and the following chapter is severe weather. In this chapter, we will examine the severe local weather produced in association with cumulonimbus clouds, namely thunderstorms and tornadoes. In Chapter 11, the focus will turn to the large tropical storms we call hurricanes.

Occurrences of severe weather have a fascination that ordinary weather phenomena cannot provide. The lightning display generated by a thunderstorm can be a spectacular event that elicits both awe and fear (Figure 10–1). Of course, hurricanes and tornadoes also attract a great deal of much-deserved attention. A single tornado outbreak or hurricane can cause billions of dollars in property damage as well as many deaths.

What's in a Name?

In Chapter 9, we examined the middle-latitude cyclones that play such an important role in causing day-to-day weather changes. Yet the use of the term "cyclone" is often confusing. For many people, the term implies only an intense storm, such as a tornado or a hurricane. When a hurricane unleashes its fury on India or Bangladesh, for example, it is usually reported in the media as a cyclone (the term denoting a hurricane in that part of the world).

Similarly, tornadoes are referred to as cyclones in some places. This custom is particularly common in portions of the Great Plains of the United States. Recall that

in the *Wizard of Oz*, Dorothy's house was carried from her Kansas farm to the land of Oz by a cyclone. Indeed, the nickname for the athletic teams at Iowa State University is the *Cyclones*. Although hurricanes and tornadoes are, in fact, cyclones, the vast majority of cyclones are not hurricanes or tornadoes. The term "cyclone" simply refers to the circulation around any low-pressure center, no matter how large or intense it is.

Tornadoes and hurricanes are both smaller and more violent than middle-latitude cyclones. Middle-latitude cyclones may have a diameter of 1600 kilometers (1000 miles) or more. By contrast, hurricanes average only 600 kilometers (375 miles) across, and tornadoes, with a diameter of just 0.25 kilometer (0.16 mile), are much too small to show up on a weather map.

The thunderstorm, a much more familiar weather event, hardly needs to be distinguished from tornadoes, hurricanes, and midlatitude cyclones. Unlike the flow of air about these storms, the circulation associated with thunderstorms is characterized by strong up-and-down movements. Winds in the vicinity of a thunderstorm do not follow the inward spiral of a cyclone, but they are typically variable and gusty.

Although thunderstorms form "on their own" away from cyclonic storms, they also form in conjunction with cyclones. For instance, thunderstorms are frequently spawned along the cold front of a midlatitude cyclone, where on rare occasions a tornado may descend from the thunderstorm's cumulonimbus tower. Hurricanes also generate widespread thunderstorm activity. Thus, thunderstorms are related in some manner to all three types of cyclones mentioned here.

Thunderstorms

Almost everyone has observed various small-scale phenomena that result from the vertical movements of relatively warm, unstable air. Perhaps you have seen a dust devil over an open field on a hot day whirling its dusty load to great heights (see Box 7–1) or maybe you have seen a bird glide effortlessly skyward on an invisible thermal of hot air. These examples illustrate the dynamic thermal instability that occurs during the development of a *thunderstorm*. A **thunderstorm** is simply a storm that generates lightning and thunder. It frequently produces gusty winds, heavy rain, and hail. A thunderstorm may be produced by a single cumulonimbus cloud and influence only a small area or it may be associated with clusters of cumulonimbus clouds covering a large area.

Thunderstorms form when warm, humid air rises in an unstable environment. Various mechanisms can trigger the upward air movement needed to create thunderstorm-producing cumulonimbus clouds. One mechanism, the unequal heating of Earth's surface, significantly contributes to the formation of *air-mass thunderstorms*. These storms are associated with the scattered puffy cumulonimbus clouds that commonly form *within* maritime tropical air masses and produce scattered thunderstorms on summer days. Such storms are usually short-lived and seldom produce strong winds or hail.

In contrast, thunderstorms in a second category not only benefit from uneven surface heating but are associated with the lifting of warm air, as occurs along a front or a mountain slope. Moreover, diverging winds aloft frequently contribute to the formation of these storms because they tend to draw air from lower levels upward beneath them. Some of the thunderstorms in this second category may produce high winds, damaging hail, flash floods, and tornadoes. Such storms are described as *severe*.

At any given time, there are an estimated 2000 thunderstorms in progress. As we would expect, the greatest proportion occurs in the tropics, where warmth, plentiful moisture, and instability are always present. About 45,000

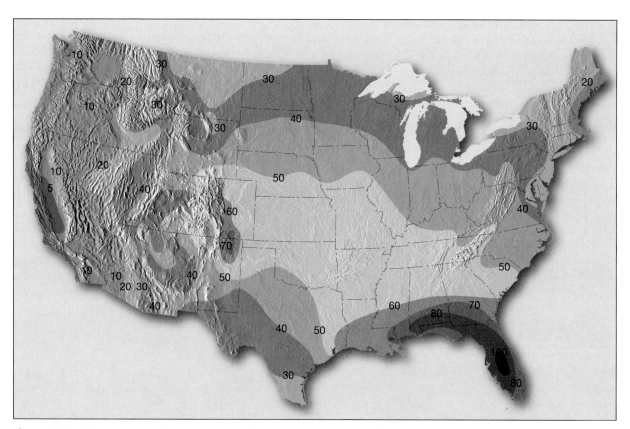

Figure 10–2 Average number of days each year with thunderstorms. The humid subtropical climate that dominates the southeastern United States receives much of its precipitation in the form of thunderstorms. Most of the Southeast averages 50 or more days each year with thunderstorms. *(Source: Environmental Data Service, NOAA)*

thunderstorms take place each day and more than 16 million occur annually around the world. The lightning from these storms strikes Earth 100 times each second.

Annually, the United States experiences about 100,000 thunderstorms and millions of lightning strikes. A glance at Figure 10–2 shows that thunderstorms are most frequent in Florida and the eastern Gulf Coast region, where activity is recorded on between 70 and 100 days each year. The region on the east side of the Rockies in Colorado and New Mexico is next, with thunderstorms occurring on 60 to 70 days annually. Most of the rest of the nation experiences thunderstorms on 30 to 50 days a year. Clearly, the western margin of the United States has little thunderstorm activity. The same is true for the northern tier of states and for Canada, where warm, moist, unstable mT air seldom penetrates.

Air-Mass Thunderstorms

In the United States, **air-mass thunderstorms** frequently occur in maritime tropical (mT) air that moves northward from the Gulf of Mexico. These warm, humid air masses contain abundant moisture in their lower levels and can be rendered unstable when heated from below or lifted along a front. Because mT air most often becomes unstable in spring and summer when it is warmed from below by the heated land surface, it is during these seasons that air-mass thunderstorms are most frequent. They also have a strong preference for mid-afternoon, when surface temperatures are highest. Because local differences in surface heating aid the growth of air-mass thunderstorms, they generally occur as scattered, isolated cells instead of being organized in relatively narrow bands or other configurations.

Stages of Development

Important field experiments that were conducted in Florida and Ohio in the late 1940s probed the dynamics of air-mass thunderstorms. This pioneering work, known as the *Thunderstorm Project*, was prompted by a number of thunderstorm-related airplane crashes. It involved the use of radar, aircraft, radiosondes, and an extensive network of surface instruments. The research produced a three-stage model of the life cycle of an air-mass thunderstorm that remains basically unchanged after more than 50 years. The three stages are depicted in Figure 10–3.

Cumulus Stage. Recall that an air-mass thunderstorm is largely a product of the uneven heating of the surface, which leads to rising currents of air that ultimately produce a cumulonimbus cloud. At first the buoyant thermals produce fair weather cumulus clouds that may exist for just minutes before evaporating into the drier surrounding air (Figure 10–4). This initial cumulus development is important because it moves water vapor from the surface to greater heights. Ultimately, the air becomes sufficiently humid that newly forming clouds do not evaporate, but instead continue to grow vertically.

The development of a cumulonimbus tower requires a continuous supply of moist air. The release of latent heat allows each new surge of warm air to rise higher than the last, adding to the height of the cloud (Figure 10–5). This phase in the development of a thunderstorm, called the **cumulus stage**, is dominated by updrafts (Figure 10–3a).

Once the cloud passes beyond the freezing level, the Bergeron process begins producing precipitation. Eventually, the accumulation of precipitation in the cloud is too great for the updrafts to support. The falling precipitation causes drag on the air and initiates a downdraft.

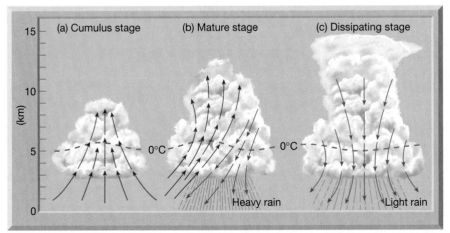

Figure 10–3 Stages in the development of a thunderstorm. During the cumulus stage, strong updrafts act to build the storm. The mature stage is marked by heavy precipitation and cool downdrafts in part of the storm. When the warm updrafts disappear completely, precipitation becomes light and the cloud begins to evaporate.

Figure 10–4 At first, buoyant thermals produce fair weather cumulus clouds that soon evaporate into the surrounding air, making it more humid. As this process of cumulus development and evaporation continues, the air eventually becomes sufficiently humid so that newly forming clouds do not evaporate but continue to grow. *(Photo by Henry Lansford/Photo Researchers, Inc.)*

The creation of the downdraft is further aided by the influx of cool, dry air surrounding the cloud, a process termed **entrainment**. This process intensifies the downdraft because the air added during entrainment is cool and therefore heavy; possibly of greater importance, it is dry. It thus causes some of the falling precipitation to evaporate (a cooling process), thereby cooling the air within the downdraft.

Mature Stage. As the downdraft leaves the base of the cloud, precipitation is released, marking the beginning of the cloud's **mature stage** (Figure 10–3b). At the surface, the cool downdrafts spread laterally and can be felt before the actual precipitation reaches the ground. The sharp, cool gusts at the surface are indicative of the downdrafts aloft. During the mature stage, updrafts exist side by side with downdrafts and continue to enlarge the cloud. When the cloud grows to the top of the unstable region, often located at the base of the stratosphere, the updrafts spread laterally and produce the characteristic anvil top. Generally ice-laden cirrus clouds make up the top and are spread downwind by strong winds aloft. The mature stage is the most active period of a thunderstorm. Gusty winds, lightning, heavy precipitation, and sometimes small hail are experienced.

Dissipating Stage. Once downdrafts begin, the vacating air and precipitation encourage more entrainment of the cool, dry air surrounding the cell. Eventually, the downdrafts dominate throughout the cloud and initiate the **dissipating stage** (Figure 10–3c). The cooling effect of falling precipitation and the influx of colder air aloft mark the end of the thunderstorm activity. Without a supply of moisture from updrafts, the cloud will soon evaporate. An interesting fact is that only a modest portion, on the order of 20 percent, of the moisture that condenses in an air-mass thunderstorm actually leaves the cloud as precipitation. The remaining 80 percent evaporates back into the atmosphere.

Figure 10–5 This developing cumulonimbus cloud became a towering August thunderstorm over central Illinois. *(Photo by E. J. Tarbuck)*

It should be noted that within a single air-mass thunderstorm there may be several individual *cells*, that is, zones of adjacent updrafts and downdrafts. When you view a thunderstorm, you may notice that the cumulonimbus cloud consists of several towers (Figure 10–5). Each tower may represent an individual cell that is in a somewhat different part of its life cycle.

To summarize, the stages in the development of an air-mass thunderstorm are as follows:

1. The *cumulus stage*, in which updrafts dominate throughout the cloud and growth from a cumulus to a cumulonimbus cloud occurs.
2. The *mature stage*, the most intense phase, with heavy rain and possibly small hail, in which downdrafts are found side by side with updrafts.
3. The *dissipating stage*, dominated by downdrafts and entrainment, causing evaporation of the structure.

Occurrence

Mountainous regions, such as the Rockies in the West and the Appalachians in the East, experience a greater number of air-mass thunderstorms than do the Plains states. The air near the mountain slope is heated more intensely than air at the same elevation over the adjacent lowlands. A general upslope movement then develops during the daytime that can sometimes generate thunderstorm cells. These cells may remain almost stationary above the slopes below.

Although the growth of thunderstorms is aided by high surface temperatures, many thunderstorms are not generated solely by surface heating. For example, many of Florida's thunderstorms are triggered by the convergence associated with sea-to-land air flow (see Figure 4–17, page 108). Many thunderstorms that form over the eastern two-thirds of the United States occur as part of the general convergence and frontal wedging that accompany passing midlatitude cyclones. Near the equator, thunderstorms commonly form in association with the convergence along the equatorial low. Most of these thunderstorms are not severe, and their life cycles are similar to the three-stage model described for air-mass thunderstorms.

Severe Thunderstorms

Severe thunderstorms are capable of producing heavy downpours and flash flooding as well as strong, gusty straight-line winds, large hail, frequent lightning, and perhaps tornadoes (see Box 10–1). For a thunderstorm to be officially classified as *severe* by the National Weather Service, it must have winds in excess of 93 kilometers (58 miles) per hour (50 knots) or produce hailstones with diameters larger than 1.9 centimeters (0.75 inch) or generate a tornado. Of the estimated 100,000 thunderstorms that occur annually in the United States, about 10 percent (10,000 storms) reach severe status.

As you learned in the preceding section, air-mass thunderstorms are localized, relatively short-lived phenomena that dissipate after a brief, well-defined life cycle. They actually extinguish themselves because downdrafts cut off the supply of moisture necessary to maintain the storm. For this reason, air-mass thunderstorms seldom if ever produce severe weather. By contrast, other thunderstorms do not quickly dissipate but instead may remain active for hours. Some of these larger, longer-lived thunderstorms attain severe status.

Why do some thunderstorms persist for many hours? A key factor is the existence of strong vertical wind shear—that is, changes in wind direction and/or speed between different heights. When such conditions prevail, the updrafts that provide the storm with moisture do not remain vertical, but become tilted. Because of this, the precipitation that forms high in the cloud falls into the downdraft rather than into the updraft as occurs in air-mass thunderstorms. This allows the updraft to maintain its strength and continue to build upward. Sometimes the updrafts are sufficiently strong that the cloud top is able to push its way into the stable lower stratosphere, a situation called *overshooting* (Figure 10–6).

Beneath the cumulonimbus tower, where downdrafts reach the surface, the denser cool air spreads out along the ground (see Box 10–2). The leading edge of this outflowing downdraft acts like a wedge, forcing warm, moist surface air into the thunderstorm. In this way the downdrafts act to maintain the updrafts, which, in turn, sustain the thunderstorm.

By examining Figure 10–6, you can see that the outflowing cool air of the downdraft acts as a "mini-cold front" as it advances into the warmer surrounding air. This outflow boundary is called a **gust front**. As the gust front moves across the ground the very turbulent air sometimes picks up loose dust and soil, making the advancing boundary visible. Frequently a *roll cloud* may form as warm air is lifted along the leading edge of the gust front (Figure 10–7). The advance of the gust front can provide the lifting needed for the formation of new thunderstorms many kilometers away from the initial cumulonimbus clouds.

Supercell Thunderstorms

Some of our most dangerous weather is caused by a type of thunderstorm called a **supercell**. Few weather phenomena are as awesome (Figure 10–8). An estimated 2000

Box 10–1 Atmospheric Hazard: Flash Floods: The Number One Thunderstorm Killer

Tornadoes and hurricanes are nature's most awesome storms. Because of this status, they are logically the focus of much well-deserved attention. Yet, surprisingly, these dreaded events are not responsible for the greatest number of storm-related deaths. That distinction is reserved for flash floods. For a recent 30-year period, the number of storm-related deaths in the United States from flooding averaged 135 per year. By contrast, tornado fatalities averaged 73 annually and hurricanes, 25.

Flash floods are local floods of great volume and short duration. The rapidly rising surge of water usually occurs with little advance warning and can destroy roads, bridges, homes, and other substantial structures (Figure 10–A). Discharges quickly reach a maximum and diminish almost as rapidly. Flood flows often contain large quantities of sediment and debris as they sweep channels clean.

Frequently flash floods result from the torrential rains associated with a slow-moving severe thunderstorm or take place when a series of thunderstorms repeatedly pass over the same location. Sometimes they are triggered by heavy rains from hurricanes and tropical storms. Occasionally floating debris or ice can accumulate at a natural or artificial obstruction and restrict the flow of water. When such temporary dams fail, torrents of water can be released as a flash flood.

Flash floods can take place in almost any area of the country. They are particularly common in mountainous terrain, where steep slopes can quickly channel runoff into narrow valleys. The hazard is most acute when the soil is already nearly saturated from earlier rains or consists of impermeable materials. A disaster in Shadydale, Ohio, demonstrates what can happen when even moderately heavy rains fall on saturated ground with steep slopes.

> On the evening of 14 June 1990, 26 people lost their lives as rains estimated to be in the range of 3 to 5 inches fell on saturated soil, which generated flood waves in streams that reached tens of feet in height, destroying near-bank residences and businesses. Preceding months of above-normal rainfall had generated soil moisture contents of near saturation. As a result, moderate amounts of rainfall caused large amounts of surface and near-surface runoff. Steep valleys with practically vertical walls channeled the floods, creating very fast, high, and steep wave crests.°

Why do so many people perish in flash floods? Aside from the factor of surprise (many are caught sleeping), people do not appreciate the power of moving water. A glance at Figure 10–A helps illustrate the force of a flood wave. Just 15 centimeters (6 inches) of fast-moving flood water can knock a person down. Most automobiles will float and be swept away in only 0.6 meter (2 feet) of

---

to 3000 supercell thunderstorms occur annually in the United States. They represent just a small fraction of all thunderstorms, but they are responsible for a disproportionate share of the deaths, injuries, and property damage associated with severe weather. Less than half of all supercells produce tornadoes, yet virtually all of the strongest and most violent tornadoes are spawned by supercells.

A supercell consists of a single, very powerful cell that at times can extend to heights of 20 kilometers (65,000 feet) and that persists for many hours. These massive clouds have diameters ranging between about 20 and 50 kilometers (12 and 30 miles).

Despite the single-cell structure of supercells, these storms are remarkably complex. The vertical wind profile may cause the updraft to rotate. For example, this could occur if the surface flow is from the south or southeast and the winds aloft increase in speed and become more westerly with height. If a thunderstorm develops in

such a wind environment, the updraft is made to rotate. It is within this column of cyclonically rotating air, called the **mesocyclone**, that tornadoes often form.°

The huge quantities of latent heat needed to sustain a supercell require special conditions that keep the lower troposphere warm and moisture-rich. Studies suggest that the existence of an inversion layer a few kilometers above the surface helps to provide this basic requirement. Recall that temperature inversions represent very stable atmospheric conditions that restrict vertical air motions. The presence of an inversion seems to aid the production of a few very large thunderstorms by inhibiting the formation of many smaller ones (Figure 10–9). The inversion prevents the mixing of warm, humid air in the lower troposphere with cold, dry air above. Consequently, surface

°More on mesocyclones can be found in the section on "Tornado Development."

Figure 10–A The disastrous nature of flash floods is illustrated by the Big Thompson River flood of July 31, 1976, in Colorado. During a four-hour period more than 30 centimeters (12 inches) of rain fell on portions of the small area drained by the river. This amounted to nearly three-quarters of the average yearly total. The flash flood in the narrow canyon lasted only a few hours, but cost 139 people their lives. Damages were estimated at more than $35 million. *(U.S. Geological Survey, Denver)*

water. *More than half of all U.S. flash flood fatalities are auto related!* Clearly, people should never attempt to drive over a flooded road. The depth of water is not always obvious. Also, the road bed may have been washed out under water. Present-day flash floods are calamities with potential for very high death tolls and huge property losses. Although efforts are being made to improve observations and warnings, flash floods remain elusive natural killers.

°"Prediction and Mitigation of Flash Floods: A Policy Statement of the American Meteorological Society." *Bulletin of the American Meteorological Society*, Vol. 74, No. 8 (Aug. 1993), p. 1586.

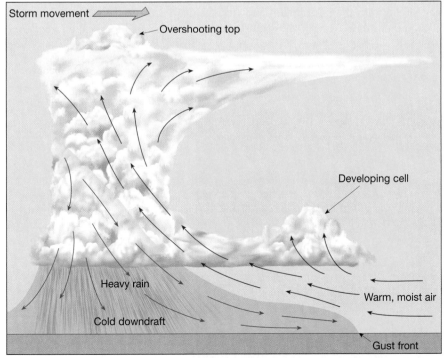

Figure 10–6 Diagram of a well-developed cumulonimbus tower showing updrafts, downdrafts, and an overshooting top. Precipitation forming in the tilted updraft falls into the downdraft. Beneath the cloud, the denser cool air of the downdraft spreads out along the ground. The leading edge of the outflowing downdraft acts to wedge moist surface air into the cloud. Eventually the outflow boundary may become a gust front that initiates new cumulonimbus development.

Figure 10–7 Roll clouds, like this one at Miles City, Montana, are sometimes produced along a gust front in an eddy between the inflow and the downdraft. *(Photo by National Science Foundation/ National Center for Atmospheric Research)*

heating continues to increase the temperature and moisture content of the layer of air trapped below the inversion. Eventually, the inversion is locally eroded by strong mixing from below. The unstable air below "erupts" explosively at these sites, producing unusually large cumulonimbus clouds. It is from such clouds with their concentrated, persistent updrafts that supercells form.

Squall Lines and Mesoscale Convective Complexes

Because the atmospheric conditions favoring the formation of severe thunderstorms often exist over a broad area, they frequently develop in groups that consist of many individual storms clustered together. Sometimes these clusters occur as elongate bands called *squall lines*.

Figure 10–8 A supercell thunderstorm. *(Photo by Howard B. Bluestein)*

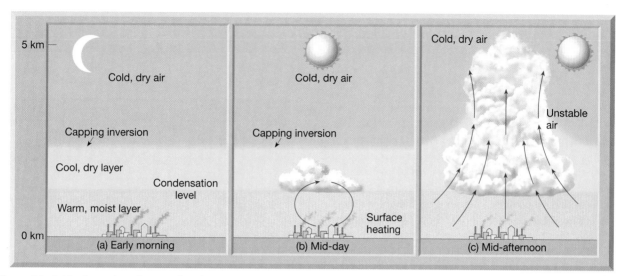

Figure 10–9 The formation of severe thunderstorms can be enhanced by the existence of a temperature inversion located a few kilometers above the surface.

At other times, the storms are organized into roughly circular clusters known as *mesoscale convective complexes.* No matter how the cells are arranged, they are not simply clusters of unrelated individual storms. Rather, they are related by a common origin, or they occur in a situation in which some cells lead to the formation of others.

Squall Lines. A **squall line** is a relatively narrow band of thunderstorms, some of which may be severe, that develops in the warm sector of a middle-latitude cyclone,

usually 100 to 300 kilometers (60 to 180 miles) in advance of the cold front. The linear band of cumulonimbus development might stretch for 500 kilometers (300 miles) or more and consists of many individual cells in various stages of development. An average squall line can last for 10 hours or more, and some have been known to remain active for more than a day. Sometimes the approach of a squall line is preceded by a *mammatus sky* consisting of dark cloud rolls that have downward pouches (Figure 10–10).

Figure 10–10 The dark overcast of a mammatus sky, with its characteristic downward bulging pouches, sometimes precedes a squall line. *(Photo by Annie Griffiths/DRK Photo)*

Box 10–2 Atmospheric Hazard: Microbursts

Beneath some thunderstorms, strong localized downdrafts known as *downbursts* sometimes occur. When downbursts are small—that is, less than 4 kilometers (2.5 miles) across—they are called *microbursts*. These straight-line concentrated bursts of wind are often produced when downdrafts are accelerated by a great deal of evaporative cooling. (Remember, the colder the air, the denser it is, and the denser the air, the faster it will "fall.") Microbursts typically last just two to five minutes. Despite their small size and short duration, microbursts represent a significant atmospheric hazard.

Upon reaching the surface, the chilled air of the microburst pushes out in all directions from the center of the downdraft, similar to a jet of water from a faucet splashing in a sink (Figure 10–B). In this manner, microbursts can produce winds in excess of 160 kilometers (100 miles) per hour. Within minutes, the source of the downdraft dissipates while the outflow at the ground continues to expand.

The violent winds of a microburst can cause a great deal of destruction. For example, in July 1993 millions of trees were uprooted by a microburst near Pak Wash, Ontario. In July 1984 a total of 11 people drowned when a microburst caused a 28-meter (90-foot)-long

Figure10–B Microburst over Stapleton Airport runway, Denver, Colorado. These powerful downward bursts of air, usually associated with thunderstorms, can prove extremely hazardous to aircraft, causing them to lose lift during critical periods of landing and takeoff. *(Courtesy of the National Science Foundation/National Center for Atmospheric Research)*

Most squall lines are not the product of forceful lifting along a cold front. Some develop from a combination of warm, moist air near the surface and an active jet stream aloft. The squall line forms when the divergence and resulting lift created by the jet stream is aligned with a strong, persistent low-level flow of warm, humid air from the south.

A squall line with severe thunderstorms can also form along a boundary called a **dryline**, a narrow zone along which there is an abrupt change in moisture. It forms when continental tropical (cT) air from the southwestern United States is pulled into the warm sector of a middle-latitude cyclone, as shown in Figure 10–11. The denser cT air acts to lift the less dense mT air with which it is converging.[†] By contrast, both cloud formation and storm development along the cold front are minimal because the front is advancing into dry cT air.

[†]Warm, dry air is more dense than warm, humid air, because the molecular weight of water vapor (H_2O) is only about 62 percent as great as the molecular weight of the mixture of gases that make up dry air.

Drylines most frequently develop in the western portions of Texas, Oklahoma, and Kansas. Such a situation is illustrated by Figure 10–12. The dryline is easily identified by comparing the dew-point temperatures on either side of the squall line. The dew points in the mT air to the east are 30° to 45°F higher than those in the cT air to the west. Much severe weather was generated as this extraordinary squall line moved eastward, including 55 tornadoes over a six-state region.

Mesoscale Convective Complexes. A **mesoscale convective complex (MCC)** consists of many individual thunderstorms organized into a large oval to circular cluster. A typical MCC is large, covering an area of at least 100,000 square kilometers (39,000 square miles). The usually slow-moving complex may persist for 12 hours or more (Figure 10–13).

MCCs tend to form most frequently in the Great Plains. When conditions are favorable, an MCC develops from a group of afternoon air-mass thunderstorms. In the evening,

sternwheeler boat to capsize on the Tennessee River. Sometimes the damage associated with these violent outflows is mistaken for tornado damage. However, wind damage from microbursts occurs in straight lines, whereas a rotational circulation pattern is usually detectable along the damage path created by a tornado.

The wind shear associated with microbursts has been responsible for a number of airplane crashes. Imagine an aircraft attempting to land and being confronted by a microburst, as in Figure 10–C. As the airplane flies into the microburst, it initially encounters a strong headwind, which tends to carry it upward. To reduce the lift, the pilot points the nose of the aircraft downward. Then, just seconds later, a tailwind is encountered on the opposite side of the microburst. Here, because the wind is moving with the airplane, the amount of air flowing over the wings and providing lift is dramatically reduced, causing the craft to suddenly lose altitude and crash.

The number of aviation deaths due to microbursts has declined significantly as understanding of the event has increased. Systems to detect the wind shifts associated with microbursts have been installed at most major airports in the United States. Moreover, pilots now receive training on how to handle microbursts during takeoffs and landings.

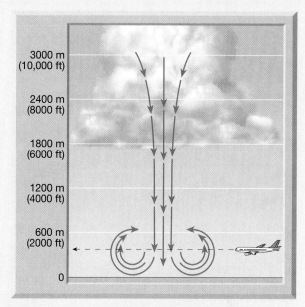

Figure10–C The arrows in this sketch represent the movement of air in a microburst, a concentrated intense downdraft formed by extraordinary evaporative cooling that creates an outward wind burst at the surface. If you could see it leave the cloud and strike the ground, it might resemble a narrow stream of water "splashing" into a sink. Because of the extremely abrupt changes in wind direction within microbursts, they are a threat to aircraft during landings and takeoffs.

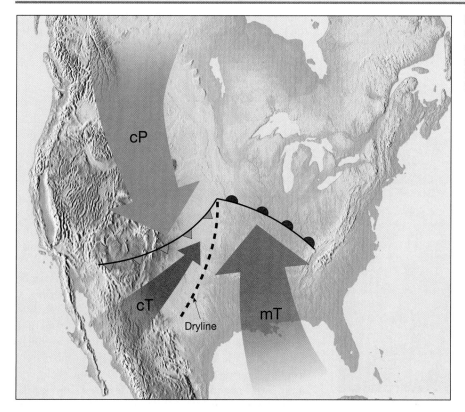

Figure 10–11 Squall-line thunderstorms frequently develop along a dryline, the boundary separating warm, dry continental tropical air and warm, moist maritime tropical air.

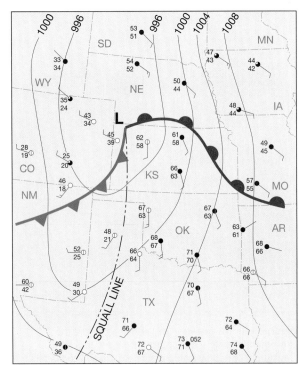

Figure 10–12 The squall line of this middle-latitude cyclone was responsible for a major outbreak of tornadoes. The squall line separates dry (cT) and very humid (mT) air. The dryline is easily identified by comparing dew-point temperatures on either side of the squall line. Dew point (°F) is the lower number at each station.

as the local storms decay the MCC starts developing. The transformation of afternoon air-mass thunderstorms into an MCC requires a strong low-level flow of very warm and

Figure 10–13 This satellite image shows a mesoscale convective complex (MCC) over the eastern Dakotas. *(NOAA)*

moist air. This flow enhances instability, which, in turn, spurs convection and cloud development. As long as favorable conditions prevail, MCCs remain self-propagating as gust fronts from existing cells lead to the formation of new powerful cells nearby. New thunderstorms tend to develop near the side of the complex that faces the incoming low-level flow of warm moist air.

Although mesoscale convective complexes sometimes produce severe weather, they are also beneficial because they provide a significant portion of the growing-season rainfall to the agricultural regions of the central United States.

Lightning and Thunder

A storm is classified as a thunderstorm only after thunder is heard. Because thunder is produced by lightning, lightning must also be present (Figure 10–14). **Lightning** is similar to the electrical shock you may have experienced on touching a metal object on a very dry day. Only the intensity is different.

During the formation of a large cumulonimbus cloud, a separation of charge occurs, which simply means that part of the cloud develops an excess negative charge, whereas another part acquires an excess positive charge. The object of lightning is to equalize these electrical differences by producing a negative flow of current from the region of excess negative charge to the region with excess positive charge or vice versa. Because air is a poor conductor of electricity (good insulator), the electrical potential (charge difference) must be very high before lightning will occur, on the order of 3000 volts per meter.

What Causes Lightning?

The origin of charge separation in clouds, although not fully understood, must hinge on rapid vertical movements within, for lightning occurs primarily in the violent mature stage of a cumulonimbus cloud. Because the formation of these tall clouds is chiefly a summertime phenomenon in the midlatitudes, it also explains why lightning is seldom observed there in the winter. Furthermore, lightning rarely occurs before the growing cloud penetrates the 5-kilometer level, where sufficient cooling begins to generate some ice crystals.

Some cloud physicists believe that charge separation occurs during the formation of ice pellets. Experimentation shows that as droplets begin to freeze, positively charged ions are concentrated in the colder regions of the droplets, whereas negatively charged ions are concentrated in the warmer regions. Thus, as the droplets freeze from the outside in, they develop a positively charged ice shell and a negatively charged interior. As the interior

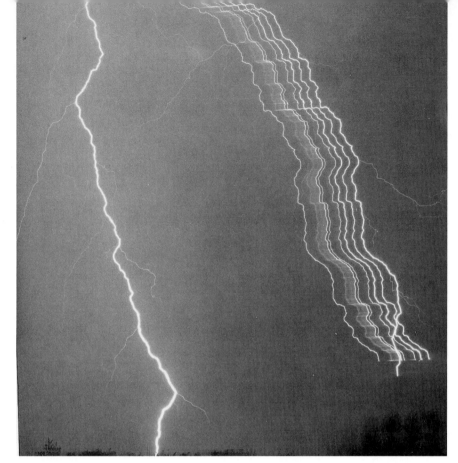

Figure 10–14 Multiple lightning stroke of a single flash as recorded by a moving-film camera. *(Courtesy of E. J. Tarbuck)*

begins to freeze, it expands and shatters the outside shell. The small positively charged ice fragments are carried upward by turbulence, and the relatively heavy droplets eventually carry their negative charge toward the cloud base. As a result, the upper part of the cloud is left with a positive charge and the lower portion of the cloud maintains an overall negative charge with small positively charged pockets (Figure 10–15).

As the cloud moves, the negatively charged cloud base alters the charge at the surface directly below by repelling negatively charged particles. Thus, the surface beneath the cloud acquires a net positive charge. These charge differences build to millions and even hundreds of millions of volts before a lightning stroke acts to discharge the negative region of the cloud by striking the positive area of the ground below, or, more frequently, the positively charged portion of that cloud, or a nearby cloud.

The Lightning Stroke

Cloud-to-ground strokes are of most interest and have been studied in detail (see Box 10–3). Moving-film cameras have greatly aided in these studies (see Figure 10–14). They show that the lightning we see as a single flash is really several very rapid strokes between the cloud and the ground. We will call the total discharge, which lasts only a few tenths of a second and appears as a bright streak, the **flash**. Individual components that make up each flash are termed **strokes**. Each stroke is separated by roughly 50 milliseconds, and there are usually three to four strokes per flash. When a lightning flash appears to flicker, it is because your eyes discern the individual strokes that make up this discharge. Moreover, each stroke consists of a downward propagating leader that is immediately followed by a luminous return stroke.

Each stroke is believed to begin when the electrical field near the cloud base frees electrons in the air immediately below, thereby ionizing the air (Figure 10–15). Once ionized, the air becomes a conductive path having a radius of roughly 10 centimeters and a length of 50 meters. This path is called a **leader**. During this electrical breakdown, the mobile electrons in the cloud base begin to flow down this channel. This flow increases the electrical potential at the head of the leader, which causes a further extension of the conductive path through further ionization. Because this initial path extends itself earthward in short, nearly invisible bursts, it is called a **step leader**. Once this channel nears the ground, the electrical field at the surface ionizes the remaining section

Figure 10–15 Discharge of a cloud via cloud-to-ground lightning. Examine this drawing carefully while reading the text.

of the path. With the path completed, the electrons that were deposited along the channel begin to flow downward. The initial flow begins near the ground.

As the electrons at the lower end of the conductive path move earthward, electrons positioned successively higher up the channel begin to migrate downward. Because the path of electron flow is continually being extended upward, the accompanying electric discharge has been appropriately named a **return stroke**. As the wave front of the return stroke moves upward, the negative charge that was deposited on the channel is effectively lowered to the ground. It is this intense return stroke that illuminates the conductive path and discharges the lowest kilometer or so of the cloud. During this phase, tens of coulombs of negative charge are lowered to the ground.‡

‡A coulomb is a unit of electrical charge equal to the quantity of charge transferred in 1 second by a steady current of 1 ampere.

The first stroke is usually followed by additional strokes that apparently drain charges from higher areas within the cloud. Each subsequent stroke begins with a **dart leader** that once again ionizes the channel and carries the cloud potential toward the ground. The dart leader is continuous and less branched than the step leader. When the current between strokes has ceased for periods greater than 0.1 second, further strokes will be preceded by a step leader whose path is different from that of the initial stroke. The total time of each flash consisting of three or four strokes is about 0.2 second.

Thunder

The electrical discharge of lightning superheats the air immediately around the lightning channel. In less than a second, the temperature rises by 8000° to 33,000°C. When air is heated this quickly, it expands explosively and produces the sound waves we hear as **thunder**. Because

Box 10–3 Atmospheric Hazard: Lightning Safety*

Lightning ranks second only to floods in the number of storm-related deaths each year. Although the number of reported lightning deaths in the United States annually averages about 85, many go unreported. About 100 people are estimated to be killed and more than 500 injured by lightning every year in the United States.

Warnings, statements, and forecasts are routinely issued for floods, tornadoes, and hurricanes but not for lightning. Why is this the case? The answer relates to the wide geographic occurrence and frequency of lightning.

The magnitude of the cloud-to-ground lightning hazard is understood better today than ever before. Lightning occurs in the United States every day in summer and nearly every day during the rest of the year. Between 1992 and 1995 the National Lightning Detection Network identified an average of 21,746,000 cloud-to-ground flashes per year. Because lightning is so widespread and strikes the ground with such great frequency, it is not possible to warn each person of every flash. For this reason, lightning can be considered the most dangerous weather hazard that many people encounter each year.

Being aware of and following proven safety guidelines can greatly reduce the risk of injury or death. Individuals are ultimately responsible for their personal safety and should take appropriate action when threatened by lightning.

No place is absolutely safe from the threat of lightning, but some places are safer than others.

- Large enclosed structures (substantially constructed buildings) tend to be much safer than smaller or open structures. The risk for lightning injury depends on whether the structure incorporates lightning protection, the types of construction materials used, and the size of the structure.

- In general, fully enclosed metal vehicles such as cars, trucks, buses, vans, and fully enclosed farm vehicles, etc., with the windows rolled up, provide good shelter from lightning. Avoid contact with metal or conducting surfaces outside or inside the vehicle.

- *Avoid* being in or near high places and open fields, isolated trees, unprotected gazebos, rain or picnic shelters, baseball dugouts, communications towers, flagpoles, light poles, bleachers (metal or wood), metal fences, convertibles, golf carts, and water (ocean, lakes, swimming pools, rivers, etc.).

- When inside a building *avoid* use of the telephone, taking a shower, washing your hands, doing dishes, or any contact with conductive surfaces with exposure to the outside such as metal door or window frames, electrical wiring, telephone wiring, cable TV wiring, plumbing, and so forth.

Safety guidelines for individuals include the following:

- Generally speaking, if individuals can see lightning and/or hear thunder they are already at risk. Louder or more frequent thunder indicates that lightning activity is approaching, thus increasing the risk for lightning injury or death. If the time delay between seeing the flash (lightning) and hearing the bang (thunder) is less than 30 seconds, the individual should be in, or seek, a safer location. Be aware that this method of ranging has severe limitations in part due to the difficulty of associating the proper thunder to the corresponding flash.

- High winds, rainfall, and cloud cover often act as precursors to actual cloud-to-ground strikes, and these should motivate individuals to take action. Many lightning casualties occur in the beginning, as the storm approaches, because people ignore these precursors. Also, many lightning casualties occur after the perceived threat has passed. Generally, the lightning threat diminishes with time after the last sound of thunder but may persist for more than 30 minutes. When thunderstorms are in the area but not overhead, the lightning threat can exist even when it is sunny, not raining, or when clear sky is visible.

- When available, pay attention to weather warning devices such as NOAA (National Oceanic and Atmospheric Administration) weather radio and/or credible lightning detection systems; however, do not let this information override good common sense.

*Material in this box is based on "Updated Recommendations for Lightning Safety—1998," in *Bulletin of the American Meteorological Society*, Vol. 80, No. 10, October 1999, pp. 2035–39.

lightning and thunder occur simultaneously, it is possible to estimate the distance to the stroke. Lightning is seen instantaneously, but the relatively slow sound waves, which travel approximately 330 meters (1000 feet) per second, reach us a little later. If thunder is heard five seconds after the lightning is seen, the lightning occurred about 1650 meters away (approximately 1 mile).

The thunder that we hear as a rumble is produced along a long lightning path located at some distance from the observer. The sound that originates along the path nearest the observer arrives before the sound that originated farthest away. This factor lengthens the duration of the thunder. Reflection of the sound waves by mountains or buildings further delays their arrival and adds to this effect. When lightning occurs more than 20 kilometers (12 miles) away, thunder is rarely heard. This type of lightning, popularly called *heat lightning*, is no different from the lightning that we associate with thunder.

Tornadoes

Tornadoes are local storms of short duration that must be ranked high among nature's most destructive forces. Their sporadic occurrence and violent winds cause many deaths each year. The nearly total destruction in some stricken areas has led many to liken their passage to bombing raids during war.

Such was the case when devastating tornadoes hit parts of Oklahoma and Kansas on Monday evening May 3, 1999 (Figure 10–16). The estimated death toll was 49 people, 44 in Oklahoma and 5 in the Wichita, Kansas, area. Altogether, 76 tornadoes occurred during this outbreak, which took place in a region appropriately called "Tornado Alley." This 160-kilometer (100-mile)-wide swath extends from northern Texas through Oklahoma and Kansas. Oklahoma officials estimated that nearly 8100 homes and businesses were damaged or destroyed. In Kansas the figure exceeded 1100. The insurance industry expected insured losses to reach or exceed $1 billion.

Tornadoes, sometimes called *twisters*, or *cyclones*, are violent windstorms that take the form of a rotating column of air, or *vortex*, that extends downward from a cumulonimbus cloud. Pressures within some tornadoes have been estimated to be as much as 10 percent lower than immediately outside the storm. Drawn by the much lower pressure in the center of the vortex, air near the ground rushes into the tornado from all directions. As the air streams inward, it is spiraled upward around the core until it eventually merges with the airflow of the parent thunderstorm deep in the cumulonimbus tower.

Because of the rapid drop in pressure, air sucked into the storm expands and cools adiabatically. If the air cools

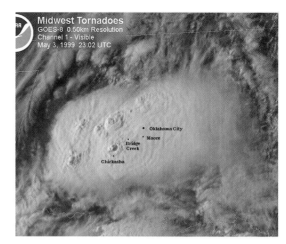

(a)

(b)

Figure 10–16 (a) Satellite image showing some of the tornado-producing storms that struck Oklahoma and Kansas on May 3, 1999. The overshooting tops of these huge clouds are clearly visible. (b) Aftermath of a violent (F5) tornado that struck Oklahoma City on May 3, 1999. The largest and most destructive of the 76 tornadoes formed about 72 kilometers (45 miles) southwest of Oklahoma City and cut a path about 0.8 kilometer (0.5 mile) wide as it moved north and east across the Oklahoma City area, staying on the ground for nearly four hours. The death toll was relatively low because of repeated warnings broadcast throughout the late afternoon and early evening. Yet, even with up to a half hour warning, many could not find adequate shelter from the tremendous power of the strongest storm. *(Photo (a) NOAA; photo (b) by Howard B. Bluestein/Photo Researchers, Inc.)*

below its dew point, the resulting condensation creates a pale and ominous-appearing cloud that may darken as it moves across the ground, picking up dust and debris (Figure 10–17). Occasionally, when the inward spiraling air is relatively dry, no condensation funnel forms because the drop in pressure is not sufficient to cause the necessary adiabatic cooling. In such cases, the vortex is only

Figure 10–17 A tornado is a violently rotating column of air in contact with the ground. The air column is visible when it contains condensation or when it contains dust and debris. Often the appearance is the result of both. When the column of air is aloft and does not produce damage, the visible portion is properly called a funnel cloud. This tornado hit in Wisconsin July 18, 1996. *(P. Lloyd/Weatherstock)*

made visible by the material that it vacuums from the surface and carries aloft.

Some tornadoes consist of a single vortex, but within many stronger tornadoes are smaller intense whirls called *suction vortices* that rotate within the main vortex (Figure 10–18). Suction vortices have diameters of only about 10 meters (30 feet) and rotate very rapidly. This structure accounts for occasional observations of virtually total destruction of one building while another one just 10 meters away suffers little damage.

Because of the tremendous pressure gradient associated with a strong tornado, maximum winds can sometimes exceed 480 kilometers (300 miles) per hour. For example, using Doppler radar observations, scientists measured wind speeds of 510 kilometers (310 miles) per hour in one of the May 3, 1999, Oklahoma City tornadoes. Reliable wind-speed

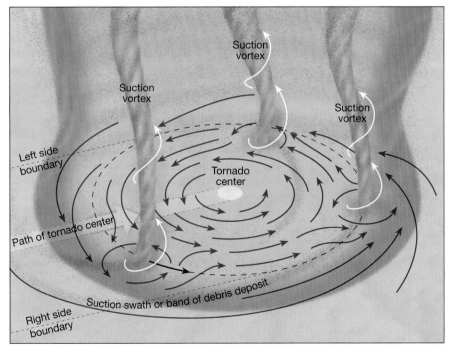

Figure 10–18 Some tornadoes have multiple suction vortices. These small and very intense vortices are roughly 10 meters (30 feet) across and move in a counterclockwise path around the tornado center. Because of this multiple vortex structure, one building might be heavily damaged and another one, just 10 meters away, might suffer little damage. *(After Fujita)*

measurements using traditional anemometers are lacking. The changes that occur in atmospheric pressure with the passage of a tornado are largely estimates and are based on a few storms that happened to pass a nearby weather station. Thus, meteorologists have had to base their tornado models on a relatively scant observational foundation. Although additional data might be gathered if shelters and instruments capable of withstanding the fury of a tornado were placed in an area, such an effort would probably not be worthwhile. Tornadoes are highly localized and randomly distributed, and the probability of placing instruments at the right site would thus be infinitesimally small.

For instance, the probability of a tornado striking a given point in the region most frequently subject to tornadoes is about once in 250 years. The development of Doppler radar, however, has increased our ability to study tornado-producing thunderstorms. As we shall see, the development of this technology is allowing meteorologists to gather important new data from a safe distance.

The Development and Occurrence of Tornadoes

Tornadoes form in association with severe thunderstorms that produce high winds, heavy (sometimes torrential) rainfall, and often damaging hail. Although hail may or may not precede a tornado, the portion of the thunderstorm adjacent to large hail is often the area where strong tornadoes are most likely to occur.

Fortunately, less than 1 percent of all thunderstorms produce tornadoes. Nevertheless, a much higher number must be monitored as potential tornado producers. Although meteorologists are still not sure what triggers tornado formation, it is apparent that they are the product of the interaction between strong updrafts in a thunderstorm and the winds in the troposphere.

Tornado Development

Tornadoes can form in any situation that produces severe weather, including cold fronts, squall lines, and tropical cyclones (hurricanes). Usually the most intense tornadoes are those that form in association with supercells. An important precondition linked to tornado formation in severe thunderstorms is the development of a mesocyclone. A *mesocyclone* is a vertical cylinder of rotating air, typically about 3 to 10 kilometers (2 to 6 miles) across, that develops in the updraft of a severe thunderstorm. The formation of this large vortex often precedes tornado formation by 30 minutes or so.

Mesocyclone formation depends on the presence of vertical wind shear. Moving upward from the surface, winds change direction from southerly to westerly and the wind speed increases. The speed wind shear (that is, stronger winds aloft and weaker winds near the surface) produces a rolling motion about a horizontal axis as shown in Figure 10–19a. If conditions are right, strong updrafts in the storm tilt the horizontally rotating air to a nearly vertical alignment (Figure 10–19b). This produces the initial rotation within the cloud interior.

At first the mesocyclone is wider, shorter, and rotating more slowly than will be the case in later stages. Subsequently, the mesocyclone is stretched vertically and narrowed horizontally causing wind speeds to accelerate in an inward vortex (just as ice skaters accelerate while spinning or as a sink full of water accelerates when it spirals down a drain).* Next, the narrowing column of rotating air stretches downward until a portion of the cloud protrudes below the cloud base to produce a very dark, slowly rotating *wall cloud*. Finally, a slender and rapidly spinning vortex emerges from the base of the wall cloud to form a *funnel cloud*. If the funnel cloud makes contact with the surface, it is then classified as a *tornado* (Figure 10–19c,d).

The formation of a mesocyclone does not necessarily mean that tornado formation will follow. Only about half of all mesocyclones produce tornadoes. The reason for this is not understood. Because this is the case, forecasters cannot determine in advance which mesocyclones will spawn tornadoes.

Tornado Climatology

Severe thunderstorms—and hence tornadoes—are most often spawned along the cold front or squall line of a middle-latitude cyclone or in association with supercell thunderstorms. Throughout the spring, air masses associated with midlatitude cyclones are most likely to have greatly contrasting conditions. Continental polar air from Canada may still be very cold and dry, whereas maritime tropical air from the Gulf of Mexico is warm, humid, and unstable. The greater the contrast, the more intense the storm.

These two contrasting air masses are most likely to meet in the central United States because there is no significant natural barrier separating the center of the country from the arctic or the Gulf of Mexico. Consequently, this region generates more tornadoes than any other part of the country or, in fact, the world. Figure 10–20, which depicts the average annual tornado incidence in the United States over a 27-year period, readily substantiates this fact.

*For more on this concept, see Box 11–1 on the conservation of angular momentum.

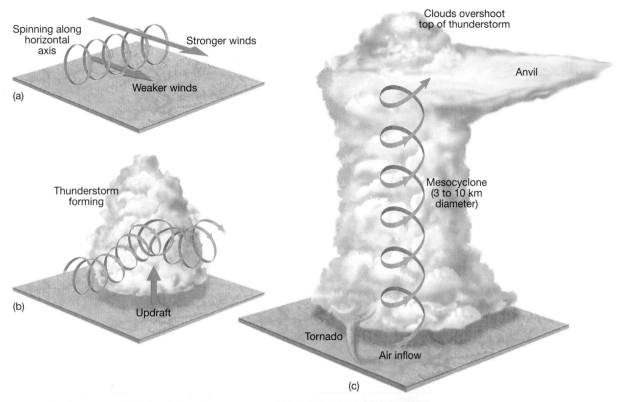

Figure 10–19 The formation of a mesocyclone often precedes tornado formation. (a) Winds are stronger aloft than at the surface (called *speed wind shear*), producing a rolling motion about a horizontal axis. (b) Strong thunderstorm updrafts tilt the horizontally rotating air to a nearly vertical alignment. (c) The mesocyclone, a vertical cylinder of rotating air, is established. (d) If a tornado develops it will descend from a slowly rotating wall cloud in the lower portion of the mesocyclone. This supercell tornado hit the Texas Panhandle in May 1996. *(Photo by Warren Faidley/Weatherstock)*

An average of about 770 tornadoes are reported annually in the United States. Still, the actual numbers that occur from one year to the next vary greatly. During a recent 45-year span, for example, yearly totals ranged from a low of 421 to a high of 1426. Tornadoes occur during every month of the year. April through June is the period of greatest tornado frequency in the United States, and December and January are the months of

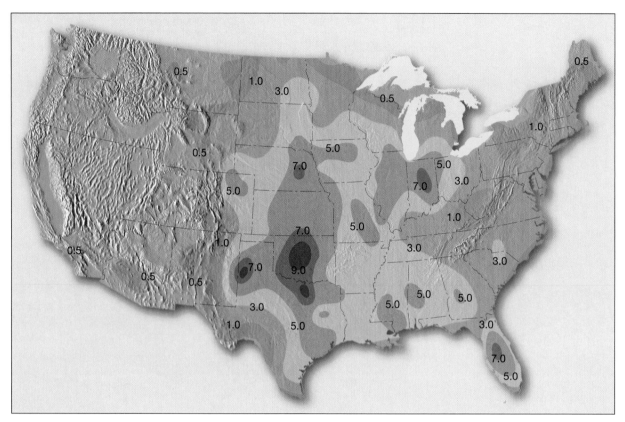

Figure 10–20 Average annual tornado incidence per 10,000 square miles (26,000 square kilometers) for a 27-year period. *(Courtesy of NOAA)*

lowest activity. Of the nearly 35,000 confirmed tornadoes reported over the contiguous 48 states during the period, an average of almost five per day occurred during May. At the other extreme, a tornado was reported only every other day in January.

Typically, 54 percent of all tornadoes take place during the spring. Fall and winter, in contrast, together account for only 19 percent (Figure 10–21). In February, when the incidence of tornadoes begins to increase, the center of maximum frequency lies over the central Gulf states. During March, this center moves eastward to the southeastern Atlantic states, where tornado frequency reaches a peak in April.

During May and June, the center of maximum frequency moves through the southern Great Plains and then to the northern Plains and Great Lakes area. This drift is due to the increasing penetration of warm, moist air while contrasting cool, dry air still surges in from the north and northwest. Thus, when the Gulf states are substantially under the influence of warm air after May, there is no cold-air intrusion to speak of, and tornado frequency drops. Such is the case across the country after June.

Winter cooling permits fewer and fewer encounters between warm and cold air masses, and tornado frequency returns to its lowest level by January.

Profile of a Tornado

The average tornado has a diameter of between 150 and 600 meters (500 to 2000 feet), travels across the landscape at approximately 45 kilometers (30 miles) per hour, and cuts a path about 26 kilometers (16 miles) long. Because many tornadoes occur slightly ahead of a cold front, in the zone of southwest winds, most move toward the northeast. The Illinois example demonstrates this fact nicely (Figure 10–22). The figure also shows that many tornadoes do not fit the description of the "average" tornado.

Of the hundreds of tornadoes reported in the United States each year, over half are comparatively weak and short-lived. Most of these small tornadoes have lifetimes of three minutes or less and paths that seldom exceed 1 kilometer (0.6 mile) in length and 100 meters (330 feet) wide. Typical wind speeds are on the order of 150 kilometers (96 miles) per hour or less. On the other end of the

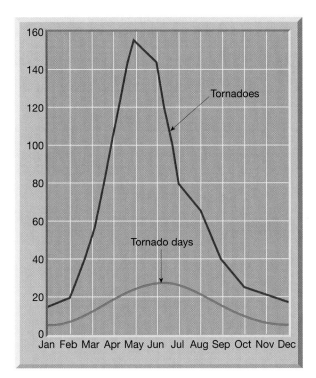

Figure 10–21 Average number of tornadoes and tornado days each month in the United States for a 27-year period. *(Courtesy of NOAA)*

tornado spectrum are the infrequent and often long-lived violent tornadoes. Although large tornadoes constitute only a small percentage of the total reported, their effects are often devastating. Such tornadoes may exist for periods in excess of three hours and produce an essentially continuous damage path more than 150 kilometers (90 miles) long and perhaps a kilometer or more wide. Maximum winds range beyond 500 kilometers (310 miles) per hour (Figure 10–23).

Tornado Destruction

Because tornadoes generate the strongest winds in nature, they have accomplished many seemingly impossible tasks, such as driving a piece of straw through a thick wooden plank and uprooting huge trees (Figure 10–24). Although it may seem impossible for winds to cause some of the fantastic damage attributed to tornadoes, tests in engineering facilities have repeatedly demonstrated that winds in excess of 320 kilometers (200 miles) per hour are capable of incredible feats.

There is a long list of documented examples. In 1931, a tornado actually carried an 83-ton railroad coach and

its 117 passengers 24 meters (80 feet) through the air and dropped them in a ditch. A year later, near Sioux Falls, South Dakota, a steel beam 15 centimeters (6 inches) thick and 4 meters (13 feet) long was ripped from a bridge, flew more than 300 meters (nearly 1000 feet) and perforated a 35-centimeter (14-inch)-thick hardwood tree. In 1970, an 18-ton steel tank was carried nearly 1 kilometer (0.6 mile) at Lubbock, Texas. Fortunately, the winds associated with most tornadoes are not this strong.

Most tornado losses are associated with a few storms that strike urban areas or devastate entire small communities. The amount of destruction wrought by such storms depends to a significant degree (but not completely) on the strength of the winds. A wide spectrum of tornado strengths, sizes, and lifetimes are observed. One commonly used guide to tornado intensity was developed by the late T. Theodore Fujita at the University of Chicago and is appropriately called the **Fujita Intensity Scale**, or simply the **F-scale** (Table 10–1). Because tornado winds cannot be measured directly, a rating on the F-scale is determined by assessing the worst damage produced by a storm. Although widely used, the F-scale is not perfect. Estimating tornado intensity based on damage alone does not take into account the structural integrity of the objects hit by a tornado. A well-constructed building can withstand very high winds, whereas a poorly built structure can suffer devastating damage from the same or even weaker winds.

The drop in atmospheric pressure associated with the passage of a tornado plays a minor role in the damage process. Most structures have sufficient venting to allow for the sudden drop in pressure. Opening a window, once thought to be a way to minimize damage by allowing inside and outside atmospheric pressure to equalize, is no longer recommended. In fact, if a tornado gets close enough to a structure for the pressure drop to be experienced, the strong winds probably will have already caused significant damage.

Although the greatest part of tornado damage is caused by violent winds, most tornado injuries and deaths result from flying debris. On the average, tornadoes cause more deaths each year than any other weather events except lightning and flash floods. For the United States, the average annual death toll from tornadoes is about 75 people. However, the actual number of deaths each year can depart significantly from the average. On April 3–4, 1974, for example, an outbreak of 148 tornadoes brought death and destruction to a 13-state region east of the Mississippi River. More than 300 people died and nearly 5500 people were injured in this worst tornado disaster in half a century.

In one statistical study that examined a 29-year period, there were 689 tornadoes that caused loss of life. This figure

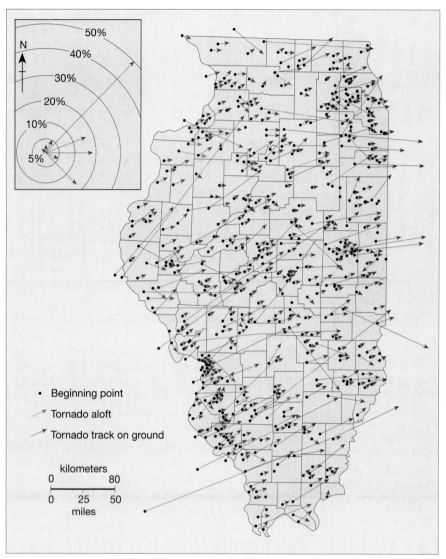

Figure 10–22 Paths of Illinois tornadoes (1916–1969). Because most tornadoes occur slightly ahead of a cold front, in the zone of southwest winds, they tend to move toward the northeast. Tornadoes in Illinois verify this. Over 80 percent exhibited directions of movement toward the northeast through east. *(After John W. Wilson and Stanley A. Changnon, Jr.,* Illinois Tornadoes, *Illinois State Water Survey Circular 103, 1971, pp. 10, 24)*

represented slightly less than 4 percent of the total 19,312 reported storms. Although the percentage of tornadoes that result in death is small, every tornado is potentially lethal. If you examine Figure 10–25, which compares tornado fatalities with storm intensities, the results are quite interesting. It is clear from this graph that the majority (63 percent) of all tornadoes are weak and that the number of storms decreases as tornado intensity increases. The distribution of tornado fatalities, however, is just the opposite. Although only 2 percent of tornadoes are classified as violent, they account for nearly 70 percent of the deaths.

If there is some question about the causes of tornadoes, there certainly is none about the destructive effects of these violent storms. A severe tornado leaves the affected area stunned and disorganized and may require a response of the magnitude demanded in war.

Tornado Forecasting

Because severe thunderstorms and tornadoes are small and short-lived phenomena, they are among the most difficult weather features to forecast precisely. Nevertheless, the prediction, the detection, and the monitoring of such storms are among the most important services provided by professional meteorologists. The timely issuance and dissemination of watches and warnings are both critical to the protection of life and property.

The Storm Prediction Center (SPC) located in Norman, Oklahoma, is part of the National Weather Service (NWS) and the National Centers for Environmental Prediction (NCEP). Its mission is to provide timely and accurate forecasts and watches for severe thunderstorms and tornadoes.

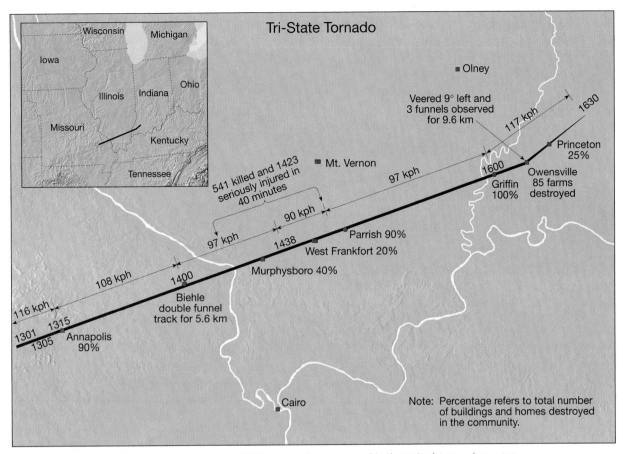

Figure 10–23 "One tornado among the more than 13,000 which have occurred in the United States since 1915 easily ranks above all others as the single most devastating storm of this type. Shortly after its occurrence on 18 March 1925, the famed Tri-State tornado was recognized as the worst on record, and it still ranks as the nation's greatest tornado disaster. The tornado remained on the ground for 219 miles. The resulting losses included 695 dead, 2027 injured, and damages equal to $43 million in 1970 dollars. This represents the greatest death toll ever inflicted by a tornado and one of the largest damage totals." *(Map and description from John W. Wilson and Stanley A. Changnon, Jr., Illinois Tornadoes, Illinois State Water Survey Circular 103, 1971, p. 32)*

Severe thunderstorm outlooks are issued several times daily. *Day 1* outlooks identify those areas likely to be affected by severe thunderstorms during the next 6 to 30 hours and *day 2* outlooks extend the forecast through the following day. Both outlooks describe the type, coverage, and intensity of the severe weather expected. Many local NWS field offices also issue severe weather outlooks that provide a more local description of the severe weather potential for the next 12 to 24 hours.

Tornado Watches and Warnings

Tornado watches alert the public to the possibility of tornadoes over a specified area for a particular time interval. Watches serve to fine-tune forecast areas already identified in severe weather outlooks. A typical watch covers an area of about 65,000 square kilometers (25,000 square miles) for a four to six hour period. A tornado watch is an important part of the tornado alert system because it sets in motion the procedures necessary to deal adequately with detection, tracking, warning, and response. Watches are generally reserved for organized severe weather events where the tornado threat will affect at least 26,000 square kilometers (10,000 square miles) and/or persist for at least three hours. Watches typically are not issued when the threat is thought to be isolated and/or short-lived.

Whereas a tornado watch is designed to alert people to the possibility of tornadoes, a **tornado warning** is issued by local offices of the National Weather Service when a tornado has actually been sighted in an area or is indicated by weather radar. It warns of a high probability of imminent danger. Warnings are issued for much smaller areas than watches, usually covering portions of a county or counties. In addition, they are in effect for much shorter periods, typically 30 to 60 minutes. Because a tornado

Figure 10–24 The force of the wind during a tornado near Wichita, Kansas in April 1991 was enough to drive this piece of metal into a utility pole. *(Photo by John Sokich/NOAA)*

warning may be based on an actual sighting, warnings are occasionally issued after a tornado has already developed. However, most warnings are issued prior to tornado formation, sometimes by several tens of minutes, based on Doppler radar data and/or spotter reports of funnel clouds or cloud-base rotation.

If the direction and the approximate speed of the storm are known, an estimate of its most probable path can be made. Because tornadoes often move erratically, the warning area is fan-shaped downwind from the point where the tornado has been spotted. Improved forecasts and advances in technology have contributed to a significant decline in tornado deaths over the last 50 years.

Additional deaths and serious injuries could be averted if more people would take adequate safety measures after a tornado warning is issued. This point was dramatically illustrated in a study made following the tornado that struck Wichita Falls, Texas, on April 10, 1979. The study revealed that 26 (60 percent) of the 43 traumatic deaths and 30 (51 percent) of the 59 serious injuries occurred to people who, despite ample warning, went to their cars to drive out of the storm's path. These people had a risk of serious or fatal injury of 23 per 1000. People who remained indoors and in stationary homes were at a relatively low risk (3 per 1000) if they took simple precautions; people in mobile homes were at greatest risk (85 per 1000†).

As noted earlier, the probability of one place being struck by a tornado, even in the area of greatest frequency, is slight. Nevertheless, although the probabilities may be small, tornadoes have provided many mathematical exceptions. For example, the small town of Codell, Kansas, was hit three years in a row—1916, 1917, and 1918—and each time on the same date, May 20! Needless to say, tornado watches and warnings should never be taken lightly.

Doppler Radar

The installation of **Doppler radar** across the United States has significantly improved our ability to track thunderstorms and issue warnings based on their potential to produce tornadoes. Conventional weather radar works by

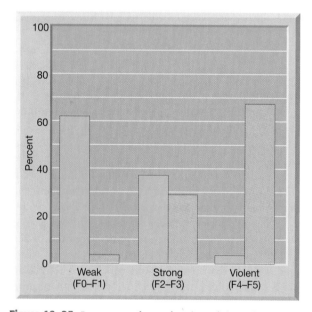

Figure 10–25 Percentage of tornadoes in each intensity category (blue) and percentage of fatalities associated with each category (red). *(From Joseph T. Schaefer et al., "Tornadoes—When, Where and How Often,"* Weatherwise, *33, no. 2 (1980), 57)*

†Roger I. Glass et al., "Injuries from the Wichita Falls Tornado: Implications for Prevention," *Science*, 207, no. 4432 (1980), 735.

Table 10–1 Fujita Intensity Scale

Scale	Wind Speed (KPH)	(MPH)	Expected Damage
F0	<116	<72	**Light Damage** Damage to chimneys and billboards; broken branches; shallow-rooted trees pushed over.
F1	116–180	72–112	**Moderate Damage** The lower limit is near the beginning of hurricane wind speed. Surfaces peeled off roofs; mobile homes pushed off foundations or overturned; moving autos pushed off the road.
F2	181–253	113–157	**Considerable Damage** Roofs torn off frame houses; mobile homes demolished; boxcars pushed over; large trees snapped or uprooted; light-object missiles generated.
F3	254–332	158–206	**Severe Damage** Roofs and some walls torn off well-constructed houses; trains overturned; most trees in forest uprooted; heavy cars lifted off ground and thrown.
F4	333–419	207–260	**Devastating Damage** Well-constructed houses leveled; structures with weak foundations blown some distance; cars thrown and large missiles generated.
F5	>419	>260	**Incredible Damage** Strong frame houses lifted off foundations and carried considerable distance to disintegrate; automobile-sized missiles fly through the air farther than 100 m; trees debarked; incredible phenomena occur.

transmitting short pulses of electromagnetic energy. A small fraction of the waves that are sent out is scattered by a storm and returned to the radar. The strength of the returning signal indicates rainfall intensity, and the time difference between the transmission and return of the signal indicates the distance to the storm.

However, to identify tornadoes and severe thunderstorms, we must be able to detect the characteristic circulation patterns associated with them. Conventional radar cannot do so except occasionally when spiral rain bands occur in association with a tornado and give rise to a hook-shaped echo.

Doppler radar not only performs the same tasks as conventional radar but also has the ability to detect motion directly (Figure 10–26). The principle involved is known as the *Doppler effect* (see Box 10–4). Air movement in clouds is determined by comparing the frequency of the reflected signal to that of the original pulse. The movement of precipitation toward the radar increases the frequency of reflected pulses, whereas motion away from the radar decreases the frequency. These frequency changes are then interpreted in terms of speed toward or away from the Doppler radar unit. It is this same principle that allows police radar to determine the speed of moving cars. Unfortunately, a single Doppler radar unit cannot detect air movements that occur parallel to it. Therefore, when a more complete picture of the winds within a cloud mass is desired, it is necessary to use two or more Doppler units.

Doppler radar can detect the initial formation and subsequent development of the mesocyclone within a severe thunderstorm that frequently precedes tornado development. Almost all (96 percent) mesocyclones produce damaging hail, severe winds, or tornadoes. Those that produce tornadoes (about 50 percent) can sometimes be distinguished by their stronger wind speeds and their sharper gradients of wind speeds. Mesocyclones can sometimes be identified within parent storms 30 minutes or more before tornado formation, and if a storm is large, at distances up to 230 kilometers (140 miles). In addition, when close to the radar, individual tornado circulations may sometimes be detected. Ever since the implementation of the national Doppler network, the average lead time for tornado warnings has increased from less than 5 minutes in the late 1980s to nearly 10 minutes today.

Doppler radar is not without problems. One concern relates to the weak tornadoes that rank at or near the bottom of the Fujita Intensity Scale (see Table 10–1). The nature of this problem has been summarized as follows:

> Presently, all tornadoes are treated in the same manner by the NWS (National Weather Service); a tornado warning is issued, and local governments usually respond by activating their emergency procedures. Doppler radar makes forecasting and detection of these tornadoes possible in real-time operations. Thus, the potential exists for numerous warnings

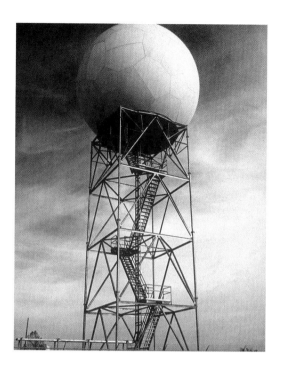

Figure 10–26 By using advanced Doppler radars, scientists are better able to estimate when and where thunderstorms will form, even in areas of seemingly clear air. Doppler radar installation (right) and advanced weather interactive processing system (left) at Sterling, Virginia, near Dulles International Airport. This is one of 115 new state-of-the-art National Weather Service facilities. *(Photos by Brownie Harris/The Stock Market)*

being issued for tornadoes which do little or no damage. This could desensitize the public to the dangers of more rare, life-threatening tornadoes. A future research goal should be the development of techniques that enable forecasting of tornado intensity.[‡]

It should also be pointed out that not all tornado-bearing storms have clear-cut radar signatures and that other

[‡]Lawrence B. Dunn, "Two Examples of Operational Tornado Warnings Using Doppler Radar Data," *Bulletin of the American Meteorological Society*, 71, no. 2 (February 1990), 152.

storms can give false signatures. Detection, therefore, is sometimes a subjective process and a given display could be interpreted in several ways. Consequently, trained observers will continue to form an important part of the warning system in the foreseeable future.

Although some operational problems exist, the benefits of Doppler radar are many. As a research tool, it is not only providing data on the formation of tornadoes, but is also helping meteorologists gain new insights into thunderstorm development, the structure and dynamics of hurricanes, and air turbulence hazards that plague aircraft. As a practical tool for tornado detection, it has significant advantages over a system that uses conventional radar.

Chapter Summary

- Although tornadoes and *hurricanes* are, in fact, cyclones, the vast majority of cyclones are not hurricanes or tornadoes. The term "cyclone" simply refers to the circulation around any low-pressure center, no matter how large or intense it is. *Thunderstorms*, storms containing lightning and thunder, are related in some manner to tornadoes, hurricanes, and midlatitude cyclones.

- Dynamic thermal instability occurs during the development of thunderstorms, which form when warm, humid air rises in an unstable environment. A number

of mechanisms, such as unequal heating of Earth's surface or lifting of warm air along a front or mountain slope, can trigger the upward air movement needed to create thunderstorm-producing cumulonimbus clouds. Severe thunderstorms produce high winds, damaging hail, flash floods, and tornadoes.

- In the United States, *air-mass thunderstorms* frequently occur in maritime tropical (mT) air that moves northward from the Gulf of Mexico. During the spring and summer, when the air is heated from below by the warmer land surface, the warm, humid air mass

Box 10–4 The Doppler Effect

We have all heard the changes in the sounds that a car, truck, or train makes as it approaches and then passes by. As the vehicle moves toward us, the sound has a higher pitch, and after the vehicle passes, the pitch drops. The faster the car, truck, or train is moving, the more the pitch drops. This phenomenon occurs not only in association with sound waves but also with electromagnetic waves. Known as the *Doppler effect*, it is named for Christian Johann Doppler, the Austrian physicist who first explained it in 1842.

Why does the wave frequency appear to shift when the wave source is moving? Figure 10–Da shows the pattern of wave crests generated by a source that is not moving. The distance between wave crests (the wavelength) is identical for each successive wave. If this were a source emitting sound, the pitch would be the same no matter where the listener was located. However, if the source were moving, the wave crests would no longer make a concentric pattern. Rather, they would become more closely spaced in the direction the source was advancing. Therefore, the wave frequency would be greater ahead of the source and lower behind the source. Figure 10–Db illustrates this pattern.

Astronomers apply this principle when they wish to determine whether a light source such as a star is approaching or retreating relative to Earth. When the lines in the star's spectrum are shifted toward wavelengths that are shorter than those observed when such a source is at rest, the star is approaching. If the star is moving away from the observer, all of the spectral lines are shifted toward longer wavelengths. In meteorology, the new generation of weather radars use the Doppler principle to probe the circulation within a cloud by monitoring the movements of raindrops.

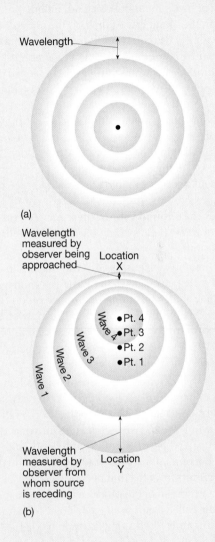

Figure 10–D The Doppler effect is the change in the observed frequency of waves produced by the motion of the wave source and/or the wave receiver. In both parts of the diagram, the circles represent wave crests that travel at a constant speed. (a) When the wave source is stationary, the distance between waves is identical for all waves that are produced. The wave frequency (the pitch of the sound) at any point in this diagram is the same. (b) Here the source is moving toward the top of the page. Whenever the source (or the receiver) moves, the wave pattern becomes distorted. Wave 1 was produced when the source was at point 1; wave 2 was created when the source was at point 2; and wave 3 was emitted at point 3. Notice that even though the frequency emitted by the source remains the same as in part (a), a listener at location *X* would experience a higher frequency (higher-pitched sound) because each wave has a shorter distance to travel and therefore arrives at location *X* more frequently than would occur if the source were not moving. Conversely, a listener at location *Y* would experience a lower frequency.

becomes unstable and thunderstorms develop. Generally three stages are involved in the development of a thunderstorm. The *cumulus stage* is dominated by rising currents of air (updrafts) and the formation of a towering cumulonimbus cloud. Falling precipitation within the cloud causes drag on the air and initiates a downdraft that is further aided by the influx of cool, dry air surrounding the cloud, a process termed *entrainment*. The beginning of the *mature stage* is marked by the downdraft leaving the base of the cloud and the release of precipitation. With gusty winds, lightning, heavy precipitation, and sometimes hail, the mature stage is the most active period of a thunderstorm. Marking the end of the storm, the *dissipating stage* is dominated by downdrafts and entrainment. Without a supply of moisture from updrafts, the cloud soon evaporates. It should be noted that within a single air-mass thunderstorm there may be several individual *cells*, that is zones of adjacent updrafts and downdrafts.

- Mountainous regions, such as the Rockies in the West and the Appalachians in the East, experience a greater number of air-mass thunderstorms than do the Plains states. Many thunderstorms that form over the eastern two-thirds of the United States occur as part of the general convergence and frontal wedging that accompany passing midlatitude cyclones.

- *Severe thunderstorms* are capable of producing heavy downpours and flash flooding as well as strong, gusty straight-line winds. They are influenced by strong vertical wind shear, that is, changes in wind direction and/or speed between different heights, and updrafts that become tilted and continue to build upward. Downdrafts from the thunderstorm cells reach the surface and spread out to produce an advancing wedge of cold air, called a *gust front*, which may form a *roll cloud* as warm air is lifted along its leading edge.

- Some of the most dangerous weather is produced by a type of thunderstorm called a *supercell*, a single, very powerful thunderstorm cell that at times may extend to heights of 20 kilometers (65,000 feet) and persist for many hours. The vertical wind profile of these cells may produce a *mesocyclone*, a column of cyclonically-rotating air, within which tornadoes sometimes form. Supercells appear to form as inversion layers are eroded locally and unstable air "erupts" from below to form unusually large cumulonimbus clouds with concentrated, persistent updrafts.

- *Squall lines* are relatively narrow, elongated bands of thunderstorms that develop in the warm sector of a middle-latitude cyclone, usually in advance of a cold front. Some develop when divergence and lifting created by an active jet stream aloft is aligned with a strong, persistent low-level flow of warm, humid air from the south. A squall line with severe thunderstorms can also form along a boundary called a *dryline*, a narrow zone along which there is an abrupt change in moisture.

- A *mesoscale convective complex* (MCC) consists of many individual thunderstorms that are organized into a large oval to circular cluster. They form most frequently in the Great Plains from groups of afternoon air-mass thunderstorms. Although MCCs sometimes produce severe weather, they also provide a significant portion of the growing-season rainfall to the agricultural regions of the central United States.

- *Thunder* is produced by *lightning*. The object of lightning is to equalize the electrical difference associated with the formation of a large cumulonimbus cloud by producing a negative flow of current from the region of excess negative charge to the region with excess positive charge, or vice versa. The origin of charge separation in clouds, although not fully understood, must hinge on rapid vertical movements within the cloud. The lightning we see as some single flashes are really several very rapid strokes between the cloud and the ground. When air is heated by the electrical discharge of lightning, it expands explosively and produces the sound waves we hear as thunder. The thunder we hear as a rumble is produced by a long lightning flash at some distance from the observer.

- *Tornadoes*, sometimes called twisters, or cyclones, are violent windstorms that take the form of a rotating column of air, or *vortex*, that extends downward from a cumulonimbus cloud. Some tornadoes consist of a single vortex, but within many stronger tornadoes are smaller intense whirls called *suction vortices* that rotate within the main vortex. Pressures within some tornadoes have been estimated to be as much as 10 percent lower than immediately outside the storm. Because of the tremendous pressure gradient associated with a strong tornado, maximum winds approach 480 kilometers (300 miles) per hour.

- Tornadoes form in association with severe thunderstorms that produce high winds, heavy rainfall, and often damaging hail. They form in any situation that produces severe weather including cold fronts, squall lines, and tropical cyclones (hurricanes). An important precondition linked to tornado formation in severe thunderstorms is the development of a mesocyclone that forms in the updraft of the thunderstorm. As the narrowing column of rotating air stretches downward, a rapidly spinning *funnel cloud* may emerge from a slowly rotating *wall cloud*. If the funnel cloud makes contact with the surface, it is then classified as a tornado.

- Severe thunderstorms, and hence tornadoes, are most often spawned along the cold front or squall line of a middle-latitude cyclone or in association with super-cell thunderstorms. Although April through June is the period of greatest tornado activity, tornadoes occur during every month of the year. The average tornado has a diameter of between 150 and 600 meters, travels across the landscape toward the northeast at approximately 45 kilometers per hour, and cuts a path about 26 kilometers long.

- Most tornado damage is caused by tremendously strong winds. One commonly used guide to tornado intensity is the *Fujita Intensity Scale*, or simply *F-scale*. A rating on the F-scale is determined by assessing the worst damage produced by a storm. Although most tornado damage is done by violent winds, most tornado injuries and deaths result from flying debris.

- Because severe thunderstorms and tornadoes are small and short-lived phenomena, they are among the most difficult weather features to forecast precisely. When necessary, the Storm Prediction Center of the National Weather Service issues severe thunderstorm outlooks several times daily. When weather conditions favor the formation of tornadoes, a *tornado watch* is issued to alert the public to the possibility of tornadoes over a specified area for a particular time interval. A *tornado warning* is issued by local offices of the National Weather Service when a tornado has been sighted in an area or is indicated by weather radar. With its ability to detect air motion within a cloud, *Doppler radar* technology has greatly advanced the accuracy of tornado warnings. Using the principle known as the *Doppler effect*, Doppler radar can identify the initial formation and subsequent development of the mesocyclone within a thunderstorm that frequently precedes tornado development.

Vocabulary Review

air-mass thunderstorm (p. 271)
cumulus stage (p. 271)
dart leader (p. 282)
dissipating stage (p. 272)
Doppler radar (p. 292)
dryline (p. 278)
entrainment (p. 272)
flash (p. 281)
Fujita Intensity Scale (F-scale) (p. 289)
gust front (p. 273)
leader (p. 281)
lightning (p. 280)
mature stage (p. 272)

mesocyclone (p. 274)
mesoscale convective complex (MCC) (p. 278)
return stroke (p. 282)
severe thunderstorm (p. 273)
squall line (p. 277)
step leader (p. 281)
stroke (p. 281)
supercell (p. 273)
thunder (p. 282)
thunderstorm (p. 270)
tornado (p. 284)
tornado warning (p. 291)
tornado watch (p. 291)

Review Questions

1. If you hear that a cyclone is approaching, should you immediately seek shelter?
2. Compare the wind speeds and the sizes of middle-latitude cyclones, tornadoes, and hurricanes.
3. Although tornadoes and hurricanes are dangerous storms, they are not responsible for the greatest number of storm-related deaths in the United States. What is the deadliest storm phenomenon? (See Box 10–1.)
4. What are the primary requirements for the formation of thunderstorms?
5. Where would you expect thunderstorms to be most common on Earth? In the United States?

6. During what season and at what time of day is air-mass thunderstorm activity greatest? Why?
7. Why does entrainment intensify thunderstorm downdrafts?
8. Describe how downdrafts in a severe thunderstorm act to maintain updrafts. What is a *gust front*?
9. Briefly describe the formation of a squall line along a dry line.
10. How is thunder produced?
11. What is heat lightning?
12. Why do tornadoes have such high wind speeds?

13. What general atmospheric conditions are most conducive to the formation of tornadoes?

14. When is the "tornado season"? Can you explain why it occurs when it does? Why does the area of greatest tornado frequency migrate?

15. Violent (F4–F5) tornadoes are only about 2 percent of the total. What percentage of tornado fatalities is associated with these strongest storms? (See Figure 10–25.)

16. Distinguish between a tornado watch and a tornado warning.

17. A vehicle with its horn sounding moves away from an observer at point *A* and toward an observer at point *B*. Should the pitch of the horn at point *B* be higher or lower than at point *A*? What effect explains the difference in pitch experienced at points *A* and *B*? (See Box 10–4.)

18. What advantages does Doppler radar have over conventional radar?

Problems

1. If thunder is heard 15 seconds after lightning is seen, about how far away was the lightning stroke?

2. Examine the upper left portion of Figure 10–22 and determine the percentage of tornadoes that exhibited directions of movement toward the E through NNE.

3. Figures 10–27 and 10–28 represent two common ways that United States tornado statistics are graphically presented to the public. Which four states experience the greatest number of tornadoes? Are these the states with the greatest tornado threat? Which map is most useful for depicting the tornado hazard in the United States? Does the map in Figure 10–20 have an advantage over either or both of these maps?

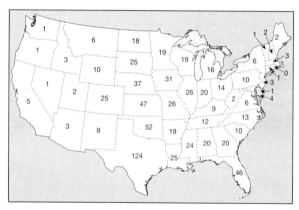

Figure 10–27 Annual average number of tornadoes by state for a 45-year period.

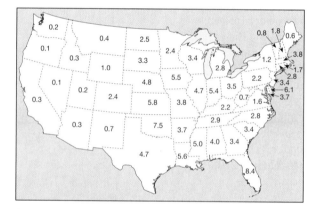

Figure 10–28 Annual average number of tornadoes per 10,000 square miles by state for a 45-year period.

Atmospheric Science Online

The following are informative and interesting Internet sites that address topics related to those presented in the chapter:

Storm Prediction Center (NWS)
* **http://www.spc.noaa.gov/index.html**

National Severe Storms Laboratory (NOAA)
* **http://www.nssl.noaa.gov/**

For direct links to these sites and others, chapter objectives and reviews, quiz questions, and topical investigations that utilize Web resources, visit *The Atmosphere, Eighth Edition* Home Page at:
* **http://www.prenhall.com/lutgens**

11

Hurricanes

This pier along the North Carolina coast was battered by the pounding surf from hurricanes Dennis and Floyd during the 1999 hurricane season. An adjacent pier house (where observer is standing) was completely destroyed. Torrential rains from these hurricanes caused record-breaking floods in North Carolina. *(Photo by Steve Helber/Associated Press)*

The whirling tropical cyclones that occasionally have wind speeds exceeding 300 kilometers per hour are known in the United States as **hurricanes**—the greatest storms on Earth. Hurricanes are among the most destructive of natural disasters. When a hurricane reaches land, it is capable of annihilating coastal areas and killing tens of thousands of people. On the positive side, however, these storms provide essential rainfall over many areas they cross. Consequently, a resort owner along the Florida coast may dread the coming of hurricane season, but a farmer in Japan may welcome its arrival.

The vast majority of hurricane-related deaths and damage are caused by relatively infrequent, yet powerful storms. Table 11–1 lists the deadliest hurricanes to strike the United States during the twentieth century. The storm that pounded an unsuspecting Galveston, Texas, in 1900 was not just the deadliest U.S. hurricane ever, but the deadliest natural disaster *of any kind* to affect the United States in the twentieth century. The Galveston tragedy occurred long before the development of weather radar, satellites, and reconnaissance aircraft. Such a great loss of life is unlikely to occur again. Nevertheless, our coasts are vulnerable. People are flocking to live near the ocean. The proportion of the U.S. population residing within 75 kilometers (45 miles) of a coast in 2010 is projected to be well in excess of 50 percent. The concentration of such large numbers of people near the shoreline means that hurricanes place millions at risk. Moreover, the potential costs of property damage are incredible.

For example, in August 1992 when Hurricane Andrew slammed into southern Florida and then coastal Louisiana, damages (especially from wind) exceeded $25 billion. It was the costliest natural disaster in U.S. history. In September 1999 Hurricane Floyd brought flooding rains, high winds, and rough seas to a large portion of the Atlantic seaboard (Figure 11–1). More than 2.5 million people evacuated their homes from Florida north to the Carolinas and beyond. It was the largest peacetime evacuation in U.S. history. Torrential rains falling on already saturated ground created devastating inland flooding. Altogether Floyd dumped more than 48 centimeters (19 inches) of rain on Wilmington, North Carolina, 33.98 cm

Table 11–1 The 20 Deadliest U.S. Hurricanes, 1900–1999

Ranking	Hurricane	Year	Category	Deaths
1.	Texas (Galveston)	1900	4	8000°
2.	Florida (Lake Okeechobee)	1928	4	1836
3.	Florida (Keys)/S. Texas	1919	4	600†
4.	New England	1938	3	600
5.	Florida (Keys)	1935	5	408
6.	Audrey (SW Louisiana/Texas)	1957	4	390
6.	NE United States	1944	3	390‡
8.	Louisiana (Grand Isle)	1909	4	350
9.	Louisiana (New Orleans)	1915	4	275
9.	Texas (Galveston)	1915	4	275
11.	Camille (Mississippi/ Louisiana)	1969	5	256
12.	Florida (Miami)/Mississippi/ Alabama/Florida (Pensacola)	1926	4	243
13.	Diane (NE United States)	1955	1	184
14.	SE Florida	1906	2	164
15.	Mississippi/Alabama/ Florida (Pensacola)	1906	3	134
16.	Agnes (NE United States)	1972	1	122
17.	Hazel (North Carolina/ South Carolina)	1954	4	95
18.	Betsy (SE Florida/SE Louisiana)	1965	3	75
18.	Floyd (North Carolina)	1999	2	75
20.	Carol (NE United States)	1954	3	60

°May actually have been as high as 10,000 to 12,000.

†Over 500 of these lost on ships at sea; 600 to 900 estimated deaths.

‡Some 344 of these lost on ships at sea.

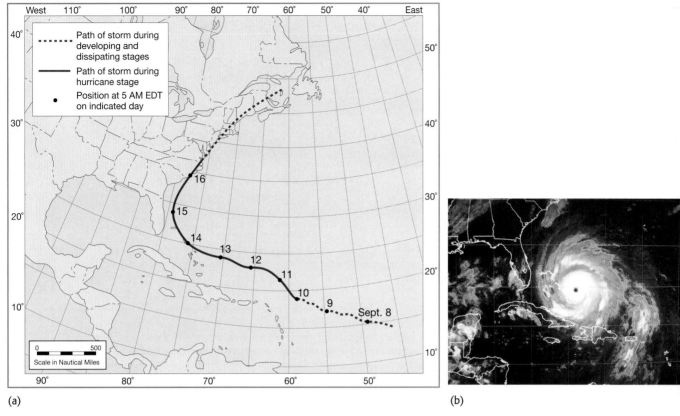

Figure 11–1 Hurricane Floyd formed from a tropical disturbance (easterly wave) that moved westward from the coast of Africa on September 2, 1999. The system became a tropical depression on September 7. The path on this map begins at that point. The hurricane ravaged portions of the Bahamas on September 13–14 and posed a serious threat to Florida. The satellite image shows the storm at this point. After weakening somewhat, Floyd eventually made landfall near Cape Fear, North Carolina, on September 16. Torrential rains were responsible for massive inland flooding. *(AP/Wide World Photos)*

(13.38 inches) in a single 24-hour span. Floyd was the deadliest hurricane to strike the U.S. mainland since Agnes in 1972 (Table 11–1). Most of the deaths were due to drowning from freshwater floods. Each hurricane season, storms like Andrew and Floyd remind us of our vulnerability to these great forces of nature.

Profile of a Hurricane

Most of us view the weather in the tropics with favor. Places like Hawaii and the islands of the Caribbean are known for their lack of significant day-to-day variations. Warm breezes, steady temperatures, and rains that come as heavy but brief tropical showers are expected. It is ironic that these relatively tranquil regions sometimes produce the most violent storms on Earth.

Most hurricanes form between the latitudes of 5° and 20° over all the tropical oceans except the South Atlantic and the eastern South Pacific (Figure 11–2). The North

Pacific has the greatest number of storms, averaging 20 per year. Fortunately for those living in the coastal regions of the southern and eastern United States, fewer than five hurricanes, on the average, develop each year in the warm sector of the North Atlantic.

These intense tropical storms are known in various parts of the world by different names. In the western Pacific, they are called *typhoons*, and in the Indian Ocean, including the Bay of Bengal and Arabian Sea, they are simply called *cyclones*. In the following discussion, these storms will be referred to as hurricanes. The term *hurricane* is derived from Huracan, a Carib god of evil.

Although many tropical disturbances develop each year, only a few reach hurricane status. By international agreement a hurricane has sustained wind speeds of at least 119 kilometers (74 miles) per hour and a rotary circulation.[*] Mature hurricanes average about 600 kilometers (375 miles) across, although they can range in

[*]*Sustained winds* are defined as the wind averaged over a one-minute interval.

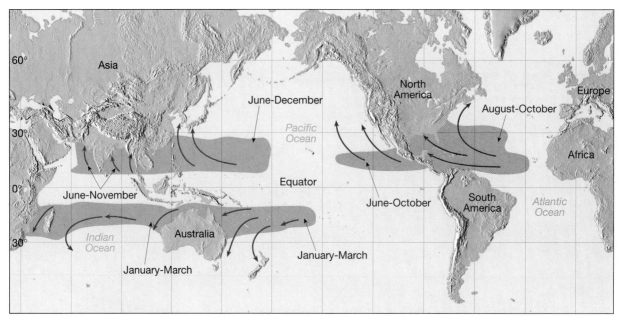

Figure 11–2 This world map shows the regions where most hurricanes form as well as their principal months of occurrence and the most common tracks they follow. Hurricanes do not develop within about 5° of the equator because the Coriolis force is too weak. Because warm surface ocean temperatures are necessary for hurricane formation, they seldom form poleward of 20° latitude nor over the cool waters of the South Atlantic and the eastern South Pacific.

diameter from 100 kilometers (60 miles) up to about 1500 kilometers (930 miles). From the outer edge of the hurricane to the center, the barometric pressure has sometimes dropped 60 millibars, from 1010 to 950 millibars. The lowest pressures ever recorded in the Western Hemisphere are associated with these storms.

A steep pressure gradient like that shown in Figure 11–3 generates the rapid, inward spiraling winds of a hurricane. As the air moves closer to the center of the storm, its velocity increases. This acceleration is explained by the law of conservation of angular momentum (see Box 11–1).

As the inward rush of warm, moist surface air approaches the core of the storm, it turns upward and ascends in a ring of cumulonimbus towers (Figure 11–4). This doughnut-shaped wall of intense convective activity surrounding the center of the storm is called the **eye wall**. It is here that the greatest wind speeds and heaviest rainfall occur. Surrounding the eye wall are curved bands of clouds that trail away in spiral fashion. Near the top of the hurricane the airflow is outward, carrying the rising air away from the storm center, thereby providing room for more inward flow at the surface.

At the very center of the storm is the **eye** of the hurricane (Figure 11–5). This well-known feature is a zone where precipitation ceases and winds subside. The graph in Figure 11–6 shows wind speeds recorded by a data buoy (a remote floating instrument package). As the eye of Hurricane Kate moved into the vicinity of the data buoy, the wind speed dropped by two-thirds, from 48 meters per second (174 kilometers or 108 miles per hour) to 16 meters per second (58 kilometers or 36 miles per hour). Then as the eye moved away, winds rapidly climbed back to 36 meters per second (128 kilometers or 79 miles per hour).

The eye offers a brief but deceptive break from the extreme weather in the enormous curving wall clouds that surround it. The air within the eye gradually descends and heats by compression, making it the warmest part of the storm. Although many people believe that the eye is characterized by clear blue skies, this is usually not the case because the subsidence in the eye is seldom strong enough to produce cloudless conditions. Although the sky appears much brighter in this region, scattered clouds at various levels are common.

Hurricane Formation and Decay

A hurricane is a heat engine that is fueled by the latent heat liberated when huge quantities of water vapor condense. The amount of energy produced by a typical hurricane in just a single day is truly immense. The release of latent heat warms the air and provides buoyancy for its upward flight. The result is to reduce the pressure near

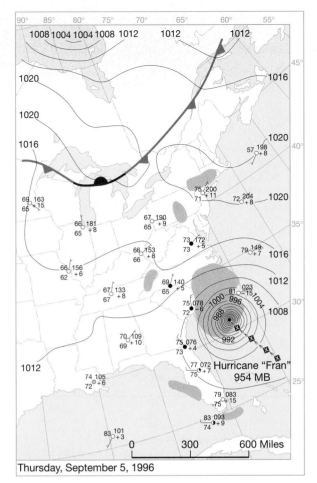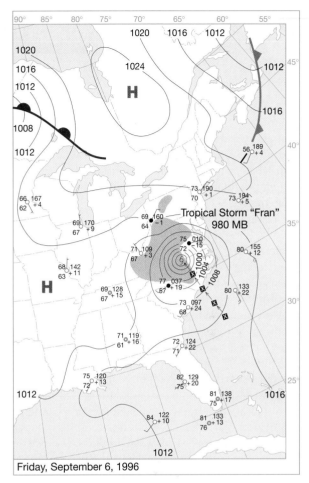

Figure 11-3 Weather maps showing Hurricane Fran at 7 A.M., EST, on two successive days, September 5 and 6, 1996. On September 5, winds exceeded 190 kph. As the storm moved inland, heavy rains caused flash floods, killed 30 people, and caused more than $3 billion in damages. The station information plotted off the Gulf and Atlantic coasts is from data buoys, which are remote floating instrument packages.

the surface, which, in turn, encourages a more rapid inflow of air. To get this engine started, a large quantity of warm, moist air is required, and a continuous supply is needed to keep it going.

Hurricanes develop most often in the late summer when ocean waters have reached temperatures of 27°C (80°F) or higher and are thus able to provide the necessary heat and moisture to the air. This ocean-water temperature requirement accounts for the fact that hurricanes do not form over the relatively cool waters of the South Atlantic and the eastern South Pacific. For the same reason, few hurricanes form poleward of 20 degrees of latitude (see Figure 11–2). Although water temperatures are sufficiently high, hurricanes do not form within 5 degrees of the equator, because the Coriolis force is too weak to initiate the necessary rotary motion.

Hurricane Formation

Many tropical storms achieve hurricane status in the western parts of oceans, but their origins often lie far to the east. There disorganized arrays of clouds and thunderstorms, called **tropical disturbances**, sometimes develop and exhibit weak pressure gradients and little or no rotation. Most of the time these zones of convective activity die out. However, occasionally tropical disturbances grow larger and develop a strong cyclonic rotation.

Several different situations can trigger tropical disturbances. They are sometimes initiated by the powerful convergence and lifting associated with the intertropical convergence zone (ITCZ). Others form when a trough from the middle latitudes intrudes into the tropics. Tropical disturbances that produce many of the strongest hurricanes that enter the western North Atlantic and threaten

Box 11–1 The Conservation of Angular Momentum

Why do winds blowing around a storm move faster near the center and more slowly near the edge? To understand this phenomenon, we must examine the *law of conservation of angular momentum*. This law states that the product of the velocity of an object around a center of rotation (axis) and the distance of the object from the axis is constant.

Picture an object on the end of a string being swung in a circle. If the string is pulled inward, the distance of the object from the axis of rotation decreases and the speed of the spinning object increases. The change in radius of the rotating mass is balanced by a change in its rotational speed.

Another common example of the conservation of angular momentum occurs when a figure skater starts whirling on the ice with both arms extended (Figure 11–A). Her arms are traveling in a circular path about an axis (her body). When the skater pulls her arms inward, she decreases the radius of the circular path of her arms. As a result, her arms go faster and the rest of her body must follow, thereby increasing her rate of spinning.

In a similar manner, when a parcel of air moves toward the center of a storm, the product of its distance and velocity must remain unchanged. Therefore, as air moves inward from the outer edge, its rotational velocity must increase.

Let us apply the law of conservation of angular momentum to the horizontal movement of air in a hypothetical hurricane. Assume that air with a velocity of 5 kilometers per hour begins 500 kilometers from the center of the storm. By the time it reaches a point 100

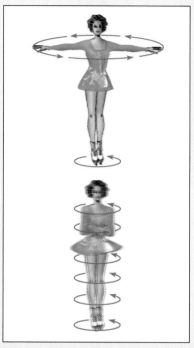

Figure 11–A When the skater's arms are extended, she spins slowly. When her arms are pulled in, she spins much faster.

kilometers from the center, it will have a velocity of 25 kilometers per hour (assuming there is no friction). If this same parcel of air were to continue to advance toward the storm's center until its radius was just 10 kilometers, it would be traveling at 250 kilometers per hour. Friction reduces these values somewhat.

North America often begin as large undulations or ripples in the trade winds known as **easterly waves**, so named because they gradually move from east to west.

Figure 11–7 illustrates an easterly wave. The lines on this simple map are not isobars. Rather, they are *streamlines*, lines drawn parallel to the wind direction used to depict surface airflow. When middle-latitude weather is analyzed, isobars are usually drawn on the weather map. By contrast, in the tropics the differences in sea-level pressure are quite small, so isobars are not always useful. Streamlines are helpful because they show where surface winds converge and diverge.

To the east of the wave axis the streamlines move poleward and get progressively closer together, indicating that the surface flow is convergent. Convergence, of course, encourages air to rise and form clouds. Therefore, the tropical disturbance is located on the east side of the wave. To the west of the wave axis, surface flow diverges as it turns toward the equator. Consequently, clear skies are the rule here.

Easterly waves frequently originate as disturbances in Africa. As these storms head westward with the prevailing trade winds, they encounter the cold Canaries current (see Figure 3–7, p. 65). If the disturbance survives

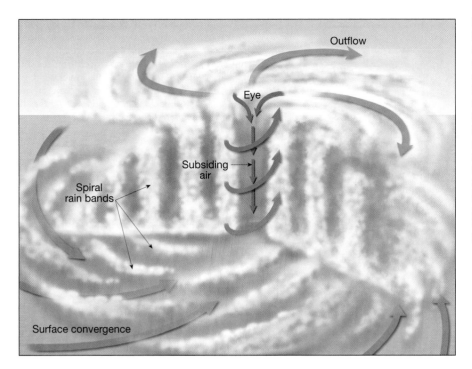

Figure 11–4 Cross section of a hurricane. Note that the vertical dimension is greatly exaggerated. The eye, the zone of relative calm at the center of the storm, is a distinctive hurricane feature. Sinking air in the eye warms by compression. Surrounding the eye is the eyewall, the zone where winds and rain are most intense. Tropical moisture spiraling inward creates rain bands that pinwheel around the storm center. Outflow of air at the top of the hurricane is important because it prevents the convergent flow at lower levels from "filling in" the storm. *(After NOAA)*

the trip across the cold stabilizing waters of the current, it is rejuvenated by the heat and moisture of the warmer water of the mid-Atlantic. From this point on, a small percentage develop into more intense and organized systems, some of which may reach hurricane status.

Even when conditions seem to be right for hurricane formation, many tropical disturbances do not strengthen. One circumstance that may inhibit further development is a temperature inversion called the *trade wind inversion*. It forms in association with the subsidence that

Figure 11–5 A satellite image of Hurricane Fran as the storm approached the Florida coast September 4, 1996. The eye is clearly visible in the center of the storm. The eye wall, the most intense part of the storm, is a doughnut-shaped wall of cumulonimbus development surrounding the eye. *(Photo by WSI Corp./Liaison Agency, Inc.)*

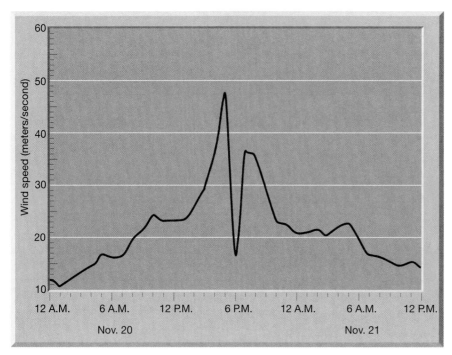

Figure 11–6 Wind speeds recorded by a data buoy in the eastern Gulf of Mexico during the passage of Hurricane Kate on November 20–21, 1985. The substantially lower wind speeds associated with the eye of the storm are clearly shown. *(After the National Hurricane Center/NOAA)*

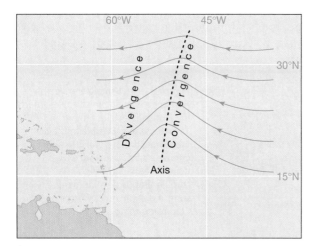

Figure 11–7 Easterly wave in the subtropical Atlantic. Streamlines show low-level airflow. To the east of the wave, axis winds converge as they move slightly poleward. To the west of the axis, flow diverges as it turns toward the equator. Tropical disturbances are associated with the convergent flow of the easterly wave. Easterly waves extend for 2000 to 3000 kilometers (1200 to 1800 miles) and move from east to west with the trade winds at rates between 15 and 35 kilometers (10 and 20 miles) per hour. At this rate, it takes an imbedded tropical disturbance a week or 10 days to move across the North Atlantic.

occurs in the region influenced by the subtropical high.† A strong inversion diminishes air's ability to rise and thus inhibits the development of strong thunderstorms. Another factor that works against the strengthening of tropical disturbances is strong upper-level winds. When present, a strong flow aloft disperses the latent heat released from cloud tops, heat that is essential for continued growth and development.

What happens on those occasions when conditions favor hurricane development? As latent heat is released from the clusters of thunderstorms that make up the tropical disturbance, areas within the disturbance get warmer. As a consequence, air density lowers and surface pressure drops, creating a region of weak low pressure and cyclonic circulation. As pressure drops at the storm center, the pressure gradient steepens. In response, surface wind speeds increase and bring additional supplies of moisture to nurture storm growth. The water vapor condenses, releasing latent heat, and the heated air rises. Adiabatic cooling of rising air triggers more condensation and the release of more latent heat, which causes a further increase in buoyancy. And so it goes.

†In the troposphere, a temperature inversion exists when temperatures in a layer of air increase with an increase in altitude rather than by decreasing with height, which is usually the case. For more on how subsidence can produce an inversion, see the section on "Inversions Aloft" in Chapter 13, p. 361.

Meanwhile, higher pressure develops at the top of the developing tropical depression.‡ This causes air to flow outward (diverge) from the top of the storm. Without this outward flow up top, the inflow at lower levels would soon raise surface pressures and thwart further storm development.

Although many tropical disturbances occur each year, only a few develop into full-fledged hurricanes. Recall that tropical cyclones are called hurricanes only when their winds reach 119 kilometers (74 miles) per hour. By international agreement, lesser tropical cyclones are given different names based on the strength of their winds. When a cyclone's strongest winds do not exceed 61 kilometers (37 miles) per hour, it is called a **tropical depression**. When sustained winds are between 61 and 119 kilometers (37 and 74 miles) per hour, the cyclone is termed a **tropical storm**. It is during this phase that a name is given (Andrew, Fran, Opal, etc.). Should the tropical storm become a hurricane, the name remains the same (see Box 11–2). Each year between 80 and 100 tropical storms develop around the world. Of them, usually half or more eventually reach hurricane status.

Hurricane Decay

Hurricanes diminish in intensity whenever they (1) move over ocean waters that cannot supply warm, moist tropical air, (2) move onto land, or (3) reach a location where the large-scale flow aloft is unfavorable. Richard Anthes describes the possible fate of hurricanes in the first category as follows:

Many hurricanes approaching the North American or Asian continents from the southeast are turned toward the northeast, away from the continents, by the steering effect of an upper-level trough. This recurvature carries the storms toward higher latitudes where the ocean temperatures are cooler and an encounter with cool, dry polar air masses is more likely. Often the tropical cyclone and a polar front interact, with cold air entering the tropical cyclone from the west. As the release of latent heat is diminished, the upper-level divergence weakens, mean temperatures in the core fall and the surface pressure rises.°

Whenever a hurricane moves onto land, it loses its punch rapidly. For example, in Figure 11–3 notice how the isobars show a much weaker pressure gradient on September 6 after Hurricane Fran moved ashore, than on

‡See Figure 6–10 and the discussion of the sea–land breeze in the section on "Pressure Gradient Force" in Chapter 6.
°*Tropical Cyclones: Their Evolution, Structure, and Effects*, Meteorological Monographs, Vol. 19, no. 41 (1982), p. 61. Boston: American Meteorological Society.

September 5 when it was over the ocean. The most important reason for this rapid demise is the fact that the storm's source of warm, moist air is cut off. When an adequate supply of water vapor does not exist, condensation and the release of latent heat must diminish.

In addition, the increased surface roughness over land results in a rapid reduction in surface wind speeds. This factor causes the winds to move more directly into the center of the low, thus helping to eliminate the large pressure differences.

Hurricane Destruction

Although the amount of damage caused by a hurricane depends on several factors, including the size and population density of the area affected and the nearshore bottom configuration, certainly the most significant factor is the strength of the storm itself.

Based on the study of past storms, the **Saffir–Simpson scale** was established to rank the relative intensities of hurricanes (Table 11–2). Predictions of hurricane severity and damage are usually expressed in terms of this scale. When a tropical storm becomes a hurricane, the National Weather Service assigns it a scale (category) number. Category assignments are based on observed conditions at a particular stage in the life of a hurricane and are viewed as estimates of the amount of damage a storm would cause if it were to make landfall without changing size or strength. As conditions change, the category of a storm is reevaluated so that public-safety officials can be kept informed. By using the Saffir–Simpson scale, the disaster potential of a hurricane can be monitored and appropriate precautions can be planned and implemented.

A rating of 5 on the scale represents the worst storm possible, and a 1 is least severe. Storms that fall into category 5 are rare. Only three such storms hit the United States in the twentieth century: Camille pounded Mississippi in 1969, a Labor Day hurricane struck the Florida Keys in 1935, and Allen hammered the southern Texas coast in 1980. Damage caused by hurricanes can be divided into three classes: (1) storm surge, (2) wind damage, and (3) inland freshwater flooding.

Storm Surge

Without question, the most devastating damage in the coastal zone is caused by the storm surge. It not only accounts for a large share of coastal property losses, but it is also responsible for 90 percent of all hurricane-caused deaths. A **storm surge** is a dome of water 65 to 80 kilometers (40 to 50 miles) wide that sweeps across the coast near

Box 11–2 Naming Tropical Storms and Hurricanes

Tropical storms are named to provide ease of communication between forecasters and the general public regarding forecasts, watches, and warnings. Tropical storms and hurricanes can last a week or longer, and two or more storms can be occurring in the same region at the same time. Thus, names can reduce the confusion about what storm is being described.

During World War II, tropical storms were informally assigned women's names (perhaps after wives and girlfriends) by U.S. Army Corps and Navy meteorologists who were monitoring storms over the Pacific. From 1950 to 1952, tropical storms in the North Atlantic were identified by the phonetic alphabet—Able, Baker, Charlie, and so forth. In 1953 the U.S. Weather Bureau (now the National Weather Service) switched to women's names.

The practice of using feminine names continued until 1978, when a list containing both male and female names was adopted for tropical cyclones in the eastern Pacific. In the same year a proposal that both male and female names be adopted for Atlantic hurricanes, beginning with the 1979 season, was accepted by the World Meteorological Organization (WMO).

The WMO has created lists of names for tropical storms over ocean areas. The names used for Atlantic, Gulf of Mexico, and Caribbean hurricanes are shown in Table 11–A. The names are ordered alphabetically and do not contain names that begin with the letters Q, U, X, Y, and Z because of the scarcity of names beginning with those letters. When a tropical depression reaches tropical storm status, it is assigned the next unused name on the list. At the beginning of the next hurricane season, names from the next list are selected even though many names may not have been used the previous season.

The names on each list are used again at the end of each six-year cycle unless a hurricane was particularly noteworthy. Names of such storms as Andrew (1992), Camille (1969), Floyd (1999), and Hugo (1989) are retired to prevent confusion when storms are discussed in future years.

Table 11–A Tropical Storm and Hurricane Names for the Atlantic, Gulf of Mexico, Caribbean Sea

1999	2000	2001	2002	2003	2004
Arlene	Alberto	Allison	Arthur	Ana	Alex
Bret	Beryl	Barry	Bertha	Bill	Bonnie
Cindy	Chris	Chantal	Cristobal	Claudette	Charley
Dennis	Debby	Dean	Dolly	Danny	Danielle
Emily	Ernesto	Erin	Edouard	Erika	Earl
Floyd	Florence	Felix	Fay	Fabian	Frances
Gert	Gordon	Gabrielle	Gustav	Grace	Gaston
Harvey	Helene	Humberto	Hanna	Henri	Hermine
Irene	Isaac	Iris	Isidore	Isabel	Ivan
Jose	Joyce	Jerry	Josephine	Juan	Jeanne
Katrina	Keith	Karen	Kyle	Kate	Karl
Lenny	Leslie	Lorenzo	Lili	Larry	Lisa
Maria	Michael	Michelle	Marco	Mindy	Matthew
Nate	Nadine	Noel	Nana	Nicholas	Nicole
Ophelia	Oscar	Olga	Omar	Odette	Otto
Philippe	Patty	Pablo	Paloma	Peter	Paula
Rita	Rafael	Rebekah	Rene	Rose	Richard
Stan	Sandy	Sebastien	Sally	Sam	Shary
Tammy	Tony	Tanya	Teddy	Teresa	Thomas
Vince	Valerie	Van	Vicky	Victor	Virginie
Wilma	William	Wendy	Wilfred	Wanda	Walter

Table 11–2 Saffir–Simpson Hurricane Scale

Scale Number (Category)	Central Pressure (Millibars)	Wind Speed (KPH)	Wind Speed (MPH)	Storm Surge (Meters)	Storm Surge (Feet)	Damage
1	≥980	119–153	74–95	1.2–1.5	4–5	*Minimal.* No real damage to building structures. Damage primarily to unanchored mobile homes, shrubbery, and trees. Also, some coastal road flooding and minor pier damage.
2	965–979	154–177	96–110	1.6–2.4	6–8	*Moderate.* Some roofing material, door, and window damage to buildings. Some trees blown down. Considerable damage to mobile homes. Coastal and low-lying escape routes flood 2 to 4 hours before arrival of the hurricane center. Small craft in unprotected anchorages break moorings.
3	945–964	178–209	111–130	2.5–3.6	9–12	*Extensive.* Some structural damage to small residences and utility buildings. Large trees blown down. Mobile homes are destroyed. Flooding near the coast destroys smaller structures with larger structures damaged by battering of floating debris. Terrain lower than 2 meters above sea level may be flooded inland 13 km or more. Evacuation of low-lying residences within several blocks of the shoreline may be required.
4	920–944	210–250	131–155	3.7–5.4	13–18	*Extreme.* Some complete roof structure failures on small residences. Extensive damage to doors and windows. Low-lying escape routes may be cut by rising water 3 to 5 hours before arrival of the hurricane center. Major damage to lower floors of structures near the shore. Terrain lower than 3 meters above sea level may be flooded, requiring massive evacuation of residential areas as far inland as 10 km.
5	<920	>250	>155	>5.4	>18	*Catastrophic.* Complete roof failure on many residences and industrial buildings. Some complete building failures. Severe window and door damage. Low-lying escape routes are cut by rising water 3 to 5 hours before arrival of the hurricane center. Major damage to lower floors of all structures located less than 5 meters above sea level and within 500 meters of the shoreline. Massive evacuation of residential areas on low ground within 8 to 16 km of the shoreline may be required.

the point where the eye makes landfall. If all wave activity were smoothed out, the storm surge would be the height of the water above normal tide level (Figure 11–8). In addition, tremendous wave activity is superimposed on the surge. We can easily imagine the damage that this surge of water could inflict on low-lying coastal areas (Figure 11–9).

In the delta region of Bangladesh, for example, most of the land is less than 2 meters (6.5 feet) above sea level. When a storm surge superimposed on normal high tide inundated that area on November 13, 1970, the official death toll was 200,000; unofficial estimates ran to 500,000. It was one of the worst natural disasters of modern times. In May 1991, a similar event again struck Bangladesh. This time the storm took the lives of at least 135,000 people and devastated coastal towns in its path.

An important process that contributes to the development of a storm surge is the piling up of ocean water by strong onshore winds. Gradually the hurricane's winds push water toward the shore, causing sea level to elevate while also churning up violent wave activity. In addition,

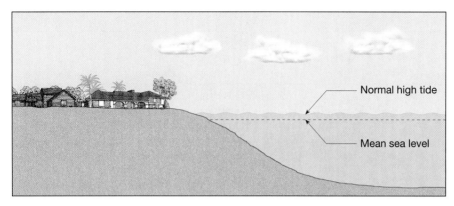

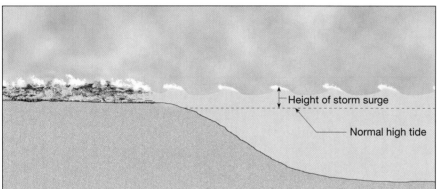

the storm's low atmospheric pressure causes the water level to rise as well.

As a hurricane advances toward the coast in the Northern Hemisphere, storm surge is always most intense on the right side of the eye where winds are blowing *toward* the shore. In addition, on this side of the storm the forward movement of the hurricane also contributes to the storm surge. In Figure 11–10, assume a hurricane with peak winds of 175 kilometers (109 miles) per hour is moving toward the shore at 50 kilometers (31 miles) per hour. In this case, the net wind speed on the right side of the advancing storm is 225 kilometers (140 miles) per hour. On the left side, the hurricane's winds are blowing opposite the direction of storm movement, so the net winds are *away* from the coast at 125 kilometers (78 miles) per hour. Along the shore facing the left side of the oncoming hurricane, the water level may actually decrease as the storm makes landfall.

Wind Damage

Destruction caused by wind is perhaps the most obvious of the classes of hurricane damage. For some structures, the force of the wind is sufficient to cause total ruin. Just read the descriptions of category 3, 4, and 5 storms in Table 11–2. Mobile homes are particularly vulnerable. In addition, the strong winds can create a dangerous barrage of flying debris. In regions with good building codes, wind damage is usually not as catastrophic as storm-surge damage. However, hurricane-force winds affect a much larger area than storm surge and can cause huge economic losses.

For example, winds associated with Hurricane Andrew produced more than $20 billion of damage in southern Florida and Louisiana (see Box 11–3). Tornadoes are spawned by most hurricanes that strike the United States and represent another aspect of wind damage. Although they can be very destructive locally, tornadoes generally account for only a small percentage of the total storm damage.

Inland Flooding

The torrential rains that accompany most hurricanes represent a third significant threat—flooding. Recall the flooding caused by Hurricane Floyd, which devastated North Carolina in September 1999. Whereas the effects of storm surge and strong winds are concentrated in coastal areas, heavy rains may affect places hundreds of kilometers from the coast for up to several days after the storm has lost its hurricane-force winds.

A well-known example of such destruction is Hurricane Agnes (1972). Although it was only a category 1

(a)

(b)

Figure 11–9 The 25-foot storm surge of Hurricane Camille at Pass Christian, Mississippi, devastated the coast. (a) The Richelieu Apartments before the hurricane. This substantial-looking three-story building was directly across the highway from the beach. (b) The same apartments after the hurricane. *(Estate of Chauncy T. Hinman)*

Figure 11–10 Winds associated with a Northern Hemisphere hurricane that is advancing toward the coast. This hypothetical storm, with peak winds of 175 kilometers (109 miles) per hour, is moving toward the coast at 50 kilometers (31 miles) per hour. On the right side of the advancing storm, the 175-kilometer-per-hour winds are in the same direction as the movement of the storm (50 kilometers per hour). Therefore, the *net* wind speed on the right side of the storm is 225 kilometers (140 miles) per hour. On the left side, the hurricane's winds are blowing opposite the direction of storm movement, so the *net* winds of 125 kilometers (78 miles) per hour are away from the coast. Storm surge will be greatest along that part of the coast hit by the right side of the advancing hurricane.

Box 11–3 Atmospheric Hazard: Hurricane Andrew

Ever since 1965 when Hurricane Betsy struck, the residents of southern Florida's vulnerable east coast waited for the next big one. In the intervening years, many powerful storms skirted the region and caused destruction elsewhere. In August 1992, the inevitable became reality.

On August 16, 1992, a tropical depression formed out in the Atlantic, closer to the west coast of Africa than to the United States. The next day, the National Hurricane Center declared that the depression had reached the status of a tropical storm and christened it Andrew (see Figure 11–B). For the next week, Andrew made its way westward, achieving hurricane status on August 22. By Sunday, the 23rd, the spiraling storm was on a collision course with South Florida.

Hurricane Andrew made landfall south of Key Biscayne, Florida, during the dark, early morning hours of August 24 (Figure 11–C). Maximum sustained surface

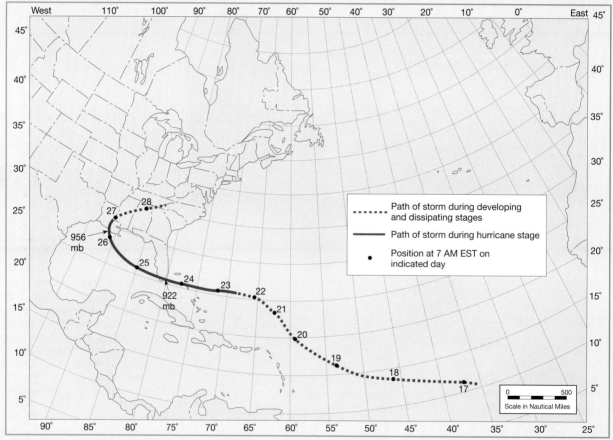

Figure11–B Path of Hurricane Andrew in August 1992.

storm on the Saffir–Simpson scale, it was one of the costliest hurricanes of the century, creating more than $2 billion in damage and taking 122 lives. The greatest destruction was attributed to flooding in the northeastern portion of the United States, especially in Pennsylvania, where record rainfalls occurred. Harrisburg received

nearly 32 centimeters (12.5 inches) in 24 hours and western Schuykill County measured more than 48 centimeters (19 inches) during the same span. Agnes's rains were not as devastating elsewhere. Prior to reaching Pennsylvania, the storm caused some flooding in Georgia, but most farmers welcomed the rain because dry conditions had been

winds were about 230 kilometers (145 miles) per hour, with gusts exceeding 280 kilometers (175 miles) per hour. With a central pressure of 922 millibars (27.23 inches), Andrew was a category 4 storm on the Saffir–Simpson scale (see Table 11–2). It was close to being a rare category 5. Andrew had the third lowest pressure in this century for a hurricane making landfall in the United States.

It took Hurricane Andrew just hours to cut its destructive path across Florida. Property damage from Andrew was primarily wind damage along a 40-kilometer (25-mile)-wide swath of destruction that was centered a few kilometers north of the town of Homestead. Fortunately, the highly developed coastline of Miami Beach was more than 27 kilometers (17 miles) from the eye wall of the storm and so did not experience sustained hurricane-force winds. Two other aspects of the storm made it less damaging than it might otherwise have been: The storm surge was small and there was little rainwater flooding because the hurricane did not linger but advanced rapidly across the region.

The storm's destructive accomplishments were awesome. In one area, the wind carried 6-meter (20-foot)-long steel and concrete beams, with roofs still attached, more than 50 meters (165 feet). Cars and boats were tossed about like toys, tens of thousands of homes and businesses were devastated, hundreds of acres of citrus groves were uprooted, and Homestead Air Force Base was leveled. Government officials who toured the area said it looked like a "war zone."

After crossing southern Florida, Andrew went on to cross the Gulf of Mexico and lashed the Louisiana coastline on August 26 as a category 3 storm. In all, Hurricane Andrew was responsible for at least 62 deaths and caused $20 billion to 30 billion in damages. It was the costliest natural disaster in U.S. history.

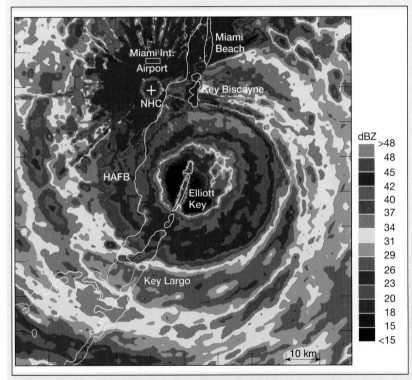

Figure 11–C Color radar image of Hurricane Andrew at 4:35 A.M., EDT, August 24, 1992. The picture is from the last full sweep of the National Weather Service's Miami radar (located at the National Hurricane Center [NHC]) before the radar was destroyed by the storm. The digitized radar imagery shows the eye centered over Elliott Key just before landfall at Homestead Air Force Base (HAFB). As Andrew traveled due west, the heaviest damage occurred in those areas affected by the eye wall (doughnut-shaped region shown in shades of red). The weather radar measures the power from the portion of the radar beam scattered back by raindrops and ice particles. The colors associated with greater reflection (i.e., red) correspond to areas with larger amounts of rain, which typically are also regions of stronger winds. Red areas in the center of the eye are caused by ground clutter from islands. (Ground clutter is the reflection of the radar beam by terrain, large structures, and rough water.) Ground clutter in the vicinity of NHC has been removed. *(Courtesy of National Hurricane Center/NOAA)*

plaguing them earlier. In fact, the value of the rains to crops in the region far exceeded the losses caused by flooding.

Another well-known example is Hurricane Camille (1969). Although this storm is best known for its exceptional storm surge and the devastation it brought to coastal areas, the greatest number of deaths associated with this storm occurred in the Blue Ridge Mountains of Virginia two days after Camille's landfall. Here many places received more than 25 centimeters (10 inches) of rain and severe flooding took more than 150 lives.

To summarize, extensive damage and loss of life in the coastal zone can result from storm surge, strong winds, and torrential rains. When loss of life occurs, it is commonly caused by the storm surge, which can devastate entire barrier islands or zones within a few blocks of the coast. Although wind damage is usually not as catastrophic as the storm surge, it affects a much larger area. Where building codes are inadequate, economic losses can be especially severe. Because hurricanes weaken as they move inland, most wind damage occurs within 200 kilometers (125 miles) of the coast. Far from the coast, a weakening storm can produce extensive flooding long after the winds have diminished below hurricane levels. Sometimes the damage from inland flooding exceeds storm-surge destruction.

Detecting and Tracking Hurricanes

North Atlantic hurricanes develop in the trade winds, which generally move these storms from east to west at about 25 kilometers (15 miles) per hour. Then almost without exception, hurricanes curve poleward and are deflected into the westerlies, which increase their forward motion up to a maximum of 100 kilometers (60 miles) per hour. Some move toward the mainland, but their irregular paths make prediction of their movement difficult.

A location only a few hundred kilometers from a hurricane—just a day's striking distance away—may experience clear skies and virtually no wind. Before the age of weather satellites, such a situation made it difficult to warn people of impending storms. The worst natural disaster in U.S. history came as a result of a hurricane that struck an unprepared Galveston, Texas, on September 8, 1900. The strength of the storm, together with the lack of adequate warning, caught the population by surprise and cost the lives of 6000 people in the city and at least 2000 more elsewhere (Figure 11–11). Fortunately, hurricanes are no longer the unheralded killers they once were.

In the United States, early warning systems have greatly reduced the number of deaths caused by hurricanes. At the same time, however, an astronomical rise has occurred in the amount of property damage. The primary reason for this trend has been the rapid population growth in coastal areas. The National Weather Service (NWS) is concerned that such population increases could set the stage for a major hurricane disaster because the evacuation of large numbers of people might require greater warning times than are presently available.

The Role of Satellites

Today, many different tools provide data that are used to detect and track hurricanes. This information is used to develop forecasts and to issue watches and warnings. The greatest single advancement in tools used for observing tropical cyclones has been the development of meteorological satellites.

Because the tropical and subtropical regions that spawn hurricanes consist of enormous areas of open ocean, conventional observations are limited. The needs for meteorological data from these vast regions are now met primarily by satellites. Even before a storm begins to develop cyclonic flow and the spiraling cloud bands so typical of a hurricane, it can be detected and monitored by satellites (Figure 11–12).

Figure 11–11 The aftermath of the Galveston hurricane of 1900. Entire blocks were swept clean, while mountains of debris accumulated around the few remaining buildings. *(UPI/Corbis)*

(a)

(b)

(c)

Figure 11–12 Satellite images showing the movement of Hurricane Allen in August 1980. Allen was a rare category 5 storm, one of just a few to strike the United States in the twentieth century. Notice how the storm diminishes in intensity when it begins to move over land. *(Courtesy of the National Environmental Satellite Data, and Information Center/NOAA)*

The advent of weather satellites has largely solved the problem of detecting tropical storms and has significantly improved monitoring. However, satellites are remote sensors, and it is not unusual for wind-speed estimates to be off by tens of kilometers per hour and for storm position estimates to have errors of tens of kilometers. It is still not possible to determine with accuracy surface-wind distributions or detailed structural characteristics. A combination of observing systems is necessary to provide the data needed for accurate forecasts and warnings.

Aircraft Reconnaissance

Aircraft reconnaissance represents a second important source of information about hurricanes. Ever since the first experimental flights into hurricanes were made in the 1940s, the aircraft and the instruments employed have become quite sophisticated. When a hurricane is within range, specially instrumented aircraft can fly directly into a threatening storm and accurately measure details of its position and current state of development. Data transmission can be made directly from an aircraft in the midst of a storm to the forecast center where input from many sources is collected and analyzed.

A major contribution to hurricane forecasting and warning programs has been an improved understanding of the structure and characteristics of these storms. Although advancements in remote sensing from satellites have been made, measurements from reconnaissance aircraft will be required for the foreseeable future to maintain the present level of accuracy for forecasts of potentially dangerous tropical storms.

Radar and Data Buoys

Radar is a third basic tool in the observation and study of hurricanes (Figure 11–13). Ever since the 1960s, a system has been in place that covers the entire Gulf of Mexico and Atlantic coastal regions. It provides continuous coverage of tropical storms within 240 kilometers (150 miles) of the coast. Details revealed by these radar units permit refinement of hurricane warnings as storms approach the mainland. These capabilities have improved significantly as new-generation Doppler radar systems have been put into operation.† These units provide additional information on wind fields and contribute to more accurate rainfall

†For a more complete discussion of this important tool, see the section "Doppler Radar" in Chapter 10.

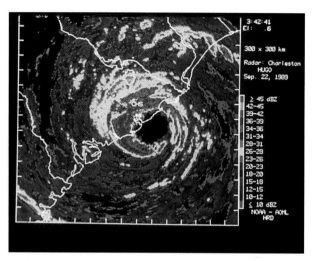

Figure 11–13 Radar image of Hurricane Hugo moving over the coast of South Carolina. Colors show rainfall intensity. Red is the most intense and blue is less intense. No rain is falling in the black areas. The rainless "hole" in the center is the eye. As expected, the heaviest rains are occurring in the eye wall. *(Courtesy of Peter Dodge, National Hurricane Center/NOAA)*

forecasts and flood warnings. Doppler radar also improves the detection of tornadoes spawned by a hurricane.

Data buoys represent a fourth method of gathering data for the study of hurricanes. These remote, floating instrument packages are positioned in fixed locations all along the Gulf Coast and Atlantic Coast of the United States. When you examine the weather maps in Figure 11–4 you can see data buoy information plotted at several offshore stations. Ever since the early 1970s, data provided by these units have become a dependable and routine part of daily weather analysis as well as an important element of the hurricane warning system. The buoys represent the only means of making nearly continuous direct measurements of surface conditions over ocean areas. The graph in Figure 11–6, which shows wind-speed changes that occurred with the passage of a hurricane, was made with data from a buoy.

Hurricane Watches and Warnings

Using input from the observational tools that were just described in conjunction with sophisticated computer models, meteorologists attempt to forecast the movements

and intensity of a hurricane. The goal is to issue timely watches and warnings.

A **hurricane watch** is an announcement aimed at specific coastal areas that a hurricane poses a possible threat, generally within 36 hours. By contrast, a **hurricane warning** is issued when sustained winds of 119 kilometers (74 miles) per hour or higher are expected within a specified coastal area in 24 hours or less. A hurricane warning can remain in effect when dangerously high water or a combination of dangerously high water and exceptionally high waves continue, even though winds may be less than hurricane force.

Two factors are especially important in the watch-and-warning decision process. First, adequate lead time must be provided to protect life and, to a lesser degree, property. Second, forecasters must attempt to keep overwarning at a minimum. This, however, is a difficult task. A policy statement from the American Meteorological Society describes the situation as follows:

> Consistent with current forecast accuracy, it is necessary to issue hurricane warnings for rather large coastal areas. Warnings issued 24 hours before hurricane landfall average 300 nautical miles (560 kilometers) in length. Normally, the swath of damage encompasses about one-third of the warned area so the ratio of affected area to warned area is about one to three. In other words, approximately two-thirds of the area is, in effect, "overwarned." Such overwarning is not only costly, but also results in a loss of credibility in the warnings.[‡]

Clearly, the decision to issue a warning represents a delicate balance between the need to protect the public on the one hand and the desire to minimize the degree of overwarning on the other.

Although many improvements in observational tools and forecast techniques have occurred in recent years, property losses and the potential for loss of life due to hurricanes continue to grow. Forecasts are improving, but not as rapidly as populations are rising in hurricane-prone areas. The result is that longer and longer lead times are needed for communities to prepare for hurricanes.

[‡]"Hurricane Detection, Tracking and Forecasting," *Bulletin of the American Meteorological Society*, 67, no. 7 (July 1993), a policy statement of an organization.

Chapter Summary

- The vast majority of hurricane deaths and damage are caused by relatively infrequent, yet powerful storms. The costliest natural disaster in U.S. history, with damages in excess of $25 billion, was Hurricane Andrew in 1992.

- Most hurricanes form between the latitudes of 5° and 20° over all tropical oceans except the South Atlantic

and eastern South Pacific. The North Pacific has the greatest number of storms, averaging 20 per year. In the western Pacific, hurricanes are called *typhoons*, and in the Indian Ocean, they are referred to as *cyclones*.

- A steep pressure gradient generates the rapid, inward spiraling winds of a hurricane. As the warm, moist air approaches the core of the storm, it turns upward and ascends in a ring of cumulonimbus towers and forms a doughnut-shaped wall called the *eye wall*. At the very center of the storm, called the *eye*, the air gradually descends, precipitation ceases, and winds subside.

- A hurricane is a heat engine fueled by the latent heat liberated when huge quantities of water vapor condense. They develop most often in late summer when ocean waters have reached temperatures of 27°C (80°F) or higher and are thus able to provide the necessary heat and moisture to the air. The initial stage of a tropical storm's life cycle, called a *tropical disturbance*, is a disorganized array of clouds that exhibits a weak pressure gradient and little or no rotation. Tropical disturbances that produce many of the strongest hurricanes that enter the western North Atlantic and threaten North America often begin as large undulations or ripples in the trade winds known as *easterly waves*.

- Each year, only a few tropical disturbances develop into full-fledged hurricanes that require minimum wind speeds of 119 kilometers per hour. When a cyclone's strongest winds do not exceed 61 kilometers per hour, it is called a *tropical depression*. When winds are between 61 and 119 kilometers per hour, the cyclone is termed a *tropical storm*. Hurricanes diminish in intensity whenever they (1) move over cool ocean waters that cannot supply warm, moist tropical air, (2) move onto land, or (3) reach a location where large-scale flow aloft is unfavorable.

- Although damages caused by a hurricane depend on several factors, including the size and population density of the area affected and the near-shore bottom configuration, the most significant factor is the strength of the storm itself. The *Saffir-Simpson* scale ranks the relative intensities of hurricanes. A 5 on the scale represents the worst storm possible, and a 1 is the least severe. Damage caused by hurricanes can be divided into three categories: (1) *storm surge*, which is most intense on the right side of the eye where winds are blowing toward the shore, occurs when a dome of water sweeps across the coast near the point where the eye makes landfall, (2) *wind damage*, and (3) *inland freshwater flooding*, which is caused by torrential rains that accompany most hurricanes.

- North Atlantic hurricanes develop in the trade winds, which generally move these storms from east to west at about 25 kilometers per hour. Today, because of early warning systems that help detect and track hurricanes, the number of deaths associated with these violent storms have been greatly reduced. Because the tropical and subtropical regions that spawn hurricanes consist of enormous areas of open oceans, meteorological data from these vast regions are provided primarily by *satellites*. Other important sources of hurricane information are *aircraft reconnaissance, radar*, and remote, floating instrument platforms called *data buoys*. Using data from these observational tools, meteorologists can issue an announcement, call a *hurricane watch*, aimed at specific coastal areas threatened by a hurricane, generally within 36 hours. By contrast, a *hurricane warning* is issued when sustained winds of 119 kilometers per hour or higher are expected within a specified coastal area in 24 hours or less. Two important factors in the watch and warning decision process are (1) adequate lead time and (2) attempting to keep overwarning at a minimum.

Vocabulary Review

easterly wave (p. 304)
eye (p. 302)
eye wall (p. 302)
hurricane (p. 300)
hurricane warning (p. 316)
hurricane watch (p. 316)

Saffir–Simpson scale (p. 307)
storm surge (p. 307)
tropical depression (p. 307)
tropical disturbance (p. 303)
tropical storm (p. 307)

Review Questions

1. Why might people in some parts of the world welcome the arrival of the hurricane season?

2. When a parcel of air approaches the center of a hurricane, how does its speed change? What law explains this change? (See Box 11–1.)

3. Which of these statements about the eye of a hurricane are true and which are false?
    a. It is typically the warmest part of the storm.
    b. It is usually characterized by clear, blue skies.
    c. It is in the eye that winds are strongest.

4. During what time of year do most of the hurricanes that affect North America form? Why is hurricane formation favored at this time?

5. Tropical storms that form near the equator do not acquire a rotary motion as cyclones of higher latitudes do. Why?

6. What are streamlines? How do streamlines indicate an easterly wave in the North Atlantic?

7. List two factors that inhibit the strengthening of tropical disturbances.

8. Which has the stronger winds, a tropical storm or a tropical depression?

9. Why does the intensity of a hurricane diminish rapidly when it moves onto land?

10. What is the purpose of the Saffir–Simpson scale?

11. Hurricane damage can be divided into three broad categories. Name them. Which one of the categories is responsible for the greatest percentage of hurricane-related deaths?

12. Great damage and significant loss of life can take place a day or more after a hurricane has moved ashore and weakened. When this occurs, what is the likely cause?

13. List four tools that provide data used to track hurricanes and develop forecasts.

14. A hurricane has slower wind speeds than a tornado, but a hurricane inflicts more total damage. How might this be explained?

15. The number of deaths in the United States attributable to hurricanes has continually declined over the last 50 years, but the number of tornado deaths has increased. Write an explanation to account for this situation.

16. Briefly describe the problem of "overwarning" that is related to the issuing of hurricane warnings.

17. Although observational tools and hurricane forecasts continue to improve, the potential for loss of life due to hurricanes is growing. Explain this apparent contradiction.

Problems

The questions that follow refer to the weather maps of Hurricane Fran in Figure 11–3.

1. On which of the two days were Fran's wind speeds probably highest? How were you able to determine this?

2. **a.** How far did the center of the hurricane move during the 24-hour period represented by these maps?
   b. At what rate did the storm move during this 24-hour span? Express your answers in miles per hour.

3. The middle-latitude cyclone shown in Figure 9–20 has an east–west diameter of approximately 1200 miles (when the 1008-mb isobar is used to define the outer boundary of the low). Measure the diameter (north–south) of Hurricane Fran on September 5. Use the 1008-mb isobar to represent the outer edge of the storm. How does this figure compare to the middle-latitude cyclone?

4. Determine the pressure gradient for Hurricane Fran on September 5. Measure from the 1008-mb isobar at Charleston, to the center of the storm. Express your answer in millibars per 100 miles.

5. The weather map in Figure 9–20 shows a well-developed middle-latitude cyclone. Calculate the pressure gradient of this storm from the 1008-mb isobar at the Wyoming–Idaho border to the center of the low. Assume the pressure at the center of the storm to be 986 mb and the distance to be 625 miles. Express your answer in millibars per 100 miles. How does this answer compare to your answer to problem 4?

Atmospheric Science Online

The following are informative and interesting Internet sites that address topics related to those presented in the chapter:

Hurricane Central: Storm 2000 (Lowe's)
- **http://www.GoPBI.com/weather/storm/**

National Hurricane Center (NOAA)
- **http://www.nhc.noaa.gov/**

For direct links to these sites and others, chapter objectives and reviews, quiz questions, and topical investigations that utilize Web resources, visit *The Atmosphere, Eighth Edition* Home Page at:
- **http://www.prenhall.com/lutgens**

Weather Analysis and Forecasting

Predicting severe weather is one of the main tasks of the forecaster. *(Photo by Richard Kaulin/Tony Stone Images)*

319

Figure 12–1 New York City following a major blizzard that hit on January 7–8, 1996. The storm paralyzed much of the Northeast. Probably no other aspect of our physical environment affects the daily lives of people more than the weather. *(Photo by Porter Gifford/Liaison Agency, Inc.)*

Each year, modern society demands more accurate weather forecasts. The desire for sound weather predictions ranges from wanting to know if the weekend's weather will permit a beach outing to NASA's need to evaluate conditions on a space shuttle launch date (Figure 12–1). Such diverse industries as airlines and fruit growers depend heavily on accurate weather forecasts. In addition, the designs of buildings, smokestacks, and many industrial facilities rely on a sound knowledge of the atmosphere. We are no longer satisfied with short-range predictions, but instead demand accurate long-range predictions. Such a question as "Will the Northeast experience an unseasonably cold winter?" has become common.

The Weather Business: A Brief Overview

In the United States, the governmental agency responsible for gathering and disseminating weather-related information is the **National Weather Service (NWS)**. The mission of the NWS is as follows:

> The National Weather Service (NWS) provides weather, hydrologic and climate forecasts and warnings for the United States, its territories, adjacent waters and ocean areas, for the protection of life and property and the enhancement of the national economy. NWS data and products form a national information database and infrastructure that can be used by other governmental agencies, the private sector, the public and the global community.

Perhaps the most important services provided by the NWS are forecasts and warnings of hazardous weather including thunderstorms, flooding, hurricanes, tornadoes, winter weather, and extreme heat. According to the federal Emergency Management Agency, 80 percent of all declared emergencies are weather related (see Box 12–1). In a similar vein, the Department of Transportation reports that more than 6000 fatalities per year can be attributed to the weather. Heat waves claim approximately 1000 lives annually in the United States. Further, in 1993, flooding of the upper Mississippi River Valley inflicted damages in excess of $15 billion and took 48 lives.

As the population grows, the economic impact from weather-related phenomena also escalates. During the years 1986 through 1995, for example, property damage due to wind, hail, snow, and tornadoes increased by 500 percent. An estimated 90 percent of the U.S. public consults weather forecasts every day. As a result, the pressure on the NWS to provide more accurate and longer range forecasts continues to grow.

To produce even a short-range forecast is an enormous task. It involves numerous steps, including collecting weather data, transmitting it, and compiling it on a global scale. These data must then be analyzed so that an accurate assessment of the current conditions can be made. From current weather patterns, various methods are used to determine the future state of the atmosphere, a task called *weather forecasting*. Although weather forecasting is a major component of this chapter, many of the procedures that are employed in this endeavor are beyond the scope of this text. Consequently, we provide only a brief overview of this important aspect of the weather business.

The final phase in the weather business is the dissemination of a wide variety of forecasts. Each of the 119 Weather Forecast Offices operated by the NWS issues regional forecasts, aviation forecasts, and warnings covering their forecast area. The *local forecast* seen on The Weather Channel is an unedited version of a forecast issued by one of these offices. Further, all the data and products (maps, charts, and forecasts) produced by the NWS are available at no cost to the general public and to private forecasting services.

The demand for highly visual forecasts containing computer-generated graphics has grown along with the use of personal computers and the Internet. Because it is outside the mission of the National Weather Service, this publicly funded entity is not the source of the animated depictions of the weather that appear on most local newscasts. Instead, the private sector has taken over this task. In addition, private forecast services customize the NWS products to create a variety of specialized weather reports that are tailored for specific audiences. In a farming community, for example, the weather reports might include frost warnings, while winter forecasts in Denver, Colorado, include the snow conditions at ski resorts. Hence, the public receives most of its weather information through local TV stations and newspapers and/or through more widely distributed sources such as The Weather Channel, CNN, and USA Today (Figure 12–2).

It is important to note that, despite the valuable role that the private sector plays in disseminating weather-related information to the public, the NWS is the *official* voice in the United States for issuing warnings during life-threatening weather situations. Two major weather centers operated by the NWS serve critical functions in this regard. The Storm Prediction Center in Norman, Oklahoma, maintains a constant vigil for severe thunderstorms and tornadoes (see Chapter 10). Hurricane watches and warnings for the Atlantic, Caribbean, Gulf of Mexico, and eastern

Figure 12–2 Weather broadcaster previewing graphics. *(Photo by Bob Daemmrich/Stock Boston)*

Pacific are issued by the National Hurricane Center/Tropical Prediction Center in Miami, Florida (see Chapter 11).

In summary, the process of providing weather forecasts and warnings throughout the United States occurs in three stages. First, to provide a picture of the current state of the atmosphere, data are collected and analyzed on a global scale. Second, the NWS employs a variety of techniques to establish the future state of the atmosphere; a process called *weather forecasting*. Third, forecasts are disseminated to the public, mainly through the private sector. The National Weather Service serves as the sole entity responsible for issuing watches and warnings of extreme weather events.

Weather Analysis

Before weather can be predicted, the forecaster must have an accurate picture of current atmospheric conditions. This enormous task, called **weather analysis**, involves collecting, transmitting, and compiling millions of pieces of observational data. Because the atmosphere is ever-changing, this job must be accomplished quickly. High-speed supercomputers have greatly aided the weather analyst.

Box 12–1 Atmospheric Hazards: Debris Flows in the San Francisco Bay Region*

When prolonged, intense rain falls on steep hillsides, the saturated soils can become unstable and move rapidly downslope. Such land movements, called *debris flows*, are capable of destroying homes, washing out roads and bridges, knocking down trees, and obstructing streams and roadways with thick deposits of mud and rocks (Figure 12–A). An especially destructive event occurred in 1982 when thousands of debris flows caused nearly $70 million in damages and took 25 lives. Since then, several serious but less severe events have occurred. As more and more people build in the hills around the Bay region, the potential impact of debris flows on life and property is increasing.

The Debris-Flow Warning System

Debris flows can begin suddenly, often with little warning. Loss of lives during the intense 1982 storm prompted the National Weather Service (NWS) and the U.S. Geological Survey (USGS) to develop a debris-flow warning system for the San Francisco Bay area.

Figure 12–A On January 25, 1997, a debris flow literally buried this one-story home in Mill Valley, California. Heavy rains from a powerful Pacific storm triggered the event. *(Justin Sullivan/AP Wide World Photos)*

During the rainy season (October through April), this warning system measures rainfall using more than 50 radio-telemetered rain gauges, called the ALERT network (Figure 12–B). Early in each rainy season, these

Gathering Data

A vast network of weather stations is required to provide enough data for a weather chart that is useful for generating even short-range forecasts. On a global scale, the **World Meteorological Organization**, which includes more than 130 nations, is responsible for gathering the needed data and for producing numerous weather maps and upper-level charts that describe the current state of the atmosphere. About 10,000 surface stations on land, 7000 ships at sea, and 3000 data buoys report the atmospheric conditions four times each day at 0000, 0600, 1200, and 1800 Greenwich Mean Time (Figure 12–3). In addition, weather satellites and radiosondes (weather balloons) are used to determine conditions aloft throughout the entire depth of the atmosphere. This network is becoming more complete each year. Nevertheless, even today, vast sections of the globe, especially large portions of the oceans, are monitored inadequately.

Once collected, the information is transmitted to three *World Meteorological Centers* located near Washington, D.C., Moscow, and Melbourne, Australia. From these world centers, the compiled data are sent to the meteorological center in each participating country. These national centers use computer-generated models to project the current state of the atmosphere into the future. From here, these somewhat generalized forecasts are disseminated to regional and local meteorological centers, where they are used to produce site-specific predictions.

In the United States, the National Weather Service operates about 119 Weather Forecast offices. In addition to their role as regional forecast centers, these offices are responsible for gathering and transmitting weather information to a central database. The Federal Aviation Administration (FAA), in cooperation with the NWS, also operates observation stations at most metropolitan airports. Together, these facilities, combined with a few hundred automated systems strategically located around the country, bring the total number of surface observation stations in the United States to over 1000.

Weather Maps: Pictures of the Atmosphere

Once this large body of data has been collected, the analyst displays it in a form that can be comprehended easily by the forecaster. This step is accomplished by placing

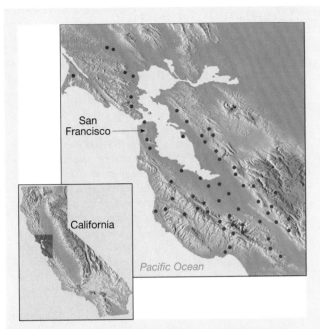

Figure 12–B San Francisco Bay region, showing hilly areas where debris flows are possible. Dots show locations of ALERT rain gauges. During storms, these gauges radio-transmit rainfall data to the U.S. Geological Survey and the National Weather Service. Scientists analyze the data to determine debris-flow danger. If danger is high, a Debris-Flow Watch or Warning is issued.

rainfall measurements, along with measurements of soil moisture from a study site in the hills south of San Francisco, are used to estimate the moisture level of soils throughout the Bay region. Soils must reach a sufficient moisture level each year before slopes become susceptible to debris flows during intense rainstorms. Once soils reach this moisture level, the USGS monitors weather forecasts and uses up-to-the-minute data from the ALERT network to determine the potential for imminent debris flows during each subsequent rainstorm. Warnings are then broadcast by the National Weather Service.

How Much Rain Is Needed to Trigger Debris Flows?

Once soils have reached sufficient moisture levels during a rainy season, it is the rainfall rate, rather than total rainfall amount, that is most important for determining whether debris flows will occur. For example, 4 inches of rain in 24 hours is generally not sufficient to trigger debris flows in the San Francisco Bay region. However, 4 inches of rain in 6 hours generally will trigger numerous debris flows. On burned slopes that have lost their anchoring vegetation, and altered slopes, such as road cuts, greater caution is needed because debris flows can be triggered by less severe rainfall conditions.

*Based on material prepared by the U.S. Geological Survey

the information on a number of **synoptic weather maps** (see Box 12–2). They are called synoptic, which means "coincident in time," because they display a synopsis of the weather conditions at a given moment. These weather charts are a symbolic representation of the state of the atmosphere. Thus, to the trained eye, a weather map is a snapshot that shows the status of the atmosphere, including data on temperature, humidity, pressure, and airflow. (Forecast maps that depict the future state of the atmosphere are also produced, a topic we will consider later in this chapter.)

Over 200 surface maps and charts covering several levels of the atmosphere are produced each day by the NWS and its forecast centers. A task once done by hand, computers are now employed to analyze and plot the data in a systematic fashion. Typically, lines and symbols are used to depict the weather patterns (Figure 12–4a). Once a map is generated, it is the analyst who fine-tunes it, correcting any errors or omissions. Figure 12–4 shows a simplified version of a surface map, as well as a 500-millibar-height contour chart covering the same time period.

In addition to the surface map, twice-daily upper-air charts are drawn at 850-, 700-, 500-, 300-, and 200-millibar

levels. Recall that on these charts, height contours (in tens of meters) instead of isobars are used to depict the pressure field. These charts also contain isotherms (equal temperature lines) shown as dashed lines labeled in degrees Celsius. This series of upper-air charts provides a three-dimensional view of the atmosphere.

Each of the regularly generated upper-air charts is valuable in predicting certain aspects of the weather. For example, the 700-millibar maps, which represent the region about 3 kilometers (2 miles) above sea level, are useful in establishing the movement of air-mass thunderstorms because these storms travel at roughly the speed of the 700-millibar winds. The 500-millibar level is located about 5.5 kilometers (18,000 feet) above sea level, where the average temperature is approximately –20°C (–4°F). Prior to the advent of the computer, when weather charts had to be drawn by hand, the 500-millibar chart was heavily relied upon to predict sites of cyclogenesis (Figure 12–4b). This was done by applying a set of "rules," which were based on past behavior, to predict how the existing weather pattern might develop. Today, high-speed computers allow forecasters to simultaneously analyze the weather at several levels throughout the depth of the atmosphere.

Figure 12–3 Data buoy used to record atmospheric conditions over a section of the global ocean. These data are transmitted via satellite to a land-based station. *(Photo by Matthew Neal McVay/Stock Boston)*

Weather Forecasting

The primary goal of the National Weather Service is to predict the future state of the atmosphere, a process called **weather forecasting**. This is a formidable task, as illustrated by the following quote.

> Imagine a system on a rotating sphere that is 8000 miles wide, consists of different materials, different gases that have different properties (one of the most important of which, water, exists in different concentrations), heated by a nuclear reactor 93 million miles away. Then, just to make life interesting, this sphere is oriented such that, as it revolves around the nuclear reactor, it is heated differently at different locations at different times of the year. Then, someone is asked to watch the mixture of gases, a fluid only 20 miles deep, that covers an area of 250 million square miles, and to predict the state of that fluid at one point on the sphere 2 days from now. This is the problem weather forecasters face.[*]

Because of the complex and highly quantitative nature of modern weather forecasting, we can only highlight the approaches used. These include, but are not limited to, traditional synoptic weather forecasting, numerical weather prediction, statistical methods, and various short-range forecasting techniques. Remember, the object of weather forecasting is not only to project the location and possible intensification of existing pressure systems, but also to identify probable sites for the formation of new storm centers.

Synoptic Weather Forecasting

Until the late 1950s, **synoptic weather forecasting** was the primary basis for making weather predictions. As the name implies, synoptic weather maps are the main tools for producing such forecasts (Figure 12–5). In this approach, various techniques are employed to extrapolate future conditions from the patterns depicted on current weather charts. Some of these techniques, which will be considered later in this section, include *persistence forecasting*, *trend forecasting*, and the *analog method*.

One of these techniques involves looking at a developing cyclonic storm and comparing it to the behavior of similar weather systems from the past. From this comparison, the forecaster makes a prediction regarding how the current weather pattern might evolve. Called the *analog method*, the forecaster attempts to match current conditions with similar well-established patterns from the past. As this practice matured, "rules of thumb"

Two of the upper-level maps, the 300- and 200-millibar charts, are located in the upper troposphere. Here at altitudes above 10 kilometers (6 miles) the temperatures can reach a chilly –60°C (–75°F). More importantly, these levels have the strongest jet streams. In order to show the airflow aloft, these maps plot *isotachs*, which are lines of equal wind speed, in addition to height contours and isotherms. Recall that the areas near the jet stream where airflow makes dramatic changes in speed are also the regions where upper-level convergence and divergence occur. Because upper-level divergence supports the development and intensification of surface cyclones, analysis of the flow at this level is a critical component of modern forecasting.

In summary, the analysis phase involves collecting and compiling millions of pieces of observational data describing the current state of the atmosphere. These data are then displayed on a number of different weather maps that show current weather patterns at selected levels throughout the depth of the atmosphere, not just at the surface.

[*]Robert T. Ryan, "The Weather Is Changing…or Meteorologists and Broadcasters, the Twain Meet," *Bulletin of the American Meteorological Society*, 63, no. 3 (March 1982), 308.

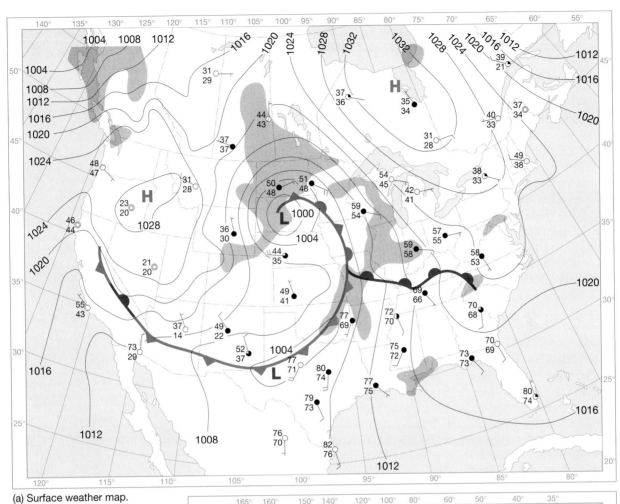

(a) Surface weather map.

Figure 12–4 Simplified synoptic weather maps. (a) Surface weather map for 7:00 A.M. Eastern Standard Time depicting a well-developed middle-latitude cyclone. (b) A 500-millibar-level map for the same time period.

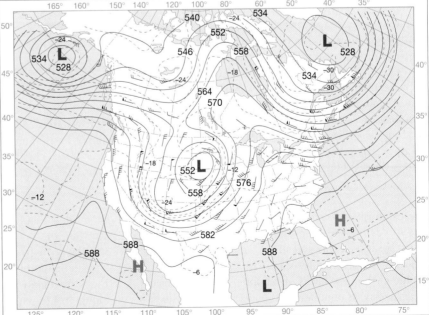

(b) 500-millibar level chart.

Box 12–2 Constructing a Synoptic Weather Chart

Production of surface weather charts first involves plotting the data from selected observing stations. By international agreement, data are plotted using the symbols illustrated in Figure 12–C. Normally, data that are plotted include temperature, dew point, pressure and its tendency, cloud cover (height, type, and amount), wind speed and direction, and weather, both current and past. These data are always plotted in the same position around the station symbol for consistent reading. Using Figure 12–C, for example, you can see that the temperature is plotted in the upper-left corner of the sample model, and it will always appear in that location. (The only exception to this arrangement is the wind arrow, because it is oriented with the direction of airflow.) A more complete weather station model and a key for decoding weather symbols are in Appendix B.

The data are plotted as shown in Figure 12–D (left). Once data have been plotted, isobars and fronts are added to the weather chart (Figure 12–D on right). Isobars are usually plotted on surface maps at intervals of 4 millibars (1004, 1008, 1012, etc.). The positions of the isobars are estimated as accurately as possible from the pressure readings available. Note in Figure 12–D (right) that the 1012-millibar isobar is about halfway between two stations that report 1010 millibars and 1014 millibars. Frequently, observational errors and other complications require the analyst to smooth the isobars so that they conform to the overall picture. Many irregularities in the pressure field are caused by local influences that have little bearing on the larger circulation depicted on the charts. Once the isobars are drawn, centers of high and low pressure are designated.

Because fronts are boundaries that separate contrasting air masses, they can often be identified on weather charts by locating zones exhibiting abrupt changes in conditions. Because several elements change

Figure 12–C A specimen station model showing the location of general data. *(Abridged from the International Code)*

across a front, all are examined so that the frontal position is located most accurately. Some of the more easily recognized changes that can aid in identifying a front on a surface chart are as follows:

were established to aid the forecaster. Applying these rules to current weather charts became the backbone of weather forecasting and still serves an important role in making short-range forecasts.

One of the drawbacks of synoptic weather forecasting, at least in its infancy, was the fact that very little was known about the winds aloft and the key role they play in cyclogenesis. Modern weather forecasting relies heavily

on models that predict changes in these wind patterns. Further, the only good model of cyclone development was the polar-front theory introduced by a group of Norwegian meteorologists during World War I. Recall that in the Norwegian model, midlatitude cyclones develop along fronts where two contrasting air masses clash. Although a very useful model for forecasting cyclone evolution, it has limited predictive value when it comes to

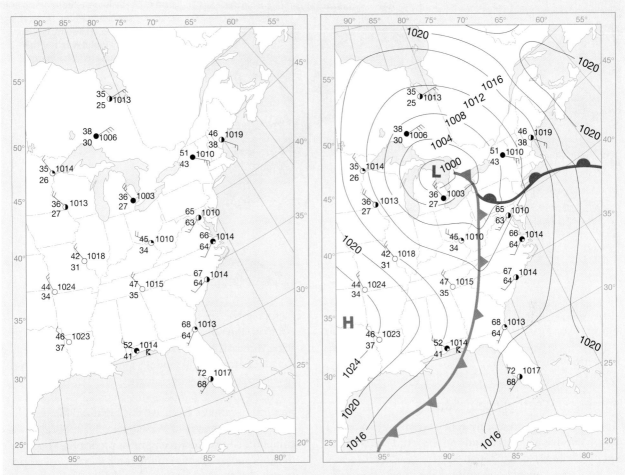

Figure 12–D Simplified weather charts. (left) Stations, with data for temperature, dew point, wind direction, wind speed, sky cover, and barometric pressure. (right) Same chart showing isobars and fronts.

1. Marked temperature contrast over a short distance.
2. Wind shift (in a clockwise direction) over a short distance by as much as 90°.
3. Humidity variations commonly occurring across a front that can be detected by examining dew-point temperatures.
4. Clouds and precipitation patterns giving clues to the position of fronts.

Notice in Figure 12–D that all the conditions listed are easily detected across the frontal zone. However, not all fronts are as easily defined as the one on our sample map. In some cases, surface contrasts on opposite sides of the front are subdued. When this happens, charts of the upper air, where flow is less complex, become an important tool for detecting fronts.

forecasting favorable sites for cyclogenesis. In other words, the Norwegian model would receive low marks as a tool for forecasting when and where a midlatitude cyclone might form a few days into the future. Thus, as other methods were developed that could more accurately predict the future state of the atmosphere, the importance of synoptic forecasting declined. This is particularly true for forecasts made for periods longer than a day or two.

Despite their limitations, synoptic weather charts are still widely used (along with radar and weather satellite images) by experienced forecasters to pinpoint the occurrence of specific events. For example, air-mass thunderstorms travel at roughly the speed of the 700-millibar winds. Knowing that the upper-level winds "steer" these storm cells, and using upper-air charts to establish their rate of movement, forecasters can predict if and when a

Figure 12–5 Two meteorologists at the National Hurricane Center/Tropical Prediction Center, in Miami, Florida, examining a synoptic weather map. *(Photo by L. Mulvehill/The Image Works)*

storm will reach a particular locale. Much of the accuracy of these forecasts depends on the experience of the forecaster. Perhaps synoptic forecasting is as much of an art as it is a science.

Numerical Weather Prediction

Modern weather forecasting relies heavily on so-called **numerical weather prediction**. The word "numerical" is somewhat misleading, for all types of weather forecasting are based on some quantitative data and therefore could fit under this heading. Numerical weather prediction relies on the fact that the gases of the atmosphere obey a number of known physical principles (see Box 12–3). Ideally, these physical laws can be used to predict the future state of the atmosphere, given the current conditions. This situation is analogous to predicting future positions of Mars, using Newton's laws of motion and knowing the planet's current position.

Numerical weather prediction employs a number of highly refined computer models that attempt to mimic the behavior of the atmosphere. In the United States, three different models, plus several variations, are used. The process begins by entering the current atmospheric parameters into a computer program containing equations that describe atmospheric motion. After billions of calculations, the supercomputers used to run the program projects how the present weather is expected to change. Because slight variations in the initial conditions can produce very different forecasts, several runs of the same program may be performed. It is up to the forecaster to select which prediction best fits the current situation. To assist in this endeavor, forecasters take into account such

factors as the season and climatic data. Forecasters also compare the output of two different models before making a forecast. Most often the final forecast is a blend of the data from both models, plus input from satellite images, radar, and other sources.

Using numerical models, the National Weather Service produces a variety of somewhat generalized forecast charts at its headquarters near Washington, D.C. Because these computer-generated forecast maps predict the condition of the atmosphere at some future time, they are called **prognostic charts**. Most numerical models are designed to generate prognostic charts that predict the flow pattern aloft. Meteorologists use this information, along with experience and "rules of thumb," to project favorable sites for cyclogenesis. These numerical models also generate forecasts for other conditions, including maximum and minimum temperatures, wind speeds, and precipitation probabilities. Even the most simplified models, however, require such a vast number of calculations that they could only be used after the advent of high-speed computers.

Generated at regular intervals, these prognostic charts are sent to the 119 weather forecasting offices of the NWS. The forecasters at these sites blend local conditions, numerical predictions, and the weather quirks of the region into a site-specific forecast. This task is complicated by the fact that various prognostic charts are available to the regional forecaster. For example, generally two different numerical models are employed to predict the minimum temperature for the day. One method works better on some days than others, and performs better in some locales than others. It is up to the forecaster to select the correct model each day. In addition, computer models do not take into account all surface conditions, including minor topographic features. The presence of snow cover or a body of water can alter local weather in ways that are not predicted on these computer-generated charts.

In summary, numerical weather-prediction models have greatly improved our ability to forecast the weather; however, the prognostic charts obtained by these techniques are somewhat general. Thus, the detailed aspects of weather must still be determined by applying traditional forecasting methods to these charts.

Statistical Methods

Statistical methods are often used in conjunction with numerical weather predictions. Past weather data are carefully examined to determine weather patterns that are good predictors of future events. Once these relations are established, current data can then be used to project future conditions.

Although this procedure can be used to make a general forecast, it is most often used to determine a single aspect of the weather—for example, to project the maximum temperature for the day. To do so, statistical data relating temperature to wind speed and direction, cloud cover, humidity, and to the season of the year are compiled. After these data are displayed on charts, a reasonable estimate of the maximum temperature for the day can be made.

Another statistical approach to weather forecasting is the **analog method**. As indicated earlier, this method employs a search of past weather records to locate ones that come close to duplicating current conditions. Once found, the sequence of current weather events should parallel those of the past. Although this seems straightforward, the method has drawbacks: no two periods of weather are identical in all respects, and there are simply too many variables to match. Today, the analog method is most often applied to long-range forecasting.

Techniques Used in Short-Range Forecasting

Perhaps the simplest forecasting technique is based on the tendency of the weather at a given site to remain unchanged for several hours, or even a day or so. **Persistence forecasts** predict that the future weather will be the same as the present conditions. For example, if it is raining now at a particular location, it might be reasonable to assume that it will still be raining in a few hours. Persistence forecasts do not account for changes that might occur in the intensity or direction of a weather system, and they cannot predict the formation or dissipation of cyclones. Because of these limitations and the rapidity with which weather systems change, persistence forecasts break down after 6 to 12 hours, or a day at most.

Another related method, called **trend forecasting**, assumes that the weather occurring upstream will persist and move on to affect the area in its path. For example, forecasters extrapolate the rate of movement of a line of thunderstorms to predict the time these storms may inflict their fury on some distant community.

Severe weather events are typically short-lived (a few hours) mesoscale (less than 100 km) phenomena that are frequently too small to appear on general prognostic charts. Included in this category are thunderstorms, tornadoes, hail storms, and microbursts (Figure 12–6). It should be obvious that forecasting and issuing warnings for these events must be done quickly and be site-specific. The techniques used for this work, often called **nowcasting**, are heavily dependent on weather radar and geostationary satellites. These weather tools are important in detecting areas of heavy precipitation or at least cloud types capable of triggering these severe conditions. Nowcasting techniques use highly interactive computers capable of integrating data from a variety of sources. Prompt forecasting of tornadic winds is one example of the importance of nowcasting techniques.

Figure 12–6 A line of thunderstorms developing across the Badlands of South Dakota. Mesoscale phenomena like these are too small to appear on computer-generated prognostic charts. Predictions of these events rely heavily on weather radar and geostationary satellites. *(Photo by Rod Planck/Dembinsky Photo Associates)*

Box 12–3 Numerical Weather Prediction

Gregory J. Carbone*

During the past several centuries, the physical laws governing the atmosphere have been refined and expressed through mathematical equations. In the early 1950s, meteorologists began using computers, which provided an efficient means of solving these mathematical equations, to forecast the weather. The goal of such numerical weather prediction is to predict changes in large-scale atmospheric flow patterns. The equations relate to many of the processes already discussed in this book. Two *equations of motion* describe how horizontal air motion changes over time, taking into account pressure gradient, Coriolis, and frictional forces. The *hydrostatic equation* describes vertical motion in the atmosphere. The *first law of thermodynamics* is used to predict changes in temperature that result from the addition or subtraction of heat or from expansion and compression of air. Two equations refer to the conservation of mass and water. Finally, the *ideal gas law* or *equation of state* shows the relationship among three fundamental variables—temperature, density, and pressure.

Weather-prediction models begin with observations describing the current state of the atmosphere. They use the equations to compute new values for each variable of interest, usually at 5–10-minute intervals, called the *time step*. Predicted values serve as the initial conditions for the next series of computations and are made for specific locations and levels of the atmosphere. Each model has a spatial resolution that describes the distance between prediction points. Solving the model's fundamental equations repeatedly predicts the future state of the atmosphere. The model output is provided to weather forecasters for fixed intervals, such as 12, 24, 36, 48, and 72 hours in the future.

Despite the sophistication of numerical weather-prediction models, most still produce forecast errors. Three factors in particular restrict their accuracy—inadequate representation of physical processes, errors in initial observations, and inadequate model resolution. Whereas the models are grounded on sound physical laws and capture the major characteristics of the atmosphere, they necessarily simplify the workings of a very complex system. The representation of land–surface processes and topography are just two examples of features that are incompletely treated in current numerical models. Errors in the initial observations fed into the computer will be amplified over time because numerical weather prediction models include many nonlinear relationships. Finally, physical conditions at all spatial scales can influence atmospheric changes. Yet the spatial resolution of numerical models is too coarse to capture many important processes. In fact, the atmospheric system moves at scales too small to ever be observed and incorporated explicitly into models.

A simple example illustrates how misrepresentation of physical processes or observation error might lead to inaccurate predictions. Figure 12–E is based on an equation used to predict future values of a given variable Y. The equation is written as

$$Y_t = (a \times Y_t) - Y_t^2$$

where Y represents the value of some variable at time t

Y_{t+1} represents the same variable at the next time step

and a represents a constant coefficient.

In summary, weather forecasts are produced using several methods. The National Weather Service uses numerical weather prediction to generate general prognostic charts for large regions. These charts are then disseminated to regional and local forecast centers that apply traditional synoptic, statistical, and short-range forecasting techniques to generate more area-specific forecasts.

Long-Range Forecasts

Long-range forecasting is an area that relies heavily on statistical averages obtained from past weather data, also referred to as *climatic data*. The National Weather Service prepares weekly, monthly, and seasonal weather outlooks. These are not weather forecasts in the usual

Notice that each predicted value serves as the initial value for the next calculation, the same way in which output from a numerical weather-prediction model provides input for subsequent computations. The solid line in Figure 12–E shows the equation solution over a number of time steps, given an initial value of (e.g., a meteorological observation), $Y_{t=0}$ = 1.5, and a coefficient value, a = 3.75. The graph illustrates how the precision of our equations describing the evolving state of the atmosphere may affect predictions. The blue line represents values of Y that result from the adjustment of a from 3.75 to 3.749. Similarly, we can demonstrate how

a very small observation error could amplify over time by adjusting $Y_{t=0}$ from 1.5 to 1.499. The red line in the graph shows how an incremental change in the initial value affects predictions. Small errors may make very little difference in the early stages of our prediction, but such errors amplify dramatically over time. Because we cannot observe many small-scale features of the atmosphere, nor incorporate all of its processes into computer models, weather forecasts have a theoretical limit.

*Professor Carbone is a faculty member in the Department of Geography at the University of South Carolina.

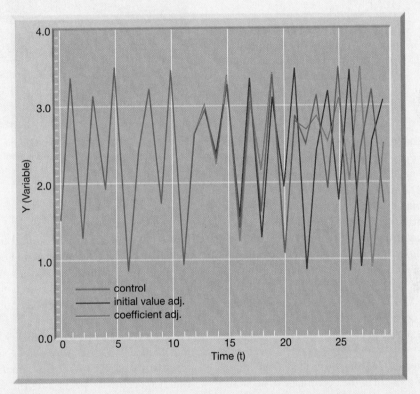

Figure 12–E Tiny errors may not significantly influence the early stages of a prediction, but with time, such errors amplify dramatically.

sense ("partly cloudy and cooler tomorrow"). They are estimates of the precipitation and temperatures that can be expected during these periods. Such projections indicate only whether the region will experience near-normal conditions or not.

The general monthly extended forecasts are produced by the following procedure. First, a mean 700-millibar contour chart is constructed for the coming

month. This requires examination of statistical records for the season of the year, and then modifying the data based on the known effects of ocean temperatures, snow cover, and other factors. Once this chart is compiled, relations between the flow aloft and the development and movement of surface weather patterns are considered in making a prediction for each segment of the United States.

A typical monthly forecast indicates the probabilities that temperature and precipitation will be above or below normal for that time of year. For example, a forecast for December may indicate that the average temperatures in the Northeast are expected to be cooler than normal, whereas those in the Midwest are expected to be warmer than normal. Furthermore, this chart may predict higher than normal precipitation for southern California and lower than normal precipitation for much of the Southeast. Despite their very general nature, the success of monthly forecasts has been disappointing.

Forecast Accuracy

Never, no matter what the progress of science, will honest scientific men who have a regard for their professional reputations venture to predict the weather.

Dominique François Arago
French physicist (1786–1853)

Some might argue that Arago's statement is still true! Nevertheless, a great deal of scientific and technological progress has been made in the two centuries since this observation. Today, accurate weather forecasts and warnings of extreme weather conditions are provided by the U.S. National Weather Service (NWS). Their forecasts are used by government agencies to protect life and property, by electric utilities and farmers, by the construction industry, travelers, airlines—in other words, nearly everyone (Figure 12–7).

How does the NWS measure the **skill** of its forecasts? In determining the occurrence of precipitation, for example, NWS forecasts are correct more than 80 percent of the time. Does this mean the NWS is doing a great job? Not necessarily. When establishing the skill of a forecaster, we need to examine more than the percentage of accurate forecasts. For example, measurable precipitation in Los Angeles is recorded only 11 days each year, on average. Therefore, the chance of rain in Los Angeles is 11 out of 365, or about 3 percent. Knowing this, a forecaster could predict no rain for every day of the year and be correct 97 percent of the time! Although the accuracy would be high, it would not indicate skill.

Any measure of forecasting skill must consider climatic data. Thus, if a forecaster is to exhibit forecasting skill, he or she must do better than forecasts that are based simply on climatic averages. In the Los Angeles example, the forecaster must be able to predict rain on at least a few of the rainy days to demonstrate forecasting skill.

Figure 12–7 Forecasters at the National Weather Service provide nearly 2 million predictions annually to the public and commercial interests. *(Photo by Lawrence Migdale/Photo Researchers, Inc.)*

The only aspect of the weather that is predicted as a percentage probability is rainfall (see Box 12–4). Here statistical data are studied to find how often precipitation occurred under similar conditions. Although the fact that precipitation will take place can be predicted with better than 80 percent accuracy, predictions of the amount and time of occurrence are not as reliable.

Just how skillful are the weather forecasts provided by the National Weather Service? In general very short-range forecasts (0 to 12 hours) have demonstrated considerable skill, especially for predicting the formation and movement of large weather systems like midlatitude cyclones. The so-called short-range forecasts (12–72 hours) are much better at predicting precipitation amounts than forecasts made only two decades ago. However, the exact distribution of precipitation is tied to mesoscale structures, such as individual thunderstorms, that cannot be predicted by the current numerical models.

Thus, a forecast could predict 2 inches of rain for an area, but one town might receive only a trace while a neighboring community is deluged with 4 inches. Predictions of maximum and minimum temperatures and wind, in contrast, are quite accurate. Medium-range forecasts (3–7 days) have also shown significant improvement in the last

Box 12–4 Precipitation Probability Forecasts*

Many of us are satisfied with a precipitation forecast that simply tells us whether rain is likely or not (Figure 12–F). In traditional precipitation forecasts like these, the inherent uncertainty is expressed using qualifiers such as "chance" and "likely." However, such vague terminology is subject to a wide range of misinterpretations by everyone. Moreover, compared to the use of probabilities to express level of uncertainty, traditional forecast terminology is imprecise and often ambiguous. To overcome these shortcomings, the National Weather Service instituted *precipitation probability forecasts* in 1965.

Probability refers to the chance that an event will occur and is represented as a number between 0 (the probability of an impossible one) and 1 (the probability of an inevitable one). Probability can also be expressed as a percentage, so that a 0.3 chance of an event occurring is expressed as 30 percent. This expresses the idea of probability in terms that people quickly understand.

In forecasting, probability is the percentage chance that at least 0.01 inch (0.025 cm) of precipitation will fall at *any point* in the area during the time period covered by the forecast. Thus, a 70 percent probability indicates a 7 in 10 chance of precipitation, and a 3 in 10 chance of no measurable precipitation at any location in the forecast area. In general, these forecasts cover 12-hour periods and moderate-sized metropolitan areas.

Unfortunately, most people interpret this forecast to mean that there is a 70 percent chance of precipitation somewhere in the forecast area, and a 30 percent chance that it will not occur anywhere in that area. This is not the case.

In actuality, this forecast states that at any point in the forecast area (for example, your home), there is a 70 percent chance of measurable precipitation.

Further, the chance of a shower occurring at a specific site is the product of two quantities: the probability that a precipitation-producing storm will develop or move into the area and the percent of the area that the storm is expected to cover.

For example, a forecaster can have a high degree of confidence that a storm will move into an area (say, 80 percent), but determine that only 40 percent of the area will be affected. Under these conditions, the forecaster will call for a 30 percent chance of precipitation (0.80 × 0.40 = 0.32, or about 30 percent). Although precipitation is nearly certain, the chance that it will affect you, wherever you are in the forecast area, is only 3 in 10. Consequently, in the summer, when storms are frequently isolated or scattered, the probability that your immediate area will have rain tends to be smaller than during the winter months when storms are more widespread.

*The information presented here is based on a National Weather Service publication "Precipitation Probability Forecasts" and an article by Allan H. Murphy et al., "Misinterpretations of Precipitation Probability Forecasts," *Bulletin of the American Meteorological Society*, 61, no. 6 (June), 695–700.

Figure 12–F Umbrellas serve as safe haven in rainy weather. *(Photo by Roy Morsch/The Stock Market)*

few decades. Large cyclonic storms like the 1996 East Coast blizzard shown in Figure 12–1 are often predicted a few days in advance. Yet, beyond day seven, the predictability of day-to-day weather using these modern methods proves no more accurate than projections made from climatic data.

The reasons for the limited range of modern forecasting techniques are many. As noted, the network of observing stations is incomplete. Not only are large areas of Earth's land–sea surface monitored inadequately, but on a global scale data from the middle and upper troposphere is meager. Moreover, the physical laws governing the atmosphere are not fully understood and the current models of the atmosphere remain incomplete.

Nevertheless, numerical weather prediction has greatly improved the forecaster's ability to project changes in the upper-level flow. When the flow aloft can be tied more closely to surface conditions, weather forecasts should show even greater skill.

A policy statement by the American Meteorological Society summarizes the predictive skill of weather forecasts:

1. Very short-range forecasts (0–12 hours). *These forecasts have shown considerable skill and utility, especially for predictions of the evolution and movement of large- and medium-sized weather systems. However, the accuracy of the forecasts decreases rapidly as the scale of the weather features decreases and the time range of the forecasts increases. Forecasting the evolution and movement of smaller-scale, short-lived, often intense weather phenomena such as tornadoes, hail storms, and flash floods is less mature than for predictions of larger-scale weather systems. The difficulty in forecasting small-scale systems is due to insufficient computational ability, inadequate observational capabilities, and limited understanding of the physical processes that are taking place during these events. General areas where these systems are likely to form can often be predicted up to 3 days in advance but the precise location that such a small-scale storm will form cannot usually be forecast reliably with much lead time. However, forecasts of small-scale features have improved markedly in regions where weather-related phenomena are generated or modulated by fixed terrain features, land–sea contrasts, and land-use characteristics.*

Despite the difficulties in predicting these small-scale phenomena, the lead time of watches and warnings has increased. For example, the lead time for tornado warnings has more than doubled in the last decade due to the improved observing systems provided by the NWS operational Doppler radar network and satellite imagery. These warnings and watches rely heavily on observing and detecting when conditions are favorable for the development of severe convection and then monitoring each storm's evolution. Forecaster interpretation of radar and satellite imagery and local spotters play a critical part in these very short-term forecasts.

2. Short-range forecasts (12–72 hours). *The accuracy of short-range forecasts (12–72 hours) has continued to increase during the past decade. Improvements in observing systems and in how the data are assimilated into the computer models have resulted in steady improvement in the ability to predict the evolution of major, larger-scale weather systems. Accurate predictions of the development and movement of large-scale weather systems and the associated day-to-day variations in temperature, precipitation,*

cloudiness, and air quality are made regularly throughout this time range.

Forecasts of how much precipitation will fall in the 36–60-hour time frame are now more accurate than 12–36-hour predictions were during the late 1970s. However, the details of precipitation patterns are often tied to smaller-scale structures such as fronts, thunderstorm outflow boundaries, and mesoscale convective systems that are still difficult for the current generation of numerical models to simulate. High-resolution computer models with more advanced physics show promise for being able to simulate these small-scale features, and this, together with the advances in observing systems, gives reason to believe that further improvement in forecasts of precipitation is likely.

3. Medium-range forecasts (3–7 days into the future). *Medium-range forecasts have shown significant improvement in the last two decades. Large-scale events like the East Coast blizzards of 1993 and 1996 are now often forecast days in advance of the first flake of snow, allowing emergency managers the opportunity to make plans to mitigate potential life-threatening situations that might develop. Three-day forecasts of major low-pressure systems that determine the general evolution of the weather are more skillful today than 36-hour forecasts were 15 years ago. In the late 1970s, day 5 forecasts of precipitation were no more accurate than climatology. Since then, skill of day 5 forecasts has more than doubled, with predictions of major cyclones now being as skillful as day 3 forecasts were a decade ago. Temperature forecasts have also improved and now show considerable skill on day 3, with the skill decreasing with time until generally only marginal skill remains by day 7. However, there is reason to believe skillful day 7 forecasts will be possible in the future given the steady improvements in computer models, observational approaches, and forecast strategies.*

4. Extended-range forecasts (week 2). *The predictability of the day-to-day weather for periods beyond day 7 is usually small. Operationally, forecasts at these time ranges have taken the form of 6–10 day mean temperature and occurrence of precipitation departures from normal. The accuracy of these 5-day mean temperature and precipitation forecasts has more than doubled since the 1970s. The accuracy of precipitation forecasts is less than that for temperature, even though the skill of both has increased at about the same rate. Advances in observing systems, computer models, and statistical techniques may allow skillful forecasts of the mean conditions for the 8–14-day period ("week 2") in the near future.*

5. Monthly and seasonal forecasts. *As a result of research over the last decade, monthly and seasonal forecasts of mean temperature and precipitation are*

now useful for specialized applications in major economic sectors, such as agricultural and energy interests, if utilized over a long period. There is reason for optimism that the utility of these long-range forecasts can be improved as computer models and statistical methods become more sophisticated. These new techniques are expected to improve monthly and seasonal outlooks, especially when the relatively strong signals associated with El Niño and La Niña events are present. Notwithstanding these advances, no verifiable skill exists or is likely to exist for forecasting day-to-day weather changes beyond two weeks. Claims to the contrary should be viewed with skepticism.*

In summary, the accuracy of weather forecasts covering short- and medium-range periods has improved steadily over the past few decades, particularly in forecasting the evolution and movement of middle-latitude cyclones. Technological developments and improved computer models, as well as an increased understanding of how the atmosphere behaves, have greatly contributed to this success. The ability to accurately predict the weather beyond day seven, however, remains poor.

Tools in Weather Forecasting

A number of technical advances have been made or are under development to improve forecast accuracy, particularly in the range of one to five days. One of the current deficiencies, limited observational data, is being addressed through the use of automated weather instruments. It has

*From the *Bulletin of the American Meteorological Society*, 79, no. 10 (October 1998), 2161–63.

been projected that about 1000 of these **Automated Surface Observing Systems (ASOS)** will be employed in places currently outside the observational network. Eventually, these devices will replace most human observers (see Box 12–5). Similarly, special radar units called *wind profilers* are being installed to measure wind speed and direction up to 10 kilometers (6 miles) above the surface. Further, these measurements can be taken every six minutes instead of every 12 hours as weather balloons currently do.

Important advances have also been made in the area of data handling. With the advent of interactive microcomputer systems, it is possible for forecasters to display, manipulate, and rapidly digest the great quantity and variety of available data. This capability allows them to combine a variety of data on the same monitor or to focus more closely on regions experiencing severe weather. As part of a major modernization program, a state-of-the-art interactive computer network called *Advanced Weather Interactive Processing Systems (AWIPS)* has been installed in National Weather Service forecast offices.

Small-scale weather phenomena, such as tornadoes and thunderstorm microbursts, cannot be *predicted* using available forecasting techniques. Therefore, the emphasis is placed on detecting and tracking these features. The capabilities of weather radar, which is a particularly valuable tool for this purpose, are steadily improving. The NEXRAD (Next Generation Weather Radar) network is an important example. The backbone of this system, Doppler radar, is discussed in Chapter 10. During the decade of the 1990s, the National Weather Service installed these advanced units in all of its modern forecasting centers (Figure 12–8). These new, or revamped,

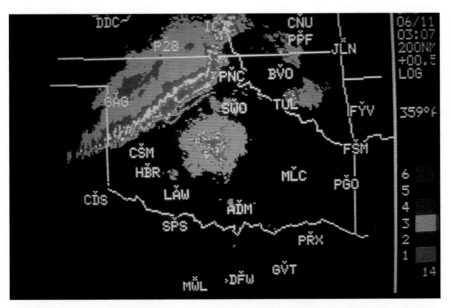

Figure 12–8 This image, taken from an advanced meteorological work station, shows the intensity of precipitation as detected by Doppler radar. The heaviest precipitation (red) occurs along a squall line located in northwestern Oklahoma. *(Photo by Howard Bluestein/Science Source/Photo Researchers, Inc.)*

Box 12–5 Automatic Weather Stations

Gong-Yuh Lin*

The photograph in Figure 12–G shows the Novalynx ALERT (Automated Local Evaluation in Real Time) weather station at Eel Point, San Clemente Island, California. The Eel Point automatic weather station is one of three stations on San Clemente Island established by the California State University at Northridge, in collaboration with the U.S. Navy Natural Resources Office, for the purpose of studying different microclimates on the island. The weather station consists of an antenna and meteorological sensors to measure solar radiation, temperature, relative humidity, wind speed and direction, and rainfall. A tipping-bucket rain gauge (not visible in the photo) is placed atop a standpipe about 6 meters above the ground. Using radio transmissions, hourly weather data can be acquired and displayed in real time on a desktop computer.

Similar automatic weather stations, equipped with additional weather sensors, are used jointly by the National Weather Service and the Federal Aviation Administration at many airports. This program, called Automated Surface Observation Systems (ASOS), substantially reduces the labor costs required for traditional manual-surface observations. Further, this system is able to broadcast computer-generated-voice weather information to pilots.

Portable weather stations of this type are also being used in the National Weather Service flood-warning program. Because of low operating costs, automatic weather stations are quickly replacing many manual facilities.

Figure 12–G Automated weather station located at Eel Point, San Clemente, California. *(Photo by E. J. Tarbuck)*

*Professor Lin is a faculty member in the Department of Geography at California State University at Northridge.

facilities have the capability to provide more timely and site-specific storm warnings.

Satellites in Weather Forecasting

Meteorology entered the space age on April 1, 1960, when the first weather satellite, TIROS 1, was launched. (TIROS stands for Television and Infra-Red Observation Satellite.) In its short life span of only 79 days, TIROS 1 transmitted thousands of images to Earth. Nine TIROS satellites had been launched by 1965. In 1964, the second-generation Nimbus (Latin for *cloud*) satellites had infrared sensors capable of "seeing" cloud coverage at night.

Another series of satellites was placed into polar orbits so that they circled Earth in a north-to-south direction (Figure 12–9a). These **polar satellites** orbit Earth at low

altitudes (a few hundred kilometers) and require only 100 minutes per orbit. By properly orienting the orbits, these satellites drift about 15° westward over Earth's surface during each orbit. Thus, they are able to obtain images of the entire Earth twice each day and coverage of a large region in only a few hours.

By 1966, **geostationary satellites** were placed in orbit over the equator (Figure 12–9b). These satellites, as their name implies, remain fixed over a point on Earth because their rate of travel keeps pace with Earth's rate of rotation. To keep a satellite positioned over a given site, however, the satellite must orbit at a greater distance from Earth's surface (about 35,000 kilometers, or 22,000 miles). At this altitude, the speed required to keep a satellite in orbit will also keep it moving in time with the rotating Earth. However, at this distance, some detail is lost on the images.

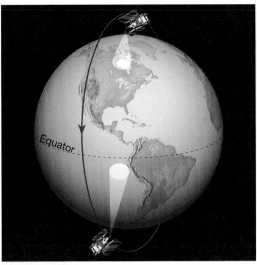
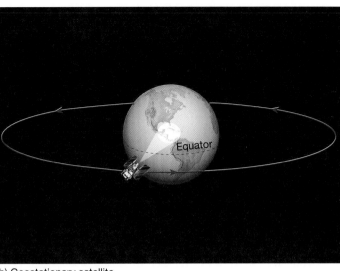

(a) Polar orbiters   (b) Geostationary satellite

Figure 12–9 Weather satellites. (a) Polar-orbiting satellites about 850 kilometers (530 miles) above Earth travel over the North and South poles. Presently, NOAA operates two TIROS satellites that scan an area farther to the west in each orbit. (b) A geostationary satellite moves west to east above the equator at a distance of 36,000 kilometers (22,300 miles) and at the same rate that Earth rotates.

What Weather Satellites Reveal

Weather satellites greatly add to our knowledge of weather patterns. Most important, they help to fill gaps in observational data, especially over the oceans. For example, examine the clarity with which the clouds outline the fronts of the wave cyclone shown in Figure 9–13. These wishbone-shaped swirls are easily traced by satellites as they migrate across even the most remote portions of our planet. Developing hurricanes are frequently spotted long before they are detected by our surface observational network.

Weather satellites are equipped to generate several types of images simultaneously, including visible, infrared, and water vapor images. *Visible images*, like the one shown in Figure 12–10, are views of the Earth the way an astronaut would see our planet from space. The primary difference is that the satellite images are usually black and white. Visible images are produced by measuring the intensity of light reflected from cloud tops and other surfaces. By contrast, *infrared images* are obtained from radiation emitted (rather than reflected) from the same objects.

Infrared images have proven very useful in determining regions of possible precipitation within a middle-latitude cyclone. Compare the visible image in Figure 12–10 with the infrared image in Figure 12–11. Note that in the visible image all the clouds exhibit the same intensity of white and appear very similar (Figure 12–10).

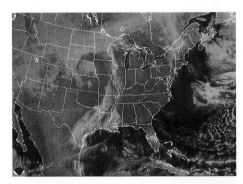

Figure 12–10 GOES 8 satellite view of cloud distribution about midday on March 8, 2000. This image records visible light, much as your eyes would see it. Compare this with the infrared image in Figure 12–11. *(Courtesy of NOAA)*

Figure 12–11 GOES 8 infrared image of the same cloud pattern shown in Figure 12–10. On this image, some of the clouds appear much whiter than others. These are the thicker, vertically developed clouds that have cold tops. A band of rain clouds can be easily seen curling from western Illinois into the Dakotas. *(Courtesy of NOAA)*

Figure 12–12 Water vapor image from the GOES 10 satellite for the morning of March 7, 2000. The greater the intensity of white, the greater the atmosphere's water vapor content. Black areas are driest. *(Courtesy of NOAA)*

In contrast, the infrared image provides a way to determine which clouds are the most probable precipitation producers (Figure 12–11). Here, warm objects appear darker and cold objects appear lighter. Because high cloud tops are colder than low cloud tops, towering cumulonimbus clouds that may generate heavy precipitation appear very white. By contrast, the lower, thinner nimbostratus clouds that produce only medium-to-light precipitation appear much darker. Thus, infrared satellite images are valuable forecasting tools.

Water vapor images provide yet another way to view our planet. Most of Earth's radiation with a wavelength of 6.7 micrometers is emitted by water vapor. Therefore, satellites equipped with detectors for this narrow band of radiation are, in effect, mapping the concentration of water vapor in the atmosphere. Bright regions in Figure 12–12 translate into regions of high water vapor concentration whereas dark areas are covered by drier air. In addition, because most fronts occur between air masses having contrasting moisture contents, water vapor images are valuable tools for locating frontal boundaries.

The third generation of weather satellites, known as Geostationary Operational Environmental Satellites (GOES), provides visible, infrared, and water vapor images for North America every half hour (Figure 12–13). (These are the same satellite images that often appear on The Weather Channel.) Such frequent observations allow meteorologists to track the movement of large weather systems that cannot be adequately followed by weather radar or polar-orbiting satellites. GOES has also been very important in monitoring the development and movement of tropical storms and hurricanes (Figure 12–14).

Measurement by Satellite

Although weather satellites are still more weather observers than weather forecasters, ingenious developments have made them more than simply TV cameras pointed at Earth from space. For example, geostationary satellites are presently used to estimate wind speeds from cloud movements. Satellites are also being equipped with instruments designed to measure temperatures at various elevations. Future satellites will be able directly or indirectly to determine wind speeds, humidity, and temperatures at various altitudes.

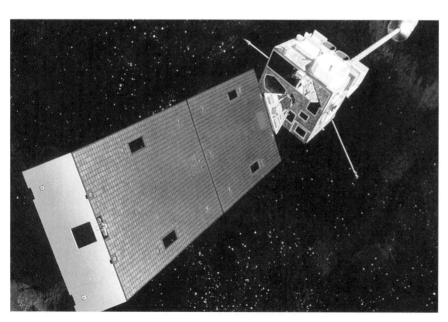

Figure 12–13 Artist's rendering of the Geostationary Operational Environmental Satellite (GOES). From a position almost 36,000 kilometers in space over the equator, GOES can monitor a substantial portion of the Western Hemisphere with good resolution, and obtain and transmit data from any point on Earth within its view. *(Courtesy of NOAA)*

Figure 12–14 Weather satellites are invaluable tools for tracking storms and gathering atmospheric data. This enhanced infrared image shows Hurricane Andrew as the storm crossed the Gulf of Mexico on August 25, 1992. *(Photo by Photri/The Stock Market)*

How can a satellite be used to measure temperatures at various heights? The technique relies on the fact that different wavelengths of energy are absorbed and radiated by different layers of the atmosphere. On a clear night, for example, the atmosphere does not absorb radiation from Earth's surface that has a wavelength of 10 micrometers (recall that this is the atmospheric window). Thus, by measuring the radiation at 10 micrometers, we have a method of determining Earth's surface temperature. Because other wavelengths indicate the altitudes within the atmosphere from which they originate, they, too, can be used to measure temperatures. This method of obtaining temperature readings aloft is simpler in principle than in practice, but with refinement it can supplement the limited supply of conventional radiosonde data.

Weather Forecasting and Upper-Level Flow

In Chapter 9, we demonstrated the strong correlation between cyclonic disturbances at the surface and the wavy flow in the westerlies aloft. Once meteorologists determine the air flow aloft for some specified time in the future, the data can be used to predict changes in surface pressure systems.

Although much is still unknown about the wavy flow of the westerlies, its most basic features are understood with some certainty. An obvious feature is seasonal change. The seasonal fluctuations in wind speeds are a consequence of the seasonal changes of the temperature gradient. The steep temperature gradient across the middle latitudes in winter corresponds to stronger flow aloft. In the cool season, the change in wind speed is reflected on upper-air charts by more closely spaced contour lines.

The position of the midlatitude jet stream also fluctuates seasonally. Its mean position migrates southward with the approach of winter and northward as summer nears. By midwinter, the jet core may penetrate as far south as central Florida. Because the paths of cyclonic systems shift with the flow aloft, we can expect the southern tier of states to encounter most of their stormy weather in the winter. During hot summer months, the storm track runs across the northern states, and some cyclones never leave Canada. The northerly storm track associated with summer also applies to Pacific storms, which move toward Alaska during the warm months, thus producing a rather long dry season for much of our West Coast. The number of cyclones generated is also seasonal, with the largest number occurring in the cooler months when temperature gradients are greatest.

In the cool season, the westerly flow goes through an irregular cyclic change. There may be periods of a week or more when the flow is nearly west to east, as shown in Figure 12–15a. These conditions bring relatively mild temperatures, and few disturbances are experienced south of the jet stream. Then, without warning, the upper flow begins to meander wildly and produces large-amplitude waves and a general north-to-south flow

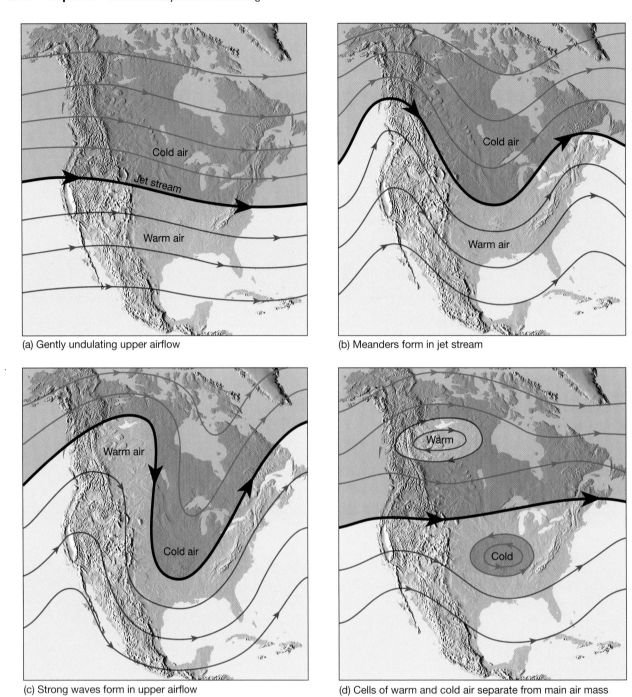

(a) Gently undulating upper airflow

(b) Meanders form in jet stream

(c) Strong waves form in upper airflow

(d) Cells of warm and cold air separate from main air mass

Figure 12–15 Cyclic changes that occur in the upper-level airflow of the westerlies. The flow, which has the jet stream as its axis, starts out nearly straight and then develops meanders that are eventually cut off. *(After J. Namais, NOAA)*

(Figure 12–15b). This change allows cold air to move southward, intensifying the temperature gradient and the flow aloft (Figure 12–15c).

During these periods, cyclonic activity dominates the weather. For a week or more, cyclonic storms redistribute large quantities of heat across the midlatitudes by moving cold air southward and warm air northward (Figure 12–15d). This redistribution eventually results in a weakened temperature gradient and a return to a flatter flow aloft and less intense weather at the surface. These cycles of alternating calm and stormy weather can last from one to six weeks.

Box 12–6 What Is "Normal"?

When we watch a weather report on TV, the person making the presentation usually lists statistics for the day. Frequently, after stating the high, low, and average temperatures, the reporter indicates whether the daily mean is above or below normal. Similarly, following a report that includes the monthly or annual precipitation total, we will be told how much of a *departure from normal* the figure represents. Sometimes the weathercaster even follows up such information with a remark informing us that the daily mean temperature or the monthly rainfall total is "much" above or below normal.

Does this mean that we are experiencing abnormal conditions? Certainly, many people would assume that the statistics are unusual or even extraordinary, because most of us perceive "normal" as implying ordinary or frequent. Such a perception arises from the common usage of the word *normal*: "Conforming, adhering to, or constituting a typical or usual standard, pattern, level or type."°

If normal does not imply ordinary or frequent, just what does it mean when applied to meteorological observations? *Normal is simply an average of a climate element over a 30-year period.* It is a standard that is used in making comparisons. A departure from normal is the difference between currently observed values and the 30-year average. The normal is usually not the most frequent value (called the *mode*) nor the value above which half the cases fall (called the *median*). It is safe to say that no matter what meteorological statistic is considered, experiencing a value that equals the normal is actually the exception and not the rule (Figure 12–H).

The World Meteorological Organization has established a standard definition for the 30-year span used to compute normals. Normals are recalculated each decade in an attempt to keep up with any climatic changes that might take place. For example, on January 1, 2001, the period for computing normals changes. Prior to that date, normals represent the period January 1, 1961, through December 31, 1990. Beginning January 1, 2001, data for the 1990s replaces the observations from the 1960s.

°From *Webster's II New Riverside University Dictionary* (Boston: Houghton Mifflin Company, 1984), p. 803.

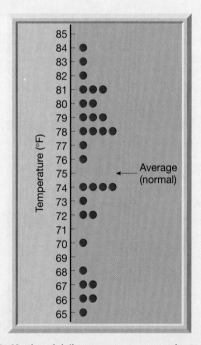

Figure 12–H Plot of daily mean temperatures for Peoria, Illinois, August 1993.

The Winter of 1977

Let us consider an example of the influence of the flow aloft on the weather over an extended period. As our example, we will examine an atypical winter (see Box 12–6). In a normal January, an upper-air ridge is situated over the Rocky Mountains and a trough extends across the eastern two-thirds of the United States (Figure 12–16). This "typical" flow pattern is believed to be influenced by the mountains. During January 1977, the normal flow pattern was greatly accentuated, as illustrated in Figure 12–16. The greater amplitude of the upper-level flow caused an almost continuous influx of cold air into the Deep South, producing record low temperatures throughout much of the eastern and central United States (Figure 12–17). Because of dwindling natural gas supplies, many industries experienced layoffs. Much of Ohio was hit so hard that four-day work weeks and massive shutdowns were ordered.

While most of the East was in a deep freeze, the westernmost states were under a strong ridge of high pressure. Generally mild temperatures and clear skies

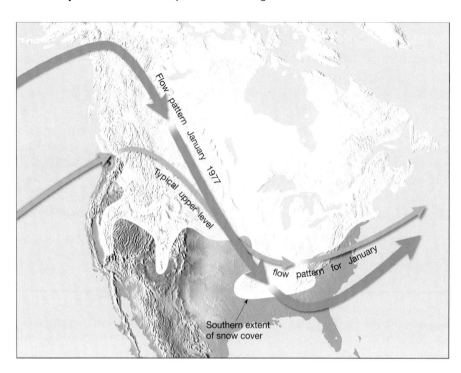

Figure 12–16 The unusually high amplitude experienced in the flow pattern of the prevailing westerlies during the winter of 1977 brought warmth to Alaska, drought to the west, and frigid temperatures to the central and eastern United States.

dominated their weather. This was no blessing, however, because the ridge of high pressure blocked the movement of Pacific storms that usually provide much needed winter precipitation. The shortage of moisture was especially serious in California, where January is the middle of its three-month rainy season.

Throughout most of the western states, the winter rain and snow that supply water for summer irrigation was far below normal. This dilemma was compounded by the fact that the previous year's precipitation had also been far below normal and many reservoirs were almost empty. Although much of the country was concerned about economic disaster caused either by a lack of moisture or frigid temperatures, the highly accentuated flow pattern channeled unseasonably warm air into Alaska. Even Fairbanks, which generally experiences

Figure 12–17 Arctic air invades the eastern United States.
(Photo by Seth Resnick, Stock Boston)

temperatures as low as –40°C (–40°F), had a mild January with numerous days above freezing.°

°For an excellent review of the winter weather of 1976–1977, see Thomas Y. Canby, "The Year the Weather Went Wild," *National Geographic*, 152, no. 6 (1977), 798–892.

Because the flow aloft fluctuates in a somewhat unpredictable manner, long-range weather prediction remains beyond the forecaster's reach. Nonetheless, attempts are being made to predict more accurately long-term changes in the upper-level flow. It is hoped that such research will allow meteorologists to answer questions such as: Will next winter be colder than recent winters? Will California experience a drought next year?

Chapter Summary

- In the United States, the government agency responsible for gathering and disseminating weather related information is the *National Weather Service* (NWS). Perhaps the most important services provided by the NWS are forecasts and warnings of hazardous weather including thunderstorms, flooding, hurricanes, tornadoes, winter weather, and extreme heat.

- The process of providing weather forecasts and warnings throughout the United States occurs in three stages. First, data are collected and analyzed on a global scale. Second, a variety of techniques are used to establish the future state of the atmosphere; a process called weather *forecasting*. Finally, forecasts are disseminated to the public, mainly through the private sector.

- Assessing current atmospheric conditions, called *weather analysis*, involves collecting, transmitting, and compiling millions of pieces of observational data. On a global scale, the *World Meteorological Organization* is responsible for gathering, plotting, and distributing weather data. Once collected, the information is distributed to three *World Meteorological Centers* near Washington, D.C., Moscow, and Melbourne, Australia.

- Initially, weather data are displayed on a synoptic weather map. A weather map shows the status of the atmosphere and includes data on temperature, humidity, pressure, and airflow. In addition to surface maps, twice-daily upper air charts depicting the pressure field are drawn at 850, 700, 500, 300, and 200-millibar levels.

- The approaches used in modern weather forecasting include traditional *synoptic weather forecasting*, *numerical weather prediction*, *statistical methods*, and various *short-range forecasting techniques*. Synoptic weather forecasting, the primary method for making weather predictions until the late 1950s, involves the analysis of synoptic weather charts, employing several empirical rules. Numerical weather prediction, used extensively in modern weather forecasting, is based

on the fact that the gases of the atmosphere obey many known physical principles. Ideally, these physical laws can be used to predict the future state of the atmosphere, when the current conditions are known. Numerical weather prediction uses a number of highly refined computer models that attempt to mimic the behavior of the atmosphere. Statistical methods, using past weather data to predict future events, are often used in conjunction with numerical weather predictions. One statistical approach, the *analog method*, examines past weather records to find situations that come close to duplicating current conditions. The simplest short-range forecasting techniques, called *persistence forecasts*, assume future weather will be the same as the present conditions. Another technique, often called *nowcasting*, uses radar and geostationary satellites to quickly forecast severe weather events, such as thunderstorms, tornadoes, hail storms, and microbursts.

- *Long range weather forecasting* is an area that relies heavily on statistical averages obtained from past weather events, also referred to as *climatic data*. Weekly, monthly, and seasonal weather outlooks prepared by the National Weather Service are not weather forecasts in the usual sense. Rather, they indicate only whether the region will experience near-normal precipitation and temperatures or not.

- Weather forecasting relies on the skill of the forecaster. Very short-range (0 to 12 hours) forecasts have demonstrated considerable skill, especially for predicting the formation and movement of large weather systems. Short-range forecasts (12 to 72 hours) of maximum and minimum temperatures and wind speeds are quite accurate. Furthermore, predicting precipitation amounts is much better than forecasts made only two decades ago. Medium-range forecasts (3–7 days into the future) have shown significant improvement in the last 20 years. However, the accuracy of day-to-day weather forecasts for periods beyond 7 days is relatively unreliable.

- Many technical advances have been made to improve forecast accuracy. *Automated Surface Observing Systems* (ASOS) are now being used in places currently outside the observational network. Interactive microcomputer systems make it possible for forecasters to display, manipulate, and rapidly digest the great quantity and variety of available data. Advanced Doppler radar networks aid the detection and tracking of small-scale weather phenomena, such as tornadoes and thunderstorms.

- Weather forecasting relies heavily on information provided by both *polar* and *geostationary weather satellites*. Their primary importance is to help to fill gaps in observational data, especially over the oceans. Weather satellites can generate several types of images, including visible, infrared, and water vapor images. Currently, infrared images (images obtained from radiation emitted rather than reflected by an object) help determine regions of possible precipitation. Future satellites will be able to detect wind speeds, humidity, and temperatures at various heights.

- For many years meteorologists have been aware of a strong correlation between cyclonic disturbances at the surface and the seasonal fluctuations in the wavy flow of the westerlies aloft. Frequently, when upper-air flow produces large-amplitude waves and a general north-to-south flow, cold air moves southward and cyclonic activity dominates the weather. By contrast, when the flow is nearly west-to-east, mild temperatures and few cyclonic disturbances are experienced south of the jet stream. Although the effects of upper-level flow on weather are well documented, the somewhat unpredictable manner of the flow aloft keeps long range weather forecasts unreliable.

Vocabulary Review

analog method (p. 329)
Automated Surface Observing System (ASOS) (p. 335)
geostationary satellite (p. 336)
long-range forecasting (p. 330)
National Weather Service (NWS), (p. 320)
nowcasting (p. 329)
numerical weather prediction (p. 328)
persistence forecasts (p. 329)
polar satellite (p. 336)

prognostic charts (p. 328)
skill (p. 332)
statistical methods (p. 329)
synoptic weather maps (p. 323)
synoptic weather forecasting (p. 324)
trend forecasting (p. 329)
weather analysis (p. 321)
weather forecasting (p. 324)
World Meteorological Organization (p. 322)

Review Questions

1. What is the mission of the National Weather Service?

2. List the three steps involved in providing weather forecasts.

3. What is meant by *weather analysis*?

4. Compare the tasks of a weather analyst with those of a weather forecaster.

5. What information is provided by a surface weather map? (See Box 12–2.)

6. Briefly describe what is involved in synoptic weather forecasting.

7. Why is the name "numerical weather forecasting" somewhat misleading?

8. Briefly describe the basis of numerical weather predictions.

9. How are prognostic charts different from synoptic weather maps?

10. What do computer-generated numerical models try to predict?

11. Describe the statistical approach called the "analog method." What are its drawbacks?

12. If it is snowing today, what could be predicted for tomorrow if persistence forecasting were employed?

13. What type of weather is typically forecast using nowcasting techniques?

14. What elements are predicted in a long-range (monthly) weather chart?

15. Give an example of why the percent of correct forecasts is not always a good measure of forecasting skill.

16. How do satellites help identify clouds that are most likely to produce precipitation?

17. What information is provided by water vapor images?

18. What advantage do geostationary satellites have over polar satellites? Name one disadvantage.

Problems

1. The map in Figure 12–18 has several weather stations plotted on it. Using the weather data for a typical day in March, which are given in Table 12–1, complete the following:

   a. On a copy of Figure 12–18, plot the temperature, wind direction, pressure, and sky coverage by using the international symbols given in Appendix B.

   b. Using Figure 12–D as a guide, complete this weather map by adding isobars at 4-millibar intervals, the cold front and the warm front, and the symbol for low pressure.

   c. Apply your knowledge of the weather associated with a middle-latitude cyclone in the spring of the year, and describe the weather conditions at each of the following locations:

   (i) Philadelphia, Pennsylvania
   (ii) Quebec, Canada
   (iii) Toronto, Canada
   (iv) Sioux City, Iowa

2. Many TV weather reports include a seven-day outlook. Tune in such a report and jot down the forecast for the last (seventh) day. Then, each day thereafter, write down the forecast for the day in question. Finally, record what actually occurred on that day. Contrast the forecast seven days ahead with what actually took place. How accurate (or inaccurate) was the seven-day forecast for the day you selected? How accurate was the five-day forecast for this day? The two-day forecast?

3. Using Box 12–4 as a guide, calculate the precipitation probability for the following situations:

   a. There is a 50 percent chance that a storm will move into the forecast area and that it will affect 20 percent of the area.

Figure 12–18 Map to accompany problem number 1.

   b. There is a 90 percent chance that a storm will move into the forecast area and that it will affect 100 percent of the area.

   c. There is a 50 percent chance that a storm will move into the forecast area and that it will affect 50 percent of the area.

Table 12–1 Weather data for a typical March day

Location	Temperature (°F)	Pressure (MB)	Wind Direction	Sky Cover (Tenths)
Wilmington, N.C.	57	1009	SW	7
Philadelphia, Pa.	59	1001	S	10
Hartford, Conn.	47	1001	SE	Sky obscured
International Falls, Minn.	–12	1008	NE	0
Pittsburgh, Pa.	52	995	WSW	10
Duluth, Minn.	–1	1006	N	0
Sioux City, Iowa	11	1010	NW	0
Springfield, Mo.	35	1011	WNW	2
Chicago, Ill.	34	985	NW	10
Madison, Wis.	23	995	NW	10
Nashville, Tenn.	40	1008	SW	10
Louisville, Ky.	40	1002	SW	5
Indianapolis, Ind.	35	994	W	10
Atlanta, Ga.	49	1010	SW	7
Huntington, W.Va.	52	998	SW	6
Toronto, Canada	44	985	E	9
Albany, N.Y.	50	998	SE	7
Savanna, Ga.	63	1012	SW	10
Jacksonville, Fla.	66	1013	WSW	10
Norfolk, Va.	67	1005	S	10
Cleveland, Ohio	49	988	SW	4
Little Rock, Ark.	37	1014	WSW	0
Cincinnati, Ohio	41	997	WSW	10
Detroit, Mich.	44	984	SW	10
Montreal, Canada	42	993	E	10
Quebec, Canada	34	999	NE	Sky obscured

Atmospheric Science Online

The following are informative and interesting Internet sites that address topics related to those presented in the chapter:

National Weather Service:
- **http://www.nws.noaa.gov/**

Weather Channel, The:
- **http://www.weather.com/homepage.html**

World Meteorological Organization:
- **http://www.wmo.ch/**

For direct links to these sites and others, chapter objectives and reviews, quiz questions, and topical investigations that utilize Web resources, visit *The Atmosphere, Eighth Edition* Home Page at:

- **http://www.prenhall.com/lutgens**

CHAPTER THIRTEEN

Air Pollution

Pudong, China on a moderately smoggy day. *(Photo by F. Hoffman/The Image Works)*

ir pollution and meteorology are linked in two ways. One concerns the influence that weather conditions have on the dilution and dispersal of air pollutants. The second connection is the reverse and deals with the effect that air pollution has on weather and climate. The first of these associations is examined in this chapter. The second and equally important relationship is discussed in Chapter 14 and is the focus of several sections and special-interest boxes.°

In Chapter 4, the concept of atmospheric stability was introduced. You learned that the stability of air plays a significant role in controlling many aspects of daily weather. In this chapter, you will see that atmospheric stability is closely related to urban air pollution. Air quality is not just a function of the quantity and types of pollutants emitted into the air; it is also closely linked to the atmosphere's ability to disperse these noxious substances. Dispersal, in turn, is related to the stability of the atmosphere.

Air pollution is a continuing threat to our health and welfare. An average adult male requires about 13.5 kilograms (30 pounds) of air each day compared with about 1.2 kilograms (2.6 pounds) of food and 2 kilograms (4.4 pounds) of water. The cleanliness of air, therefore, should certainly be as important to us as the cleanliness of our food and water.

A Brief Historical Perspective

Air is never perfectly clean. Many natural sources of air pollution have always existed (Figure 13–1). Ash from volcanic eruptions, salt particles from breaking waves,

°See Box 3–4 "How Cities Influence Temperature" and Box 13–1 "Air Pollution Changes the Climate of Cities." In addition, see the section on "Inadvertent Weather Modification: Urban-Induced Precipitation," in Chapter 5 and the discussion of "Country Breeze" in Chapter 7.

pollen and spores released by plants, smoke from forest and brush fires, and windblown dust are all examples of "natural air pollution" (Figure 13–2). Ever since people have been on Earth, however, they have added to the frequency and intensity of some of these natural pollutants, especially the last two. For example, the dust storm in Figure 13–3 occurred when strong winds raised dry soil from plowed farm fields.

With the discovery of fire came an increased number of accidental as well as intentional burnings. Even today, in many parts of the world, fire is used to clear land for agricultural purposes (the so-called slash-and-burn method), filling the air with smoke and reducing visibility. When people clear the land of its natural vegetative cover for whatever purpose, soil is exposed and blown into the air. Yet when we consider the air in a modern-day industrial city, these human-accentuated forms of pollution, although significant, may seem minor by comparison.

Air Pollution: Not a New Problem

Although some types of air pollution are relatively recent creations, others have been around for centuries. Smoke pollution, for example, plagued London for centuries. Because of the odor and smoke produced by the burning of coal, King Edward I made the following proclamation in 1300: "Be it known to all within the sound of my voice, whosoever shall be found guilty of burning coal shall suffer the loss of his head." Unfortunately, one Londoner did not heed the king's warning and paid the extreme price for his misdeed. As far as is known, however, this is the only case of capital punishment resulting from an air pollution violation!

The ban on burning coal led to the use of an alternative fuel—wood. The extensive wood burning soon dramatically reduced English forests, and coal consumption

Figure 13–1 Forest fires triggered by lightning are one of many natural sources of air pollution. *(Photo by Kent and Donna Dennen/Photo Researchers, Inc.)*

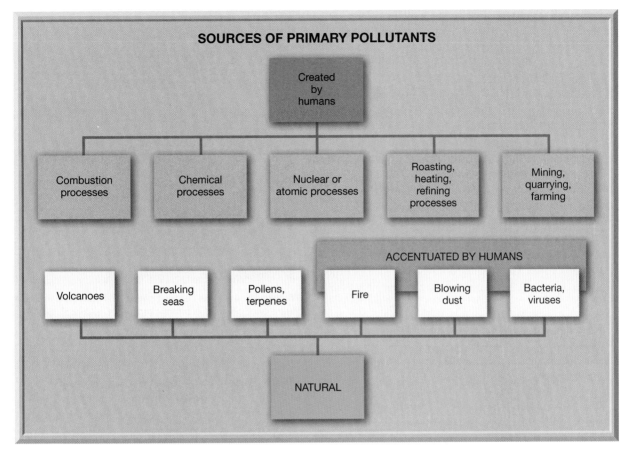

SOURCES OF PRIMARY POLLUTANTS

Created by humans

Combustion processes

Chemical processes

Nuclear or atomic processes

Roasting, heating, refining processes

Mining, quarrying, farming

ACCENTUATED BY HUMANS

Volcanoes

Breaking seas

Pollens, terpenes

Fire

Blowing dust

Bacteria, viruses

NATURAL

Figure 13–2 Sources of primary pollution. *(After Reid A. Bryson and John E. Kutzbach,* Air Pollution, *Commission on College Geography Resource Paper No. 6, Figure 2, p. 8.)*

again increased despite royal disapproval. Thus, in 1661, when John Evelyn wrote *Fumifugium, or the Inconvenience of Aer and Smoak of London Dissipated, together with some Remidies Humbly Proposed*, the problem of foul air still plagued Londoners. In his book, Evelyn noted that a traveler, although many miles from London,

Figure 13–3 An example of natural air pollution that has been accentuated by human activities. This dust storm near Elkhart, Kansas, in May 1937 occurred because the soil was plowed and vulnerable to strong winds. It was because of storms like this that portions of the Great Plains were called the "Dust Bowl" in the 1930s. *(Photo reproduced from the collection of the Library of Congress)*

"sooner smells than sees the city to which he repairs." In fact, London continued to have severe air pollution problems well into the twentieth century. It was only after a devastating smog disaster in 1952 that truly decisive action was taken to clean the air.

London, however, has not monopolized the air pollution scene. With the coming of the Industrial Revolution, many cities began to experience "big-time" air pollution. Instead of just simply accelerating natural sources, people found many new ways to pollute the air (Figure 13–2) and many new things with which to pollute it. In the mid- to late nineteenth century, the population of many American and European cities swelled as people sought work in the growing numbers of new foundries and steel mills. As a result, the urban environment became increasingly fouled by the fumes of industry. In *Hard Times*, Charles Dickens vividly describes the scene in a late nineteenth-century factory town:

> It was a town of machinery and tall chimneys out of which interminable serpents of smoke trailed themselves forever and ever, and never got uncoiled. It had a black canal in it, and a river that ran purple with ill-smelling dye.

It is clear that poor air quality was not the only environmental pollution that plagued these places! However, it should be noted that the rapid rise in urban air pollution was not necessarily viewed with great alarm. Rather, chimneys belching forth smoke and soot were a symbol of growth and prosperity (Figure 13–4). The following quotation from an 1880 speech by the well-known lawyer and orator Robert Ingersoll, for example, is reported to have elicited great cheering and cries of "Good! Good!" from the audience: "I want the sky to be filled with the smoke of American industry and upon that cloud of smoke will rest forever the bow of perpetual promise. That is what I am for." With the rapid growth of the world's population and accelerated industrialization, the quantities of atmospheric pollutants increased drastically.

Some Historic Episodes

The first major air pollution disaster to be studied in depth occurred in the Meuse valley of Belgium. Here for five days in December 1930, a blanket of smog hung in the valley, killing 63 people and causing 6000 to become ill. Since the 1930s, many air pollution episodes have demonstrated the devastating effect that dirty air can have on life and property. In October 1948, Donora, Pennsylvania, had such an experience. The grime that settled from the air coated houses, streets, and sidewalks, so that pedestrians and autos actually left distinct

Figure 13–4 Stacks belching smoke and soot such as these were once a sign of economic prosperity. *(Marc St. Gil EPA-Documerica)*

footprints and tire tracks. Almost 6000 of the town's 14,000 inhabitants became ill, and 20 died.

One of the most tragic air pollution episodes ever occurred in London in December 1952. More than 4000 people died as a result of this five-day ordeal. The people who suffered most were those with respiratory and heart problems, primarily the elderly. Extreme air pollution darkened London again in 1953 and 1962 and affected New York City in 1953, 1963, and 1966. Since these events, the passage of legislation, the development of regulations and standards, and the advancement of control technology have reduced the frequency and severity of such episodes. Nevertheless, health authorities are equally concerned about the slow and subtle effects on our lungs and other organs of air pollution levels that are much lower but that are present every day year after year.

Sources and Types of Air Pollution

Air pollutants are airborne particles and gases that occur in concentrations that endanger the health and well-being of organisms or disrupt the orderly functioning of the environment. Pollutants can be grouped into two categories: primary and secondary. **Primary pollutants** are emitted directly from identifiable sources (Table 13–1). They pollute the air immediately upon being emitted. **Secondary pollutants**, in contrast, are produced in the atmosphere when certain chemical reactions take place among primary

Table 13–1 Air Pollution Source Categories

Category	Comments
Point sources	Includes factories and electric power plants
Mobil sources	Not only includes cars and trucks but also lawn mowers, airplanes, and anything else that moves and pollutes the air
Biogenic sources	This category includes all nonanthropogenic (not human-generated) sources. Examples include trees and other vegetation, microbial activity, oil and gas seeps, etc.
Area sources	Small and individual sources such as dry cleaners and degreasing operations.

pollutants. The chemicals that make up smog are important examples. In some cases, the impact of primary pollutants on human health and the environment is less severe than the effects of the secondary pollutants they form.

Primary Pollutants

Figure 13–5 depicts the major primary pollutants and the sources that produce them. When the various sources are examined, the significance of the transportation category is obvious. It accounts for nearly half of our pollution (by weight). In addition to highway vehicles, this category includes trains, ships, and airplanes. Still, the tens of millions of cars and trucks on U.S. roads are, without a doubt, the greatest contributors in this category (Figure 13–6). What follows is a brief survey and description of the major primary pollutants.

Particulate Matter. *Particulate matter* (PM) is the general term used for a mixture of solid particles and liquid droplets found in the air. Some particles are large or dark enough to be seen as soot or smoke. Others are so small they can be detected only with an electron microscope. These particles, which come in a wide range of sizes (*fine*

particles are less than 2.5 micrometers in diameter, and coarser-size particles are larger than 2.5 micrometers), originate from many different stationary and mobile sources as well as from natural sources. Fine particles ($PM_{2.5}$) result from fuel combustion from motor vehicles, power generation, and industrial facilities, as well as from residential fireplaces and wood stoves. Coarse particles (PM_{10}) are generally emitted from sources such as vehicles traveling on unpaved roads, materials handling, and crushing and grinding operations, as well as windblown dust. Some particles are emitted directly from their sources, such as smokestacks and cars. In other cases, gases such as sulfur dioxide interact with other compounds in the air to form fine particles.

Particulates are frequently the most obvious form of air pollution because they reduce visibility and leave deposits of dirt on the surfaces with which they come in contact (see Box 13–1). In addition, particulates may carry any or all of the other pollutants dissolved in or absorbed on their surfaces.

Originally, total suspended particulate (TSP) was the indicator used to represent this category. It included all particles up to 45 micrometers in diameter. In 1987, the

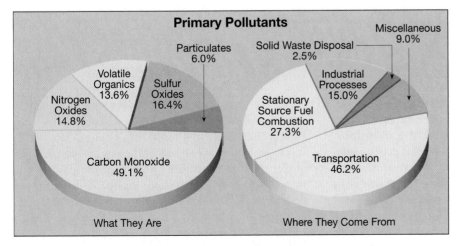

Primary Pollutants

What They Are
- Nitrogen Oxides 14.8%
- Volatile Organics 13.6%
- Sulfur Oxides 16.4%
- Particulates 6.0%
- Carbon Monoxide 49.1%

Where They Come From
- Solid Waste Disposal 2.5%
- Miscellaneous 9.0%
- Industrial Processes 15.0%
- Stationary Source Fuel Combustion 27.3%
- Transportation 46.2%

Figure 13–5 Major primary pollutants and their sources. Percentages are calculated on the basis of weight. *(Data from the U.S. Environmental Protection Agency)*

Figure 13–6 This crowded freeway in San Diego, California, reminds us that the transportation category is a major contributor to air pollution. Emissions from an individual vehicle are generally low, relative to the smokestack image many people associate with air pollution. But in numerous cities across the country, the personal automobile is the single greatest polluter, when emissions from millions of vehicles are added together. Driving a private car is probably a typical person's most polluting daily activity. *(Photo by Jerry Schad/Photo Researchers, Inc.)*

U.S. Environmental Protection Agency (EPA) set new standards that related only to particles smaller than 10 micrometers (identified as PM_{10}). Ten years later, in 1997, the EPA revised its standards for particulate matter again so that in the future they will be based on $PM_{2.5}$. This change was in response to a large amount of research that analyzed the health effects of particulates.

Inhalable particulate matter includes both fine and coarse particles. These particles can accumulate in the respiratory system and are associated with numerous health effects. Exposure to coarse particles is primarily associated with the aggravation of respiratory conditions, such as asthma. Fine particles are most closely associated with such health effects as increased hospital admissions and emergency room visits for heart and lung disease, increased respiratory symptoms and disease, decreased lung function, and even premature death.

Sensitive groups that appear to be at greatest risk to such effects include the elderly, individuals with cardiopulmonary disease, such as asthma, and children. In addition to health problems, particulate matter is the major cause of reduced visibility in many parts of the United States. Airborne particles also can cause damage to paints and building materials.

Sulfur Dioxide. *Sulfur dioxide* (SO_2) is a colorless and corrosive gas that originates largely from the combustion of sulfur-containing fuels, primarily coal and oil. Important sources include power plants, smelters, petroleum refineries, and pulp and paper mills. Once SO_2 is in the air, it is frequently transformed into sulfur trioxide (SO_3), which reacts with water vapor or water droplets to form sulfuric acid (H_2SO_4). Very tiny particles act as a medium on which the acidic sulfate ion (SO_4^{-2}) is carried over long distances in the atmosphere. When it is "washed out" of the air or deposited on surfaces, it contributes to a serious environmental problem known as *acid precipitation*. This issue is the subject of a later section.

High concentrations of SO_2 can result in temporary breathing impairment for asthmatic children and adults who are active outdoors. Short-term exposures of asthmatic individuals to elevated SO_2 levels while at moderate exertion may result in reduced lung function that may be accompanied by such symptoms as wheezing, chest tightness, or shortness of breath. Other effects that have been associated with longer-term exposures to high concentrations of SO_2, in conjunction with high levels of particulate matter, include respiratory illness and aggravation of existing cardiovascular disease.

Nitrogen Oxides. *Nitrogen oxides* are gases that form during the high-temperature combustion of fuel when nitrogen in the fuel or the air reacts with oxygen. Power plants and motor vehicles are the primary sources. These gases also form naturally when certain bacteria oxidize nitrogen-containing compounds.

The initial product formed is nitric oxide (NO). When NO oxidizes further in the atmosphere, nitrogen dioxide (NO_2) forms. Commonly, the general term NO_x is used to describe these gases. Although NO_x forms naturally, its concentration in cities is 10 to 100 times higher than in rural areas. Nitrogen dioxide has a distinctive reddish, brown color that frequently tints polluted city air and reduces visibility. When concentrations are high, NO_2 can also contribute to lung and heart problems. When air is humid, NO_2 reacts with water vapor to form nitric acid (HNO_3). Like sulfuric acid, this corrosive substance also contributes to the acid-rain problem. Moreover, because nitrogen oxides are highly reactive gases, they play an important part in the formation of smog.

Box 13–1 Air Pollution Changes the Climate of Cities

In Box 3–4 you saw that air pollution in cities contributes to the heat island by inhibiting the loss of longwave radiation at night. In Chapter 5 you learned that pollutants may have a "cloud seeding" effect that increases precipitation in and downward of cities. These influences, however, are not the only ways in which pollutants influence urban climate.

The blanket of particulates over most large cities significantly reduces the amount of solar radiation reaching the surface. In some cities, the overall reduction in the receipt of solar energy is 15 percent or more, whereas short wavelength ultraviolet is decreased by up to 30 percent. This weakening of incoming solar energy is variable. During air pollution episodes, the decrease is much greater than for periods when air quality is good (Figure 13–A). Furthermore, particulates are most effective in reducing solar radiation near the ground when the Sun angle is low. This occurs because the length of the path through the polluted air increases as the Sun angle drops. Thus, for a given quantity of particulate matter, solar energy will be reduced by the largest percentage at high-latitude cities and during the winter.

When compared to surrounding rural areas, the relative humidity in cities is generally from 2 to 8 percent lower. One reason is that cities are hotter. Remember from Chapter 4 that as air temperature increases, capacity also rises and relative humidity drops. A second reason is that less water vapor is supplied to city air by evaporation from the surface. Evaporation is reduced in cities because rain water rapidly runs off, frequently into subsurface storm sewers.

Although relative humidity tends to be lower in cities, the occurrences of clouds and fogs are greater. What is the cause of this apparent paradox? It is likely that the large quantities of condensation nuclei produced by human activities in urban areas led to this greater incidence of clouds and fogs. When hygroscopic (water-seeking) nuclei are plentiful, water vapor readily condenses on them, even when the air is not quite saturated.

Figure 13–A An air pollution episode at Shenyang, China. It is not difficult to understand why the amount of solar radiation reaching the surface is reduced in cities. *(Photo by F. Hoffmann/The Image Works)*

Volatile Organic Compounds. *Volatile organic compounds* (VOC for short), also called *hydrocarbons*, encompass a wide array of solid, liquid, and gaseous substances that are composed exclusively of hydrogen and carbon. Large quantities occur naturally, with methane (CH_4) being the most abundant. Methane, however, does not interact chemically with other substances and has no negative health effects. In cities, the incomplete combustion of gasoline in motor vehicles is the principal source of reactive VOCs. Although some hydrocarbons from other sources are cancer-causing agents, most of the VOCs in city air do not by themselves appear to pose significant environmental problems. However, as we shall see in a later discussion, when VOCs react with certain other pollutants (especially nitrogen oxides), noxious secondary pollutants result.

Carbon Monoxide. *Carbon monoxide* (CO) is a colorless, odorless, and poisonous gas produced by incomplete burning of carbon in fuels. It is the most abundant primary pollutant, with about two-thirds of the nationwide emissions coming from transportation sources, mainly highway vehicles.

Although CO is quickly removed from the atmosphere, it can nevertheless be dangerous. Carbon monoxide enters the bloodstream through the lungs and reduces oxygen delivery to the body's organs and tissues. Because it cannot be seen, smelled, or tasted, CO can have an effect on people without their realizing it. In small amounts, it causes drowsiness, slows reflexes, and impairs judgment. If concentrations are sufficiently high, CO can cause death. Carbon monoxide poses a serious health hazard where concentrations can reach high levels as in poorly ventilated tunnels and underground parking facilities.

Lead. *Lead (Pb)* is very dangerous because it accumulates in the blood, bones, and soft tissues. It can impair the functioning of many organs. Even at low doses, lead exposure is associated with damage to the nervous systems of young children.

In the past, automotive sources were the major contributor of lead emissions to the atmosphere because lead was added to gasoline as a way to prevent engine knock. Ever since the EPA-mandated phaseout of leaded gasoline, lead concentrations in the air of U.S. cities have shown a dramatic decline (Table 13–2). Occasional violations of the lead air-quality standard still occur near large industrial sources such as lead smelters.

Secondary Pollutants

Recall that secondary pollutants are not emitted directly into the air, but form in the atmosphere when reactions take place among primary pollutants. The sulfuric acid described earlier is one example of a secondary pollutant. After the primary pollutant, sulfur dioxide, is emitted into the atmosphere, it combines with oxygen to produce sulfur trioxide, which then combines with water to create this irritating and corrosive acid.

Table 13–2 Air Quality and Emissions Trends, 1988–1997

Pollutant	Percent Decrease in Concentrations	Percent Decrease in Emissions
Carbon monoxide (CO)	38	25
Lead (Pb)	67	44
Nitrogen dioxide (NO_2)	14	1
Ozone (O_3)	16	*
Particulate matter (PM_{10})	26	12
Sulfur dioxide (SO_2)	39	12

*Ozone is not emitted directly into the atmosphere but rather is a secondary pollutant.

Air pollution in urban and industrial areas is often termed **smog**. The word was coined in 1905 by Harold A. Des Veaux, a London physician, and was created by combining the words "smoke" and "fog." Des Veaux's term was indeed an apt description of London's principal air pollution threat, which was associated with the products of coal burning coupled with periods of high humidity.

Today, however, smog is used as a synonym for general air pollution and does not necessarily imply the smoke–fog combination. Therefore, when greater clarity is desired, we sometimes find the word "smog" preceded by such modifiers as "London-type," "classical," "Los Angeles-type," or "photochemical." The first two modifiers refer to the original meaning of the word, and the last two to air quality problems created by secondary pollutants.

Many reactions that produce secondary pollutants are triggered by strong sunlight and so are called **photochemical reactions**. One common example occurs when nitrogen oxides absorb solar radiation, initiating a chain of complex reactions. When certain volatile organic compounds are present, the result is the formation of a number of undesirable secondary products that are very reactive, irritating, and toxic. Collectively, this noxious mixture of gases and particles is called *photochemical smog*. One of the substances it usually contains is called PAN (peroxyacetyl nitrate), which damages vegetation and irritates the eyes. The *major* component in photochemical smog is ozone. Recall from Chapter 1 that ozone is formed by natural processes in the stratosphere. However, when produced near Earth's surface, ozone is considered a pollutant (see Box 13–2).

The negative effects of ozone are well documented. Short-term exposure to elevated levels causes eye and lung irritation. Moreover, there is mounting evidence of chronic effects from longer term or recurring exposures to more moderate levels. Ozone also lowers crop yields, retards tree growth, and damages ornamental plants and

Box 13–2 Ozone: Good or Bad?

In any examination of atmospheric environmental issues, ozone is likely to be given a prominent place. Sometimes, however, there is confusion among the general public and media about the role of ozone in the atmosphere. The reason for this confusion is the fact that ozone is related to more than one environmental issue.

In Chapter 1 we learned that ozone occurs naturally in the region of the atmosphere known as the stratosphere. Here, 10 to 50 kilometers above Earth's surface, ozone absorbs ultraviolet radiation from the Sun. Because the wavelengths absorbed in the stratosphere are harmful to life at the surface, ozone functions as a protective shield. Without this shield, potentially lethal intensities of ultraviolet radiation would reach the surface. We also learned that ozone-destroying chemicals called chlorofluorocarbons (CFCs) threaten the layer of stratospheric ozone. Therefore, in this context, ozone is considered a desirable atmospheric component that is critical to life on Earth and should be preserved.

In contrast to the foregoing view of ozone as a beneficial atmospheric component, many people living in big cities view the presence of ozone in the atmosphere with alarm. Rather than wanting to "save the ozone," these individuals want to see ozone removed from the air. Are these urban dwellers referring to the *same* ozone that functions to shield life from damaging ultraviolet radiation? The answer is *no*.

Whereas ozone forms and exists naturally in the stratosphere, it is produced as a pollutant in the troposphere. Photochemical (Los Angeles-type) smog is a major environmental problem in many big cities, especially during the summer months. Ozone is the major component of this noxious mixture of gases and particles forming as a result of reactions among pollutants emitted by motor vehicles and industries. During air pollution episodes, ozone levels are frequently used as an index to the poor quality of urban air. The higher the ozone concentrations, the poorer the air quality. Therefore, in this context, ozone is an undesirable pollutant that threatens human health.

In conclusion, the answer to the question posed as the title to this box must be yes. Ozone is both good and bad. Whereas ozone in the upper atmosphere is beneficial to life by shielding the planet from harmful ultraviolet radiation, high concentrations at ground level are a major health and environmental concern.

---

shrubs. It also damages materials such as rubber, some plastics, and paint. Its harsh, biting odor is a distinctive characteristic of photochemical smog.

Because the reactions that create ozone are stimulated by strong sunlight, the formation of this pollutant is limited to daylight hours. Peaks occur in the afternoon following a series of hot, sunny, calm days. As we might expect, ozone levels are highest during the warmer summer months. The "ozone season" varies from one part of the country to another. Although May through October is typical, areas in the Sunbelt of the American South and Southwest may experience problems throughout the year. By contrast, northern states have shorter ozone seasons, such as May through September for North Dakota.

Trends in Air Quality

Although considerable progress has been made in controlling air pollution, the quality of the air we breathe still remains a serious public health problem (see Box 13–3).

Economic activity, population growth, meteorological conditions, and regulatory efforts to control emissions all influence the trends in air pollutant emissions. Up until the 1950s, the greatest influences on emissions were related to the economy and population growth. Emissions grew as the economy and population increased. Emissions fell in periods of economic recession. For example, dramatic declines in emissions in the 1930s were due to the Great Depression (Figure 13–7). Emissions also increase as a result of shifts in the demand for various products. For instance, the tremendous upsurge in demand for gasoline following World War II increased emissions associated with petroleum refining and on-road vehicles.

In the 1950s, the states issued air pollution statutes generally targeted toward smoke and particulate emissions. It was not until the passage of the federal Clean Air Act as amended in 1970 that major strides were made in reducing air pollution. This legislation created the Environmental Protection Agency (EPA) and charged it with establishing air quality and emissions standards.

Box 13–3 Indoor Air Pollution

The primary focus of this chapter is urban and regional air pollution. However, the indoor environment should not be overlooked as a place where air pollution can be a significant problem. In fact, it is becoming increasingly clear that the air we breathe indoors may pose a greater health risk than the air we breathe outdoors. More than 100 dangerous substances, in concentrations 10 to 40 times greater than outdoors, occur in many American homes. The health risks of indoor air pollution are magnified because people spend 70 to 90 percent or more of their time indoors. The U.S. Environmental Protection Agency (EPA) has placed indoor air pollution at the top of a list of 18 sources of environmental cancer risks.

According to health officials and the EPA, three of the most dangerous indoor air pollutants are cigarette smoke, radioactive radon gas, and formaldehyde. The hazards of cigarette smoke are well known. Radon gas occurs in the soil, rock, and water around us and is known to cause lung cancer. It is a radioactive product of uranium and can reach high levels in some houses, depending on the local geology and house construction.

Formaldehyde is a part of many materials commonly found in homes and other buildings. Plywood, fiberboard, and paneling are three examples. In addition to being a possible carcinogen, formaldehyde can cause chronic breathing problems, dizziness, rashes, headaches and more. Many other materials found in modern homes are sources of indoor airborne contaminants. Examples include cleaning products, carpet adhesives, aerosol sprays, and mothballs.

Indoor air pollution may be more of a problem today than it was in the past. One reason is the trend toward greater energy efficiency, which has tended to make houses and office buildings more airtight. When such structures are not adequately ventilated, pollutants become more concentrated and occupants can suffer (Figure 13–B). When acute, this is called "sick building syndrome." The EPA estimates that as many as 20 percent of all U.S. buildings may be "sick," costing the nation billions of dollars in reduced productivity and absenteeism.

In developing countries, indoor air pollution can be severe in the dwellings of many poor people. In such settings, the daily activities of cooking and heating with fuels such as coal, wood, and dung produce unhealthy levels of certain gases and particulate matter. The World Health Organization estimates that about two-thirds of the world's population, primarily in rural areas, burn these traditional fuels, often in stoves with little or no ventilation. The extremely high pollutant levels that sometimes result can have major detrimental health effects.

In summary, indoor air pollution is a serious issue. Whether it is a modern office building or a primitive rural dwelling, the indoor environment may contain a host of airborne pollutants. Some are chemical effluents associated with modern technology; others are particles of soot from burning coal or dung. Addressing the problem involves a number of actions. In industrialized countries, people must be made aware of and then pay attention to their total indoor exposure. This will involve regulating pollution sources and improving ventilation to dilute unavoidable pollutants. In the developing world, the use of cleaner fuels and properly vented stoves is a key to a healthier indoor environment.

Figure 13–B For some people indoor air pollution can be a significant risk. When buildings are not adequately ventilated, pollutants can build up. In recent years, comparative-risk studies performed by the EPA have consistently ranked indoor air pollution among the top five environmental risks to public health. *(Photo by Jon Feingersh/The Stock Market)*

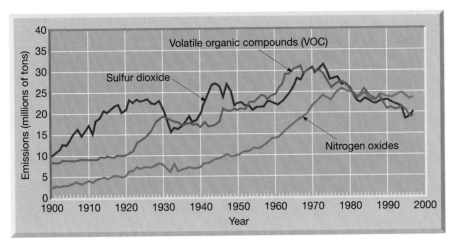

Figure 13–7 Trend in national emissions 1900–1997. Prior to the post-1970 period, economic activity and population growth were major factors influencing the emissions of air pollutants. Emissions grew as the economy and population increased and emissions declined during economic downturns. For example, dramatic declines in emissions in the 1930s were due to the Great Depression. Since 1970, much of the downward trend in emissions is due to the Clean Air Act. ((*After U.S. E.P.A., Office of Air Quality Planning and Standards*)

The Clean Air Act of 1970 mandated the setting of standards for four of the primary pollutants—particulates, sulfur dioxide, carbon monoxide, and nitrogen oxides—as well as the secondary pollutant ozone. At the time, these five pollutants were recognized as being the most widespread and objectionable. Today, with the addition of lead, they are known as the *criteria pollutants* and are covered by the National Ambient Air Quality Standards (Table 13–3). The primary standard for each pollutant shown in Table 13–3 is based on the highest level that can be tolerated by humans

Table 13–3 National Ambient Air Quality Standards

Pollutant	Standard Value	
Carbon Monoxide (CO)		
8-Hour average	9 ppm°	(10 mg/m³)†
1-Hour average	35 ppm	(40 mg/m³)
Nitrogen Dioxide (NO$_2$)		
Annual arithmetic mean	0.053 ppm	(100 µg/m³)°°
Ozone (O$_3$)		
1-Hour average	0.12 ppm	(235 µg/m³)
8-Hour average	0.08 ppm	(157 µg/m³)
Lead (Pb)		
Quarterly average		1.5 µg/m³
Particulate <10 micrometers (PM$_{10}$)		
Annual arithmetic mean		50 µg/m³
24-Hour average		150 µg/m³
Particulate <2.5 micrometers (PM$_{2.5}$)		
Annual arithmetic mean		15 µg/m³
24-Hour average		65 µg/m³
Sulfur Dioxide (SO$_2$)		
Annual arithmetic mean	0.03 ppm	(80 µg/m³)
24-Hour average	0.14 ppm	(365 µg/m³)
3-Hour average	0.50 ppm	(1300 µg/m³)

°ppm, parts per million.

†mg/m³, milligrams per cubic meter of air. A milligram is one-thousandth of a gram.

°°µg/m³, micrograms per cubic meter. A microgram is one-millionth of a gram.

Source: U.S. Environmental Protection Agency, Office of Air Quality Planning and Standards.

without noticeable ill effects, minus a 10 to 50 percent margin for safety.

For some of the pollutants, both long-term and short-term levels are set. Short-term levels are designed to protect against acute effects, whereas the long-term standards were established to guard against chronic effects. *Acute* refers to pollutant levels that may be life-threatening within a period of hours or days. *Chronic* pollutant levels cause gradual deterioration of a variety of physiological functions over a span of years. It should be pointed out that standards are established using human health criteria and not according to their impact on other species or on atmospheric chemistry.

By the late 1990s, approximately 107 million people in the United States resided in counties that did not meet one or more air quality standards (Figure 13–8). It is clear why the EPA describes ozone as our "most pervasive ambient air pollution problem." As Figure 13–8 shows, the number of people living in counties that exceeded the ozone standard is greater than the total number of those living in counties affected by the other five pollutants.

The fact that air quality standards have not yet been met in a large number of places does not mean that progress has not been made. The United States has made significant strides in reducing air pollution. In

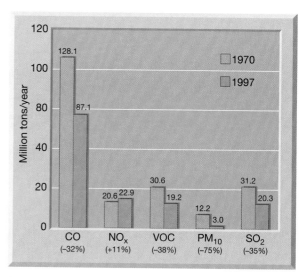

Figure 13–9 Comparison of 1970 and 1997 emissions. The 1997 total is about 31 percent lower than 1970. During this time span, the U.S. population increased 31 percent and vehicle miles traveled went up 127 percent.

1997, emissions of the five major primary pollutants shown in Figure 13–5 totaled about 152 million tons. By contrast, in 1970, when the first Clean Air Act became law, the same five pollutants totaled about 223 million tons. The 1997 total is about 31 percent lower than 1970 (Figure 13–9). This progress is also shown in Figure 13–7, where downward trends in all pollutants except nitrogen oxides are evident. This improvement in air quality has been achieved during a time when urban growth has been substantial. However, methods of control have not been as effective as expected in upgrading urban air quality.

An important reason for the slower than expected progress in bettering air quality is related to growth. For example, *on a per car basis*, emissions of primary pollutants have dramatically improved. However, at the same time this was occurring, the U.S. population increased by 31 percent and vehicle miles traveled went up by 127 percent.

In other words, pollution controls have improved air quality, but the positive effects have been partly offset by an increase in the number of vehicles on the road. This is borne out in Los Angeles, where, despite significant progress in reducing emissions, the city still faces substantial air quality problems. Since the 1950s, when serious efforts to alleviate air pollution began, the population of the area has more than tripled and the number of motor vehicles has more than quadrupled.

Late in 1990, Congress passed the Clean Air Act Amendments. The lengthy technical legislation is broad

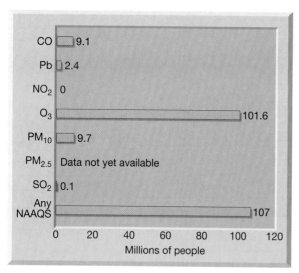

Figure 13–8 Number of people living in counties with air quality concentrations above the levels of the National Ambient Air Quality Standards (NAAQS) in 1997. For example, 9.1 million people live in counties where carbon monoxide (CO) concentrations exceed the national standard. Despite substantial progress in reducing emissions, there were still approximately 107 million people nationwide who lived in counties with monitored air quality levels above the primary national standards.

in scope and organized into several different parts called *titles*. The amendments impose the following:

1. Tighter controls on air quality, including more stringent standards on the emission of sulfur dioxide produced by coal combustion.
2. Lower limits on automobile emissions, which include the use of cleaner-burning fuels.
3. Greater constraints on hazardous air pollutants, which are aimed at greatly reducing urban smog.
4. Controls on water quality degradation from acid rain.
5. Clean air research that authorizes the continuation of the National Acid Precipitation Assessment Program.
6. Limits on and the eventual abolishment of chlorofluorocarbons and other ozone-depleting compounds.
7. Data collection on greenhouse gases that contribute to global climate change.

Regulations and standards regarding these titles are periodically established and revised. For example, in December 1999 the EPA announced new standards that require lower nitrogen oxide emissions from motor vehicles and reduced sulfur levels in gasoline. Recall that nitrogen oxides play a significant role in the formation of photochemical smog. Sulfur in gasoline diminishes the effectiveness of catalytic converters (devices that reduce pollution from tailpipes). By significantly reducing sulfur, the catalytic converter becomes more effective, and tailpipe emissions improve.

Meteorological Factors Affecting Air Pollution

Certainly, the most obvious factor influencing air pollution is the quantity of contaminants emitted into the atmosphere. Still, experience shows that even when emissions remain relatively steady for extended periods, we often find wide variations in air quality from one day to the next. Indeed, when air pollution episodes occur, they are not generally the result of a drastic increase in the output of pollutants; instead, they occur because of changes in certain atmospheric conditions.

Perhaps you have heard the phrase, "The solution to pollution is dilution." To a significant degree, it is true. If the air into which the pollution is released is not dispersed, the air will become more toxic. Two of the most important atmospheric conditions affecting the dispersion of pollutants are (1) the strength of the wind and (2) the stability of the air. These factors are critical because they determine how rapidly pollutants are diluted by mixing with the surrounding air after leaving the source.

Wind as a Factor

The manner in which wind speed influences the concentration of pollutants is shown in Figure 13–10. Assume that a burst of pollution leaves the stack every second. If the wind speed were 10 meters per second, the distance between each pollution "cloud" would be 10 meters. If the wind is reduced to 5 meters per second, the distance between "clouds" will be 5 meters. Consequently, because of the direct effect of wind speed, the concentration of pollutants is twice as great with the 5 meters per second wind as with the 10 meters per second wind. It is easy to understand why air pollution problems seldom occur when winds are strong, but rather are associated with periods when winds are weak or calm.

A second aspect of wind speed influences air quality. The stronger the wind, the more turbulent the air. Thus, strong winds mix polluted air more rapidly with the surrounding air, thereby causing the pollution to be more dilute. Conversely, when winds are light, there is little turbulence and the concentration of pollutants remains high.

The Role of Atmospheric Stability

Whereas wind speed governs the amount of air into which pollutants are initially mixed, atmospheric stability determines the extent to which vertical motions will mix the pollution with cleaner air above. The vertical distance between Earth's surface and the height to which convectional movements extend is called the **mixing depth**. Generally, the greater the mixing depth, the better the air quality. When the mixing depth is several kilometers, pollutants are mixed through a large volume of cleaner air and dilute rapidly. When the mixing depth is shallow, pollutants are confined to a much smaller volume of air and concentrations can reach unhealthy levels.

When air is stable, convectional motions are suppressed and mixing depths are small. Conversely, an unstable atmosphere promotes vertical air movements and greater mixing depths. Because heating of Earth's surface by the Sun enhances convectional movements, mixing depths are usually greater during the afternoon hours. For the same reason, mixing depths during the summer months are typically greater than during the winter months.

Temperature inversion represents a situation in which the atmosphere is very stable and the mixing depth is significantly restricted. Warm air overlying cooler air acts as a lid and prevents upward movement, leaving the pollutants trapped in a relatively narrow zone near the ground. This effect is dramatically illustrated

Figure 13–10 Effect of wind speed on the dilution of pollutants. The concentration of pollutants increases as wind speed decreases.

by the photograph in Figure 13–11. Most of the air pollution episodes cited earlier were linked to the occurrence of temperature inversions.

Surface Temperature Inversions. Solar heating can result in high surface temperatures during the late morning and afternoon that increase the environmental lapse rate and render the lower air unstable. During nighttime hours, however, just the opposite situation may occur; temperature inversions, which result in very stable atmospheric conditions, can develop close to the ground. These surface inversions form because the ground is a more effective radiator than the air above.

Figure 13–11 Air pollution in downtown Los Angeles. Temperature inversions act as lids to trap pollutants below. *(Photo by Ted Spiegel/Black Star)*

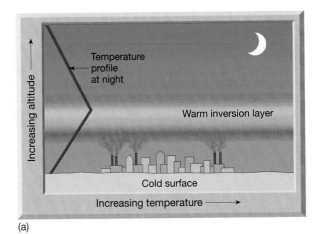

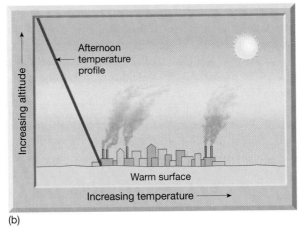

Figure 13–12 (a) A generalized temperature profile for a surface inversion. (b) Temperature-profile changes after the Sun has heated the surface.

This being the case, radiation from the ground to a clear night sky causes more rapid cooling at the surface than higher in the atmosphere. Consequently, the coldest air is found next to the ground, yielding a vertical temperature profile resembling the one shown in the upper portion of Figure 13–12. Once the Sun rises, the ground is heated and the inversion disappears.

Although usually rather shallow, surface inversions may be very deep in regions where the land surface is uneven. Because cold air is denser than warm air, the chilled air near the surface gradually drains from the uplands and slopes into adjacent lowlands and valleys. As might be expected, this deeper surface inversion will not dissipate as quickly after sunrise. Thus, although valleys are often preferred sites for manufacturing because they afford easy access to water transportation, they are also more likely to experience relatively thick surface inversions that, in turn, will have a negative effect on air quality.

Inversions Aloft. Many extensive and long-lived air pollution episodes are linked to temperature inversions that develop in association with the sinking air that characterizes centers of high air pressure (anticyclones). As the air sinks to lower altitudes, it is compressed and so its temperature rises. Because turbulence is almost always present near the ground, this lowermost portion of the atmosphere is generally prevented from participating in the general subsidence. Thus, an inversion develops aloft between the lower turbulent zone and the subsiding warmer layers above (Figure 13–13).

The air pollution that plagues Los Angeles is frequently related to inversions associated with the subsiding eastern portion of the subtropical high in the North Pacific. In addition, the adjacent cool waters of the Pacific Ocean and the mountains surrounding the city compound the problem. When winds move cool air from the Pacific into Los Angeles, the warmer air that is pushed aloft creates or strengthens an inversion aloft that acts as an effective lid. Because the surrounding mountains keep the smog from moving farther inland, air pollution is trapped in the basin until a change in weather brings relief. Clearly, the geographic setting of a place can significantly contribute to air quality problems. The Los Angeles area is an excellent example.

In summary, we have seen that when the wind is strong and an unstable environmental lapse rate prevails, the diffusion of pollutants is rapid and high pollution concentrations will not occur except perhaps near a major source. In contrast, when an inversion exists and winds are

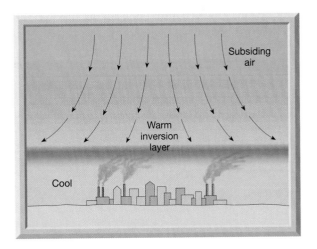

Figure 13–13 Inversions aloft frequently develop in association with slow-moving centers of high pressure, where the air aloft subsides and warms by comparison. The turbulent surface zone does not subside as much. Thus, an inversion often forms between the lower turbulent zone and the subsiding layers above.

light, diffusion is inhibited and high pollution concentrations are to be expected in areas where there are sources. Air pollution is especially acute in urban areas experiencing frequent and prolonged temperature inversions.

Acid Precipitation

As a consequence of burning large quantities of fossil fuels, primarily coal and petroleum products, about 43 million tons of sulfur and nitrogen oxides are released into the atmosphere each year in the United States. The major sources of these emissions include power-generating plants, industrial processes, such as ore smelting and petroleum refining, and motor vehicles of all kinds. Through a series of complex chemical reactions, some of these pollutants are converted into acids that then fall to Earth's surface as rain or snow. Another portion is deposited in dry form and subsequently converted into acid after coming in contact with precipitation, dew, or fog.

In 1852, the English chemist Angus Smith coined the term *acid rain* to refer to the effect that industrial emissions had on precipitation in the British Midlands. A century and a half later, this phenomenon is not only the focus of research for many environmental scientists but also a topic having substantial international political importance. Although Smith clearly realized that acid rain caused environmental damage, large-scale effects were not recognized until the middle part of the twentieth century. Eventually, widespread public concern in the late 1970s led to significant government-sponsored studies of the problem. At the turn of the twenty-first century, such research activities continue to examine this still unresolved environmental problem.

Extent and Potency of Acid Precipitation

Rain is naturally somewhat acidic. When carbon dioxide from the atmosphere dissolves in water, it becomes weak carbonic acid. Small amounts of other naturally occurring acids also contribute to the acidity of precipitation. It was once thought that unpolluted rain has a pH of about 5.6 on the pH scale (Figure 13–14). However, studies in uncontaminated remote areas have shown that precipitation usually has a pH closer to 5. Unfortunately, in most areas within several hundred kilometers of large centers of human activity, precipitation has much lower pH values. This rain or snow is called **acid precipitation**.

Widespread acid rain has been known in northern Europe and eastern North America for some time. Studies have also shown that acid rain occurs in many other regions, including western North America, Japan, China, Russia, and South America. In addition to local pollution sources, a portion of the acidity found in the northeastern United States and eastern Canada originates hundreds of kilometers away in industrialized regions to the south and southwest. This situation occurs because many pollutants remain in the atmosphere for periods as long as five days, during which time they may be transported great distances.

One contributing factor is, of all things, a technology that is used to reduce pollution in the immediate vicinity of a source. Taller chimney stacks improve local air quality by releasing pollutants into the stronger and more persistent winds that exist at greater heights (Figure 13–15). Although such stacks enhance dilution and dispersion, they also promote the long-distance transport of these unwanted emissions. In this way, individual stack plumes with pollution concentrations considered too dilute to be a direct health or environmental threat locally contribute

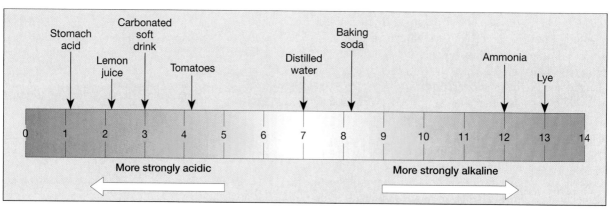

Figure 13–14 The *pH scale* is a common measure of the degree of acidity or alkalinity of a solution. The scale ranges from 0 to 14, with a value of 7 denoting a solution that is neutral. Values below 7 indicate greater acidity, whereas numbers above 7 indicate greater alkalinity. The pH values of some familiar substances are shown on the diagram. Although distilled water is neutral (pH 7), rainwater is naturally acidic. It is important to note that the pH scale is logarithmic; that is, each whole number increment indicates a tenfold difference. Thus, pH 4 is 10 times more acidic than pH 5 and 100 times (10×10) more acidic than pH 6.

Figure 13–15 Tall stacks improve local air quality by releasing pollutants at greater heights where winds are stronger. However, such stacks promote the long-distance transport of pollutants that may ultimately contribute to acid precipitation far from their source. *(Photo by Sandra Baker/Liaison Agency, Inc.)*

to interregional pollution problems. Unfortunately, because atmospheric processes in eastern North America lead to a thorough mixing of pollutants, it is not yet possible to distinguish clearly between the relative impact of distant sources compared with local sources.

Effects of Acid Precipitation

For many years, acid precipitation had not been directly linked to any adverse effects on human health. This is no longer true. A growing body of evidence shows that acid aerosols affect people's health. Exposure may impair the ability of the upper respiratory tract and the deep parts of the lungs to clear themselves of harmful particles. A link has also been made with the incidence of bronchitis among children.

Beyond these possible impacts on health, the damaging effects of acid rain on the environment are believed to be considerable in some areas and imminent in others. The best-known effect of acid precipitation is the lowering of pH in thousands of lakes and streams in Scandinavia and eastern North America. Accompanying this have been substantial increases in dissolved aluminum that is leached from the soil by the acidic water and that, in turn, is toxic to fish. Consequently, some lakes are virtually devoid of fish, whereas others are approaching this condition. Furthermore, ecosystems are characterized by many interactions at many levels of organization, which means that evaluating the effects of acid precipitation on these complex systems is difficult and expensive, and far from complete.

Even within small areas, the effects of acid precipitation can vary significantly from one lake to another. Much of this variation is related to the nature of the soil and

rock materials in the area surrounding the lake. Because minerals such as calcite in some rocks and soils can neutralize acid solutions, lakes surrounded by such materials are less likely to become acidic. In contrast, lakes that lack this buffering material can be severely affected. Even so, over a period of time, the pH of lakes that have not yet been acidified may drop as the buffering material in the surrounding soil becomes depleted.

In addition to the thousands of lakes that can no longer support fish, research indicates that acid precipitation may also reduce agricultural crop yields and impair the productivity of forests. Acid rain not only harms the foliage but also damages roots and leaches nutrient minerals from the soil (Figure 13–16). Finally, acid precipitation is known to promote the corrosion of metals and contributes to the destruction of stone structures (Figure 13–17).

Figure 13–16 Damage to forests by acid precipitation is well documented in Europe and eastern North America. These trees in the Great Smoky Mountains have been injured by acid-laden clouds. *(Photo by Doug Locke/Dembinsky Photo Associates)*

Figure 13–17 Acid rain accelerates the chemical weathering of stone monuments and structures. *(Photo by Don & Pat Valenti/DRK Photo)*

In summary, acid precipitation involves the delivery of acidic substances through the atmosphere to Earth's surface. These compounds are introduced into the air as by-products of combustion and industrial activity. The atmosphere is both the avenue by which offending compounds travel from sources to the sites where they are deposited and the medium in which the combustion products are transformed into acidic substances. In addition to its detrimental impact on aquatic systems, acid precipitation has a number of other harmful effects.

Acid precipitation is a complex and multifaceted issue. Because current knowledge regarding many aspects of the problem is incomplete, additional research continues to be carried out. Yet it has been clear for some time that our understanding is sufficient to take corrective actions as well. One important step occurred with the passage of the Clean Air Act of 1990, which requires a substantial reduction in U.S. emissions of sulfur dioxide and modest cuts in emissions of nitrogen oxides. Canada has agreed to make comparable cuts in its emissions of these air pollutants. Thanks to such controls, precipitation in much of the eastern United States and Canada is less acid than before. Monitoring has shown that improvements of from 15 to 25 percent are common.

Chapter Summary

- Air pollution and weather are linked in two ways. One concerns the influence that weather conditions have on the dilution and dispersal of air pollutants. The second connection is the reverse and deals with the effect that air pollution has on weather and climate.

- Air is never perfectly clean. Volcanic ash, salt particles, pollen and spores, smoke and windblown dust are all examples of "natural air pollution." Although some types of air pollution are recent creations, others, such as London's infamous smoke pollution, have been around for centuries. One of the most tragic air pollution episodes ever occurred in London in December 1952 when more than four thousand people died.

- *Air pollutants* are airborne particles and gasses that occur in concentrations that endanger the health and well-being of organisms or disrupt the orderly functioning of the environment. Pollutants can be grouped into two categories: (1) *primary pollutants*, which are emitted directly from identifiable sources, and (2) *secondary pollutants*, which are produced in the atmosphere when certain chemical reactions take place among primary pollutants. The major primary pollutants include, particulate matter (PM), sulfur dioxide, nitrogen oxides, volatile organic compounds (VOCs), carbon monoxide, and lead. Atmospheric sulfuric acid is one example of a secondary pollutant. Air pollution in urban and industrial areas is often called *smog*. *Photochemical smog*, a noxious mixture of gases and particles, is produced when strong sunlight triggers *photochemical reactions* in the atmosphere. A major component of photochemical smog is *ozone*.

- Although considerable progress has been made in controlling air pollution, the quality of the air we breathe remains a serious public health problem. Economic activity, population growth, meteorological conditions, and regulatory efforts to control emissions, all influence the trends in air pollution. The *Clean Air Act of 1970* mandated the setting of standards for four of the primary pollutants—particulates, sulfur dioxide, carbon monoxide, and nitrogen—as well as the secondary pollutant ozone. In 1997, the emissions of the five major primary pollutants in the United States were about 31 percent lower than 1970. In 1990, Congress passed the *Clean Air Act Amendments*, which further tightened controls on air quality. Regulations and standards regarding the provisions of the Clean Air Act Amendments of 1990 are periodically established and revised.

- The most obvious factor influencing air pollution is the quantity of contaminants emitted into the atmosphere. However, when air pollution episodes take place, they are not generally the result of a drastic increase in the output of pollutants; instead, they occur because of changes in certain atmospheric conditions. Two of the most important atmospheric conditions affecting the dispersion of pollutants are (1) the strength of the wind and (2) the stability of the air. The direct effect of wind speed is to influence the concentration of pollutants. Atmospheric stability determines the extent to which vertical motions will mix the pollution with cleaner air above the surface layers. The vertical distance between Earth's surface and the height to which convectional movements extend is called the *mixing depth*. Generally the greater the mixing depth, the better the air quality. *Temperature inversions* represent a situation in which the atmosphere is very stable and the mixing depth is significantly restricted. When an inversion exists and winds are light, diffusion is inhibited and high pollution concentrations are to be expected in areas where pollution sources exist. *Surface temperature inversions* form because the ground is a more effective radiator than the air above. *Inversions aloft* are associated with sinking air that characterizes centers of high air pressure.

- In most areas within several hundred kilometers of large centers of human activity, the pH value of precipitation is much lower than the usual value found in unpopulated areas. This acidic rain or snow, formed when sulfur and nitrogen oxides produced as by-products of combustion and industrial activity are converted into acids during complex atmospheric reactions, is called *acid precipitation*. The atmosphere is both the avenue by which offending compounds travel from sources to the sites where they are deposited and the medium in which the combustion products are transformed into acidic substances. Beyond possible impacts on health, the damaging effects of acid precipitation on the environment include the lowering of pH in thousands of lakes in Scandinavia and eastern North America. Besides producing water that is toxic to fish, acid precipitation has also detrimentally altered complex ecosystems.

Vocabulary Review

acid precipitation (p. 362)
air pollutants (p. 350)
mixing depth (p. 359)
photochemical reaction (p. 354)

primary pollutant (p. 350)
secondary pollutant (p. 350)
smog (p. 354)
temperature inversion (p. 359)

Review Questions

1. List two ways in which meteorology and air pollution are linked.

2. What is the difference between a primary pollutant and a secondary pollutant?

3. Relate the proper primary pollutant to each statement.
   a. These are also known as hydrocarbons.
   b. Colorless, corrosive gas that originates from burning coal and oil.
   c. The most prominent sources of this colorless, odorless, poisonous gas are motor vehicles.
   d. This gas has a distinctive reddish-brown color.

4. Consult Figure 13–5 to answer the following.
   a. Which source category is responsible for the most pollution?
   b. What is the single greatest air pollutant by weight?

5. What was the original meaning of the term "smog"? What is the current meaning of this term?

6. What triggers a photochemical reaction?

7. What is the major component in photochemical smog?

8. During what part of the day is ozone formation at its peak? When is the "ozone season"?

9. The average motor vehicle today emits *much* less pollution than 30 years ago. Why have the positive effects of this sharp reduction *not* been as great as we might have expected?

10. Why are air pollution problems more acute when winds are weak or calm?

11. How do temperature inversions influence air pollution?

12. Describe the formation of a surface inversion and compare it with an inversion that occurs aloft.

13. How does the geographic setting of Los Angeles contribute to the air pollution episodes it experiences?

14. How much more acidic is a substance with a pH of 4 compared with a substance with a pH of 6? (See Figure 13–14.)

15. How has the building of tall smokestacks contributed to interregional air pollution problems?

16. List some possible environmental effects of acid precipitation.

Atmospheric Science Online

The following are informative and interesting Internet sites that address topics related to those presented in the chapter:

U.S. Environmental Protection Agency (EPA)
- **http://www.epa.gov/**

Office of Air and Radiation (EPA)
- **http://www.epa.gov/oar/oarhome.html**

For direct links to these sites and others, chapter objectives and reviews, quiz questions, and topical investigations that utilize Web resources, visit *The Atmosphere, Eighth Edition* Home Page at:
- **http://www.prenhall.com/lutgens**

The Changing Climate

Aerial view of Greenland. Glaciers cover about 10 percent of Earth's land area. By contrast, during the recent ice age, glacial ice covered up to three times as much land as today. *(Photo by Dave Bartruff/Stock Boston)*

In Chapter 1 we characterized climate as an aggregate of weather. You learned that climate consists not only of average atmospheric values but also involves the variability of elements and the occurrence of extreme events. This is the first of two chapters that focus on *climate*. In this chapter, we examine how climate changes and why. In the next chapter, we take a tour of Earth's major climates, from steamy equatorial rain forests to the frigid poles.

The Climate System

To understand and appreciate climate, it is important to realize that climate involves more than just the atmosphere:

> The atmosphere is the central component of the complex, connected, and interactive global environmental system upon which all life depends. Climate may be broadly defined as the long-term behavior of this environmental system. To understand fully and to predict changes in the atmospheric component of the climate system, one must understand the sun, oceans, ice sheets, solid earth, and all forms of life.[*]

Indeed, we must recognize that there is a **climate system** that includes the atmosphere, hydrosphere, solid Earth, biosphere, and cryosphere. (The *cryosphere* is the ice and snow that exist at Earth's surface.) The climate system *involves the exchanges of energy and moisture that occur among the five spheres.* These exchanges link the atmosphere to the other spheres so that the whole functions as an extremely complex interactive unit. The major components of the climate system are shown in Figure 14–1.

[*]The American Meteorological Society and the University Corporation for Atmospheric Research, "Weather and the Nation's Well-Being," *Bulletin of the American Meteorological Society*, 73, no. 12 (December 1991) 2038.

Figure 14–1 The climate system involves the complex interactions that occur among the atmosphere, hydrosphere, lithosphere, cryosphere, and biosphere. All of the components are represented in this scene in Alaska's Denali National Park. *(Photo by Carr Clifton Photography)*

Changes to the climate system do not occur in isolation. Rather, when one part of it changes, the other components also react. This well-established relationship will be demonstrated often as we study climate change and world climates.

Is Our Climate Changing?

Not too many years ago the concept of *climate change* seemed to have little but academic importance, for the problems most often investigated related to the remote past. "What caused the Ice Age?" was (and is) a major question. Today, however, climate change is a major topic in the scientific community and among the general public as well. What generated this current interest in past and future climates? We can point to the following:

1. Detailed reconstructions of past climates show that the climate has varied on all time scales, from decades to millions of years. This suggests that climate in the future is more likely to differ from the present than to stay the same.
2. Research focused on human activities and their effect on the environment has demonstrated that we are inadvertently changing the climate.
3. There is observational evidence that world climate has become more variable.

Clearly, much of the new attention to climate change comes from the realization that it may adversely affect people in the near future. Further, we have learned that a knowledge of past climates helps our understanding of potential future shifts in climate.

How Do We Detect Climate Change?

High-technology and precision instrumentation are now available to study the composition and dynamics of the atmosphere. But such tools are recent inventions and therefore have been providing data for only a short time span. To understand fully the behavior of the atmosphere and to anticipate future climate change, we must somehow discover how climate has changed over broad expanses of time.

Instrumental records go back only a couple of centuries at best, and the further back we go, the more incomplete and unreliable the data become. To overcome this lack of direct measurements, scientists must decipher and reconstruct past climates by using indirect evidence.

Such evidence is found in sea-floor sediment, oxygen isotope ratios in fossil shells and glacial ice, old soils, tree-growth rings, and even historical documents. Scientists analyze these phenomena, which respond to and reflect changing atmospheric conditions. In the following discussion we will briefly look at some of these techniques. But keep in mind that these reconstructions may capture no more than the most general features of climate.

Among the most interesting and important techniques for analyzing Earth's climate history on a scale of hundreds to thousands of years is the study of ocean-floor sediments and oxygen isotope analysis.

Evidence from Sea-Floor Sediment

Most sea-floor sediments contain the remains of organisms that once lived near the sea surface (the ocean–atmosphere interface). When such near-surface organisms die, their shells slowly settle to the floor of the ocean, where they become part of the sedimentary record. These sea-floor sediments are useful recorders of worldwide climate change because the numbers and types of organisms living near the sea surface change with the climate:

> We would expect that in any area of the ocean/atmosphere interface the average annual temperature of the surface water of the ocean would approximate that of the contiguous atmosphere. The temperature equilibrium established between surface seawater and the air above it should mean that…changes in climate should be reflected in changes in organisms living near the surface of the deep sea.…When we recall that the sea-floor sediments in vast areas of the ocean consist mainly of shells of pelagic foraminifers, and that these animals are sensitive to variations in water temperature, the connection between such sediments and climatic change becomes obvious.°

Thus, in seeking to understand climatic change, scientists have become increasingly interested in the huge reservoir of data concealed in sea-floor sediments. Since the late 1960s the United States has been involved in major international projects. Presently the Ocean Drilling Program uses a specially designed research vessel, the *JOIDES Resolution*, that is capable of drilling into the ocean floor and collecting cores of deep-sea sediments (Figure 14–2). The sediment cores have proved to be excellent sources of useful data that have greatly expanded our understanding of past climates.

°Richard F. Flint, *Glacial and Quaternary Geology* (New York: John Wiley & Sons, 1971), p. 718.

Figure 14–2 The *JOIDES Resolution*, drilling ship of the Ocean Drilling Program. During cruises, holes are drilled deep into the sea floor. The cores of sediment and rock that are recovered represent millions of years of Earth history and are used by scientists to study many aspects of Earth science, including changes in global climate. The ship can drill in water depths up to 8200 meters (about 27,000 feet) and can deploy as much as 9100 meters (about 30,000 feet) of drill pipe. *(Photo courtesy of the Ocean Drilling Program)*

Evidence from Oxygen Isotope Analysis

The second technique, **oxygen isotope analysis**, is based on precise measurement of the ratio between two isotopes of oxygen: ^{16}O, which is the most common, and the heavier ^{18}O. A molecule of H_2O can form from either ^{16}O or ^{18}O. But the lighter isotope, ^{16}O, evaporates more readily from the oceans. Because of this, precipitation (and hence the glacial ice that it may form) is enriched in ^{16}O. This leaves a greater concentration of the heavier isotope, ^{18}O, in the ocean water. Thus, during periods when glaciers are extensive, more of the lighter ^{16}O is tied up in ice, so the concentration of ^{18}O in seawater increases. Conversely, during warmer interglacial periods when the amount of glacial ice decreases dramatically, more ^{16}O is returned to the sea, so the proportion of ^{18}O relative to ^{16}O in ocean water also drops. Now, if we had some ancient recording of the changes of the $^{18}O/^{16}O$ ratio, we could determine when there were glacial periods and therefore when the climate grew cooler.

Fortunately, we do have such a recording. As certain microorganisms secrete their shells of calcium carbonate ($CaCO_3$), the prevailing $^{18}O/^{16}O$ ratio is reflected in the composition of these hard parts. When the organisms die, their hard parts settle to the ocean floor, becoming part of the sediment layers there. Consequently, periods of glacial activity can be determined from variations in the oxygen isotope ratio found in shells of certain microorganisms buried in deep-sea sediments.

The $^{18}O/^{16}O$ ratio also varies with temperature. Thus, more ^{18}O is evaporated from the oceans when temperatures are high, and less is evaporated when temperatures are low. Therefore, the heavy isotope is more abundant in the precipitation of warm eras and less abundant during colder periods. Using this principle, scientists studying the layers of ice and snow in glaciers have been able to produce a record of past temperature changes (see Box 14–1).

Evidence from Other Sources

Several other methods have been used to gain insight into past climates. Because climate has a major effect on soil development and the growth of vegetation, the study of buried soils (*paleosols*), the analysis of the yearly growth rings of trees, and the study of pollen contained in sediments have been used to infer past climates. Figure 14–3 shows an example of tree rings.

Historical documents sometimes contain helpful information. Although it might seem that such records should readily lend themselves to climate analysis, such is not the case. Most manuscripts were written for purposes other than climate description. Furthermore, writers understandably neglected periods of relatively stable atmospheric conditions and mention only droughts, severe storms, memorable blizzards, and other extremes. Nevertheless, records of crops, floods, and the migration of people have furnished useful evidence of the possible influences of changing climate.

Even modern instrument records can be problematic:

Climatic records are not readily subjected to objective study. In the first place, weather observers are subject to human failings, and even small errors affect calculations that may involve equally small trends. The

Box 14–1 Climate Change Recorded in Glacial Ice

Vertical cores taken from the Greenland and Antarctic ice sheets are important sources of data about climate change during and following the most recent cycle of glaciation. Scientists collect samples with a drilling rig, like a small version of an oil drill. A hollow shaft follows the drill head into the ice, and an ice core is extracted. In this way, cores that sometimes exceed 2000 meters (6500 feet) in length and may represent more than 200,000 years of climate history are acquired for study (Figure 14–A).

The ice provides a detailed record of changing air temperatures and snowfall. Air bubbles trapped in the ice record variations in atmospheric composition. Changes in carbon dioxide and methane are linked to fluctuating temperatures. The cores also include atmospheric fallout such as windblown dust, volcanic ash, pollen, and modern-day pollution.

Past temperatures are determined by *oxygen isotope analysis*. Using this technique, scientists are able to produce a record of past temperature changes. A portion of such a record is shown in Figure 14–B.

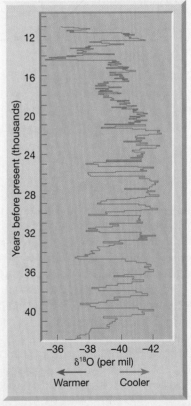

Figure 14–B Temperature variations as revealed by fluctuations in the $^{18}O/^{16}O$ ratio in a portion of a Greenland ice core. An increase in ^{18}O indicates an increase in air temperature. Levels of ^{18}O fall when cooler temperatures prevail.

Figure 14–A Scientists at the National Ice Core Laboratory in Denver, Colorado, examine an ice core sample. Faint lines in the sample are annual dust layers deposited in summer months. Using these layers, the ice cores can be dated much like dating trees by studying their rings. *(Photo by Ken Abbot/National Ice Core Laboratory)*

exposure and height above ground of instruments also materially affect results. Removal of a weather station to a new location practically destroys the value of its records for purposes of studying climatic change. But even if a station remains in the same location for a century, the changes in vegetation, drainage, surrounding buildings, and atmospheric pollution are likely to produce a greater effect on climatic records

Figure 14–3 Each year a growing tree produces a layer of new cells beneath the bark. If the tree is felled and the trunk examined (or if a core is taken, to avoid cutting the tree), each year's growth can be seen as a ring. Because the amount of growth (thickness of a ring) depends upon precipitation and temperature, tree rings are useful records of past climates. *(Photo by Stephen J. Krasemann/DRK Photo)*

than any true climatic changes. Thus, very careful checking and comparison of climatic records are necessary to detect climatic fluctuations.[*]

Natural Causes of Climate Change

A great variety of hypotheses have been proposed to explain climate change. Several have gained wide support, only to lose it and then sometimes to regain it again. Some explanations are controversial. This is to be expected, because planetary atmospheric processes are so large-scale and complex that they cannot be reproduced physically in laboratory experiments. Rather, climate and its changes must be simulated mathematically (modeled) using powerful computers. Although such models are sophisticated tools for climate research, they cannot yet approach the actual complexity of the atmosphere. Computer models are powerful and essential aids, but climate forecasts based on such simulations are still fraught with uncertainty.

In this section we examine several current hypotheses that have earned serious consideration from the scientific community. These describe "natural" mechanisms of climatic change, causes that are unrelated to human activities:

- Plate tectonics (rearranging Earth's continents, moving them closer or farther from the equator and the poles).

[*]Howard J. Critchfield, *General Climatology*, 3rd ed. (Englewood Cliffs, N.J.: Prentice Hall, 1974), p. 376.

- Volcanic activity (changing the reflectivity of the atmosphere and reducing the solar radiation that reaches the surface).
- Variations in Earth's orbit (the natural, cyclic change in our planet's orbit, axial tilt, and wobble).
- Solar variability (Does the Sun vary in its radiation output? Do sunspots affect the output?)

A later section examines human-made climatic changes, including the effect of rising carbon dioxide levels caused primarily by our burning of fossil fuels.

As you read this section, you will find that more than one hypothesis may explain the same climatic change. In fact, several mechanisms may interact to shift climate. Also, no single hypothesis can explain climate change on all time scales. A proposal that explains variations over millions of years generally cannot explain fluctuations over hundreds of years. If our atmosphere and its changes ever become fully understood, we will probably see that climate change is caused by many of the mechanisms discussed here, plus new ones yet to be proposed.

Plate Tectonics and Climate Change

Over the past few decades a revolutionary idea has emerged from the science of geology: **plate tectonics theory**. This theory now has gained nearly universal acceptance in the scientific community. It states that the outer portion of Earth is made up of several vast rigid slabs, called *plates*, which move in relation to one another over a weak plastic rock layer below. They move with incredible slowness, at only a few centimeters a year.

Most of the largest plates include an entire continent plus a lot of seafloor. Thus, as plates ponderously grind along, the continents also change position. Not only does this theory allow geologists to understand and explain many processes and features of Earth's continents and oceans, but it also provides the climatologist with a probable explanation for some hitherto unexplainable climate changes.

For example, glacial evidence in the present-day warm areas of Africa, Australia, South America, and India indicate that these regions experienced an ice age about 250 million years ago. This finding puzzled scientists for many years. How could the climate in these presently warm latitudes once have been frigid like Greenland and Antarctica?

Until the plate-tectonics theory was proved, no reasonable explanation existed. Today scientists realize that the areas containing these ancient glacial features were joined as a single "supercontinent" that was located toward the South Pole (Figure 14–4a). Later, as the plates spread apart, portions of the landmass, each moving on a different plate, slowly migrated toward their present locations.

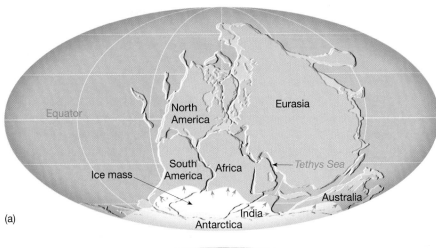

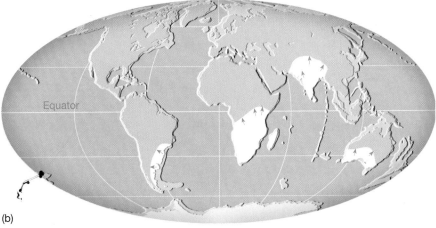

Figure 14-4 (a) The supercontinent Pangaea showing the area covered by glacial ice 300 million years ago. (b) The continents as they are today. The white areas indicate where evidence of old ice sheets exists.

Thus, large fragments of glaciated terrain have ended up in widely scattered subtropical locations (Figure 14–4b).

It is now understood that during the geologic past, plate movements accounted for many other dramatic climate changes as landmasses shifted in relation to one another and moved to different latitudes. Changes in oceanic circulation must also have occurred, altering the transport of heat and moisture and, hence, the climate as well.

Because the rate of plate movement is so slow, appreciable changes in the positions of the continents occur only over *great* spans of geologic time. Thus, climatic changes brought about by plate movements are extremely gradual and happen on a scale of millions of years. As a result, the theory of plate tectonics is not useful for explaining climate variations that occur on shorter time scales, such as tens, hundreds, or thousands of years. Other explanations must be sought to explain these changes.

Volcanic Activity and Climate Change

The idea that explosive volcanic eruptions might alter Earth's climate was first proposed many years ago. It is still regarded as a plausible explanation for some aspects of climatic variability. Explosive eruptions emit huge quantities of gases and fine-grained debris into the atmosphere. The greatest eruptions are sufficiently powerful to inject material high into the stratosphere, where it spreads around the globe and remains for many months or even years.

The basic premise is that this suspended volcanic material will filter out a portion of the incoming solar radiation, which in turn will lower temperatures in the troposphere. More than two hundred years ago Benjamin Franklin used this idea to argue that material from the eruption of a large Icelandic volcano could have reflected sunlight back to space and therefore might have been responsible for the unusually cold winter of 1783–1784.

Perhaps the most notable cool period linked to a volcanic event is the "year without a summer" that followed the 1815 eruption of Mount Tambora in Indonesia (see Box 14–2). Similar, although apparently less dramatic, effects were associated with other great explosive volcanoes, including Indonesia's Krakatoa in 1883.

Three major volcanic events have provided considerable data and insight regarding the impact of volcanoes on

Box 14–2 The Year Without a Summer

The graph in Figure 14–C allows us to compare the volume of volcanic debris extruded during some well-known eruptions, beginning with Mt. Vesuvius in A.D. 79. The eruption of the volcano named Tambora is clearly the largest of modern times. During April 7–12, 1815, this nearly 4000-meter-high (13,000 foot) Indonesian volcano violently ejected over 100 cubic kilometers (24 cubic miles) of volcanic debris.

Although the Tambora eruption occurred in an isolated part of the world, stratospheric circulation spread its influence far and wide. The impact of the volcanic aerosols on climate is believed to have been widespread in the Northern Hemisphere. According to one researcher, "The extreme cold that prevailed during the spring and summer of 1816 in some regions of the world represents one of the most unusual climatic episodes that has occurred since the advent of instrumental weather observations."[*] The effects were especially severe in New England, where 1816 came to be known as the "year without a summer."

From May through September 1816, an unprecedented series of cold spells affected the northeastern United States and adjacent portions of Canada. The result was a late spring, a cold summer, and an early fall. There was heavy snow in June and frost in July and August. Across the Northeast, crops were killed by the cold. Temperatures in New England averaged up to 3.5°C (6°F) below normal in June and 1–2°C (2–3.5°F) below normal in August. Although temperatures this cold had occurred before, there had never been such a protracted span of cold since recordkeeping began. The temperature reductions may seem modest, but they took place in a region where even a small drop in minimum temperatures can mean severe frost.

New England and adjacent Canada were not the only areas to experience a "year without a summer." As one writer observed, "Although the New England farmer considered it a local tragedy, the abnormal weather was widespread throughout the Northern Hemisphere. In England it was almost as cold as in the United States, and 1816 was a famine year there, as it was in France and Germany."[**]

The unusual meteorological events of 1816, which followed the massive 1815 eruption of Tambora, are regarded by many as a spectacular example of the influence of explosive volcanism on climate. Although the effects were relatively short-lived, this geologic event had a significant impact on both the atmosphere and humanity.

[*]Kevin Hamilton, "Early Canadian Weather Observers and the 'Year Without a Summer,'" *Bulletin of the American Meteorological Society*, 67, no. 5 (May 1986), 524.
[**]Patrick Hughes, *American Weather Stories* (Washington, D.C.: National Oceanic and Atmospheric Administration, 1976), p. 43.

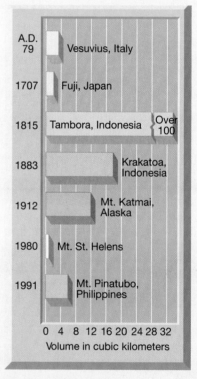

Figure 14–C Approximate volume of volcanic debris emitted during some well-known eruptions. The 1815 eruption of Tambora, the largest-known eruption in historic time, ejected over 100 times more ash than did Mount St. Helens in 1980.

global temperatures. The eruptions of Washington State's Mount St. Helens in 1980, the Mexican volcano El Chichón in 1982, and the Philippines' Mount Pinatubo

in 1991 have given scientists an opportunity to study the atmospheric effects of volcanic eruptions with the aid of more sophisticated technology than had been available in the past. Satellite images and remote-sensing instruments allowed scientists to monitor closely the effects of the clouds of gases and ash that these volcanoes emitted.

Mount St. Helens. When Mount St. Helens erupted, there was immediate speculation about the possible effects on our climate. Could such an eruption cause our climate to change? There is no doubt that the large quantity of volcanic ash emitted by the explosive eruption had significant local and regional effects for a short period (Figure 14–5). Still, studies indicated that any longer-term lowering of hemispheric temperatures was negligible. The cooling was so slight, probably less than 0.1°C(0.2°F), that it could not be distinguished from other natural temperature fluctuations.

El Chichón. Two years of monitoring and studies following the 1982 El Chichón eruption indicated that its cooling effect on global mean temperature was greater than that of Mount St. Helens, on the order of 0.3 to 0.5°C (0.5 to 0.9°F). The eruption of El Chichón was *less explosive* than the Mount St. Helens blast, so why did it have a greater impact on global temperatures? The reason is that the material emitted by Mount St. Helens was largely fine ash that settled out in a relatively short time. El Chichón, on the other hand, emitted far greater quantities of sulfur dioxide gas (an estimated 40 times more) than Mount St. Helens. This gas combines with

water vapor in the stratosphere to produce a dense cloud of tiny sulfuric-acid particles. The particles, called *aerosols*, take several years to settle out completely. They lower the troposphere's mean temperature because they reflect solar radiation back into space.

We now understand that volcanic clouds that remain in the stratosphere for a year or more are composed largely of sulfuric-acid droplets and not of dust, as was once thought. Thus, the volume of fine debris emitted during an explosive event is not an accurate criterion for predicting the global atmospheric effects of an eruption.

Mount Pinatubo. The Philippines volcano, Mount Pinatubo, erupted explosively in June 1991, injecting 25 million to 30 million tons of sulfur dioxide into the stratosphere. The event provided scientists with an opportunity to study the climatic impact of a major explosive volcanic eruption using NASA's spaceborne Earth Radiation Budget Experiment. During the next year the haze of tiny aerosols increased albedo and lowered global temperatures by 0.5°C (0.9°F).

It may be true that the impact on global temperature of eruptions like El Chichón and Mount Pinatubo is relatively minor, but many scientists agree that the cooling produced could alter the general pattern of atmospheric circulation for a limited period. Such a change, in turn, could influence the weather in some regions. Predicting

Figure 14–5 When Mount St. Helens erupted on May 18, 1980, huge quantities of volcanic ash were blown into the atmosphere. The satellite image (at right) was taken less than 8 hours after the eruption. The ash cloud has already spread as far as western Montana. Volcanic ash has little long-term impact on global climate because it settles quickly from the air. A more significant factor affecting climate is the quantity of sulfur dioxide gas emitted during an eruption. *(Photos courtesy of the U.S. Geological Survey, and the National Environmental Satellite Data, and Information Center/NOAA)*

or even identifying specific regional effects still presents a considerable challenge to atmospheric scientists.

The preceding examples illustrate that the impact on climate of a single volcanic eruption, no matter how great, is relatively small and short-lived. Therefore, if volcanism is to have a pronounced impact over an extended period, many great eruptions, closely spaced in time, need to occur. If this happens, the stratosphere would be loaded with enough gases and volcanic dust to seriously diminish the amount of solar radiation reaching the surface. Because no such period of explosive volcanism is known to have occurred in historic times, it is most often mentioned as a possible contributor to such prehistoric climatic shifts as the Ice Age. However, as we shall see in the following section, there is convincing evidence that other mechanisms can better explain Ice Age climates. At present, there is little support in the scientific community for the view that explosive volcanism can contribute significantly to an ice age.

Orbital Variations

"Variations in the earth's orbit influence climate by changing the seasonal and latitudinal distribution of incoming solar radiation."[*] Proposals that link orbital variations and climate change have been around since early in the nineteenth century. However, credit for developing the modern theory that relates Earth motions and climate change is given to the Yugoslavian astronomer Milutin Milankovitch (1879–1954). He formulated a comprehensive mathematical model based on the following elements:

1. Variations in the shape (**eccentricity**) of Earth's orbit about the Sun.
2. Changes in **obliquity**—changes in the angle that Earth's axis makes with the plane of Earth's orbit.
3. **Precession**—the wobbling of Earth's axis, like a spinning top that is winding down.

The three motions together are called **Milankovitch cycles**. We will now look at each.

Orbital Eccentricity. Although variations in the distance between Earth and Sun are of minor significance in understanding current seasonal temperature differences, they may play an important role in producing global climate changes on a time scale of thousands of years. A difference of only 3 percent exists between aphelion, which occurs

about July 4 in the middle of the Northern Hemisphere summer, and perihelion, which takes place in the midst of the Northern Hemisphere winter about January 3.

This small difference in distance means that Earth receives about 6 percent more solar energy in January than in July. Such is not always the case, however. The shape of Earth's orbit changes during a cycle that astronomers say takes between 90,000 and 100,000 years: It stretches into a longer ellipse and then returns to a more circular shape (Figure 14–6a). When the orbit is most elliptical, the amount of radiation received at closest approach (perihelion) would be on the order of 20 to 30 percent greater than at aphelion. This would most certainly result in a substantially different climate from what we now have.

Change in Axial Tilt. In Chapter 2 the inclination of Earth's axis to the plane of its orbit was shown to be the most significant cause for seasonal temperature change. At present the angle that Earth's axis makes with the plane of its orbit is about 23.5°. But this angle changes. During a cycle that averages about 41,000 years, the tilt of the axis varies between 22.1 and 24.5° (Figure 14–6b). Because this angle varies, the severity of the seasons must also change. The smaller our tilt, the smaller the temperature difference between winter and summer.

It is believed that such a reduced seasonal contrast could promote the growth of ice sheets. Because winters could be warmer, more snow would fall because the capacity of air to hold moisture increases with temperature. Conversely, summer temperatures would be cooler, meaning that less snow would melt. The result could be the growth of ice sheets.

Precession. Like a partly rundown top, Earth is wobbling slowly as it spins on its axis. At present, the axis points toward the star Polaris (often called the North Star). However, about the year A.D. 14,000, the axis will point toward the bright star Vega, which will then become the North Star (Figure 14–6c). Because the period of precession is about 26,000 years, Polaris will once again be the North Star by the year 28,000.

As a result of this cyclical wobble of the axis, a climatically significant change must take place. When the axis is tilted toward Vega in about 12,000 years, the orbital positions at which the winter and summer solstices occur will be reversed. Consequently, the Northern Hemisphere will experience winter near aphelion (when Earth is farthest from the Sun), and summer will occur near perihelion (when our planet is closest to the Sun). Thus, seasonal contrasts will be greater because winters will be colder and summers will be warmer than at present.

---

[*]John Imbrie and John Z. Imbrie, "Modeling the Climatic Response to Orbital Variations," *Science*, 207, no. 4434 (1980), 943.

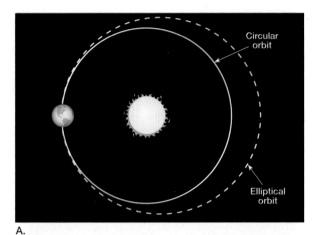

A.

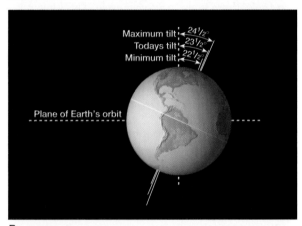

B.

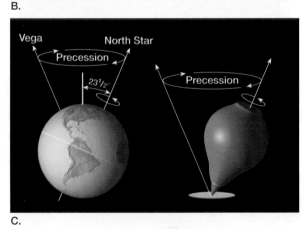

C.

Figure 14–6 Orbital variations. **A.** The shape of Earth's orbit changes during a cycle that spans about 100,000 years. It gradually changes from nearly circular to one that is more elliptical and then back again. This diagram greatly exaggerates the amount of change. **B.** Today the axis of rotation is tilted about 23.5° to the plane of Earth's orbit. During a cycle of 41,000 years, this angle varies from 21.5° to 24.5°. **C.** Precession. Earth's axis wobbles like that of a spinning top. Consequently, this axis points to different spots in the sky during a cycle of about 26,000 years.

Milankovitch Cycles. Using these factors, Milankovitch calculated variations in insolation and the corresponding surface temperature of Earth back into time in an attempt to correlate these changes with the climate fluctuations of the Ice Age. In explaining climate changes that result from these three variables, it should be pointed out that they cause little or no variation in the total annual solar energy reaching the ground. Instead, their impact is felt because they change the degree of contrast between the seasons.

Ever since Milankovitch's pioneering work, the time scales for orbital and insolational changes have been recalculated several times. Past errors have been corrected and measurements have been made with greater precision. Over the years, the Milankovitch cycles, as they are known, have become widely accepted, then largely rejected, and now, in light of recent investigations, have been shown to be a viable explanation for some aspects of climate change.

Among the studies that added credibility to Milankovitch cycles is one that examined deep-sea sediments.° Through oxygen isotope analysis and statistical analyses of climatically sensitive microorganisms, the study established a chronology of temperature change going back 450,000 years. This time scale of climate change was then compared to astronomical calculations of eccentricity, obliquity, and precession to determine if a correlation did indeed exist. (Note that the study was not aimed at identifying or evaluating the *mechanisms* by which the climate is modified by the three orbital variables. The goal simply was to see whether past changes in climate and the orbital variables corresponded.)

Although the study was involved and mathematically complex, its conclusions were straightforward. The authors found that major variations in climate over the past several hundred thousand years were closely associated with changes in the geometry of Earth's orbit. Cycles of climate change were shown to correspond closely with the periods of obliquity, precession, and orbital eccentricity. More specifically, they stated: "It is concluded that changes in the earth's orbital geometry are the fundamental cause of the succession of Quaternary ice ages."°°

Also, the study went on to predict the future trend of climate, toward a cooler climate and extensive glaciation in the Northern Hemisphere. But there are two qualifications: (1) that the prediction apply only to the *natural*

°J. D. Hays, John Imbrie, and N. J. Shackleton, "Variations in the Earth's Orbit: Pacemaker of the Ice Ages," *Science*, 194, no. 4270 (1976), 1121–1132.
°°J. D. Hays, et al., p. 1131. The term "Quaternary" refers to the period on the geologic time scale that encompasses the last 1.8 million years.

component of climate change and ignore any human influence and (2) that it be a forecast of *long-term trends* because it must be linked to factors that have periods of 20,000 years and longer. Thus, even if the prediction is correct, it contributes little to our understanding of climate changes over briefer periods of tens to hundreds of years because the cycles are too long for this purpose. Since the time of this study, subsequent research has supported its basic conclusions, namely that

> Orbital variations remain the most thoroughly examined mechanism of climatic change on time scales of tens of thousands of years and are by far the clearest case of a direct effect of changing insolation on the lower atmosphere of Earth.°

If the Milankovitch cycles indeed explain alternating glacial–interglacial periods, a question immediately arises: Why have glaciers been absent throughout most of Earth's history? Prior to plate-tectonics theory, there was no widely accepted answer. In fact, this question was a major obstacle for the supporters of Milankovitch's hypothesis. Today we have a plausible answer. Because glaciers can form only on the continents, landmasses must exist somewhere in the higher latitudes before an ice age can commence. Long-term temperature fluctuations are not great enough to create widespread glacial conditions in the tropics. Thus, many now suggest that ice ages have occurred only when Earth's shifting crustal plates carried the continents from tropical latitudes to more poleward positions.

Solar Variability and Climate

Among the most persistent hypotheses of climate change have been those based on the idea that the Sun is a variable star and that its output of energy varies through time. The effect of such changes would seem direct and easily understood: Increases in solar output would cause the atmosphere to warm, and reductions would result in cooling. This notion is appealing because it can be used to explain climate change of any length or intensity. However, no major *long-term* variations in the total intensity of solar radiation have yet been measured outside the atmosphere. Such measurements were not even possible until satellite technology became available. Now that it is possible, we will need many years of records before we begin to sense how variable (or invariable) energy from the Sun really is.

Several proposals for climate change, based on a variable Sun, relate to sunspot cycles. The most conspicuous

and best-known features on the surface of the Sun are the dark blemishes called **sunspots** (Figure 14–7). Sunspots are huge magnetic storms that extend from the Sun's surface deep into the interior. Moreover, these spots are associated with the Sun's ejection of huge masses of particles that, on reaching Earth's upper atmosphere, interact with gases there to produce auroral displays (see Figure 1–21).

Along with other solar activity, the numbers of sunspots seem to increase and decrease in a regular way, creating a cycle of about 11 years. The graph in Figure 14–8 shows the annual number of sunspots, beginning in the early 1700s. However, this pattern does not always occur. There have been periods when the Sun was essentially free of sunspots. In addition to the well-known 11-year cycle, there is also a 22-year cycle. This longer cycle is based on the fact that the magnetic polarities of sunspot clusters reverse every successive 11 years.

Interest in possible Sun–climate effects has been sustained by an almost continuous effort to find correlations on time scales ranging from days to tens of thousands of years. Two widely debated examples are briefly described here.

Sunspots and Temperature. Studies indicate prolonged periods when sunspots have been absent or nearly so. Moreover, these events correspond closely with cold periods in Europe and North America. Conversely, periods characterized by plentiful sunspots have correlated well with warmer times in these regions.

Referring to these matches, some scientists have suggested that such correlations make it appear that changes on the Sun are an important cause of climate change. But other scientists seriously question this notion. Their hesitation stems in part from subsequent investigations using different climate records from around the world that failed to find a significant correlation between solar activity and climate. Even more troubling is that no testable physical mechanism exists to explain the purported effect.

Sunspots and Drought. A second possible Sun–climate connection, on a time scale different from the preceding example, relates to variations in precipitation rather than temperature. An extensive study of tree rings revealed a recurrent period of about 22 years in the pattern of droughts in the western United States. This periodicity coincides with the 22-year magnetic cycle of the Sun mentioned earlier.

Commenting on this possible connection, a panel of the National Research Council pointed out that:

> No convincing mechanism that might connect so subtle a feature of the sun to drought patterns in limited regions has yet appeared. Moreover, the cyclic pattern

°National Research Council, *Solar Variability, Weather, and Climate* (Washington, D.C.: National Academy Press, 1982), p. 7.

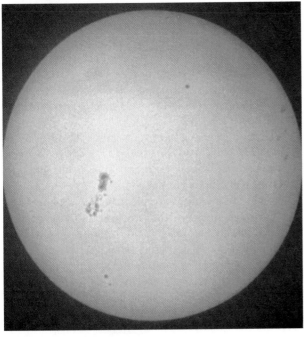 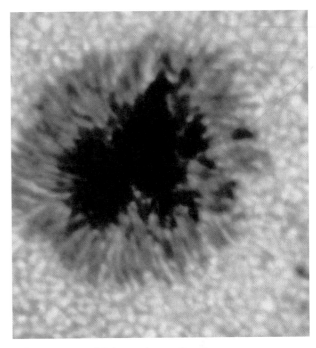

Figure 14–7 Large sunspot group (left) on the solar disk. *(Celestron 8 photo courtesy of Celestron International)* Sunspots (right) having visible umbra (dark central area) and penumbra (lighter area surrounding umbra). *(Courtesy of National Optical Astronomy Observatories)*

of droughts found in tree rings is itself a subtle feature that shifts from place to place within the broad region of the study.[*]

Possible connections between solar variability and climate would be much easier to determine if researchers could identify physical linkages between the Sun and the lower atmosphere. But despite much research, no connection between solar variations and weather has ever been

[*]*Solar Variability, Weather, and Climate* (Washington, D.C.: National Academy Press, 1982), p. 7.

well established. Apparent correlations have almost always faltered when put to critical statistical examination or when tested with different data sets. As a result, the subject has been characterized by ongoing controversy and debate.

Human Impact on Global Climate

So far we have examined four potential causes of climate change, each of them natural. In this section we discuss how humans may contribute to global climate change. One impact largely results from the addition of carbon

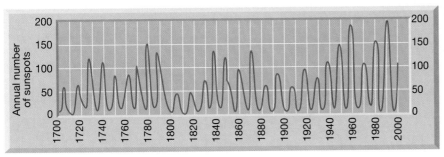

Figure 14–8 Mean annual sunspot numbers.

dioxide and other greenhouse gases to the atmosphere. A second impact is related to the addition of human-generated aerosols to the atmosphere.

One often hears that human influence on regional and global climate began with the onset of the modern industrial period, but this probably is not so. There is good evidence that people have been modifying the environment over extensive areas for thousands of years. The use of fire and the overgrazing of marginal lands by domesticated animals have both reduced the abundance and distribution of vegetation. By altering ground cover, we have modified such important climatological factors as surface albedo, evaporation rates, and surface winds. Commenting on this aspect of human-induced climate modification, the late astronomer Carl Sagan noted: "In contrast to the prevailing view that only modern humans are able to alter climate, we believe it is more likely that the human species has made a substantial and continuing impact on climate since the invention of fire."[*]

Carbon Dioxide, Trace Gases, and Climate Change

In Chapter 1 you learned that carbon dioxide (CO_2) represents only about 0.036 percent of the gases that make up clean, dry air. Nevertheless, it is a very significant component

[*]Carl Sagan et al., "Anthropogenic Albedo Changes and the Earth's Climate," *Science*, 206, no. 4425 (1980), 1367.

meteorologically. Carbon dioxide is influential because it is transparent to incoming short-wavelength solar radiation, but it is not transparent to some of the longer-wavelength outgoing Earth radiation. A portion of the energy leaving the ground is absorbed by atmospheric CO_2. This energy is subsequently reemitted, part of it back toward the surface, thereby keeping the air near the ground warmer than it would be without CO_2.

Thus, along with water vapor, carbon dioxide is largely responsible for the *greenhouse effect* of the atmosphere. Carbon dioxide is an important heat absorber, and it follows logically that any change in the air's CO_2 content could alter temperatures in the lower atmosphere.

Earth's tremendous industrialization of the past two centuries has been fueled—and still is fueled—by burning fossil fuels: coal, natural gas, and petroleum (Figure 14–9). Combustion of these fuels has added great quantities of carbon dioxide to the atmosphere.

The use of coal and other fuels is the most prominent means by which humans add CO_2 to the atmosphere, but it is not the only way. The clearing of forests also contributes substantially because CO_2 is released as vegetation is burned or decays. Deforestation is particularly pronounced in the tropics, where vast tracts are cleared for ranching and agriculture or subjected to inefficient commercial logging operations. According to U.N. estimates, the destruction of tropical forests exceeded 15 million hectares (38 million acres) per year during the 1990s.

(a)

(b)

Figure 14–9 (a) Paralleling the rapid growth of industrialization, which began in the nineteenth century, has been the combustion of fossil fuels, which has added great quantities of carbon dioxide to the atmosphere. *(Photo by Bruce Forster/Tony Stone Images)* (b) Energy consumption in the United States, 1999. Fossil fuels (petroleum, coal, and natural gas) represent about 86 percent of the total. *(Data from U.S. Department of Energy)*

Coal 25%

Petroleum 41%

Natural Gas 20%

Nuclear 9%

Hydroelectric 4%

Others 1%

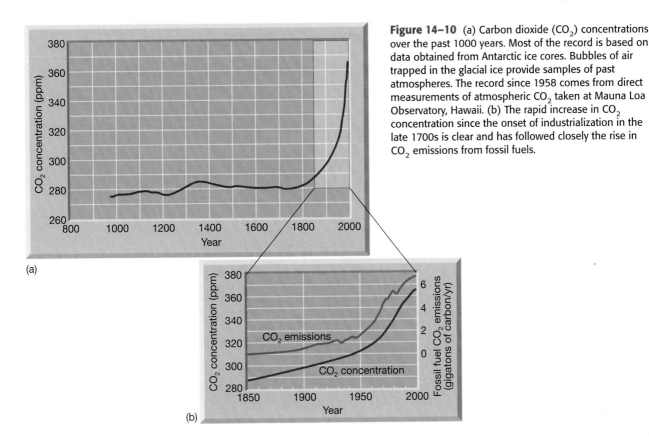

Figure 14–10 (a) Carbon dioxide (CO_2) concentrations over the past 1000 years. Most of the record is based on data obtained from Antarctic ice cores. Bubbles of air trapped in the glacial ice provide samples of past atmospheres. The record since 1958 comes from direct measurements of atmospheric CO_2 taken at Mauna Loa Observatory, Hawaii. (b) The rapid increase in CO_2 concentration since the onset of industrialization in the late 1700s is clear and has followed closely the rise in CO_2 emissions from fossil fuels.

Although some of the excess CO_2 is taken up by plants or is dissolved in the ocean, it is estimated that 45 to 50 percent remains in the atmosphere. Figure 14–10a shows CO_2 concentrations over the past thousand years based on ice-core records and (since 1958) measurements taken at Mauna Loa Observatory, Hawaii. The rapid increase in CO_2 concentration since the onset of industrialization is obvious and has closely followed the increase in CO_2 emissions from burning fossil fuels. Figure 14–10b shows this parallel growth.

We might expect that global temperature has increased as a result of the growing carbon dioxide level. In fact, since the late 1800s the mean global temperature has risen by about 0.3 to 0.6°C (0.5 to 1.1°F) and by about 0.2 to 0.3°C (0.4 to 0.5°F) over the last 45 years. Moreover, recent years have been among the warmest since the mid-1800s despite the cooling effect of the 1991 Mount Pinatubo eruption (Figure 14–11). Are these temperature trends caused by human activities or would they have occurred anyway? Scientists are cautious, but now they seem convinced that "the observed trend in global mean temperature over the past 100 years is unlikely to be natural in origin," and that the data "point towards a human influence on global climate."[*]

[*]Intergovernmental Panel on Climate Change, *Climate Change 1995: The Science of Climate Change*, New York: Cambridge University Press, 1996, p. 4.

If fossil-fuel use continues to increase at projected rates, current estimates are that the atmosphere's CO_2 content will approach 400 ppm by the year 2010 and 600 ppm during the second half of the twenty-first century. With such an increase, the greenhouse effect would become much more dramatic and measurable than in the past. The most realistic models predict an increase in the mean global surface temperature of about 2.5°C (4.5°F). A change of this magnitude would be unprecedented in human history. It would approach the warming that has taken place since the peak of the most recent glacial stage 18,000 years ago, but it would occur *much more rapidly*.

Carbon dioxide is not the only gas contributing to a possible global increase in temperature. In recent years atmospheric scientists have come to realize that the industrial and agricultural activities of people are causing a buildup of several trace gases that may also play a significant role. The substances are called *trace gases* because their concentrations are so much smaller than that of carbon dioxide. The trace gases that appear to be most important are methane (CH_4), nitrous oxide (N_2O), and chlorofluorocarbons (CFCs). These gases absorb wavelengths of outgoing radiation from Earth that would otherwise escape into space. Although individually their impact is modest, taken together the effects of these trace gases may be as great as CO_2 in warming the troposphere (see Box 14–3).

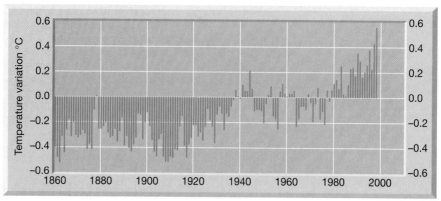

Figure 14–11 Annual average global temperature variations for the period 1860–1998. The basis for comparison is the average for the 1961–90 period (the 0.0 line on the graph). Each narrow bar on the graph represents the departure of the global mean temperature from the 1961–1990 average for one year. For example, the global mean temperature for 1862 was more than 0.5°C (1°F) *below* the 1961–90 average, whereas the global mean for 1998 was more than 0.5°C above. (Specifically, 1998 was 0.56°C warmer.) The bar graph clearly indicates that there can be *significant variations from year to year*. But the graph also shows a trend. Estimated global mean temperatures have been above the 1961–90 average every year since 1978. Also, the three warmest years in the 138-year record shown here were 1995, 1997, and 1998. The very strong 1997–1998 El Niño episode contributed substantially to the recordbreaking temperatures observed in 1997 and 1998. *(After G. Bell, et al. "Climate Assessment for 1998," Bulletin of the American Meteorological Society, Vol. 80, No. 5, May 1999, p. 54)*

Sophisticated computer models show that the warming of the lower atmosphere caused by CO_2 and trace gases will not be the same everywhere. Rather, the temperature response in polar regions could be two to three times greater than the global average. One reason is that the polar troposphere is very stable, which suppresses vertical mixing and thus limits the amount of surface heat that is transferred upward. In addition, an expected reduction in sea ice would also contribute to the greater temperature increase. This topic will be explored more fully later in this section.

How Do Aerosols Influence Climate?

Increasing the levels of carbon dioxide and other greenhouse gases in the atmosphere is the most direct human influence on global climate. But it is not the only impact. Global climate is also affected by human activities that contribute to the atmosphere's aerosol content. Recall that *aerosols* are the tiny, often microscopic, liquid and solid particles that are suspended in the air. Unlike cloud droplets, aerosols are present even in relatively dry air. Atmospheric aerosols are composed of many different materials, including soil, smoke, sea salt, and sulfuric acid. Natural sources are numerous and include such phenomena as dust storms and volcanoes.

Presently the human contribution of aerosols to the atmosphere *equals* the quantity emitted by natural sources. Most human-generated aerosols come from the sulfur dioxide emitted during the combustion of fossil fuels and as a consequence of burning vegetation to clear agricultural land. Chemical reactions in the atmosphere convert the sulfur dioxide into sulfate aerosols, the same material that produces acid precipitation.

How do aerosols affect climate? Aerosols act directly by reflecting sunlight back to space and indirectly by making clouds "brighter" reflectors. The second effect relates to the fact that many aerosols (such as those composed of salt or sulfuric acid) attract water and thus are especially effective as cloud condensation nuclei. The large quantity of aerosols produced by human activities (especially industrial emissions) trigger an increase in the number of cloud droplets that form within a cloud. A greater number of small droplets increases the cloud's brightness—that is, more sunlight is reflected back to space.

By reducing the amount of solar energy available to the climate system, aerosols have a net cooling effect. Studies indicate that the cooling effect of human-generated aerosols could offset a portion of the global warming caused by the growing quantities of greenhouse gases in the atmosphere. Unfortunately, the magnitude and extent of the cooling effect of aerosols is highly uncertain. This uncertainty is a significant hurdle in advancing our understanding of how humans alter Earth's climate.

It is important to point out some significant differences between global warming by greenhouse gases and aerosol cooling. After being emitted, greenhouse gases such as carbon dioxide remain in the atmosphere for many decades. By contrast, aerosols released into the

Box 14–3 Other Greenhouse Gases and Climate Change

Methane is among the gases that contribute to the greenhouse effect. It is present in much smaller amounts than CO_2, but its significance is greater than its relatively small concentration of about 1.7 ppm (parts per million) would indicate. The reason is that methane is 20 to 30 times more effective than CO_2 at absorbing infrared radiation emitted by Earth.

Methane is produced by *anaerobic* bacteria in wet places where oxygen is scarce (anaerobic means "without air," specifically oxygen). Such places include swamps, bogs, wetlands, and the guts of termites and grazing animals like cattle and sheep. Methane is also generated in flooded paddy fields ("artificial swamps") used for growing rice. Mining of coal and drilling for oil and natural gas are other sources because methane is a product of their formation (Figure 14–D).

The concentration of methane in the atmosphere is believed to have about doubled since 1800, an increase that has been in step with the growth in human population. This relationship reflects the close link between methane formation and agriculture. As population has risen, so have the number of cattle and rice paddies.

Nitrous oxide, sometimes called "laughing gas," is also building in the atmosphere, although not as rapidly as methane. The increase is believed to result primarily from agricultural activity. When farmers use nitrogen fertilizers to boost crop yield, some of the nitrogen enters the air as nitrous oxide. This gas is also produced by high-temperature combustion of fossil fuels. Although the annual release into the atmosphere is small, the lifetime of a nitrous oxide molecule is about 150 years! If the use of nitrogen fertilizers and fossil fuels grows at projected rates, nitrous oxide may make a contribution to greenhouse warming that approaches half that of methane.

Unlike methane and nitrous oxide, chlorofluorocarbons (CFCs) are not naturally present in the atmosphere. As you learned in Chapter 1, CFCs are manufactured chemicals with many uses (air-conditioning, for example) that have gained notoriety because they are responsible for ozone depletion in the stratosphere. The role of CFCs in global warming is less well known. CFCs are very effective greenhouse gases. They were not developed until the 1920s and were not used in great quantities until the 1950s, but they already contribute to the greenhouse effect at a level equal to methane. Although the Montreal Protocol represents strong corrective action, CFC levels will *not* drop rapidly (see the section on the Montreal Protocol in Chapter 1). CFCs remain in the atmosphere for decades, so even if all CFC emissions were to stop immediately, the atmosphere would not be free of them for many years.

Carbon dioxide from the burning of fossil fuels and deforestation is clearly the most important single cause for the projected global greenhouse warming. However, as this box has shown, it is not the only contributor. When the effects of all human-generated greenhouse gases other than CO_2 are added together and projected into the future, their collective impact significantly increases the impact of CO_2 alone.

Figure 14–D The formation of methane is associated with the stagnant, oxygen-poor water in swamps and flooded rice paddies ("artificial swamps"). These paddies are in India's Ganges lowlands. *(Photo by George Holton/Photo Researchers, Inc.)*

troposphere remain there for only a few days or, at most, a few weeks before they are "washed out" by precipitation. Because of their short lifetime in the troposphere, aerosols are distributed unevenly over the globe. As expected, human-generated aerosols are concentrated near the areas that produce them, namely industrialized regions that burn fossil fuels and land areas where vegetation is burned.

Because their lifetime in the atmosphere is short, the effect of aerosols on today's climate is determined by the amount emitted during the preceding couple of weeks. By contrast, the carbon dioxide released into the atmosphere remains for much longer spans and thus influences climate for many decades.

Climate-Feedback Mechanisms

Climate is a very complex interactive physical system. Thus, when any component of the climate system is altered, scientists must consider many possible outcomes. These possible outcomes are called **climate-feedback mechanisms**. They complicate climate-modeling efforts and add greater uncertainty to climate predictions.

What climate-feedback mechanisms are related to carbon dioxide and other greenhouse gases? The most important mechanism is that warmer surface temperatures increase evaporation rates. This in turn increases the water vapor in the atmosphere. Remember that water vapor is an even more powerful absorber of radiation emitted by Earth than is carbon dioxide. Therefore, with more water vapor in the air, the temperature increase caused by carbon dioxide and the trace gases is reinforced.

Recall that the temperature increase at high latitudes may be two to three times greater than the global average. This assumption is based in part on the likelihood that the area covered by sea ice will decrease as surface temperatures rise. Because ice reflects a much larger percentage of incoming solar radiation than does open water, the melting of the sea ice would replace a highly reflecting surface with a relatively dark surface (Figure 14–12).

The result would be a substantial increase in the solar energy absorbed at the surface. This in turn would feed back to the atmosphere and magnify the initial temperature increase created by higher levels of greenhouse gases.

So far the climate-feedback mechanisms discussed have magnified the temperature rise caused by the buildup of carbon dioxide. Because these effects reinforce the initial change, they are called **positive-feedback mechanisms**. However, other effects must be classified as **negative-feedback mechanisms** because they produce results that are just the opposite of the initial change and tend to offset it.

One probable result of a global temperature rise would be an accompanying increase in cloud cover due to the higher moisture content of the atmosphere. Most clouds are good reflectors of solar radiation. At the same time, however, they are also good absorbers and emitters of radiation emitted by Earth. Consequently, clouds produce two opposite effects. They are a negative-feedback mechanism because they increase albedo and thus diminish the amount of solar energy available to heat the atmosphere. On the other hand, clouds act as a positive-feedback mechanism by absorbing and emitting radiation that would otherwise be lost from the troposphere (see Figure 3–12).

Which effect, if either, is stronger? Atmospheric modeling shows that the negative effect of a higher albedo is dominant. Therefore, the net result of an increase in cloudiness should be a decrease in air temperature. The magnitude of this negative feedback, however, is not believed to be as great as the positive feedback caused by added moisture and decreased sea ice. Thus, although

Figure 14–12 Sea ice in the Matha Strait, Antarctica. A reduction in sea ice would act as a positive feedback mechanism because surface albedo would decrease and the amount of solar energy absorbed at the surface would increase. *(Photo by Kim Heacox/DRK Photo)*

increases in cloud cover may partly offset a global temperature increase, climate models show that the ultimate effect of the projected increase in CO_2 and trace gases will still be a temperature increase.

The problem of global warming caused by human-induced changes in atmospheric composition continues to be one of the most studied aspects of climate change. Although no models yet incorporate the full range of potential factors and feedbacks, the scientific consensus is that the increasing levels of atmospheric carbon dioxide and trace gases will lead to a warmer planet with a different distribution of climate regimes.

Some Possible Consequences of a Greenhouse Warming

What consequences can be expected if the carbon dioxide content of the atmosphere reaches a level that is twice what it was early in the twentieth century? Because the climate system is so complex, predicting the distribution of particular regional changes is very speculative. It is not yet possible to pinpoint specifics, such as where or when it will become drier or wetter. Nevertheless, plausible scenarios can be given for larger scales of space and time. As computers grow in power and as data improve, scientists will gradually develop models that provide more specific and reliable results.

As noted, the magnitude of the temperature increase will not be the same everywhere. The temperature rise will probably be smallest in the tropics and increase toward the poles. As for precipitation, the models indicate that some regions will experience significantly more precipitation and runoff. However, others will experience a decrease in runoff (due to reduced precipitation or greater evaporation caused by higher temperatures).

Water Resources and Agriculture

Such changes could profoundly alter the distribution of the world's water resources and hence affect the productivity of agricultural regions that depend on rivers for irrigation water. For example, a 2°C (3.6°F) warming and 10 percent precipitation decrease in the region drained by the Colorado River could diminish the river's flow by 50 percent or more. Because the present flow of the river barely meets current demand for irrigation agriculture, the negative effect would be serious (Figure 14–13). Many other rivers are the basis for extensive irrigated agriculture, and the projected reduction of their flow could have equally grave consequences. In contrast, large precipitation increases in other areas would increase the flow of some rivers and bring more frequent destructive floods.

Harder to estimate is the effect on nonirrigated crops that depend on direct rainfall and snowfall for moisture. Some places will no doubt experience productivity loss due to a decrease in rainfall or increase in evaporation. Still, these losses may be offset by gains elsewhere. Warming in the higher latitudes could lengthen the growing season, for instance. This in turn could allow expansion of agriculture into areas presently unsuited to crop production.

Sea-Level Rise

Another impact of a human-induced global warming is a probable rise in sea level. How is a warmer atmosphere related to a global rise in sea level? The most obvious

Figure 14–13 It is not yet possible to specify the magnitude and location of particular climate changes that may result from greenhouse warming. Many consequences are possible. Decreased rainfall and increased evaporation rates could diminish the flow of certain rivers and force the abandonment of some presently productive irrigated farmland. *(Photo by E. J. Tarbuck)*

connection, the melting of glaciers, is important, but *not* the most significant. Far more significant is that a warmer atmosphere causes an increase in ocean volume due to thermal expansion. Higher air temperatures warm the adjacent upper layers of the ocean, which in turn causes the water to expand and the sea level to rise.

Research indicates that sea level has risen between 10 and 25 centimeters (4 and 8 inches) over the past century and that the trend will continue at an accelerated rate. Some models indicate that the rise may approach or even exceed 50 centimeters (20 inches) by the end of the twenty-first century (Figure 14–14a). Such a change may seem modest, but scientists realize that any rise in sea level along a *gently* sloping shoreline, such as the Atlantic and Gulf coasts of the United States, will lead to significant erosion and severe, permanent inland flooding (Figure 14–15). If this happens, many beaches and wetlands will be eliminated and coastal civilization would be severely disrupted.

Because rising sea level is a gradual phenomenon, it may be overlooked by coastal residents as an important contributor to shoreline erosion problems. Rather, the blame may be assigned to other forces, especially storm activity. Although a given storm may be the immediate cause, the magnitude of its destruction may result from the relatively small sea-level rise that allowed the storm's power to cross a much greater land area (Figure 14–16).

As mentioned, a warmer climate will cause glaciers to melt. In fact, a portion of the 10- to 25-centimeter (4- to 8-inch) rise in sea level over the past century is attributed to the melting of alpine glaciers. This contribution is projected to continue through the twenty-first century, as shown in Figure 14–14b. Of course, if the Greenland and Antarctic ice sheets were to experience a significant increase in melting, it would lead to a much greater rise in sea level and a major encroachment by the sea in coastal zones. It should be emphasized, however, that a significant melting of major ice sheets, although possible at some future date, is *not* expected during the next century.

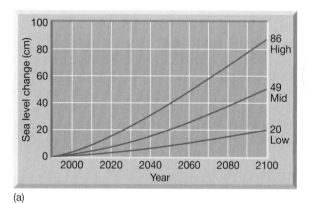

(a)

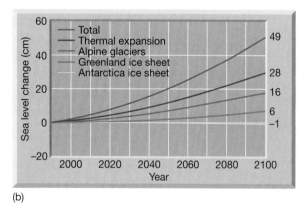

(b)

Figure 14–14 (a) High, middle, and low projections of global sea-level rise, 1990–2100. (b) Projected individual contributions to global sea-level change, 1990–2100 for the middle projection in part (a). Thermal expansion is responsible for 28 centimeters of the sea-level rise. Melting alpine glaciers and ice caps contribute another 16 centimeters. The impact shown for Antarctica indicates that this massive ice sheet is projected to grow larger during this period.

New Weather Patterns

Atmospheric scientists also expect that weather patterns will change with global warming. Potential weather changes may include:

1. Although uncertain, a higher frequency and greater intensity of hurricanes because of warmer ocean temperatures.
2. Shifts in the paths of large-scale cyclonic storms, which in turn would affect the distribution of precipitation and the occurrence of severe storms, including tornadoes.
3. More intense heat waves and droughts in some regions and fewer such events in other places.

The impact on climate of an increase in atmospheric CO_2 and trace gases is obscured by many unknowns and uncertainties. The changes will probably be gradual environmental shifts, imperceptible from year to year. Nevertheless, the effects, accumulated over decades, will have powerful economic, social, and political consequences.

Policymakers are confronted with responding to the risks posed by emissions of greenhouse gases in the face of significant scientific uncertainties. However, they are also faced with the fact that climate-induced environmental changes cannot be reversed quickly, if at all, owing to the lengthy time scales associated with the climate system. Addressing this issue, the Intergovernmental Panel on Climate Change states:

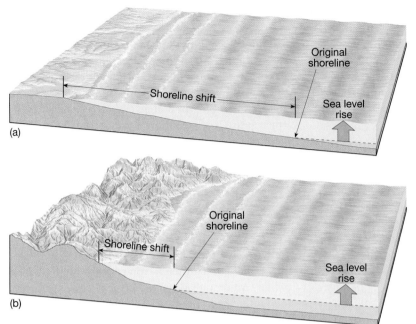

Figure 14–15 The slope of a shoreline is critical to determining the degree to which sea-level changes will affect it. (a) When the slope is gentle, small changes in the sea level cause a substantial shift. (b) The same sea-level rise along a steep coast results in only a small shoreline shift.

Uncertainty does not mean that a nation or the world community cannot position itself better to cope with the broad range of possible climate changes or protect against potentially costly future outcomes. Delaying such measures may leave a nation or the world poorly prepared to deal with adverse changes and may increase the possibility of irreversible or very costly consequences. Options for adapting to change or mitigating change that can be justified for other reasons today (e.g., abatement of air and water pollution) and make society more flexible or resilient to anticipated adverse effects of climate change appear particularly desirable.[*]

Figure 14–16 As sea level gradually rises, the shoreline retreats, and structures that were once thought to be safe from wave attack are exposed to the force of the sea. *(Photo by Kenneth Hasson)*

[*]Intergovernmental Panel on Climate Change, *Climate Change 1995: Impacts, Adaptations and Mitigation of Climate Change: Scientific–Technical Analysis*. New York: Cambridge University Press, 1996, p. 23. *Note:* The Intergovernmental Panel on Climate Change was jointly established by the World Meteorological Organization and the United Nations Environment Program to evaluate and make recommendations regarding the many aspects of global climate change.

Chapter Summary

- The *climate system* includes the atmosphere, hydrosphere, solid Earth, biosphere, and cryosphere (the ice and snow that exists at Earth's surface). The system involves the exchanges of energy and moisture that occur among the five spheres.

- Current interest in past and future climates is due to several factors. Detailed reconstructions of past climates show that the climate has varied on all time scales from decades to millions of years. Also, research focused on human activities and their effect on the environment has demonstrated that we are inadvertently changing climate. And finally, there is observational evidence that world climate has become more variable.

- Techniques for analyzing Earth's climate history on a scale of hundreds to thousands of years include evidence from *sea-floor sediments* and *oxygen isotope analysis*. Sea-floor sediments are useful recorders of worldwide climate change because the numbers and types of organic remains included in the sediment are indicative of past sea surface temperatures. Using oxygen isotope analysis, scientists can use the $^{18}O/^{16}O$ ratio found in the shells of microorganisms in sediment and layers of ice and snow to detect past temperatures. Other methods for determining past climates use buried soils (*paleosols*), the analysis of the growth rings of trees, the study of pollen contained in sediments, and information contained in historical documents.

- Several explanations have been formulated to explain climate change. Current hypotheses for the "natural" mechanisms (causes unrelated to human activities) of climatic change include (1) plate tectonics, rearranging Earth's continents closer or farther from the equator, (2) volcanic activity, reducing the solar radiation that reaches the surface, (3) variations in Earth's orbit, involving changes in the shape of the orbit (*eccentricity*), angle that Earth's axis makes with the plane of its orbit (*obliquity*), and/or the wobbling of the axis (*precession*), and (4) changes in the Sun's output associated with *sunspots*.

- Humans have been modifying the environment for thousands of years. By altering ground cover with the use of fire and the overgrazing of land, people have modified such important climatological factors as surface albedo, evaporation rates, and surface winds. Along with water vapor, carbon dioxide is largely responsible for the *greenhouse effect* of the atmosphere. Therefore, by adding carbon dioxide and other trace gases (methane, nitrous oxide, and chlorofluorocarbons) to the atmosphere, humans may be significantly contributing to global warming.

- Global climate is also affected by human activities that contribute to the atmosphere's *aerosol* (tiny, often microscopic, liquid and solid particles that are suspended in air) content. By reflecting sunlight back to space, aerosols have a net cooling effect. The effect of aerosols on today's climate is determined by the amount emitted during the preceding couple of weeks, while carbon dioxide remains for much longer spans and influences climate for many decades.

- When any component of the climate system is altered, scientists must consider the many possible outcomes, called *climate-feedback mechanisms*. Changes that reinforce the initial change are called *positive-feedback mechanisms*. For example, warmer surface temperatures cause an increase in evaporation, which further increases temperature as the additional water vapor absorbs more radiation emitted by Earth. On the other hand, *negative-feedback mechanisms* produce results that are the opposite of the initial change and tend to offset it. An example would be the negative effect that increased cloud cover has on the amount of solar energy available to heat the atmosphere.

- Because the climate system is so complex, predicting specific regional changes that may occur as the result of increased levels of carbon dioxide in the atmosphere is highly speculative. However, some possible consequences of greenhouse warming include (1) altering the distribution of the world's water resources and therefore the productivity of agricultural regions that depend on rivers for irrigation, (2) a probable rise in sea level, and (3) a change in weather patterns, such as a higher frequency and greater intensity of hurricanes and shifts in the paths of large-scale cyclonic storms.

Vocabulary Review

climate system (p. 368)
climate-feedback mechanism (p. 384)
eccentricity (p. 376)
Milankovitch cycles (p. 376)
negative-feedback mechanism (p. 384)
obliquity (p. 376)

oxygen isotope analysis (p. 370)
plate-tectonics theory (p. 372)
positive-feedback mechanism (p. 384)
precession (p. 376)
sunspot (p. 378)

Review Questions

1. List the five parts of the climate system.

2. Planetary-scale atmospheric processes are so vast and complex that they cannot be reproduced in laboratory experiments. How do scientists study such processes?

3. List and describe some methods that scientists use to gain insight into the climates of the past.

4. How does plate-tectonics theory help explain the previously "unexplainable" glacial features in present-day Africa, South America, and Australia?

5. Can plate tectonics account for short-term climate changes? Explain.

6. The volcanic eruptions of El Chichón in Mexico and Mount Pinatubo in the Philippines had measurable short-term effects on global temperatures. Describe and briefly explain these effects.

7. What geologic event was believed to be responsible for the "year without a summer"? (See Box 14–2.)

8. List and describe each of the three variables that link Earth's motions (Milankovitch cycles) and climate change.

9. Do recent studies of seafloor sediments tend to confirm or refute the importance of Milankovitch cycles? What do these studies predict for the future?

10. List two examples of possible climate change linked to solar variability. Are these Sun–climate connections widely accepted?

11. Why has the carbon dioxide level of the atmosphere been rising for more than 130 years?

12. How are temperatures in the lower atmosphere likely to change as carbon dioxide levels continue to increase?

13. Aside from carbon dioxide, what other trace gases are contributing to a future global temperature change?

14. What are the main sources of human-generated aerosols? What effect do these aerosols have on temperatures in the troposphere? How long do aerosols remain in the lower atmosphere before they are removed?

15. What are climate-feedback mechanisms? Give some examples.

16. List four possible consequences of a greenhouse warming.

Atmospheric Science Online

The following are informative and interesting Internet sites that address topics related to those presented in the chapter:

Global Change Data and Information System:
- **http://www.gcdis.usgcrp.gov/**

Global Change Master Directory (NASA):
- **http://gcmd.gsfc.nasa.gov/pointers/glob_warm.html**

For direct links to these sites and others, chapter objectives and reviews, quiz questions, and topical investigations that utilize Web resources, visit *The Atmosphere, Eighth Edition* Home Page at:
- **http://www.prenhall.com/lutgens**

15

World
Climates

The rocky Pacific coastline at Olympic National Park,
Washington. This area belongs to a climate category called
marine west coast. As the name implies, the ocean exerts a
strong influence on the climate here. *(Photo by Richard J.
Green/Photo Researchers, Inc.)*

Previous chapters examined the spatial and seasonal variations of weather and climate. Now we turn to the *combined* effects of these variations in different parts of the world. The varied nature of Earth's surface (oceans, mountains, plains, ice sheets) and the many interactions that occur among atmospheric processes give every location on our planet a distinctive (sometimes unique) climate. However, we cannot describe the climatic character of countless locales; that would require many volumes.

Instead, our purpose is to introduce you to the *major climate regions* of the world. We will examine large areas and zoom in on particular places to illustrate the characteristics of these major climate regions. In addition, for those regions that are probably unfamiliar to you (the tropical, desert, and polar realms), we briefly describe the natural landscape. Keep in mind that this chapter is a general summary of world climates, using some specific examples.

In Chapter 1 we mentioned the common misconception that climate is only "the average state of the atmosphere." Although averages are certainly important to climate descriptions, variations and extremes must also be included to accurately portray the character of an area.

Temperature and precipitation are the most important elements in climate descriptions because they have the greatest influence on people and their activities and also have an important impact on the distribution of vegetation and the development of soils. Nevertheless, other factors are also important for a complete climate description. When possible, some of these factors are introduced into our discussion of world climates.

Climate Classification

The worldwide distribution of temperature, precipitation, pressure, and wind is, to say the least, complex. Because of the many differences from place to place and time to time, it is unlikely that any two places that are more than a very short distance apart can experience identical weather. The virtually infinite variety of places on Earth makes it apparent that the number of different climates must be extremely large. Having such a diversity of information to investigate is not unique to the study of the atmosphere. It is a problem basic to all science. (Consider astronomy, which deals with billions of stars, and biology, which studies millions of complex organisms.) To cope with such variety, we must devise some means of *classifying* the vast array of data to be studied. By establishing groups of items that have common characteristics, order and manageability are introduced. Bringing order to large quantities of information not only aids comprehension and understanding but also facilitates analysis and explanation.

The first attempt at climate classification probably was made by the ancient Greeks, who divided each hemisphere into three zones: *torrid*, *temperate*, and *frigid* (Figure 15–1). The basis of this simple scheme was Earth–Sun relationships. The boundaries were the four astronomically important parallels of latitude: the Tropic of Cancer (23.5° north), the Tropic of Capricorn (23.5° south), the Arctic Circle (66.5° north), and the Antarctic Circle (66.5° south). Thus, the globe was divided into winterless climates and summerless climates and an intermediate type that had features of the other two.

Few other attempts were made until the beginning of the twentieth century. Since then, many climate-classification schemes have been devised. Remember that the classification of climates (or of anything else) is not a natural phenomenon but the product of human ingenuity. The value of any particular classification system is determined largely by its *intended use*. A system designed for one purpose may not work well for another.

In this chapter we use a classification devised by the German climatologist Wladimir Köppen (1846–1940). As a tool for presenting the general world pattern of climates, the **Köppen classification** has been the best-known and most used system for decades. It is widely accepted for many reasons. For one, it uses easily obtained data: mean monthly and annual values of temperature and precipitation. Furthermore, the criteria are unambiguous, relatively simple to apply, and divide the world into climate regions in a realistic way.

Köppen believed that the distribution of natural vegetation was the best expression of overall climate. Consequently,

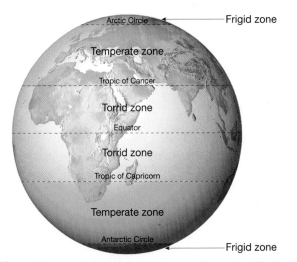

Figure 15–1 Probably the first attempt at climate classification was made by the ancient Greeks. They divided each hemisphere into three zones. The winterless *torrid* zone was separated from the summerless *frigid* zone by the *temperate* zone, which had features of the other two.

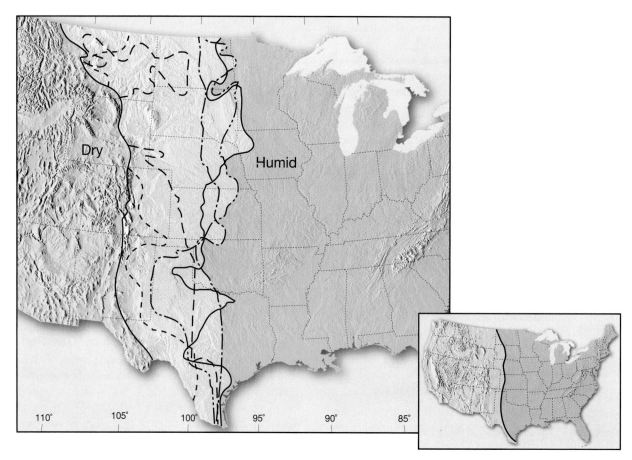

Figure 15–2 Yearly fluctuations in the dry–humid boundary during a 5-year period. The small inset shows the average position of the dry–humid boundary.

the boundaries he chose were largely based on the limits of certain plant associations. He recognized five principal climate groups, each designated by a capital letter:

A *Humid tropical*. Winterless climates; all months having a mean temperature above 18°C (64°F).

B *Dry*. Climates where evaporation exceeds precipitation; there is a constant water deficiency.

C *Humid middle-latitude*, *mild winters*; the average temperature of the coldest month is below 18°C (64°F) but above –3°C (27°F).

D *Humid middle-latitude*, *severe winters*; the average temperature of the coldest month is below –3°C (27°F), and the warmest monthly mean exceeds 10°C (50°F).

E *Polar*. Summerless climates; the average temperature of the warmest month is below 10°C (50°F).

Notice that four of these major groups (A, C, D, E) are defined on the basis of *temperature*. The fifth, the B group, has *precipitation* as its primary criterion.

Each of the five groups is further subdivided by using the criteria and symbols presented in Table 15–1.

A strength of the Köppen system is the relative ease with which boundaries are determined. However, these boundaries cannot be viewed as fixed. On the contrary, all climate boundaries shift their positions from one year to the next (Figure 15–2). The boundaries shown on climate maps are simply average locations based on data collected over many years. Thus, a climate boundary should be regarded as a broad transition zone and not a sharp line (see Box 15–1).

The world distribution of climates, according to the Köppen classification, is shown in Figure 15–3. We will refer you to this map several times as we examine Earth's climates in the following pages.

Climate Controls: A Summary

If Earth's surface were completely homogeneous, the map of world climates would be simple. It would look much like the ancient Greeks must have pictured it—a series of

Table 15-1 Köppen system of climate classification*

			Letter Symbol
1st	**2nd**	**3rd**	
A			Average temperature of the coldest month is 18°C or higher
	f		Every month has 6 cm of precipitation or more
	m		Short dry season; precipitation in driest month less than 6 cm but equal to or greater than $10 - R/25$ (R is annual rainfall in cm)
	w		Well-defined winter dry season; precipitation in driest month less than $10 - R/25$
	s		Well-defined summer dry season (rare)
B			Potential evaporation exceeds precipitation. The dry-humid boundary is defined by the following formulas: (Note: R is the average annual precipitation in cm, and T is the average annual temperature in °C) $R < 2T + 28$ when 70% or more of rain falls in warmer 6 months $R < 2T$ when 70% or more of rain falls in cooler 6 months $R < 2T + 14$ when neither half year has 70% or more of rain
	S		Steppe The BS–BW boundary is 1/2 the dry–humid boundary
	W		Desert
		h	Average annual temperature is 18°C or greater
		k	Average annual temperature is less than 18°C
C			Average temperature of the coldest month is under 18°C and above –3°C
	w		At least ten times as much precipitation in a summer month as in the driest winter month
	s		At least three times as much precipitation in a winter month as in the driest summer month; precipitation in driest summer month less than 4 cm
	f		Criteria for w and s cannot be met
		a	Warmest month is over 22°C; at least 4 months over 10°C
		b	No month above 22°C; at least 4 months over 10°C
		c	One to 3 months above 10°C
D			Average temperature of coldest month is –3°C or below; average temperature of warmest month is greater than 10°C
	s		Same as under C
	w		Same as under C
	f		Same as under C
		a	Same as under C
		b	Same as under C
		c	Same as under C
		d	Average temperature of the coldest month is –38°C or below
E			Average temperature of the warmest month is below 10°C
	T		Average temperature of the warmest month is greater than 0°C and less than 10°C
	F		Average temperature of the warmest month is 0°C or below

*When classifying climate data using Table 15–1, you should first determine whether the data meet the criteria for the E climates. If the station is not a polar climate, proceed to the criteria for B climates. If your data do not fit into either the E or B groups, check the data against the criteria for A, C, and D climates, in that order.

latitudinal bands girdling the globe in a symmetrical pattern on each side of the equator (Figure 15–1). Such is not the case, of course. Earth is not a homogeneous sphere, and many factors disrupt the symmetry just described.

At first glance the world climate map (Figure 15–3) reveals what appears to be a scrambled or even haphazard pattern, with similar climates located in widely separated parts of the world. A closer examination shows that

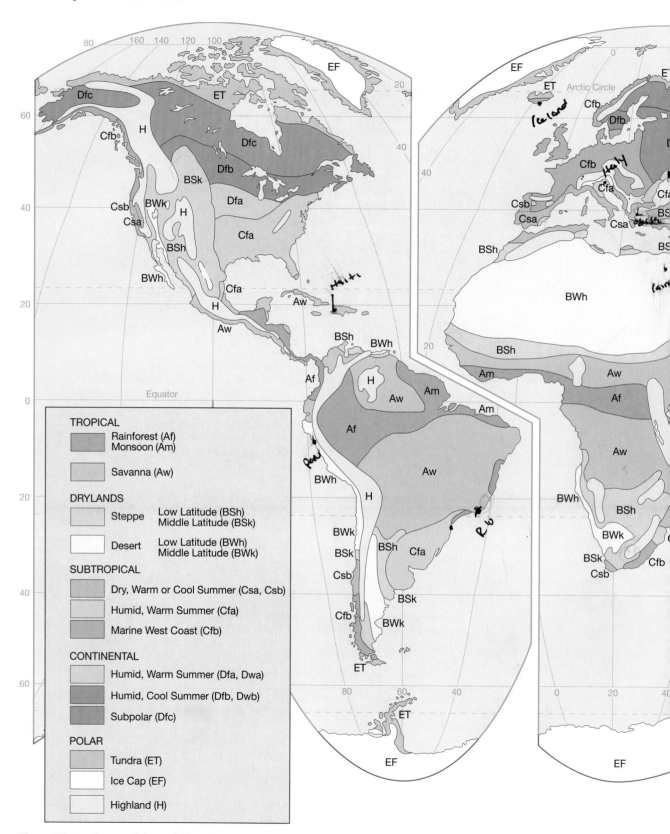

Figure 15–3 Climates of the world based on the Köppen classification.

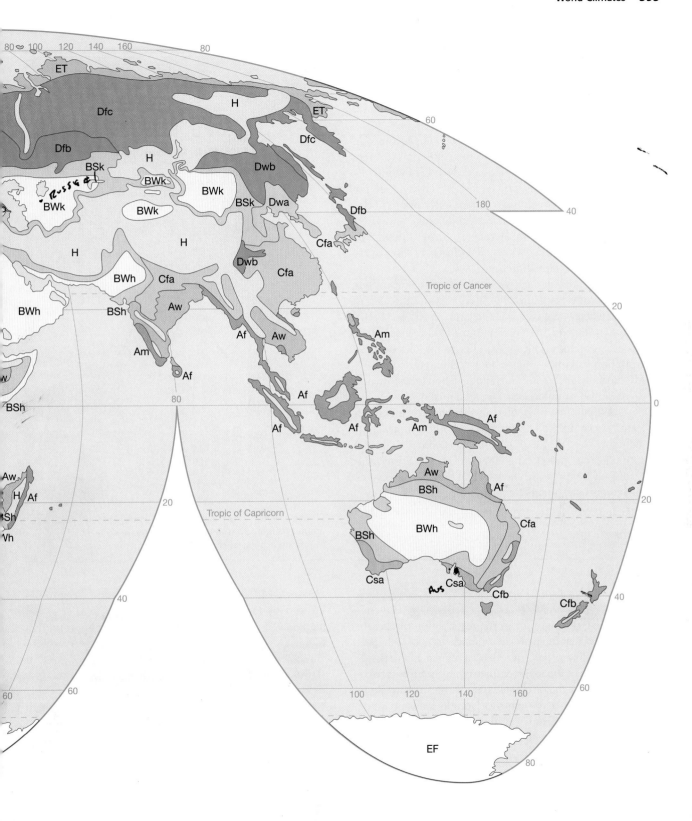

Box 15–1 Climate Diagrams

Throughout this chapter climate diagrams accompany climate descriptions. They are useful tools for presenting the basic data needed in the study of world climates. Figure 15–A shows a typical climate diagram for St. Louis. It has 12 columns, one for each month of the year. A temperature scale is on the left side, and a red line connects monthly mean temperatures. A precipitation scale is on the right side, and blue bars show average monthly precipitation totals.

Just a glance at a climate diagram reveals whether the annual temperature range is high or low and clearly shows the seasonal distribution of precipitation. Is the station in the Northern or Southern Hemisphere? Is it near the equator? Does it experience a monsoon precipitation regime or one more characteristic of a Mediterranean climate? Such information is basic in any discussion or comparison of world climates.

The diagrams presented in this chapter include a location map in the background and other information including the latitude, longitude, and climatic classification of the station. This, however, is not always the case. Only temperature and precipitation data are essential to construct a climatic diagram.

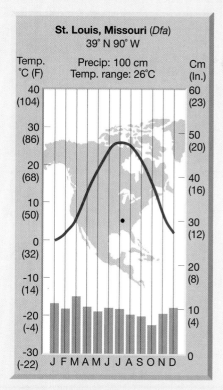

Figure 15–A Climate diagram for St. Louis, Missouri. Such a graph is valuable because it displays precise details of important aspects of climate for a specific place.

although they may be far apart, similar climates generally have similar latitudinal and continental positions. This consistency suggests an order in the distribution of climate elements and that the pattern of climates is not by chance. Indeed, the climate pattern reflects a regular and dependable operation of the major climate controls. So before we examine Earth's major climates, it will be worthwhile to review each of the major controls: latitude, land and water, geographic position and prevailing winds, mountains and highlands, ocean currents, and pressure and wind systems.

Latitude

Fluctuations in the amount of solar radiation received at Earth's surface represent the single greatest cause of temperature differences. Although variations in such factors as cloud coverage and the amount of dust in the air may be locally influential, seasonal changes in Sun angle and length of daylight are the most important factors controlling the global temperature distribution. Because all places situated along the same parallel of latitude have identical Sun angles and lengths of daylight, variations in the receipt of solar energy are largely a function of latitude. Moreover, because the vertical rays of the Sun migrate annually between the Tropic of Cancer and the Tropic of Capricorn, there is a regular latitudinal shifting of temperatures. Temperatures in the tropical realm are consistently high because the vertical rays of the Sun are never far away. As one moves farther poleward, however, greater seasonal fluctuations in the receipt of solar energy are reflected in larger annual temperature ranges.

Land and Water

The distribution of land and water must be considered second in importance only to latitude as a control of temperature. Recall that water has a greater *heat capacity* than rock and soil. Thus, land heats more rapidly and to higher temperatures than water, and it cools more rapidly and to lower temperatures than water. Consequently, variations in air temperatures are much greater over land than over water. This differential heating of Earth's surface has led to climates being divided into two broad classes—marine and continental.

Marine climates are considered relatively mild for their latitude because the moderating effect of water produces summers that are warm but not hot and winters that are cool but not cold. In contrast, **continental climates** tend to be much more extreme. Although a marine station and a continental station along the same parallel in the middle latitudes may have similar annual mean temperatures, the annual temperature *range* will be far greater at the continental station.

The differential heating and cooling of land and water can also have a significant effect on pressure and wind systems and hence on seasonal precipitation distribution as well. High summer temperatures over the continents can produce low-pressure areas that allow the inflow of moisture-laden maritime air. Conversely, the high pressure that forms over the chilled continental interiors in winter causes a reverse flow of dry air toward the oceans.

Geographic Position and Prevailing Winds

To understand fully the influence of land and water on the climate of an area, the position of that area on the continent and its relationship to the prevailing winds must be considered. The moderating influence of water is much more pronounced along the windward side of a continent, for here the prevailing winds may carry the maritime air masses far inland. On the other hand, places on the lee side of a continent, where the prevailing winds blow from the land toward the ocean, are likely to have a more continental temperature regime.

Mountains and Highlands

Mountains and highlands play an important part in the distribution of climates. This impact may be illustrated by examining western North America. Because prevailing winds are from the west, the north–south trending mountain chains are major barriers. They prevent the moderating influence of maritime air masses from reaching far inland. Consequently, although stations may lie within a few hundred kilometers of the Pacific, their temperature regime is essentially continental.

Also, these topographic barriers trigger orographic rainfall on their windward slopes, often leaving a dry rain shadow on the leeward side. Similar effects may be seen in South America and Asia, where the towering Andes and the massive Himalayan system are major barriers. For comparison, Western Europe lacks a mountain barrier to obstruct the free movement of maritime air masses from the North Atlantic. As a result, moderate temperatures and sufficient precipitation mark the entire region.

Where extensive, highlands create their own climatic regions. Because of the drop in temperature with increasing altitude, areas like the Tibetan Plateau, the Altiplano of Bolivia, and the uplands of East Africa are cooler and drier than their latitudinal locations alone would indicate.

Ocean Currents

The effect of ocean currents on the temperatures of adjacent land areas can be significant. Poleward-moving currents, such as the Gulf Stream and Kuro Siwo currents in the Northern Hemisphere and the Brazilian and East Australian currents in the Southern Hemisphere, cause air temperatures to be warmer than would be expected for their latitudes. This influence is especially pronounced in the winter. Conversely, the Canaries and California currents in the Northern Hemisphere and the Peruvian and Benguela currents south of the equator reduce the temperatures of bordering coastal zones. In addition, the chilling effect of these cold currents acts to stabilize the air masses moving across them. The result is marked aridity and often considerable advection fog.

Pressure and Wind Systems

The world distribution of precipitation shows a close relationship to the distribution of Earth's major pressure and wind systems. Although the latitudinal distribution of these systems does not generally take the form of simple "belts," it is still possible to identify a zonal arrangement of precipitation from the equator to the poles (see Figure 7–19).

In the realm of the equatorial low, the convergence of warm, moist, and unstable air makes this zone one of heavy rainfall. In the regions dominated by the subtropical highs, general aridity prevails, creating major deserts. Farther poleward, in the middle-latitude zone dominated by the irregular subpolar low, the influence of the many traveling cyclonic disturbances again increases precipitation. Finally, in polar regions, where temperatures are low and the air can hold only small quantities of moisture, precipitation totals decline.

The seasonal shifting of the pressure and wind belts, which follows the movement of the Sun's vertical rays, significantly affects areas in intermediate positions. Such regions are alternately influenced by two different pressure and wind systems. A station located poleward of the equatorial low and equatorward of the subtropical high, for example, will experience a summer rainy period as the low migrates poleward, and a wintertime drought as the high moves equatorward. This latitudinal shifting of pressure belts is largely responsible for the seasonality of precipitation in many regions.

World Climates—An Overview

The remainder of this chapter is a tour of world climates. Our tour is organized like this:

- Beginning along the equator, we visit the *wet tropics* (the *A* climates), studying their temperature and precipitation characteristics, which foster the great rain forests.
- North and south of the wet tropics is the *tropical wet and dry* climate (still an *A* climate), including the monsoon areas.
- North and south of the tropical wet and dry areas are the *dry* climates (the *B* climates). Dominating large parts of the subtropics and extending into the interiors of continents in the middle latitudes, deserts and steppes cover nearly one-third of Earth's land surface.
- Moving poleward from the dry subtropical realm, we visit regions exhibiting the *humid middle-latitude* climate (one of the *C* climates). These prevail on the eastern side of continents between 25 and 40° latitude, the southeastern United States being an example.
- On the windward coasts of continents, we next visit *marine west coast* climates (also *C* climates), like that of Western Europe.
- We complete our look at *C* climates with a visit to *dry-summer subtropical* or *Mediterranean* climates, like those of Italy, Spain, and parts of California.
- In the Northern Hemisphere, where the continents extend into the middle and high latitudes, *C* climates give way to *D* climates called *humid continental*. These are "breadbasket" areas hospitable to growing grains and meat that feed much of the world.
- Verging on the polar climates are the *subarctic* climates. These vast zones of coniferous forest in Canada and Siberia are known for their long and bitter winters.

- Around the poles we visit the *polar* climates (the *E* climates). These are summerless areas of tundra and permafrost or permanent ice sheets.
- Finally, we visit some cool places that are not necessarily near the poles but are chilled by their high elevation. The *highland* climates even occur on mountaintops near the equator and are characteristic of the Rockies, Andes, Himalayas, and other mountain regions.

We begin our tour with the *wet tropics*.

The Wet Tropics (*Af, Am*)

In the wet tropics constant high temperatures and year-round rainfall combine to produce the most luxuriant vegetation in any climatic realm: the **tropical rain forest** (Figure 15–4). Unlike the forests that we North Americans are accustomed to, the tropical rain forest is made up of broadleaf trees that remain green throughout the year. In addition, instead of being dominated by a few species, these forests are characterized by many. It is not uncommon for hundreds of different species to inhabit a single square kilometer of the forest. As a consequence, the individuals of a single species are often widely spaced.

Standing on the shaded floor of the forest looking upward, one sees tall, smooth-barked, vine-entangled trees, the trunks branchless in their lower two-thirds, forming an almost continuous canopy of foliage above. A closer look reveals a three-level structure. Nearest the ground, perhaps 5 to 15 meters (16 to 50 feet) above, the narrow crowns of rather slender trees are visible. Rising above these relatively short components of the forest, a more continuous canopy of foliage occupies the height range of 20 to 30 meters (65 to 100 feet). Finally, visible through an occasional opening in the second level, a third level may be seen at the very top of the forest. Here the crowns of the trees tower 40 meters (130 feet) or more above the forest floor.

Much sunlight streams through the highest level to those levels below, but little light penetrates to the ground. As a result, plant foliage is relatively sparse on the dimly lit forest floor. Anywhere considerable light does make its way to the ground, as along riverbanks or in human-made clearings, an almost impenetrable growth of tangled vines, shrubs, and short trees exists. The familiar term **jungle** is used to describe such sites.

The environment of the wet tropics just described covers almost 10 percent of Earth's land area (see Box 15–2). Figure 15–3 shows that *Af* and *Am* climates form a discontinuous belt astride the equator that typically

Figure 15–4 Unexcelled in luxuriance and characterized by hundreds of different species per square kilometer, the tropical rain forest is a broadleaf evergreen forest that dominates the wet tropics. Lambir Hills, Sarawak, Malaysia. *(Photo by Michael J. Doolittle/Rainbow)*

extends 5 to 10° into each hemisphere. The poleward margins are most often marked by diminishing rainfall, but occasionally decreasing temperatures mark the boundary. Because of the general decrease in temperature with height in the troposphere, this climate region is restricted to elevations below 1000 meters (3300 feet). Consequently, the major interruptions near the equator are mostly cooler highland areas.

Also note that the rainy tropics tend to have greater north–south extent along the eastern side of continents (especially South America) and along some tropical coasts. The greater span on the eastern side of a continent is due primarily to its windward position on the weak western side of the subtropical high, a zone dominated by neutral or unstable air. In other cases, as along the eastern side of Central America, the coast, backed by interior highlands, intercepts the flow of trade winds. Orographic uplift thus greatly enhances the rainfall total.

Data for some representative stations in the wet tropics are shown in Table 15–2 and Figure 15–5a and b. A brief examination of the numbers reveals the most obvious features that characterize the climate in these areas:

1. Temperatures usually average 25°C (77°F) or more each month. Not only is the annual mean high, but the annual range is also very small (note the flat temperature curves in the graphs in Figure 15–5a and b).

2. The total precipitation for the year is high, often exceeding 200 centimeters (80 inches).

3. Although rainfall is not evenly distributed throughout the year, tropical rain-forest stations are generally wet in all months. If a dry season exists, it is very short.

Temperature Characteristics

Because places with an *Af* or *Am* designation lie near the equator, the reason for their uniform temperature rhythm is clear: Insolation is consistently intense. The Sun's rays are always near to vertical, and changes in day length are slight throughout the year. Therefore, seasonal temperature variations are minimal. The small difference that exists between the warmest and coolest months often reflects changes in cloud cover rather than in the position of the Sun. In the case of Belém, Brazil, for example, you can see that the highest temperatures occur during the months when rainfall (and hence cloud cover) is least.

A striking characteristic of temperatures in the wet tropics is that daily temperature variations greatly exceed

Box 15–2 Laterites and the Clearing of the Rain Forest

Laterites are thick red soils that form in the wet tropics and subtropics. They are the end product of extreme chemical weathering. Because lush tropical rain forests have lateritic soils, they should make great agricultural soils, right? Unfortunately, this is not the case. In fact, just the opposite is true: Laterites are among the poorest soils for farming. Why?

Because laterites develop under conditions of high temperature and heavy rainfall, they are severely *leached*. Leaching leads to low fertility because plant nutrients are removed by the large volume of downward-percolating rainwater. Therefore, even though the vegetation may be dense and luxuriant, the soil itself

contains few available nutrients. Most of the nutrients that support the rain forest are locked up in the trees themselves.

In fact, the rain forest recycles its own nutrients. As trees die and decompose, the roots of the living trees quickly absorb the nutrients before they can be leached from the soil. Thus, the nutrients are continuously recycled as trees die and decompose.

Therefore, when forests are cleared to provide land for farming or to harvest the timber, most of the nutrients are removed as well (Figure 15–B). What remains is a soil that contains little to nourish anything, including crops that people try to grow.

Figure 15–B Clearing the Amazon rain forest in Surinam. The thick lateritic soil is highly leached. *(Photo by Wesley Bocxe/Photo Researchers)*

seasonal differences. Whereas annual temperature ranges in the wet tropics rarely exceed 3°C (6°F), daily temperature ranges are from two to five times greater. Thus, there is a greater variation between day and night than there is seasonally. It is interesting that monthly and daily mean temperatures in the tropics are no greater than those in many U.S. cities during the summer. For example, the highest temperature recorded at Jakarta, Indonesia, over a 78-year period and at Belém, Brazil, over a 20-year period has been only 36.6°C (98°F), compared

with extremes of 40.5°C (105°F) at Chicago and 41.1°C (106°F) at New York City.

The uniqueness of the wet tropical temperature regime is its day-in and day-out, month-in and month-out regularity. Although the thermometer may not indicate abnormal or extreme conditions, the warm temperatures combined with the high humidity and meager winds make apparent temperatures particularly high. The reputation of the wet tropics as being oppressive and monotonous is mostly well deserved.

Figure 15–C This ancient temple at Angor Wat, Cambodia, was built of bricks made of laterite. *(Photo by R. Ian Lloyd/The Stock Market)*

The clearing of rain forests not only removes the supply of plant nutrients but also leads to accelerated erosion. When vegetation is present, its roots anchor the soil while its leaves and branches provide a canopy that protects the ground from the full force of the frequent heavy rains. The removal of the vegetation also exposes the ground to strong direct sunlight. When baked by the Sun, laterites can harden to a bricklike consistency and become practically impenetrable by water and crop roots. In only a few years, lateritic soils in a freshly cleared area may no longer be cultivatable.

The term *laterite* is derived from the Latin word *latere*, meaning "brick," and was first applied to the use of this material for brick making in India and Cambodia. Laborers simply excavated the soil, shaped it, and allowed it to harden in the Sun. Ancient but still well-preserved structures built of laterite remain standing today in the wet tropics (Figure 15–C). Such structures have withstood centuries of weathering because all of the original soluble materials were already leached from the soil by chemical weathering. Laterites are therefore virtually insoluble and thus very stable.

Precipitation Characteristics

The regions dominated by *Af* and *Am* climate normally receive from 175 to 250 centimeters (68 to 98 inches) of rain each year. But a glance at the data in Table 15–2 reveals more variability in rainfall than in temperature, both seasonally and from place to place. The rainy

Table 15–2 Data for wet tropical stations													
	J	**F**	**M**	**A**	**M**	**J**	**J**	**A**	**S**	**O**	**N**	**D**	**YR**
Singapore 1°21'N; 10 m													
Temp. (°C)	26.1	26.7	27.2	27.6	27.8	28.0	27.4	27.3	27.3	27.2	26.7	26.3	27.1
Precip. (mm)	285	164	154	160	131	177	163	200	122	184	236	306	2282
Belém, Brazil 1°18'S; 10 m													
Temp. (°C)	25.2	25.0	25.1	25.5	25.7	25.7	25.7	25.9	25.7	26.1	26.3	25.9	25.7
Precip. (mm)	340	406	437	343	287	175	145	127	119	91	86	175	2731
Douala, Cameroon 4°N; 13 m													
Temp. (°C)	27.1	27.4	27.4	27.3	26.9	26.1	24.8	24.7	25.4	25.9	26.5	27.0	26.4
Precip. (mm)	61	88	226	240	353	472	710	726	628	399	146	60	4109

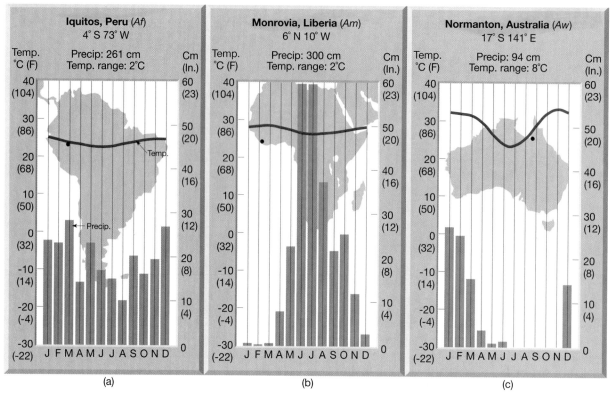

Figure 15–5 By comparing these three climate diagrams, the primary differences among the *A* climates can be seen. (a) Iquitos, the *Af* station, is wet throughout the year. (b) Monrovia, the *Am* station, has a short, dry season. (c) As is true for all *Aw* stations, Normanton has an extended dry season and a greater annual temperature range than the others.

nature of the equatorial realm is partly related to the extensive heating of the region and the consequent thermal convection. In addition, this is the zone of the converging trade winds, often referred to as the **intertropical convergence zone**, or simply the **ITCZ**. The thermally induced convection, coupled with convergence, leads to widespread ascent of the warm, humid, unstable air. Conditions near the equator are thus ideal for precipitation.

Rain typically falls on more than half of the days each year. In fact, at some stations, three-quarters of the days experience rain. There is a marked daily regularity to the rainfall at many places. Cumulus clouds begin forming in late morning or early afternoon. The buildup continues until about 3 or 4 p.m., the time when temperatures are highest and thermal convection is at a maximum; then the cumulonimbus towers yield showers. Figure 15–6, showing the hourly distribution of rainfall at Kuala Lumpur, Malaysia, exemplifies this pattern.

The cycle is different at many marine stations, with the rainfall maximum occurring at night. Here the environmental lapse rate is steepest and hence instability is greatest during the dark hours instead of in the afternoon. The environmental lapse rate steepens at night because the radiation heat loss from the air at heights of 600 to 1500 meters (2000 to 5000 feet) is greater than near the surface, where the air continues to be warmed by conduction and low-level turbulence created when air is heated by the warm water.

Portions of the rainy tropics are wet throughout the year. According to the Köppen scheme, at least 6 centimeters (2.3 inches) of rain falls each month. Yet extensive areas (those having the *Am* designation) are characterized by a brief dry season of one or two months. Despite the short dry season, the annual precipitation total in *Am* regions closely corresponds to the total in areas that are wet year-round (*Af*). Because the dry period is too brief to deplete the supply of soil moisture, the rain forest is maintained.

The seasonal pattern of precipitation in wet tropical climates is complex and not yet fully understood. Month-to-month variations are, at least in part, caused by the seasonal migration of the ITCZ, which follows the migration of the vertical rays of the Sun.

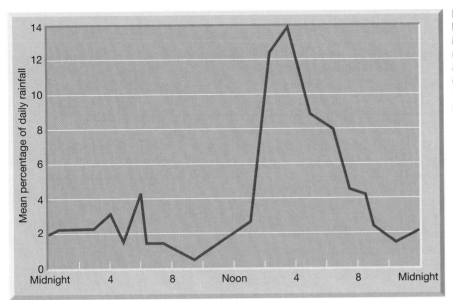

Figure 15-6 The distribution of rainfall by hour of the day at Kuala Lumpur, Malaysia, with its midafternoon maximum, illustrates the typical pattern at many wet tropical stations. *(After Ooi Jin-bee, "Rural Development in Tropical Areas,"* Journal of Tropical Geography, *12, no. 1 [1959])*

Tropical Wet and Dry (*Aw*)

Between the rainy tropics and the subtropical deserts lies a transitional climatic region referred to as **tropical wet and dry**. Along its equatorward margin, the dry season is short, and the boundary between *Aw* and the rainy tropics is difficult to define. Along the poleward side, however, the dry season is prolonged, and conditions merge into those of the semiarid realm.

Here in the tropical wet and dry climate the rain forest gives way to the **savanna**, a tropical grassland with scattered deciduous trees (Figure 15-7). In fact, *Aw* is often called the *savanna climate*. This name may not be

Figure 15-7 The tropical savanna in Kenya with its stunted, drought-resistant trees scattered amid a grassland may have resulted from seasonal burnings by native human populations. Serengeti National Park, Tanzania. *(Photo by Stan Osolinski/Dembinsky Photo Associates)*

Table 15–3 Data for tropical wet and dry stations

	J	F	M	A	M	J	J	A	S	O	N	D	YR
Calcutta, India 22°32'N; 6 m													
Temp. (°C)	20.2	23.0	27.9	30.1	31.1	30.4	29.1	29.1	29.9	27.9	24.0	20.6	26.94
Precip. (mm)	13	24	27	43	121	259	301	306	290	160	35	3	1582
Cuiaba, Brazil 15°30'S; 165 m													
Temp. (°C)	27.2	27.2	27.2	26.6	25.5	23.8	24.4	25.5	27.7	27.7	27.7	27.2	26.5
Precip. (mm)	216	198	232	116	52	13	9	12	37	130	165	195	1375

appropriate technically, because some ecologists doubt that these grasslands are climatically induced. It is believed that woodlands once dominated this zone and that the savanna grasslands developed in response to seasonal burnings by native populations.

Temperature Characteristics

The temperature data in Table 15–3 show only modest differences between the wet tropics and the tropical wet and dry regions. Because of the somewhat higher latitude of most *Aw* stations, annual mean temperatures are slightly lower. In addition, the annual temperature range is a bit greater, varying from 3°C (5°F) to perhaps 10°C (20°F). The daily temperature range, however, still exceeds the annual variation.

Because seasonal fluctuations in humidity and cloudiness are more pronounced in *Aw* areas, daily temperature ranges vary noticeably during the year. Generally, they are small during the rainy season, when humidity and cloud cover are at a maximum, and large during droughts, when clear skies and dry air prevail. Furthermore, because of the more persistent summertime cloudiness, many *Aw* stations experience their warmest temperatures at the end of the dry season, just prior to the summer solstice. Thus, in the Northern Hemisphere, March, April, and May are often warmer than June and July.

Precipitation Characteristics

Because temperature regimes among the *A* climates are similar, the primary factor that distinguishes the *Aw* climate from *Af* and *Am* is precipitation. *Aw* stations typically receive 100 to 150 centimeters (40 to 60 inches) of rainfall each year, an amount often appreciably less than in the wet tropics. The most distinctive characteristic of this climate, however, is the *markedly seasonal character of the rainfall*—wet summers followed by dry winters. This is clearly shown by the climate diagram in Figure 15–5c.

The alternating wet and dry periods are due to the latitudinal position of the *Aw* climate region. It lies between the intertropical convergence zone, with its sultry weather and convective thundershowers, and the stable, subsiding air of the subtropical highs. Following the spring equinox, the ITCZ and the other wind and pressure belts all shift poleward as they migrate with the vertical rays of the Sun (Figure 15–8). With the advance of the ITCZ into a region, the summer rainy season commences with weather patterns typical of the wet tropics. Later, with the retreat of the ITCZ back toward the equator, the subtropical high advances into the region and brings with it intense drought.

During the dry season, the landscape grows parched and nature seems to become dormant as water-stressed trees shed their leaves and the abundant tall grasses turn brown and wither. The dry season's duration depends primarily on distance from the ITCZ. Typically, the farther an *Aw* station is from the equator, the shorter the period of ITCZ control and the longer the locale will be influenced by the stable subtropical high. Consequently, the higher the latitude, the longer the dry season and the shorter the wet period.

The movement of the ITCZ is essential to understanding rainfall distribution in the tropics. This is clear from Table 15–4, which shows precipitation data for six African stations. Notice that at Malduguri (farthest north) and Francistown (farthest south), there are single rainfall maxima that occur when the ITCZ reaches its most poleward positions. Between these stations, double maxima represent the passage of the ITCZ on its way to and from these extreme locations. It is important to remember that these statistics represent long-term averages and that on a year-to-year basis the movement of the ITCZ is far from regular. There is, nevertheless, no doubt that in the tropics, rainfall follows the Sun.

The Monsoon

In much of India, Southeast Asia, and portions of Australia the alternating periods of rainfall and drought characteristic of the *Aw* precipitation regime are associated with the **monsoon**. The term is derived from the Arabic word

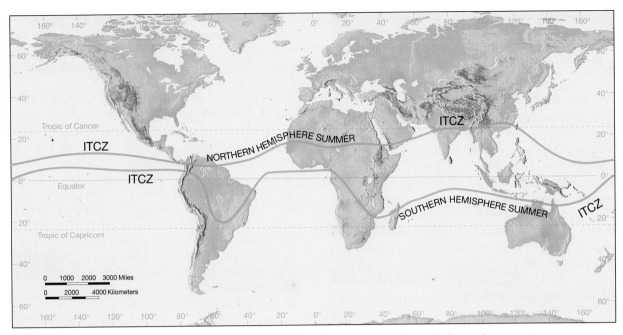

Figure 15–8 The seasonal migration of the ITCZ strongly influences precipitation distribution in the tropics.

mausim, which means "season" and typically refers to wind systems that have a pronounced seasonal reversal of direction. During the summer, conditions are conducive to rainfall because humid, unstable air moves from the oceans toward the land. In winter this reverses, and a dry wind, having its origin over the continent, blows toward the sea.

This monsoonal circulation system develops partly in response to the differences in annual temperature variations between continents and oceans. In principle, the processes associated with the monsoon are similar to those described in connection with the land and sea breeze (Chapter 7) except that the scales, in both time and space, are much larger.

During spring in the Northern Hemisphere, an irregular area of thermally induced low pressure gradually develops over the interior of southern Asia. It is further strengthened by the poleward advance of the ITCZ (Figure 15–8). Thus, the *summertime* circulation is from the higher pressure over the ocean toward the lower pressure over the continent. As winter approaches, winds reverse

Table 15–4 Rainfall regimes and the movement of the ITCZ in Africa											
	J	**F**	**M**	**A**	**M**	**J**	**J**	**A**	**S**	**O**	**N**	**D**
Malduguri, Nigeria, 11°51'N											
Precip. (mm)	0	0	0	7.6	40.6	68.6	180.3	**220.9**	106.6	17.7	0	0
Yaounde, Cameroon, 3°53'N											
Precip. (mm)	22.8	66.0	147.3	170.1	**195.6**	152.4	73.7	78.7	213.4	**294.6**	116.8	22.9
Kisangani, Zaire, 0°26'N											
Precip. (mm)	53.3	83.8	**177.8**	157.5	137.2	114.3	132.0	165.1	182.9	**218.4**	198.1	83.8
Luluabourg, Zaire, 5°54'S											
Precip. (mm)	137.2	142.2	**195.6**	193.0	83.8	20.3	12.7	58.4	116.8	165.1	**231.1**	226.0
Zomba, Malawi, 15°23'S											
Precip. (mm)	274.3	**289.6**	198.1	76.2	27.9	12.7	5.1	7.6	17.8	17.8	134.6	**279.4**
Francistown, Botswana, 21°13'S											
Precip. (mm)	**106.7**	78.7	71.1	17.8	5.1	2.5	0	0	0	22.9	58.4	86.4

direction as the ITCZ migrates southward and a deep anticyclone develops over the chilled continent. By *midwinter* dry winds blow from the continent southward to converge on Australia and southern Africa.

The *Cw* Variant

An examination of Figure 15–3 reveals areas adjacent to the wet and dry tropics in southern Africa, South America, northeastern India, and China that are designated *Cw*. Although *C* has been substituted for *A*, indicating that these regions are subtropical instead of tropical, the *Cw* climate is nevertheless just a variant of *Aw*, for the only major difference is somewhat lower temperatures. In Africa and South America, *Cw* climates are highland extensions of *Aw*. Because they occupy elevated sites, the warmer temperatures of the adjacent wet and dry tropics are reduced. In India and China, *Cw* areas are middle-latitude extensions of the tropical monsoon realm. In some cases, especially in India, the *Cw* areas are barely poleward enough to have winter temperatures below those of the *A* climates.

The Dry Climates (*B*)

The dry regions of the world cover some 42 million square kilometers (nearly 16.5 million square miles), or about 30 percent, of Earth's land surface. No other climatic group covers so large a land area (Figure 15–9).

The most characteristic feature of dry climates, aside from their meager yearly rainfall, is that precipitation is very unreliable. Generally, the smaller the mean annual rainfall, the greater its variability. As a result, yearly rainfall averages are often misleading.

For example, during one seven-year period, Trujillo, Peru, had an average rainfall of 6.1 centimeters per year (2.4 inches). Yet a closer look reveals that during the first 6 years and 11 months of the period, the station received a scant 3.5 centimeters (1.4 inch) (an annual average of slightly more than 0.5 centimeter). Then during the twelfth month of the seventh year, 39 centimeters (15.2 inches) of rain fell, 23 centimeters (9 inches) of it during a three-day span.

This extreme case illustrates the irregularity of rainfall in most dry regions. Also, there are usually more years when rainfall totals are below the average than years when they are above. As the foregoing example shows, the occasional wet period tends to lift the average (see Box 15–3).

What Is Meant by "Dry"?

It is important to realize that dryness is relative and simply refers to any *water deficiency*. Thus, climatologists define a **dry climate** as one in which the *yearly precipitation is less than the potential water loss by evaporation*. Dryness, then, is not only related to annual rainfall. It is also a function of evaporation, which in turn depends closely on temperature. As temperatures climb, potential evaporation also increases.

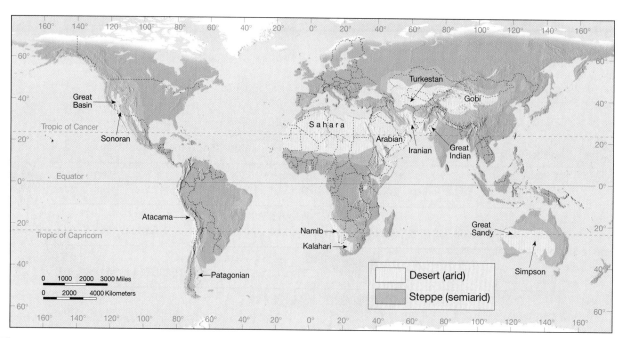

Figure 15–9 Arid and semiarid climates cover about 30 percent of Earth's land area. These dry (*B*) climates constitute the single largest climatic group.

Box 15–3 Common Misconceptions About Deserts

Deserts are hot, lifeless, sand-covered landscapes shaped largely by the force of wind. The preceding statement summarizes the image of arid regions that many people hold, especially those living in more humid places. Is it an accurate view? The answer is no. Although there are clearly elements of reality in such an impression, it is a generalization that contains a number of misconceptions (Figure 15–D).

One common fallacy about deserts is that they are lifeless or almost lifeless. Although reduced in amount and different in character, both plant and animal life are usually present. Desert plants all have one characteristic in common: They have developed adaptations that make them highly tolerant of drought. Many have waxy leaves, stems, or branches or a thickened cuticle (outermost protective layer) to reduce water loss. Others have very small leaves or no leaves at all. Also, the roots of some species often extend to great depths in order to tap the moisture found there, whereas others produce a shallow but widespread root system that enables them to absorb great amounts of moisture quickly from the infrequent desert downpours. Thus, although widely dispersed and providing little ground cover, plants of many kinds flourish in the desert.

A second widely held belief about the world's dry lands is that they are always hot. This fact seems to be reinforced by commonly quoted temperature statistics. The highest accepted temperature record for the United States as well as the entire Western Hemisphere is 57°C (134°F). This long-standing record was set at Death Valley, California, on July 10, 1913. The nearly 59°C (137°F) reading in Azizia, Libya, in North Africa's Sahara Desert on September 13, 1922, is the world record. Despite these remarkably high figures, cold temperatures are also experienced in desert regions.

For example, the average daily minimum in January at Phoenix, Arizona, is 17°C (35°F), just barely above freezing. At Ulan Bator in Mongolia's Gobi Desert, the average *high* temperature in January is only –19°C (–2°F)! Unlike any other climate group, the *B* climates are defined by precipitation criteria instead of temperature. Therefore, dry climates are found from the tropics to the high middle latitudes. Consequently, although tropical deserts lack a cold season, deserts in the middle latitudes do experience seasonal temperature changes.

The last two commonly held misconceptions about the world's deserts are more geologic than climatic. One mistaken assumption is that they consist of mile after mile of drifting sand. It is true that sand accumulations do exist in some areas and may be striking features, but they represent only a small percentage of the total desert area. For example, in the Sahara, the world's largest desert, accumulations of sand cover only one-tenth of its area. The sandiest of all deserts is the Arabian, one-third of which consists of sand.

The final mistaken assumption is the seemingly logical idea that wind is the most important agent of erosion in deserts. Although wind is relatively more significant in dry areas than anywhere else, most erosional landforms in deserts are created by running water. When the rains come, they usually take the form of thunderstorms. Because the heavy rain associated with these storms cannot all soak in, rapid runoff results. Without a thick vegetative cover to protect the ground, erosion is great.

Figure 15–D Snow blankets the rocky ground of the Sonoran Desert in southern Arizona. *(Photo by Jack W. Dykinga & Associates)*

For example, 25 centimeters (10 inches) of precipitation may be sufficient to support forests in northern Scandinavia, where evaporation is slight into the cool, humid air and a surplus of water remains in the soil. However, the same amount of rain falling on Nevada or Iran supports only a sparse vegetative cover because evaporation into the hot, dry air is great. Clearly, no specific amount of precipitation can define a dry climate.

To establish a meaningful boundary between dry and humid climates, the Köppen classification uses formulas that involve three variables: (1) average annual precipitation, (2) average annual temperature, and (3) seasonal distribution of precipitation.

The use of average annual precipitation is obvious. Average annual temperature is used because it is an index of evaporation; the amount of rainfall defining the humid–dry boundary is greater where mean annual temperatures are high, and less where temperatures are low. The third variable, seasonal distribution of precipitation, is also related to this idea. If rain is concentrated in the warmest months, loss to evaporation is greater than if it is concentrated in the cooler months. Thus, considerable differences exist in the precipitation amounts received at various stations in the *B* climates.

Table 15–5 summarizes these differences. For example, if a station with an annual mean of 20°C (68°F) and a summer wet season ("winter dry") receives less than 680 millimeters (26.5 inches) of precipitation per year, it is classified as *dry*. If the rain falls primarily in winter ("summer dry"), however, the station must receive only 400 millimeters (15.6 inches) or more to be considered humid. If the precipitation is more evenly distributed, the figure defining the humid–dry boundary is between the other two.

Within the regions defined by a general water deficiency, there are two climatic types: **arid** or **desert** (**BW**) and **semiarid** or **steppe** (**BS**). Figure 15–10 presents climatic diagrams for both types. Stations a and b

are in the subtropics and c and d are in the middle latitudes. Deserts and steppes have many features in common; their differences are primarily a matter of degree. The semiarid is a marginal and more humid variant of the arid and represents a transition zone that surrounds the desert and separates it from the bordering humid climates. The arid–semiarid boundary is commonly set at one-half the annual precipitation separating dry regions from humid. Thus, if the humid–dry boundary happens to be 40 centimeters, the steppe–desert boundary will be 20 centimeters.

Subtropical Desert (BWh) and Steppe (BSh)

The heart of low-latitude dry climates lies along the vicinity of the Tropic of Cancer and the Tropic of Capricorn. A glance at Figure 15–9 reveals a virtually unbroken desert environment stretching for more than 9300 kilometers (nearly 6000 miles) from the Atlantic coast of North Africa to the dry lands of northwestern India. In addition to this single great expanse, the Northern Hemisphere contains another much smaller area of subtropical desert (*BWh*) and steppe (*BSh*) in northern Mexico and the southwestern United States. In the Southern Hemisphere, dry climates dominate Australia. Almost 40 percent of that continent is desert, and much of the remainder is steppe.

In addition, arid and semiarid areas exist in southern Africa and make a rather limited appearance in coastal Chile and Peru. The distribution of this dry subtropical realm is primarily a consequence of the subsidence and market stability of the subtropical highs.

Precipitation. Within subtropical deserts the scanty precipitation is both infrequent and erratic. Indeed, no well-defined seasonal precipitation pattern exists in subtropical deserts. The reason is that these areas are too far poleward to be influenced by the ITCZ and too far equatorward to benefit from the frontal and cyclonic precipitation of the middle latitudes. Even during summer, when daytime heating produces a steep environmental lapse rate and considerable convection, clear skies still rule. In this case, subsidence aloft prevents the lower air, with its modest moisture content, from rising high enough to penetrate the condensation level.

The situation is different in the semiarid transitional belts surrounding the desert. Here a seasonal rainfall pattern becomes better defined. As shown by the data for Dakar in Table 15–6, stations located on the equatorward side of low-latitude deserts have a brief period of relatively heavy rainfall during the summer, when the ITCZ is farthest poleward. The rainfall regime should

Table 15–5 Average annual precipitation (mm) at *BS*-humid boundary

Average Annual Temp (°C)	Summer Dry Season	Even Distribution	Winter Dry Season
5	100	240	380
10	200	340	480
15	300	440	580
20	400	540	680
25	500	640	780
30	600	740	880

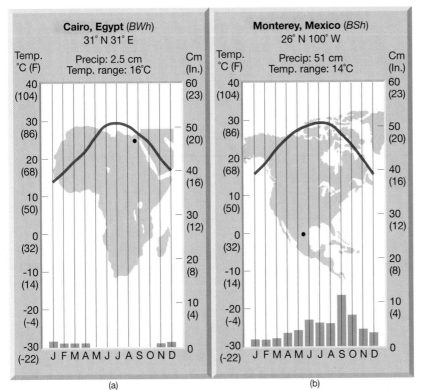

(a)

(b)

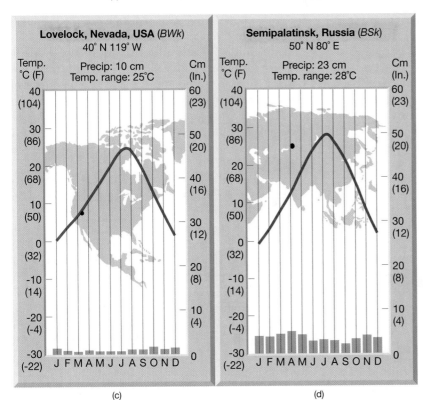

(c)

(d)

Figure 15–10 Climate diagrams for representative arid and semiarid stations. Cairo and Lovelock are classified as deserts; Monterey and Semipalatinsk are steppes. Stations (a) and (b) are in the subtropics, whereas (c) and (d) are middle-latitude sites.

Table 15–6 Data for subtropical steppe and desert stations

	J	F	M	A	M	J	J	A	S	O	N	D	YR
Marrakech, Morocco, 31°37'N; 458 m													
Temp (°C)	11.5	13.4	16.1	18.6	21.3	24.8	28.7	28.7	25.4	21.2	16.5	12.5	19.9
Precip. (mm)	28	28	33	30	18	8	3	3	10	20	28	33	242
Dakar, Senegal, 14°44'N; 23 m													
Temp. (°C)	21.1	20.4	20.9	21.7	23.0	26.0	27.3	27.3	27.5	27.5	26.0	25.2	24.49
Precip. (mm)	0	2	0	0	1	15	88	249	163	49	5	6	578
Alice Springs, Australia, 23°38'S, 570 m													
Temp. (°C)	28.6	27.8	24.7	19.7	15.3	12.2	11.7	14.4	18.3	22.8	25.8	27.8	20.8
Precip. (mm)	43	33	28	10	15	13	8	8	8	18	30	38	252

look familiar, for it is similar to that found in the adjacent wet and dry tropics, except that the amount is less and the drought lasts longer.

For steppe areas on the poleward margins of the tropical deserts, the precipitation regime is reversed. As the data for Marrakech illustrate (Table 15–6), the cool season is the period when nearly all precipitation falls. At this time of year middle-latitude cyclones often take more equatorward routes and so bring occasional periods of rain.

Temperature. The keys to understanding temperatures in the desert environment are humidity and cloud cover. The cloudless sky and low humidity allow an abundance of solar radiation to reach the ground during the day and permit the rapid exit of terrestrial radiation at night. As would be expected, relative humidities are low throughout the year. Relative humidities between 10 and 30 percent are typical at midday for interior locations.

Regarding cloud cover, desert skies are almost always clear (Figure 15–11). In the Sonoran Desert region of Mexico and the United States, for example, most stations receive nearly 85 percent of the possible sunshine. Yuma, Arizona, averages 91 percent for the year, with a low of 83 percent in January and a high of 98 percent in June. The Sahara has an average winter cloud cover of about 10 percent, which in summer drops to a mere 3 percent.

During summer, the desert surface heats rapidly after sunrise, because the clear skies permit almost all the solar energy to reach the surface, as we saw in preceding examples. By midafternoon ground-surface temperatures may approach 90°C (nearly 200°F)! Under such circumstances, it is not surprising that the world's highest temperatures are recorded in the subtropical deserts. Nor is it unexpected that the daily maximums at many stations during the hot season are consistently close to the absolute maximum (the highest temperature ever recorded at a station). At Abadán, Iran, the average daily

maximum in July is a scorching 44.7°C (112°F), or only 8.3°C (15°F) lower than the record high. Phoenix, Arizona, is not much better, recording an average July maximum of 40.5°C (105°F) compared with an absolute maximum of 47.7°C (118°F).

A contributing factor to the high ground and air temperatures is that little energy from insolation goes to power evaporation. Thus, almost all the energy goes to heating the surface. In contrast, humid regions are less likely to have such extreme ground and air temperatures, for more energy from solar radiation goes to evaporate water and less remains to heat the ground.

At night, temperatures typically drop rapidly, partly because the water vapor content of the air is fairly low. Ground-surface temperature is also a factor, however. Recall from the discussion of radiation in Chapter 2 that the higher the temperature of a radiating body, the faster it loses heat. Thus, applied to a desert setting, such environments not only heat up quickly by day but also cool rapidly at night.

Consequently, low-latitude deserts in the interiors of continents have the greatest daily temperature ranges on Earth. Daily ranges from 15 to 25°C (27 to 45°F) are common, and they occasionally reach even higher values. The highest daily temperature range ever recorded was at In Salah, Algeria, in the Sahara. On October 13, 1927, this station experienced a 24-hour range of 55.5°C (100°F), from 52.2 to –3.3°C (126 to 26°F).

Because most areas of *BWh* and *BSh* are poleward of the *A* climates, annual temperature ranges are the highest among the tropical climates. During the low-Sun period, averages are below those in other parts of the tropics, with monthly means of 16 to 24°C (60 to 75°F) being typical. Still, temperatures during the summer are higher than those in the humid tropics. Consequently, annual means at many subtropical desert and steppe stations are similar to those in the *A* climates.

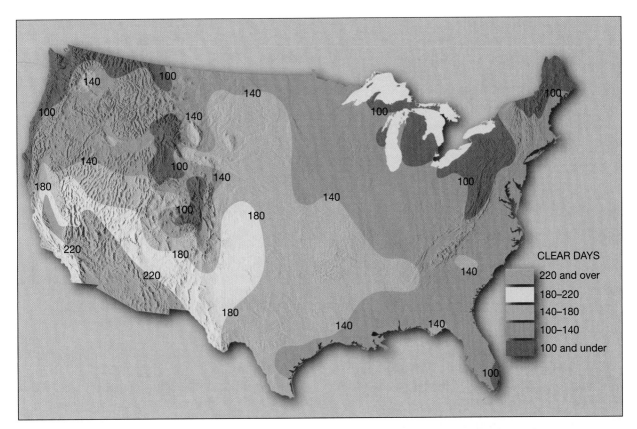

Figure 15–11 Average annual number of clear days. With few exceptions, desert skies are typically cloudless and hence receive a very high percentage of the possible sunlight. This is strikingly illustrated by the southwest desert.

West Coast Subtropical Deserts

Where subtropical deserts are found along the west coasts of continents, cold ocean currents have a dramatic influence on the climate. The principal west coast deserts are the Atacama in Peru and Chile and the Namib in southern and southwestern Africa. Other areas include portions of the Sonoran Desert in Baja, California, and coastal areas of the Sahara in northwestern Africa.

These areas deviate considerably from the general image we have of subtropical deserts. The most obvious effect of the cold current is reduced temperatures, as exemplified by the data for Lima, Peru, and Port Nolloth, South Africa (Table 15–7). Compared with other stations at similar latitudes, these places have lower annual mean temperatures and subdued annual and daily ranges. Port Nolloth, for example, has an annual mean of only 14°C (57°F) and an annual range of just 4°C (7.2°F). Contrast this with Durban, on the opposite side of South Africa, which has a yearly mean of 20°C (68°F) and an annual range twice that at Port Nolloth.

Although these stations are adjacent to the oceans, their yearly rainfall totals are among the lowest in the

Table 15–7 Data for west coast tropical desert stations

	J	F	M	A	M	J	J	A	S	O	N	D	YR
Port Nolloth, South Africa, 29°14′S; 7 m													
Temp. (°C)	15	16	15	14	14	13	12	12	13	13	15	15	14
Precip. (mm)	2.5	2.5	5.1	5.1	10.2	7.6	10.2	7.6	5.1	2.5	2.5	2.5	63.4
Lima, Peru, 12°02′S; 155 m													
Temp. (°C)	22	23	23	21	19	17	16	16	16	17	19	21	19
Precip. (mm)	2.5	T	T	T	5.1	5.1	7.6	7.6	7.6	2.5	2.5	T	40.5

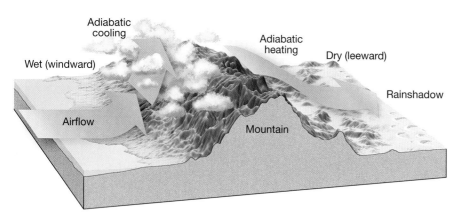

Figure 15–12 Mountains frequently contribute to the aridity of middle-latitude deserts and steppes by creating a rain shadow. Orographic lifting leads to precipitation on the windward slopes. By the time air reaches the leeward side of the mountains, much of the moisture has been lost.

world. The aridity of these coasts is intensified because the lower air is chilled by the cold offshore waters and hence further stabilized. In addition, the cold currents cause temperatures often to reach the dew point. As a result, these areas are characterized by high relative humidity and much advection fog and dense stratus cloud cover.

The point is that not all subtropical deserts are sunny, hot places with low humidity and cloudless skies. Indeed, the presence of cold currents causes west coast subtropical deserts to be relatively cool, humid places, often shrouded by low clouds or fog.

Middle-Latitude Desert (BWk) and Steppe (BSk)

Unlike their low-latitude counterparts, middle-latitude deserts and steppes are not controlled by the subsiding air masses of the subtropical highs. Instead, these dry lands exist principally because of their position in the deep interiors of large landmasses, far removed from the main moisture source—the oceans. In addition, the presence of high mountains across the paths of the prevailing winds further separates these areas from water-bearing maritime air masses (Figure 15–12). In North America the Coast ranges, Sierra Nevada, and Cascades are the foremost barriers; in Asia the great Himalayan chain prevents the summertime monsoon flow of moist air off the Indian Ocean from reaching far into the interior (Figure 15–13).

A glance at Figure 15–9 reveals that middle-latitude desert and steppe climates are most widespread in North America and Eurasia. The Southern Hemisphere lacks extensive land areas in the middle latitudes, so it

Figure 15–13 Scene in Nevada's Great Basin Desert. The aridity of this region is enhanced greatly by the rain shadow created by the Sierra Nevada. *(Photo by Charlie Ott/Photo Researchers, Inc.)*

Table 15–8 Data for middle-latitude steppe and desert stations

	J	F	M	A	M	J	J	A	S	O	N	D	YR
Ulan Bator, Mongolia, 47°55'N; 1311 m													
Temp. (°C)	−26	−21	−13	−1	6	14	16	14	9	−1	−13	−22	−3
Precip. (mm)	1	2	3	5	10	28	76	51	23	7	4	3	213
Denver, Colorado, 39°32'N; 1588 m													
Temp. (°C)	0	1	4	9	14	20	24	23	18	12	5	2	11
Precip. (mm)	12	16	27	47	61	32	31	28	23	24	16	10	327

has a much smaller area of *BWk* and *BSk*. It is found only at the southern tip of South America in the rain shadow of the towering Andes.

Like subtropical deserts and steppes, the dry regions of the middle latitudes have meager and unreliable precipitation. Unlike the dry lands of the low latitudes, however, these more poleward regions have much lower winter temperatures and hence lower annual means and higher annual ranges of temperature.

The data in Table 15–8 illustrate this point nicely. The data also reveal that rainfall is most abundant during the warm months. Although not all *BWk* and *BSk* stations have a summer precipitation maximum, most do because in winter high pressure and cold temperatures tend to dominate the continents. Both factors oppose precipitation. In summer, however, conditions are somewhat more conducive to cloud formation and precipitation because the anticyclone disappears over the heated continent, and higher surface temperatures and greater mixing ratios prevail.

Humid Middle-Latitude Climates with Mild Winters (*C*)

The Köppen classification recognizes two groups of humid middle-latitude climates. One group has mild winters (the *C* climates) and the other experiences severe winters (the *D* climates). The following three sections pertain to the *C*-type mild winter group. Figure 15–14 presents climate diagrams of the three types of *C* climates.

Humid Subtropical Climate (*Cfa*)

Humid subtropical climates are found on the eastern sides of the continents, in the 25 to 40° latitude range. They dominate the southeastern United States and other similarly situated areas: all of Uruguay and portions of Argentina and southern Brazil in South America, eastern China and southern Japan in Asia, and the eastern coast of Australia.

In the summer a visitor to the humid subtropics would experience hot, sultry weather of the type expected in the rainy tropics. Daytime temperatures are generally in the

lower 30s (°C) (high 80s °F), but it is not uncommon for the thermometer to reach into the upper 30s (90s °F) or even 40 (more than 100°F) on many afternoons. Because the mixing ratio and relative humidity are high, the night brings little relief. An afternoon or evening thunderstorm is possible because these areas experience them between 40 and 100 days each year, the majority during summer.

The primary reason for the tropical summer weather in *Cfa* regions is the dominating influence of maritime tropical air masses. During the summer months this warm, moist, unstable air moves inland from the western portions of the oceanic subtropical anticyclone. As the maritime tropical (*mT*) air passes over the heated continent, it becomes increasingly unstable, giving rise to the common convectional showers and thunderstorms.

As summer turns to autumn, the humid subtropics lose their similarity to the rainy tropics. Although winters are mild, frosts are common in higher-latitude *Cfa* areas and occasionally plague the tropical margins. The winter precipitation is also different in character. Some is in the form of snow, and most is generated along fronts of the frequent middle-latitude cyclones that sweep over these regions.

Because the land surface is colder than the maritime air, the air becomes chilled in its lower layers as it moves poleward. Consequently, convectional showers are rare, for the stabilized *mT* air masses produce clouds and precipitation only when forced to rise.

The data for two humid subtropical stations in Table 15–9 serve to summarize the general characteristics of the *Cfa* climate. Yearly precipitation usually exceeds 100 centimeters (39 inches), and the rainfall is well distributed throughout the year. Summer normally brings the most precipitation, but there is considerable variation. In the United States, for example, precipitation in the Gulf states is very evenly distributed. But as one moves poleward or toward the drier western margins, much more falls in summer. Some coastal stations have rainfall maximums in late summer or autumn when tropical cyclones or their remnants visit the area. In Asia the well-developed monsoon circulation favors a summer precipitation maximum (see the Chinese station in Figure 15–14a).

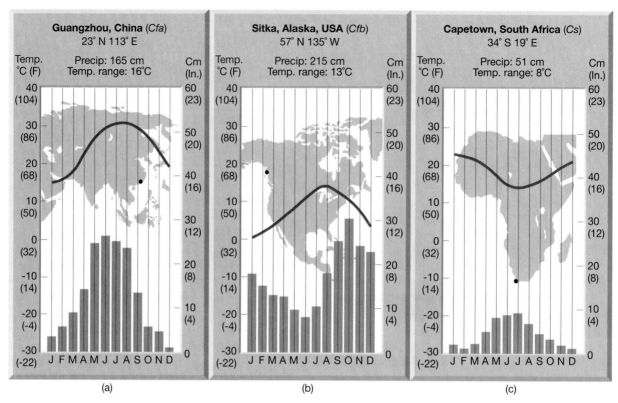

Figure 15–14 Each of these climate diagrams represents a type of C climate: (a) humid subtropical, (b) marine west coast, and (c) dry-summer subtropical.

Temperature figures show that summer temperatures are comparable to the tropics and that winter values are markedly lower. This is to be expected, because the higher-latitude position of the subtropics experiences a wider variation in Sun angle and day length, plus occasional (even frequent) invasions of continental polar (*cP*) air masses during winter.

The Marine West Coast Climate (*Cfb*)

On the western (windward) side of continents from about 40 to 65° north and south latitude is a climate region dominated by the onshore flow of oceanic air. The prevalence of maritime air masses means mild winters, cool summers, and ample rainfall throughout the year. In North America this **marine west coast climate** (*Cfb*) extends from near the U.S.–Canada border northward as a narrow belt into southern Alaska. A similar slender strip occurs in South America along the coast of Chile. In both instances high mountains parallel the coast and prevent the marine climate from penetrating far inland. The largest area of *Cfb* climate is in Europe, for here there is no mountain barrier blocking the movement of cool maritime air from the North Atlantic. Other locations include most of New Zealand as well as tiny slivers of South Africa and Australia.

Table 15–9 Data for humid subtropical stations

	J	F	M	A	M	J	J	A	S	O	N	D	YR
New Orleans, Louisiana, 29°59'N; 1 m													
Temp. (°C)	12	13	16	19	23	26	27	27	25	21	15	13	20
Precip. (mm)	98	101	136	116	111	113	171	136	128	72	85	104	1371
Buenos Aires, Argentina, 34°35'S; 27 m													
Temp. (°C)	24	23	21	17	14	11	10	12	14	16	20	22	17
Precip. (mm)	104	82	122	90	79	68	61	68	80	100	90	83	1027

Table 15–10	Data for marine west coast stations											
	J	**F**	**M**	**A**	**M**	**J**	**J**	**A**	**S**	**O**	**N**	**D**	**YR**
Vancouver, British Columbia, 49°11'N; 0 m												
Temp. (°C)	2	4	6	9	13	15	18	17	14	10	6	4	10
Precip. (mm)	139	121	96	60	48	51	26	36	56	117	142	156	1048
London, UK, 51°28'N; 5 m												
Temp. (°C)	4	4	7	9	12	16	18	17	15	11	7	5	10
Precip. (mm)	54	40	37	38	46	46	56	59	50	57	64	48	595

The data for representative marine west coast stations in Table 15–10 and Figure 15–14b reveal no pronounced dry period, although monthly precipitation drops during the summer. The reduced summer rainfall is due to the poleward migration of the oceanic subtropical highs. Although the areas of marine west coast climate are too far poleward to be dominated by these dry anticyclones, their influence is sufficient to cause a decrease in warm-season rainfall.

A comparison of precipitation data for London and Vancouver (Table 15–10) also demonstrates that coastal mountains have a significant influence on yearly rainfall. Vancouver has about two and a half times that of London. In settings like Vancouver's, the precipitation totals are higher not only because of orographic uplift but also because the mountains slow the passage of cyclonic storms, allowing them to linger and drop a greater quantity of water.

The ocean is near, so winters are mild and summers relatively cool. Therefore, a low annual temperature range is characteristic of the marine west coast climate. Because cP air masses generally drift eastward in the zone of the westerlies, periods of severe winter cold are rare. The western edge of North America is especially sheltered from incursions of frigid continental air by the high mountains that intervene between the coast and the source regions for cP air masses. Because no such mountain barrier exists in Europe, cold waves there are somewhat more frequent.

The ocean's control of temperatures can be further demonstrated by a look at temperature gradients (changes in temperature per unit distance). Although this climate encompasses a wide latitudinal span, temperatures change much more abruptly moving inland from the coast than they do in a north–south direction. The transport of heat from the oceans more than offsets the latitudinal variation in the receipt of solar energy. For example, in both January and July the temperature change from coastal Seattle to inland Spokane, a distance of about 375 kilometers (230 miles), is equal to the variation between Seattle and Juneau, Alaska. Juneau is about 11° of latitude, or roughly 1200 kilometers (750 miles), north of Seattle.

The Dry-Summer Subtropical (Mediterranean) Climate *(Csa, Csb)*

The **dry-summer subtropical climate** is typically located along the west sides of continents between latitudes 30 and 45°. Situated between the marine west coast climate on the poleward side and the subtropical steppes on the equatorward side, this climate region is transitional in character. It is the only humid climate that has a pronounced winter rainfall maximum, a feature that reflects its intermediate position (Figure 15–14c).

In summer the region is dominated by the stable eastern side of the oceanic subtropical highs. In winter, as the wind and pressure systems follow the Sun equatorward, it is within range of the cyclonic storms of the polar front. Thus, during the course of a year, these areas alternate between being a part of the dry tropics and being an extension of the humid middle latitudes. Although middle-latitude changeability characterizes the winter, tropical constancy describes the summer.

As was the case for the marine west coast climate, mountain ranges limit the dry-summer subtropics to a relatively narrow coastal zone in both North and South America. Because Australia and southern Africa barely extend to the latitudes where dry-summer climates exist, the development of this climatic type is limited on these continents as well.

Because of the arrangement of the continents and their mountain ranges, inland development occurs only in the Mediterranean basin (Figure 15–15). Here the zone of subsidence extends far to the east in summer; in winter the sea is a major route of cyclonic disturbances. Because the dry-summer climate is particularly extensive in this region, the name **Mediterranean climate** is often used as a synonym.

Temperature. Two types of Mediterranean climate are recognized and are based primarily on summertime temperatures. The cool summer type *(Csb)*, as exemplified by San Francisco and Santiago, Chile (Table 15–11), is limited to coastal areas. Here the cooler

Figure 15–15 Rolling landscape with olive trees in Tuscany (Italy). The dry-summer subtropical climate is especially well developed in the Mediterranean region. *(Photo by Grilly Bernard/Tony Stone Images)*

summer temperatures one expects on a windward coast are further intensified by cold ocean currents.

The data for Izmir, Turkey, and Sacramento illustrate the features of the warm summer type (*Csa*). At both places winter temperatures are not very different from those in the *Csb* type. But Sacramento, in the Central Valley of California, is removed from the coast, and Izmir is bordered by the warm waters of the Mediterranean; consequently, summer temperatures are noticeably higher. As a result, annual temperature ranges are also higher in *Csa* areas.

Precipitation. Yearly precipitation within the dry-summer subtropics ranges between about 40 and 80 centimeters (16 and 31 inches). In many areas such amounts mean that a station barely escapes being classified as semiarid. As a result, some climatologists refer to the dry-summer climate as *subhumid* instead of humid. This is especially true along the equatorward margins because rainfall totals increase in the poleward direction. Los Angeles, for example, receives 38 centimeters (15 inches) of precipitation annually, whereas San Francisco, 400 kilometers (250 miles) to the north, receives 51 centimeters (20 inches) per year. Still farther north, at Portland, Oregon, the yearly rainfall average is over 90 centimeters (35 inches).

Humid Continental Climates with Severe Winters (D)

The *C* climates just described have mild winters. By contrast, *D* climates experience severe winters. In this and the following section, two types of *D* climates are discussed—the humid continental and the subarctic. Climate diagrams of representative locations are shown in Figure 15–16.

This climate region is dominated by the polar front and thus is a battleground for tropical and polar air masses. No other climate experiences such rapid nonperiodic changes in the weather. Cold waves, heat waves, blizzards, and heavy downpours are all yearly events in the humid continental realm (see Box 15–4).

Humid Continental Climate *(Dfa)*

The **humid continental climate** (*Dfa*), as its name implies, is a land-controlled climate. It is the product of broad continents located in the middle latitudes. Because continentality is a basic feature, this climate does not occur in the Southern Hemisphere, where the middle-latitude zone is dominated by the ocean. Instead, it is confined to central and eastern North America and Eurasia in the latitude range 40–50°N.

Table 15–11 Data for dry-summer subtropical stations

	J	F	M	A	M	J	J	A	S	O	N	D	YR
San Francisco, California, 37°37'N; 5 m													
Temp. (°C)	9	11	12	13	15	16	17	17	18	16	13	10	14
Precip. (mm)	102	88	68	33	12	3	0	1	5	19	40	104	475
Sacramento, California, 38°35'N; 13 m													
Temp. (°C)	8	10	12	16	19	22	25	24	23	18	12	9	17
Precip. (mm)	81	76	60	36	15	3	0	1	5	20	37	82	416
Izmir, Turkey, 38°26'N; 25 m													
Temp. (°C)	9	9	11	15	20	25	28	27	23	19	14	10	18
Precip. (mm)	141	100	72	43	39	8	3	3	11	41	93	141	695
Santiago, Chile, 33°27'S; 512 m													
Temp. (°C)	19	19	17	13	11	8	8	9	11	13	16	19	14
Precip. (mm)	3	3	5	13	64	84	76	56	30	13	8	5	360

It may at first seem unusual that a continental climate should extend eastward to the margins of the ocean. However, the prevailing atmospheric circulation is from the west, so deep and persistent incursions of maritime air from the east are unlikely.

Temperature. Both winter and summer temperatures in the *Dfa* climate are relatively severe. Consequently, annual temperature ranges are great. A comparison of the stations in Table 15–12 illustrates this. July means are generally near and often above 20°C (68°F), so

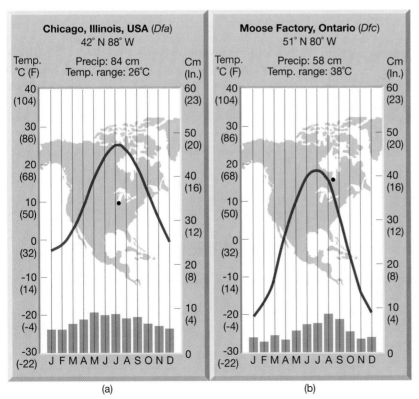

Figure 15–16 *D* climates are associated with the interiors of large landmasses in the mid-to-high latitudes of the Northern Hemisphere. Although winters can be harsh in Chicago's humid continental (*Dfa*) climate (a), the subarctic environment (*Dfc*) of Moose Factory is more extreme (b).

Box 15–4   Atmospheric Hazard: Understanding Drought*

*D*rought is a period of abnormally dry weather that persists long enough to produce a significant hydrologic imbalance such as crop damage or water supply shortages. Drought severity depends upon the degree of moisture deficiency, its duration, and the size of the affected area.

During the summer of 1999 parts of the eastern United States were suffering the effects of a yearlong drought. Four mid-Atlantic states experienced the driest summer since record keeping began in 1894. Crops were lost or stunted, and the water volume in many reservoirs and rivers dropped to seriously low levels. The drought, which began in 1998, included a heat wave and a rainfall deficit of 20 to 45 centimeters (8 to 18 inches) in many affected areas.

Although natural disasters such as floods and hurricanes usually generate more attention, droughts can be just as devastating and carry a bigger price tag. On the average, droughts cost the United States $6 to $8 billion annually compared to $2.4 billion for floods and $1.2 to $4.8 billion for hurricanes. Direct economic losses from a severe drought in 1988 were estimated at $40 billion.

Many people consider drought a rare and random event, yet it is actually a normal, recurring feature of climate. It occurs in virtually all climate zones, although its characteristics vary from region to region. The concept of drought differs from that of aridity. Drought is a temporary happening, whereas aridity describes regions where low rainfall is a permanent feature of the climate.

Drought is different from other natural hazards in several ways. First, it occurs in a gradual, "creeping" way, making its onset and end difficult to determine. The effects of drought accumulate slowly over an extended time span and sometimes linger for years after the drought has ended. Second, there is not a precise and universally accepted definition of drought. This adds to the confusion about whether or not a drought is actually occurring and, if it is, its severity. Third, drought seldom produces structural damages, so its social and economic effects are less obvious than damages from other natural disasters.

No single definition of drought works in all circumstances. The various definitions reflect many perspectives that use a number of different variables. Most definitions pertain to a specific region because of differences in climate characteristics. Thus, it is usually difficult to apply a definition that was developed for one region to another.

Definitions reflect four basic approaches to measuring drought: meteorological, agricultural, hydrological, and socioeconomic. *Meteorological drought* deals with the degree of dryness based on the departure of precipitation from normal values and the duration of the dry period. *Agricultural drought* is usually linked to a deficit of soil moisture. A plant's need for water depends on prevailing weather conditions, biological characteristics of the particular plant, its stage of growth, and various soil properties. *Hydrological drought* refers to deficiencies in surface and subsurface water supplies. It is measured as streamflow and as lake, reservoir, and groundwater levels (Figure 15–E). There is a time lag between the onset of dry conditions and a drop in streamflow, or the lowering of lakes, reservoirs, or groundwater levels. So, hydrological measurements are not the earliest indicators of drought. *Socioeconomic drought* is a reflection of what happens when a physical water shortage affects people. Socioeconomic drought occurs when the demand for an economic good exceeds supply as a result of a shortfall in water supply. For example, drought can result in significantly reduced hydroelectric power production, which in turn may require conversion to more expensive fuels and/or significant energy shortfalls.

There is a sequence of impacts associated with meteorological, agricultural, and hydrological drought (Figure 15–F). When meteorological drought begins, the agricultural sector is usually the first to be affected because of

Figure 15–E One indicator of hydrological drought is significantly diminished streamflow. Continuous records of streamflow are collected by the U.S. Geological Survey at more than 7000 gauging stations in the United States. This station is on the Rio Grande, south of Taos, New Mexico. *(Photo by E. J. Tarbuck)*

its heavy dependence on soil moisture. Soil moisture is rapidly depleted during extended dry periods. If precipitation deficiencies continue, those dependent on rivers, reservoirs, lakes, and groundwater may be affected.

When precipitation returns to normal, meteorological drought comes to an end. Soil moisture is replenished first, followed by streamflow, reservoirs and lakes, and finally groundwater. Thus, drought impacts may diminish rapidly in the agricultural sector because of its reliance on soil moisture but linger for months or years in other sectors that depend on stored surface or subsurface water supplies. Groundwater users, who are often the last to be affected following the onset of meteorological drought, may also be the last to experience a return to normal water levels. The length of the recovery period depends upon the intensity of the meteorological drought, its duration, and the quantity of precipitation received when the drought ends.

The impacts suffered because of drought are the product of both the meteorological event and the vulnerability of society to periods of precipitation deficiency. As demand for water increases as a result of population growth and regional population shifts, future droughts can be expected to produce greater impacts whether or not there is any increase in the frequency or intensity of meteorological drought.

*Based in part on material prepared by the National Drought Mitigation Center (http://enso.unl.edu/ndmc/).

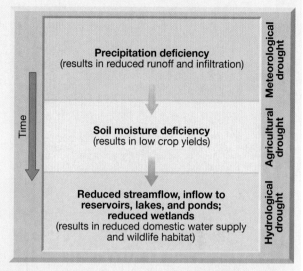

Figure 15–F Sequence of drought impacts. After the onset of meteorological drought, agriculture is affected first, followed by reductions in streamflow and water levels in lakes, reservoirs, and underground. When meteorological drought ends, agricultural drought ends as soil moisture is replenished. It takes a considerably longer time span for hydrological drought to end.

summertime temperatures, although lower in the north than in the south, are not markedly different.

This is illustrated by the temperature distribution map of the eastern United States in Figure 15–17a. The summer map has only a few widely spaced isotherms, indicating a weak summer temperature gradient. The winter map, however, shows a stronger temperature gradient (Figure 15–17b). The decrease in midwinter values with increasing latitude is appreciable. Confirming this, Table 15–12 reveals that the temperature change

Table 15–12 Data for humid continental stations													
	J	**F**	**M**	**A**	**M**	**J**	**J**	**A**	**S**	**O**	**N**	**D**	**YR**
Omaha, Nebraska, 41°18'N; 330 m													
Temp. (°C)	–6	–4	3	11	17	22	25	24	19	12	4	–3	10
Precip. (mm)	20	23	30	51	76	102	79	81	86	48	33	23	652
New York City, 40°47'N; 40 m													
Temp. (°C)	–1	–1	3	9	15	21	23	22	19	13	7	1	11
Precip. (mm)	84	84	86	84	86	86	104	109	86	86	86	84	1065
Winnipeg, Canada, 49°54'N; 240 m													
Temp. (°C)	–18	–16	–8	3	11	17	20	19	13	6	–5	–13	3
Precip. (mm)	26	21	27	30	50	81	69	70	55	37	29	22	517
Harbin, Manchuria, 45°45'N; 143 m													
Temp. (°C)	–20	–16	–6	6	14	20	23	22	14	6	–7	–17	3
Precip. (mm)	4	6	17	23	44	92	167	119	52	36	12	5	557

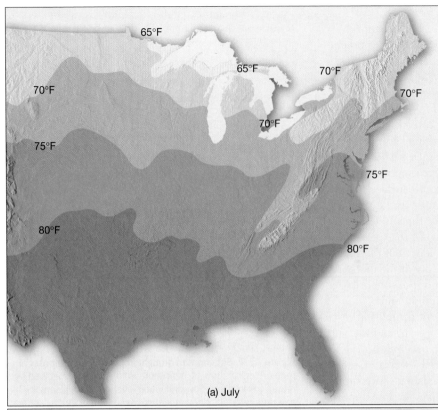

(a) July

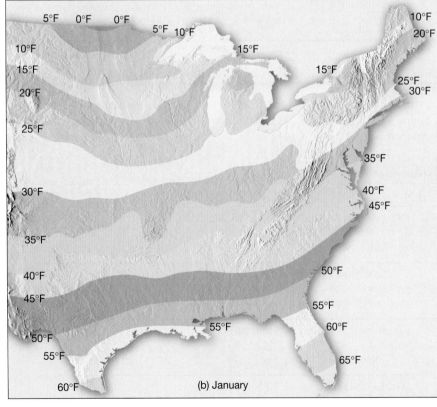

(b) January

Figure 15–17 (a) During the summer months, north–south temperature variations in the eastern United States are small—that is, the temperature gradient is weak. (b) In winter, however, north–south temperature contrasts are sharp.

between Omaha and Winnipeg is more than twice as great in winter as in summer.

Because of the steeper winter temperature gradient, shifts in wind direction during the cold season often result in sudden large temperature changes. Such is not the case in summer, for temperatures throughout the region are more uniform.

Annual temperature ranges also vary within this climate, generally increasing from south to north and from the coast toward the interior. Comparing the data for Omaha and Winnipeg illustrates the first situation, and comparing New York City and Omaha illustrates the second.

Precipitation. Records of the four stations in Table 15–12 reveal the general precipitation pattern for *Dfa* climates. A summer maximum occurs at each station. But it is weakly defined at New York City, because the East Coast is more accessible to maritime air masses throughout the year. For the same reason, New York City also has the highest total of the four stations. Harbin, Manchuria, on the other hand, shows the most pronounced summer maximum, followed by a winter drought. This is characteristic of most eastern Asian stations in the middle latitudes and reflects the powerful control of the monsoon.

Another pattern revealed by the data is that precipitation generally decreases toward the continental interior and from south to north, primarily because of increasing distance from the sources of *mT* air. Furthermore, the more northerly stations are also influenced for a greater part of the year by drier polar air masses.

Wintertime precipitation is chiefly associated with the passage of fronts connected with traveling middle-latitude cyclones. Part of this precipitation is snow, the proportion increasing with latitude. Although precipitation is often considerably less during the cold season, it is usually more conspicuous than the greater amounts that fall during summer. An obvious reason is that snow remains on the ground, often for extended periods, and rain, of course, does not. Moreover, summer rains are often in the form of relatively short convective showers, whereas winter snows are more prolonged.

The Subarctic Climate *(Dfc, Dfd)*

North of the humid continental climate and south of the polar tundra is an extensive **subarctic climate**. It covers broad, uninterrupted expanses in North America (western Alaska to Newfoundland) and in Eurasia (Norway to the Pacific coast of Russia). It is often called the **taiga** climate, for it closely corresponds to the northern coniferous forest region of the same name (Figure 15–18). Although scrawny, the spruce, fir, larch, and birch trees in the taiga represent the largest stretch of continuous forest on Earth.

Temperature. The subarctic is well illustrated by the climate diagram in Figure 15–16b and by the data for Yakutsk, Russia, and Dawson, Yukon Territory in Table 15–13. Here in the source regions of continental polar (*cP*) air masses, the outstanding feature is the dominance of winter. Not only is it long, it is bitterly cold. Winter minimum temperatures are among the lowest recorded

Figure 15–18 The northern coniferous forest is also called the *taiga*. Denali National Park, Alaska. (Photo by Tom Bean/DRK Photo)

Table 15–13	Data for subarctic stations											
	J	**F**	**M**	**A**	**M**	**J**	**J**	**A**	**S**	**O**	**N**	**D**	**YR**
Yakutsk, Russia, 62°05'N; 103 m												
Temp. (°C)	–43	–37	–23	–7	7	16	20	16	6	–8	–28	–40	–10
Precip. (mm)	7	6	5	7	16	31	43	38	22	16	13	9	213
Dawson, Yukon, Canada, 64°03'N; 315 m												
Temp. (°C)	–30	–24	–16	–2	8	14	15	12	6	–4	–17	–25	–5
Precip. (mm)	20	20	13	18	23	33	41	41	43	33	33	28	346

outside the ice caps of Greenland and Antarctica. In fact, for many years the world's coldest temperature was attributed to Verkhoyansk in east-central Siberia, where the temperature dropped to –68°C (–90°F) on February 5 and 7, 1892. Over a 23-year period this same station had an average monthly minimum of –62°C (–80°F) during January. Although exceptional, these temperatures illustrate the extreme cold that envelops the taiga in winter.

In contrast, subarctic summers are remarkably warm, despite their short duration. When compared to regions farther south, however, this short season must be characterized as cool; for despite the many hours of daylight, the Sun never rises very high in the sky, so solar radiation is not intense. The extremely cold winters and the relatively warm summers of the taiga combine to produce the greatest annual temperature ranges on Earth. Yakutsk holds the distinction of having the greatest average temperature range in the world, 63°C (113°F). As the data for Dawson show, the North American subarctic is less severe.

Precipitation. Because these far northerly continental interiors are the source regions of *cP* air masses, only limited moisture is available throughout the year. Precipitation totals are therefore small, seldom exceeding 50 centimeters (20 inches). By far the greatest precipitation comes as rain from scattered summer convectional showers. Less snow falls than in the humid continental climate to the south, yet there is the illusion of more. The reason is simple: No melting occurs for months at a time, so the entire winter accumulation (up to 1 meter) is visible all at once. Furthermore, during blizzards, high winds swirl the dry, powdery snow into high drifts, giving the false impression that more snow is falling than is actually the case. So although snowfall is not excessive, a visitor to this region could leave with that impression.

The Polar Climates (*E*)

According to the Köppen classification, **polar climates** are those in which the mean temperature of the warmest month is below 10°C (50°F). Two types are recognized: the tundra

climate (*ET*) and the ice cap climate (*EF*). Climate diagrams of representative stations are presented in Figure 15–19.

Just as the tropics are defined by their year-round warmth, so the polar realm is known for its enduring cold, with the lowest annual means on the planet. Because polar winters are periods of perpetual night, or nearly so, temperatures are understandably bitter. During the summer, temperatures remain cool despite the long days, because the Sun is so low in the sky that its oblique rays produce little warming. In addition, much solar radiation is reflected by the ice and snow or used in melting the snow cover. In either case, energy that could have warmed the land is lost. Although cool, summer temperatures are still much higher than those experienced during the severe winter months. Consequently, annual temperature ranges are extreme.

Although polar climates are classified as humid, precipitation is generally meager, with many nonmarine stations receiving less than 25 centimeters (10 inches) annually. Evaporation, of course, is also limited. The scanty precipitation is easily understood in view of the temperature characteristics of the region. The amount of water vapor in the air is always small because low mixing ratios must accompany low temperatures. In addition, steep lapse rates are not possible. Usually precipitation is most abundant during the warmer summer months, when the air's moisture content is highest.

The Tundra Climate (*ET*)

The **tundra climate** on land is found almost exclusively in the Northern Hemisphere. It occupies the coastal fringes of the Arctic Ocean, many Arctic islands, and the ice-free shores of northern Iceland and southern Greenland. In the Southern Hemisphere no extensive land areas exist in the latitudes where tundra climates prevail. Consequently, except for some small islands in the southern oceans, the *ET* climate occupies only the southwestern tip of South America and the northern portion of the Palmer Peninsula in Antarctica.

The 10°C (50°F) summer isotherm that marks the equatorward limit of the tundra also marks the poleward

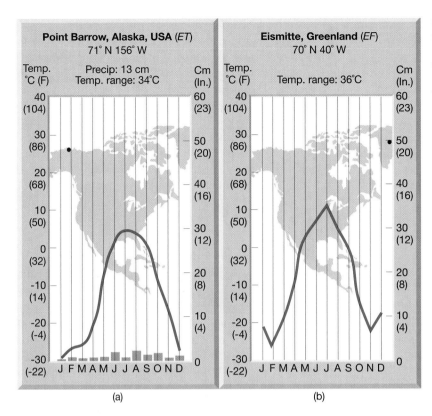

limit of tree growth. Thus, the tundra is a treeless region of grasses, sedges, mosses, and lichens (Figure 15–20). During the long cold season, plant life is dormant, but once the short, cool summer commences, these plants mature and produce seeds with great rapidity.

Because summers are cool and short, the frozen soils of the tundra generally thaw to depths of less than a meter. Consequently, the subsoil remains permanently frozen as **permafrost**. Permafrost blocks the downward movement of water and results in poorly drained, boggy soils that make the summer tundra landscape difficult to traverse (Figure 15–21).

The data for Point Barrow, Alaska (Table 15–14), on the shores of the frozen Arctic Ocean, exemplify the most common type of *ET* station, where continentality prevails. The combination of high latitude and continentality

Figure 15–20 The tundra in bloom north of Nome, Alaska. It is a region almost completely devoid of trees. Bogs and marshes are common, and plant life frequently consists of mosses, low shrubs, and flowering herbs. *(Photo by Fred Bruemmer/DRK Photo)*

makes winters severe, summers cool, and annual temperature ranges great. Yearly precipitation is small, with a modest summertime maximum.

Although Point Barrow represents the most common type of tundra setting, the data for Angmagssalik, Greenland (Table 15–14), reveal that some *ET* stations are different. Summer temperatures at both stations are equivalent, yet winters at Angmagssalik are much warmer and the annual precipitation is eight times greater than at Point Barrow. The reason is Angmagssalik's location on the southeastern coast of Greenland, where there is considerable marine influence. The warm North Atlantic Drift keeps winter temperatures relatively warm, and

maritime polar (*mP*) air masses supply moisture throughout the year. Because winters are less severe at stations like Angmagssalik, annual temperature ranges are much smaller than at stations like Point Barrow, where continentality is a major control.

Note that tundra climates are not entirely confined to the high latitudes. The summer coolness of this climate is also found at higher elevations as one moves equatorward. Even in the tropics, you can find *ET* climates if you go high enough. When compared with the Arctic tundra, however, winter temperatures in these lower-latitude counterparts become milder and less distinct from summer, as the data for Cruz Loma, Ecuador (Table 15–14), illustrate.

The Ice Cap Climate *(EF)*

The **ice cap climate**, designated by Köppen as *EF*, has no monthly mean above 0°C (32°F). Because the average temperature for all months is below freezing, the growth of vegetation is prohibited and the landscape is one of permanent ice and snow. This climate of perpetual frost covers a surprisingly large area of more than 15.5 million square kilometers (6 million square miles), or about 9 percent of Earth's land area, and aside from scattered occurrences in high mountain areas, it is largely confined to the ice sheets of Greenland and Antarctica.

Average annual temperatures are extremely low. For example, the annual mean at Eismitte, Greenland (Table 15–14), is –29°C (–20°F); at Byrd Station, Antarctica, –21°C (–6°F); and at Vostok, the Russian Antarctic Meteorological Station, –57°C (-71°F). Vostok has also experienced the lowest temperature ever recorded, –88.3°C (–127°F), on August 24, 1960.

In addition to latitude, the primary reason for such temperatures is the presence of permanent ice. Ice has a very high albedo, reflecting up to 80 percent of the meager sunlight that strikes it. The energy that is not reflected is used largely to melt the ice and so is not available for raising the temperature of air.

Another factor at many *EF* stations is elevation. Eismitte, at the center of the Greenland ice sheet, is almost 3000 meters above sea level (10,000 feet), and much of Antarctica is even higher. Thus, the permanent ice and high elevations further reduce the already low temperatures of the polar realm.

The intense chilling of the air close to the ice sheet means that strong surface-temperature inversions are common. Near-surface temperatures may be as much as 30°C (54°F) colder than air just a few hundred meters above. Gravity pulls this cold, dense air downslope, often producing strong winds and blizzard conditions. Such air movements, called *katabatic winds*, are an important

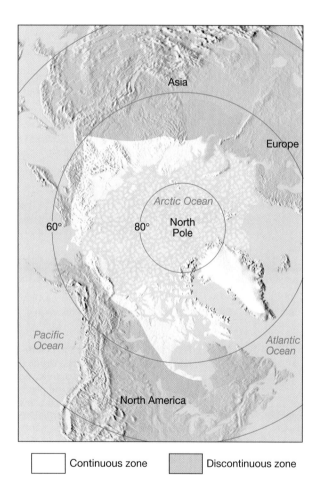

Figure 15–21 Distribution of permafrost in the Northern Hemisphere. More than 80 percent of Alaska and about 50 percent of Canada are underlain by permafrost. Two zones are recognized. In the continuous zone, the only ice-free areas are beneath deep lakes or rivers. In the higher-latitude portions of the discontinuous zones, there are only scattered islands of thawed ground. Moving southward, the percentage of unfrozen ground increases until all the ground is unfrozen. *(After the U.S. Geological Survey)*

Table 15–14 Data for polar stations

	J	F	M	A	M	J	J	A	S	O	N	D	YR
Eismitte, Greenland, 70°53'N; 2953 m													
Temp. (°C)	−42	−47	−40	−32	−24	−17	−12	−11	−11	−36	−43	−38	−29
Precip. (mm)	15	5	8	5	3	3	3	10	8	13	13	25	111
Angmagssalik, Greenland, 65°36'N; 29 m													
Temp. (°C)	−7	−7	−6	−3	2	6	7	7	4	0	−3	−5	0
Precip. (mm)	57	81	57	55	52	45	28	70	72	96	87	75	775
Point Barrow, Alaska, 71°18'N; 9 m													
Temp. (°C)	−28	−28	−26	−17	−8	0	4	3	−1	−9	−18	−24	−12
Precip. (mm)	5	5	3	3	3	10	20	23	15	13	5	5	110
Cruz Loma, Ecuador, 0°08'S; 3888 m													
Temp. (°C)	6.1	6.6	6.6	6.6	6.6	6.1	6.1	6.1	6.1	6.1	6.6	6.6	6.4
Precip. (mm)	198	185	241	236	221	122	36	23	86	147	124	160	1779

aspect of ice cap weather at many locations. Where the slope is sufficient, these gravity-induced air movements can be strong enough to flow in a direction that is opposite the pressure gradient.

Highland Climates

It is well known that mountain climates are distinctly different from those in adjacent lowlands. Sites with **highland climates** are cooler and usually wetter. The world climate types already discussed consist of large, relatively homogeneous regions. But highland climates are characterized by a great diversity of climatic conditions over small areas. Because large differences occur over short distances, the pattern of climates in mountainous areas is a complex mosaic, too complicated to depict on a world map.

In North America, highland climates characterize the Rockies, Sierra Nevada, Cascades, and the mountains and interior plateaus of Mexico. In South America, the Andes create a continuous band of highland climate that extends for nearly 8000 kilometers (5000 miles). The greatest span of highland climates stretches from western China, across southern Eurasia, to northern Spain, from the Himalayas to the Pyrenees. Highland climates in Africa occur in the Atlas Mountains in the north and in the Ethiopian Highlands in the east.

The best-known climate effect of increased altitude is lower temperatures. Greater precipitation due to orographic lifting is also common at higher elevations. Despite the fact that mountain stations are colder and often wetter than locations at lower elevations, highland climates are often very similar to those in adjacent lowlands in terms of seasonal temperature cycles and precipitation distribution. Figure 15–22 illustrates this relationship.

Phoenix, at an elevation of 338 meters (1109 feet), lies in the desert lowlands of southern Arizona. By contrast, Flagstaff is located at an altitude of 2100 meters (about 7000 feet) on the Colorado Plateau in northern Arizona. When summer averages climb to 34°C (93°F) in Phoenix, Flagstaff is experiencing a pleasant 19°C (66°F), a full 15°C (27°F) cooler. Although the temperatures at each city are quite different, the annual march of temperature for both is similar. Both experience minimum and maximum monthly means in the same months. When precipitation data are examined, both places have a similar seasonal pattern, but the amounts at Flagstaff are higher in every month. In addition, much of Flagstaff's winter precipitation is snow, whereas it only rains in Phoenix.

Because topographic variations are pronounced in mountains, every change in slope with respect to the Sun's rays produces a different microclimate. In the Northern Hemisphere, south-facing slopes are warmer and dryer because they receive more direct sunlight than do north-facing slopes and deep valleys. Wind direction and speed in mountains can be highly variable and quite different from the movement of air aloft or over adjacent plains. Mountains create various obstacles to winds. Locally, winds may be funneled through valleys or forced over ridges and around mountain peaks. When weather conditions are fair, mountain and valley breezes are created by the topography itself.

We know that climate strongly influences vegetation, which is the basis for the Köppen system. Thus, where there are vertical differences in climate, we should expect a vertical zonation of vegetation as well. Ascending a mountain can let us view dramatic vegetation changes that otherwise might require a poleward journey of thousands of kilometers (Figure 15–23). This occurs because altitude duplicates, in some respects, the influence of latitude on temperature

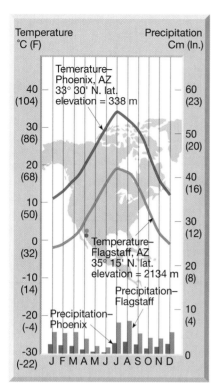

Figure 15–22 Climate diagrams for two Arizona stations illustrate the general influence of elevation on climate. Flagstaff is cooler and wetter because of its position on the Colorado Plateau, nearly 1800 meters (6000 feet) higher than Phoenix.

and hence on vegetation types. However, we know other factors, such as slope orientation, exposure, winds, and orographic effects also play a role in controlling the climate of highlands. Consequently, although the concept of vertical life zones applies on a broad regional scale, the details within an area vary considerably. Some of the most obvious variations result from differences in rainfall or the receipt of solar radiation on opposite sides of a mountain.

To summarize, *variety* and *changeability* best describe highland climates. Because atmospheric conditions fluctuate with altitude and exposure, a nearly limitless variety of local climates occurs in mountainous regions. The climate in a protected valley is very different from that of an exposed peak. Conditions on windward slopes contrast

(a)

(b)

Figure 15–23 (a) Only scanty drought-tolerant natural vegetation can survive in the hot, dry climate of southern Arizona, near Phoenix. *(Photo by Charlie Ott Photography/Photo Researchers, Inc.)*. (b) The natural vegetation associated with the cooler, wetter highlands near Flagstaff, Arizona, is much different from the desert lowlands. *(Photo by Larry Ulrich/DRK Photo)*

sharply with those on the leeward sides. Slopes facing the Sun are unlike those that lie mainly in the shadows.

Chapter Summary

- Climate is more than "the average state of the atmosphere" because a complete climate description should also include variations and extremes to accurately portray the total character of an area. The most important elements in climate descriptions are *temperature* and *precipitation* inasmuch as they have the greatest

influence on people and their activities and also have an important impact on the distribution of vegetation and the development of soils.

- Perhaps the first attempt at climate classification was made by the ancient Greeks, who divided each hemisphere into three zones: *torrid*, *temperate*, and *frigid*.

Many climate-classification schemes have been devised. The classification of climates is the product of human ingenuity and its value is determined largely by its intended use.

- For decades, a climate classification devised by Wladimir Köppen (1846–1940) has been the best-known and most used tool for presenting the world pattern of climates. The *Köppen classification* uses easily obtained data: mean monthly and annual values of temperature and precipitation. Furthermore, the criteria are unambiguous, simple to apply, and divide the world into climate regions in a realistic way. Köppen believed that the distribution of natural vegetation was the best expression of an overall climate. Consequently, the boundaries he chose were largely based on the limits of certain plant associations. Köppen recognized five principal climate groups, each designated with a capital letter: *A (humid tropical), B (dry), C (humid middle-latitude, mild winters), D (humid middle-latitude, severe winters),* and *E (polar).* Four groups *(A, C, D, E)* are defined by temperature. The fifth, the *B* group, has precipitation as its primary criterion.

- Order exists in the distribution of climate elements and the pattern of climates is not by chance. The world's climate pattern reflects a regular and dependable operation of the major *climate controls.* The major controls of climate are (1) *latitude* (variations in the receipt of solar energy and temperature differences are largely a function of latitude), (2) *land/water influence* (*marine climates* are generally mild, whereas *continental climates* are typically more extreme), (3) *geographic position and prevailing winds* (the moderating effect of water is more pronounced along the windward side of a continent), (4) *mountains and highlands* (mountain barriers prevent maritime air masses from reaching far inland, trigger orographic rainfall, and where they are extensive, create their own climatic regions), (5) *ocean currents* (poleward-moving currents cause air temperatures to be warmer then otherwise would be expected), and (6) *pressure and wind systems* (the world distribution of precipitation is closely related to the distribution of Earth's major pressure and wind systems).

- Situated astride the equator, the *wet tropics (Af, Am)* exhibit constant high temperatures and year-round rainfall which combine to produce the most luxuriant vegetation in any climatic realm—the *tropical rain forest.* Temperatures in these regions usually average 25°C (77°F) or more each month and the daily temperature variations characteristically exceed seasonal differences. Precipitation in *Af* and *Am* climates is normally from 175 to 250 centimeters (68 to 98 inches)

per year and is more variable than temperature, both seasonally and from place to place. Thermally induced convection coupled with convergence along the *intertropical convergence zone* (ITCZ) leads to widespread ascent of the warm, humid, unstable air and ideal conditions for cloud formation and precipitation.

- The *tropical wet and dry (Aw)* climate region is a transitional zone between the rainy tropics and the subtropical steppes. Here, the rain forest gives way to the *savanna,* a tropical grassland with scattered deciduous trees. Only modest temperature differences exist between the wet tropics and the tropical wet and dry regions. The primary factor that distinguishes the *Aw* climate from *Af* and *Am* is precipitation. In *Aw* regions the precipitation is typically between 100 and 150 centimeters (40 to 60 inches) per year and exhibits some seasonal character—wet summers followed by dry winters. In much of India, southeast Asia, and portions of Australia, these alternating periods of rainfall and drought are associated with the *monsoon,* wind systems with a pronounced seasonal reversal of direction. The *Cw* climate, which is subtropical instead of tropical, is a variant of *Aw.*

- Dry regions of the world cover about 30 percent of Earth's land area. Other than their meager yearly rainfall, the most characteristic feature of dry climates is that precipitation is very unreliable. Climatologists define a *dry climate* as one in which the yearly precipitation is less than the potential water loss by evaporation. To define the boundary between dry and humid climates, the Köppen classification uses formulas that involve three variables: (1) average annual precipitation, (2) average annual temperature, and (3) seasonal distribution of precipitation.

- The two climates defined by a general water deficiency are (1) *arid* or *desert (BW)*, and *semiarid* or *steppe (BS).* The differences between deserts and steppes are primarily a matter of degree. Semiarid climates are a marginal and more humid variant of arid climates that represent transitional zones that surround deserts and separate them from the bordering humid climates. Under the strong influence of the subtropical highs, the heart of the subtropical desert *(BWh)* and steppe *(BSh)* climates lies in the vicinity of the Tropic of Cancer and the Tropic of Capricorn. Within subtropical deserts, the scanty precipitation is both infrequent and erratic. In the semiarid transitional belts surrounding the desert, a seasonal rainfall pattern becomes better defined. Due to cloudless skies and low humidities, low-latitude deserts in the interiors of continents have the greatest daily temperature ranges on Earth. Where subtropical deserts are found along the west coasts of continents, cold ocean currents produce cool, humid conditions, often shrouded by low clouds or fog. Unlike their low-altitude counterparts, middle-latitude deserts (*BWk*) and steppes (*BSk*) are not controlled by the subsiding air masses of the subtropical highs. Instead, these lands exist principally because of their position in the deep interiors of large landmasses.

- *Humid middle-latitude climates with mild winters (C climates)* occur where the average temperature of the coldest month is less than 18°C (64°F) but above –3°C (27°F). Several C climate subgroups exist. *Humid subtropical climates (Cfa)* are on the eastern sides of the continents, in the 25- to 40-degree latitude range. Because of the dominating influence of maritime tropical air masses, summer weather within these regions is hot and sultry, and winters are mild. In North America, the *marine west coast climate (Cfb)* extends from near the United States–Canada border northward as a narrow belt into southern Alaska. The prevalence of maritime air masses means mild winters, cool summers, and ample rainfall throughout the year. *Dry-summer subtropical (Mediterranean) climates (Csa, Csb)* are typically found along the west sides of continents between latitudes 30 and 45°. In summer, the regions are dominated by the stable eastern sides of the oceanic subtropical highs. In winter, as the wind and pressure systems follow the Sun equatorward, they are within range of the cyclonic storms of the polar front.

- *Humid continental climates with severe winters (D climates)* experience severe winters. The average temperature of the coldest month is –3°C (27°F) or below and the average temperature of the warmest month exceeds 10°C (50°F). *Humid continental climates (Dfa)* are land controlled and do not occur in the Southern Hemisphere. They are confined to central and eastern North America and Eurasia in the latitude range 40 to 50°N. Both winter and summer temperatures in *Dfa* climates can be characterized as severe and annual temperature ranges are large. Precipitation is generally greater in summer and generally decreases toward

the continental interior and from south to north. Wintertime precipitation is chiefly associated with the passage of fronts connected with traveling middle-latitude cyclones. *Subarctic climates (Dfc, Dfd)*, often called *taiga* climates because they correspond to the northern coniferous forests of the same name, are situated north of the humid continental climates and south of the polar tundras. The outstanding feature of subarctic climates is the dominance of winter. By contrast, summers in the subarctic are remarkably warm, despite their short duration. The greatest annual temperature ranges on Earth occur here.

- *Polar climates (ET, EF)* are those in which the mean temperature of the warmest month is below 10°C (50°F). Annual temperature ranges are extreme, with the lowest annual means on the planet. Although polar climates are classified as humid, precipitation is generally meager, with many nonmarine stations receiving less than 25 centimeters (10 inches) annually. Two types of polar climates are recognized. Found almost exclusively in the Northern Hemisphere, the *tundra climate (ET)*, marked by the 10°C (50°F) summer isotherm at its equatorward limit, is a treeless region of grasses, sedges, mosses, and lichens with permanently frozen subsoil, called *permafrost*. The *ice cap climate (EF)* does not have a single monthly mean above 0°C. Consequently, the growth of vegetation is prohibited, and the landscape is one of permanent ice and snow.

- *Highland climates* are characterized by a great diversity of climatic conditions over a small area. In North America, highland climates characterize the Rockies, Sierra Nevada, Cascades, and the mountains and interior plateaus of Mexico. Although the best known climatic effect of increased altitude is lower temperatures, greater precipitation due to orographic lifting is also common. Variety and changeability best describe highland climates. Because atmospheric conditions fluctuate with altitude and exposure to the Sun's rays, a nearly limitless variety of local climates occur in mountainous regions.

Vocabulary Review

arid or desert (*BW*) (p. 408)
continental climate (p. 397)
dry climate (p. 406)
dry-summer subtropical climate (p. 415)
highland climate (p. 425)
humid continental climate (p. 416)
humid subtropical climate (p. 413)
ice cap climate (p. 424)
intertropical convergence zone (ITCZ) (p. 402)

jungle (p. 398)
Köppen classification (p. 391)
marine climate (p. 397)
marine west coast climate (p. 414)
Mediterranean climate (p. 415)
monsoon (p. 405)
permafrost (p. 423)
polar climate (p. 422)
savanna (p. 402)

semiarid or steppe (*BS*) (p. 408)
subarctic climate (p. 421)
taiga (p. 421)

tropical rain forest (p. 398)
tropical wet and dry (p. 403)
tundra climate (p. 422)

Review Questions

1. Why is classification often a necessary task in science?
2. What climate data are needed to classify a climate using the Köppen scheme?
3. Should climate boundaries, such as those shown on the world map in Figure 15–3, be regarded as fixed? Explain.
4. List the major climate controls and briefly describe their influence.
5. How does the tropical rain forest differ from a typical middle-latitude forest?
6. Distinguish between jungle and tropical rain forest.
7. Explain each of the following characteristics of the wet tropics:
   a. This climate is restricted to elevations below 1000 meters.
   b. Mean monthly and annual temperatures are high and the annual temperature range is low.
   c. This climate is rainy throughout the year, or nearly so.
8. Why are the wet tropics considered oppressive and monotonous?
9. What is the difference between *Af* areas and *Am* areas?
10. Laterite soils are found in association with tropical rain forests. Because these soils support the growth of lush natural vegetation, are they also excellent for growing crops? Why? (See Box 15–2.)
11. a. What primary factor distinguishes *Aw* climates from *Af* and *Am*?
    b. How is this difference reflected in the vegetation?
12. Describe the influence of the ITCZ and the subtropical high on the precipitation regime in the *Aw* climate.
13. In which of the following climates is the annual rainfall likely to be more consistent from year to year: *BSh*, *Aw*, *BWh*, or *Af*? In which of these climates is the annual rainfall most variable from year to year? Explain your answers.
14. In the dry (*B*) climates, there are usually more years when rainfall totals are below the average than above. Explain and give an example.
15. List and briefly discuss the four common misconceptions about deserts (see Box 15–3).
16. The amount of precipitation that defines the humid–dry boundary is variable. Why?
17. What is the primary reason (control) for the existence of the dry subtropical realm (*BWh* and *BSh*)?
18. a. Describe and explain the seasonal distribution of precipitation for a *BSh* station on the poleward side of a tropical desert and a *BSh* station on the equatorward side.
    b. If both stations barely meet the requirements for steppe climates (that is, with only a little more rainfall, both stations would be considered humid), which station would probably have the lower rainfall total? Explain.
19. Why do ground and air temperatures reach such high values in subtropical deserts?
20. Subtropical deserts, such as the Atacama and Namib, deviate considerably from the general image that we have of deserts. In what ways are these deserts not "typical" and why?
21. What is the primary cause of middle-latitude deserts and steppes?
22. Why are desert and steppe areas uncommon in the middle latitudes of the Southern Hemisphere?
23. Describe and explain the differences between summertime and wintertime precipitation in the humid subtropics (*Cfa*).
24. Why is the marine west coast climate (*Cfb*) represented by only slender strips of land in North and South America, and why is it very extensive in Western Europe?
25. How do temperature gradients (north–south versus east–west) reveal the strong oceanic influence along the West Coast of North America?
26. In this chapter the dry-summer subtropics were described as transitional. Explain why.
27. What other name is given to the dry-summer subtropical climate?
28. Why are summer temperatures cooler at San Francisco than at Sacramento (Table 15–11)?
29. Why is the humid continental climate confined to the Northern Hemisphere?
30. Why do coastal stations like New York City experience primarily continental climatic conditions?
31. Using the four stations shown in Table 15–12, describe the general pattern of precipitation in humid continental climates.

32. In the dry-summer subtropics precipitation totals increase with an increase in latitude, but in the humid continental climates the reverse is true. Explain.

33. Although generally characterized by small precipitation totals, subarctic and polar climates are considered humid. Explain.

34. Although snowfall in the subarctic climate is relatively scant, a wintertime visitor might leave with the impression that snowfall is great. Explain.

35. Describe and explain the annual temperature range one should expect in the realm of the taiga.

36. Although polar regions experience extended periods of sunlight in the summer, temperatures remain cool. Explain.

37. What is the significance of the 10°C (50°F) summer isotherm?

38. Why is the tundra landscape characterized by poorly drained, boggy soils?

39. Why are winter temperatures higher and the annual precipitation greater in Angmagssalik, Greenland, than at Point Barrow, Alaska?

40. The tundra climate is not confined solely to high latitudes. Under what circumstances might the *ET* climate be found in more equatorward locations?

41. Where are *EF* climates developed most extensively?

42. Besides the effect of latitude, what other factor(s) contribute to the extremely low temperatures that characterize the *EF* climate? Explain.

43. The Arizona cities of Flagstaff and Phoenix are relatively close to one another, yet have contrasting climates (see Figure 15–22). Briefly explain why the differences occur.

Problems

1. Use Table 15–1 to determine the appropriate classification for stations a, b, and c below.

2. Using the maps in Figure 15–17, determine the approximate January and July temperature gradients between the southern tip of mainland Florida and the point where the Minnesota–North Dakota border touches Canada. Assume the distance to be 3100 kilometers (1900 miles). Express your answers in °C per 100 kilometers or °F per 100 miles.

			J	F	M	A	M	J	J	A	S	O	N	D	YR
a.	Temp.	(°C)	−18.7	−18.1	−16.7	−11.7	−5.0	0.6	5.3	5.8	1.4	−4.2	−12.3	−15.8	−7.5
	Precip.	(mm)	8	8	8	8	15	20	36	43	43	33	13	12	247
b.	Temp.	(°C)	24.6	24.9	25.0	24.9	25.0	24.2	23.7	23.8	23.9	24.2	24.2	24.7	24.4
	Precip.	(mm)	81	102	155	140	133	119	99	109	206	213	196	122	1675
c.	Temp.	(°C)	12.8	13.9	15.0	16.1	17.2	18.8	19.4	22.2	21.1	18.8	16.1	13.9	15.9
	Precip.	(mm)	53	56	41	20	5	0	0	2	5	13	23	51	269

Atmospheric Science Online

The following are informative and interesting Internet sites that address topics related to those presented in the chapter:

American Association of State Climatologists:
• **http://www.ncdc.noaa.gov/ol/climate/aasc.html**

National Climatic Data Center:
• **http://www.ncdc.noaa.gov/**

For direct links to these sites and others, chapter objectives and reviews, quiz questions, and topical investigations that utilize Web resources, visit *The Atmosphere, Eighth Edition* Home Page at:
• **http://www.prenhall.com/lutgens**

CHAPTER SIXTEEN

Optical Phenomena of the Atmosphere

Rainbow over spruce trees of the taiga north of Alaska Range.
(Photo by Michael Giannechini/Photo Researchers, Inc.)

One of the most spectacular and intriguing of natural phenomena must surely be the rainbow. Its splash of colors has been the focus of poets and artists alike, not to mention every amateur photographer within reach of a camera. In addition to rainbows, many other optical phenomena, such as halos, coronas, and mirages, are common in our atmosphere. In this chapter, we consider how the most familiar of these displays occur. By learning about these spectacles, and by knowing when and where to look for them, you should become better able to identify each type. It is hoped that these fascinating displays will be witnessed more frequently as a result of this study.

Nature of Light

The light that forms the array of colors that constitute the rainbow, or the deep blue color of the sky, originates as white (visible) light from the Sun. It is the interaction of white sunlight with our atmosphere that creates the numerous optical phenomena that take place in the sky.

In Chapter 2, we considered some properties of light and how they contribute to occurrences like the blue sky and the red color of sunset. Here we examine other properties of light and describe how light interacts with the gases of the atmosphere, as well with ice crystals and water droplets, to generate still other optical phenomena. We consider four basic properties of light: reflection, refraction, diffraction, and interference. The sections that follow consider reflection and refraction; diffraction and interference are considered later, in the section on coronas.

Reflection

Light traveling through the emptiness of outer space travels at a uniform speed and in a straight line. When light rays encounter a transparent material, such as a piece of glass, however, some rays bounce off the surface of the glass, whereas others are transmitted at a slower velocity through the glass. The rays that bounce back from the surface of the glass are said to be *reflected*. It is reflected light that allows you to see yourself in a mirror. The image that you see in a mirror originates as light that first reflected off you toward the mirror and then was bounced from the silvered surface of the mirror back to your eyes. When light rays are reflected, they always bounce off the reflecting surface at the same angle at which they meet that surface (Figure 16–1). This principle is called the **law of reflection**. It states that the angle of incidence (incoming ray) is equal to the angle of reflection (outgoing ray).

Although the angle of incidence always equals the angle of reflection, not all objects are perfectly smooth.

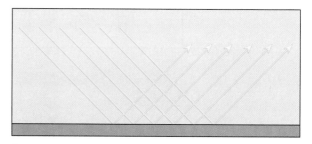

Figure 16–1 Reflection of light by a smooth surface.

Consequently, when light encounters a rough surface, the rays will strike the surface at different angles, which tends to scatter the light rays (Figure 16–2). When light is reflected from a rough surface, the image that you see is generally distorted or in some cases appears as multiple images. For example, the reflected image of the Sun when viewed on a rough ocean surface is not a circular disc, but rather a long narrow band, as shown in Figure 16–3. This bright band of light is formed because some of the sunlight that strikes each wave is reflected toward the viewer's eyes. What we are really seeing is not one image of the sun, but multiple, distorted images of a single light source. Even something as smooth-appearing as a page in this book is sufficiently rough to disperse the light in all directions, making it possible to see the print from any direction. In contrast, if this page were perfectly smooth and all the light was approaching from a specific direction, you would have to move your head to a position exactly opposite the light in order to read the print.

A type of reflection that is important to our discussion is **internal reflection**. Internal reflection occurs when light that is traveling through a transparent material, such as water, reaches the opposite surface and is reflected back into the transparent material. You can easily demonstrate this phenomenon by using a glass of water. Hold the glass of water directly overhead and look up through the water. You should be able to see clearly through the water, for very little internal reflection results when light strikes perpendicular to a surface. Keeping the glass overhead, move

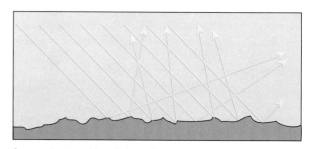

Figure 16–2 Incident light striking a rough surface is diffused.

Figure 16–3 This long, narrow band of sunlight is really multiple, distorted images of the Sun reflected from the ocean surface. Deception Pass Bridge, Washington. *(Photo by Peter Skinner/Photo Researchers, Inc.)*

it sideways so that you look up at it at an angle. Notice how the underside of the surface of the water takes on the appearance of a silvered mirror. What you are observing is the total internal reflection that occurs when the light strikes the surface at an angle greater than 48° from the vertical. Internal reflection is an important factor in the formation of optical phenomena, such as rainbows. In this instance, sunlight entering the raindrops strikes the opposite surface and is reflected back toward an observer.

Refraction

When light strikes a transparent material such as water, the rays that are not reflected are transmitted through the water and are subjected to another well-known effect, called *refraction*.

Refraction is the bending of light as it passes obliquely from one transparent medium to another. You have undoubtedly noticed this phenomenon as you have stood in a swimming pool and observed your apparently bent limbs. Obviously, it had to be the light that played such a trick on your eyes.

Refraction is caused because the velocity of light varies, depending on the material that transmits it. In a vacuum, radiation travels at 3.0×10^{10} centimeters per second; and when it travels through air, its speed is slowed only slightly. However, in such substances as water, ice, or glass, its speed is slowed considerably.°

°The speed of light is a constant; however, as it passes through various substances, this energy is delayed because of interaction with the electrons. The speed of light as it travels between these intervening electrons is the same as in a vacuum.

When light enters a transparent material perpendicular to its surface, only the velocity is affected. However, when light encounters a transparent medium at some angle other than 90°, the light rays are bent, as shown in Figure 16–4. Why light is refracted (bent) can best be demonstrated by an analogy. Imagine how an automobile responds should the driver fall asleep as the car nears a curve to the left. As the auto leaves the highway, the right front wheel will encounter the dirt shoulder before the left wheel does. Because of the soft nature of the dirt shoulder, the right wheel will slow while the left wheel, which is still on the pavement, will continue at the same rate of speed. The result will be a sudden turn of the auto toward the right, which will, it is hoped, waken the driver. Now if we can imagine that the path of the auto represents the path of light rays, and the pavement and softer shoulder represent air and water, respectively, we can see how light bends as it goes from air to water. As the light enters the water and is slowed, its path is diverted toward a line extending perpendicularly from the water's surface (Figure 16–4). Should the light pass from water into air, the bending will be in the opposite direction, that is, away from the perpendicular.

Recall that light bends because of a change in velocity. Thus, it follows that the greater the difference in the velocity at which the light travels through the materials involved, the greater the bending. Because light travels only slightly slower in air than in a vacuum, the amount of refraction in air is small; hence, air is said to have a small index of refraction. Water, in contrast, has a much larger refractive index; light will bend quite noticeably as it passes from air to water. The angle at which the light intersects the surface also affects the angle of refraction.

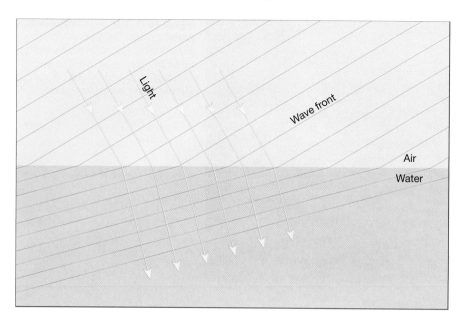

Figure 16–4 Refraction (bending) occurs as light waves pass from one material to another. The light will change direction because one part of the wave front slows before the other part.

The bending of light caused by refraction is responsible for a number of common optical illusions. These optical happenings result because our brain perceives bent light as if it has traveled to our eyes along a straight path. Try to imagine "looking down" a bent light ray to view an object located around a corner. If you could see the object, your brain would place that object out from the corner in "plain sight." On occasion, you do see things that are "around the corner." One example is our view of the setting Sun. Several minutes after the Sun has actually slipped below the horizon, it still appears to us as a full disc. We will soon provide an explanation for this occurrence.

An illustration of how refraction causes optical illusions is found in Figure 16–5. Here we see how the refraction of light produces the apparent bending of a pencil that is immersed in water. The solid lines in this figure show the actual path taken by the light, whereas the dashed lines indicate how we perceive those same light rays. As we look down at the pencil, the point appears closer to the surface than it actually is. Again, this situation occurs because our brain perceives that light as coming along the straight path indicated by the dashed line rather than along the actual bent path. Because all the light coming from the submerged portion of the pencil is bent similarly, this portion of the pencil appears nearer the surface. Therefore, where the pencil enters the water, it appears to be bent upward toward the surface of the water.

In addition to the abrupt bending of light as it passes obliquely from one transparent substance to another, light will also gradually bend as it traverses a material of varying density. As the density of a material changes, so does the velocity of light. Within Earth's atmosphere, for

example, the density of air usually increases Earthward. The results of this gradual density change are an equally gradual slowing and bending of light rays. These rays acquire a direction of curvature that has the same orientation as Earth's curvature. This bending is responsible for the apparent displacement of the position of the stars, Moon, and Sun. When the Sun (or other celestial bodies) is near the horizon, this effect is particularly

Figure 16–5 The pencil appears bent because the eye perceives light as if it were traveling along the dashed line rather than along the solid line, which represents the path of the refracted light.

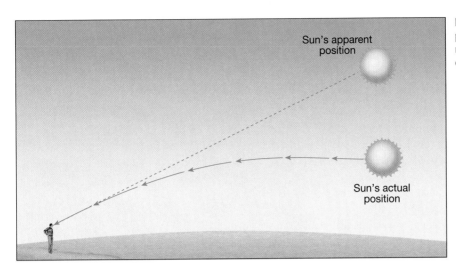

Figure 16–6 Light is refracted as it passes through the atmosphere, resulting in an apparent displacement of the position of the Sun.

great, which explains why we can see the Sun for a few minutes after it has set below the horizon.° Figure 16–6 illustrates this situation. Recall that it is our inability to perceive light as bending that places the apparent position of the Sun above the horizon.

Mirages

One of the most interesting optical events common to our atmosphere is the **mirage**. Although this phenomenon is most often associated with desert regions, it can actually be experienced anywhere (see Box 16–1). One type of mirage occurs on very hot days when the air near the ground is much less dense than the air aloft. As noted, a change in the density of air is accompanied by a gradual bending of the light rays. When light is traveling through air that is less dense near the surface, the rays will develop a curvature in a direction opposite to Earth's curvature. As we can see in Figure 16–7, this direction of bending will cause the light reflected from a distant object to approach the observer from below eye level. Consequently, because the brain perceives the light as following a straight path, the object appears below its original position and is often inverted, as is the palm tree in Figure 16–7. The palm tree will appear inverted when the rays that originate near the top of the tree are bent more than those that originate near the base of the tree.

In the classic desert mirage, a lost and thirsty wanderer encounters a mirage consisting of an oasis of palm trees and a shimmering water surface on which he can

see a reflection of the palms. The palm trees are real, but the water and the reflected palms are part of the mirage. Light traveling to the observer through the cooler air above produces the image of the actual tree. The reflected image of the palms is produced, as stated earlier, from the light that traveled downward from the trees and was gradually bent upward as it traveled through the hot (less dense) air near the ground. The image of water is produced in the same manner as the reflection of the palms. Light that traveled downward from the sky is bent upward to generate the mirage of water. Such desert mirages are called **inferior mirages** because the images appear *below* the true location of the observed object.

In addition to the "desert mirage," another common type of mirage occurs when the air near the ground is much cooler than the air aloft. Consequently, this effect is observed most frequently in polar regions or over cool ocean surfaces. When the air near the ground is substantially colder than the air aloft, the light rays bend with a curvature having the same direction as Earth's curvature. As shown in Figure 16–8, this effect allows ships to be seen where ordinarily Earth's curvature would block them from view. This phenomenon is often referred to as **looming** because sometimes the refraction of light is so great that the object appears suspended above the horizon. In contrast to a desert mirage, looming is considered a **superior mirage** because the image is seen *above* its true position.

In addition to the rather easily explained inferior and superior mirages, a number of much more complex variations have been observed. They occur when the atmosphere develops a temperature profile in which rapid temperature changes are observed with height. Under these conditions, each thermal layer acts like a glass lens. Because each layer will bend the light rays somewhat differently, the size and shape of the objects observed

°Because it takes over eight minutes for solar radiation to reach Earth, we see the Sun in the position it was located about eight minutes earlier. This situation does not affect the apparent displacement of the Sun caused by atmospheric bending of solar radiation.

Box 16–1 Are Highway Mirages Real?

You have undoubtedly seen a mirage while traveling along a highway on a hot summer afternoon. The most common highway mirages are in the form of "wet areas" that appear on the pavement ahead only to disappear as you approach (Figure 16–A). Because these "wet areas" always disappear as a person gets closer, many people believe they are optical illusions. This is not the case. Highway mirages, as well as all other types of mirages, are as real as the images observed in a mirror. As can be seen in Figure 16–A, highway mirages can be photographed. They are not "tricks played by the mind."

What causes the "wet areas" that appear on dry pavement? On hot summer days, the layer of air near Earth's surface is much warmer than the air aloft. Sunlight traveling from a region of colder (more dense) air into the warmer (less dense) air near the surface bends in a direction opposite to Earth's curvature (see Figure 16–7). As a consequence, light rays that began traveling downward from the sky are refracted upward and appear to the observer to have originated on the pavement ahead. What appears to the traveler as water is really just an inverted image of the sky. This can be verified by careful observation. The next time you view a highway mirage, look closely at any vehicle ahead of you at about the same distance as the "wet" area. Below the vehicle you should be able to see an inverted image of

Figure 16–A Classic highway mirage on a South Dakota highway. *(Photo by Tom Bean, DRK Photo)*

it. Such an image is produced in the same manner as the inverted image of the sky.

It is interesting to note that when the Sun and Moon are low on the horizon, they seem to be much larger than when they are overhead. This phenomenon is a true optical illusion that has nothing to do with the refraction of light. As meteorologist Craig Bohren stated: "The Moon illusion results from refraction by the mind, mirages from refraction by the atmosphere."

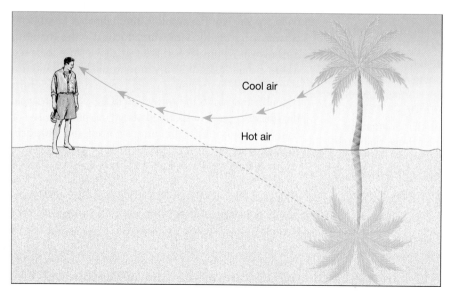

Figure 16–7 Light travels more rapidly in the hot air near the surface. Thus, as downward-directed rays enter this warm zone, they are bent upward so that they reach the observer from below eye level.

Cool air

Hot air

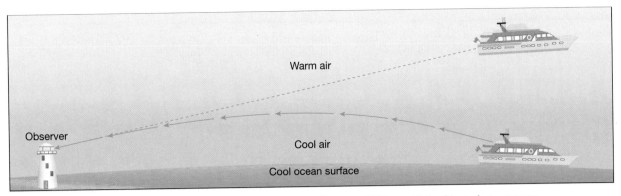

Figure 16–8 As light enters a cool layer of air, it will slow and bend downward. This results in objects that appear to loom above their true position.

through these thermal layers are greatly distorted. You may have observed an analogous sight if you have ever entered a "House of Mirrors" at the county fair. Here one of the mirrors makes you look taller, whereas others stretch the image of part of your body and compress other portions. Mirages are capable of similarly distorting objects and occasionally will even form a mountainlike image over a barren ice cap or over an open ocean.

One mirage that changes the apparent size of an object is called **towering**. As the name implies, towering results in a much larger object. An interesting type of towering is called the **Fata Morgana**. It is named for the legendary sister of King Arthur, who was credited with the magical power of being able to create towering castles out of thin air. This optical phenomenon is most frequently observed in coastal areas, where sharp temperature contrasts are common. In addition to generating magical castles, the Fata Morgana probably explains the towering mountains that were observed by early explorers of the north polar region but that never materialized.

Rainbows

Probably the most spectacular and best known of all optical phenomena that occur in our atmosphere is the **rainbow** (Figure 16–9). An observer on the ground sees the rainbow as an arch-shaped array of colors that trail across a large segment of the sky. Although the clarity of the colors varies with each rainbow, the observer can usually discern six rather distinct bands of color. The outermost band of the bow is always red and blends gradually to orange, yellow, green, blue, and eventually ends with an innermost band of violet. Typically, these spectacular splashes of color are seen when the observer is situated with the Sun on one side and a rain shower occurring in the opposite part of the sky. However, a fine mist of water droplets generated by a waterfall or lawn sprinkler can also generate a miniature rainbow.

Like other optical phenomena, rainbows have been used by people as a means of predicting the weather. A well-known weather proverb illustrates this point:

> Rainbow in the morning, sailors take warning.
> Rainbow at night, sailors delight.

This bit of weather lore relies on the fact that weather systems in the midlatitudes usually move from west to east. Remember that an observer must be positioned with his back to the Sun and facing the rain in order to see the rainbow. When a rainbow is seen in the morning, the Sun is located to the east of the observer and the raindrops that are responsible for its formation must therefore be located to the west. In the early evening, the opposite situation exists—the rain clouds are located to the east of the observer. Thus, we predict the advance of foul weather when the rainbow is seen in the morning because the rain is located to the west of the observer and is traveling toward him. Conversely, when the rainbow is seen late in the day, the rain has already passed. Although this famous proverb does have a scientific basis, a small break in the clouds, which lets the sunshine through, can generate a late-afternoon rainbow. In this situation, a rainbow may certainly be followed shortly by more rainfall.

On those occasions when a rather spectacular rainbow is visible, an observer will occasionally be treated to a view of a dimmer secondary rainbow. The secondary bow will be visible about 8° above the primary bow and will suspend a larger arc across the sky (Figure 16–9).* The secondary bow also has a slightly narrower band of colors than the primary rainbow, and the colors are in reverse order. Red makes up the innermost band of the secondary rainbow, and violet the outermost.

Although primary and secondary rainbows are produced in an almost identical manner, for clarity we consider

---

*For reference, the diameter of the Sun is equal to about 0.5° of arc.

Figure 16–9 Double rainbow over Rock Creek Valley, Montana. *(Photo by Larry Ulrich/DRK Photo)*

first the formation of the primary bow. It should be apparent that sunlight and water droplets are needed for the generation of a rainbow. Also, let us not forget the observer, who must be located between the Sun and the rain.

To understand how raindrops disperse sunlight to generate the primary rainbow, recall our discussion of refraction. Remember that as light passes obliquely from the atmosphere to water, its speed is slowed, which causes it to be refracted (bent). In addition, each color of light travels at a different velocity in water; consequently, each color will be bent at a slightly different angle. Violet-colored light, which interacts most with the intervening material, travels at the slowest rate and is therefore refracted the most, whereas red light travels most rapidly and is therefore bent the least. Thus, when sunlight, which consists of all colors, enters water, the effect of refraction is to separate it into colors according to their velocity. Sir Isaac Newton is credited with demonstrating the concept of color separation, using a prism.

Light that is transmitted through a prism is refracted twice, once as it passes from the air into the glass and again as it leaves the prism and reenters the air. Newton noted that when light is refracted twice, as by a prism, the separation of sunlight into its component colors is quite noticeable (Figure 16–10). We refer to this separation of colors by refraction as **dispersion**.

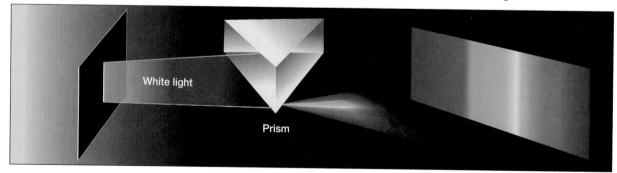

Figure 16–10 The spectrum of colors is produced when sunlight is passed through a prism and each wavelength of light is bent differently.

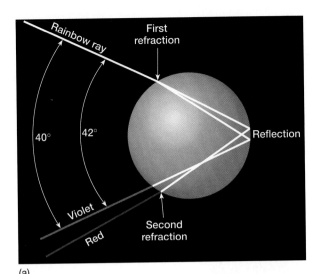

(a)

(b)

Figure 16–11 The formation of the primary rainbow. (a) Color separation results when sunlight is refracted and reflected by a raindrop to produce rainbow rays. (b) The curved shape of the rainbow results because the rainbow rays always travel toward the observer at an angle of 42° from the path of the sunlight.

When a rainbow forms, water drops act as a prism, dispersing sunlight into the spectrum of colors we see (Figure 16–11). On impacting with the droplet, the sunlight is refracted, with violet light bent the most and red the least. On reaching the opposite side of the droplet, the rays are reflected and exit the droplet on the same side they entered. After leaving the droplet, further refraction increases the dispersion already produced and accounts for the complete color separation.

The angle between the incident sunlight and the dispersed colors that constitute the rainbow is 42° for red and 40° for violet. The other colors—orange, yellow, green, and blue—are dispersed at intermediate angles. Although each droplet disperses the full spectrum of colors, an observer will see only one color from any single raindrop. For example, if green light from a particular droplet reaches an observer's eye, the violet light from the droplet will pass over his head and the red light will fall toward the ground in front of him. Consequently, each observer sees his or her "own" rainbow generated by a different set of droplets and different sunlight from that which produces another person's rainbow.

The curved shape of the rainbow results because the rainbow rays always travel toward the observer at an angle between 40° and 42° from the path of the sunlight. Consequently, when an observer looks upward at 42° from the path of the sunlight, he or she will see the color red. When the observer looks to either side at an angle of 42° the color red will also be visible. In any direction at an angle of 42° from the path of the Sun's rays, droplets will be directing red light toward the observer. Thus, we experience a 42° semicircle of color across the sky that we identify as the arch shape of the rainbow. Because an observer in an airplane can also look downward at an angle of 42°, under ideal conditions, he or she can see the rainbow as a full circle. In contrast, if the Sun is higher than 42° above the horizon, an Earthbound observer will not see a rainbow. So if you live in the midlatitudes, do not look for a rainbow during a summer rain shower occurring at midday.

As stated earlier, the secondary rainbow is generated in much the same way as the primary bow. The main difference is that the dispersed light that constitutes the secondary bow is reflected twice within a raindrop before it exits, as shown in Figure 16–12. The extra reflection results in a 50° angle for the dispersion of the color red (about 8° higher than the primary rainbow) and a reverse order of the colors.

In addition, the extra reflection accounts for a dimmer and therefore less frequently observed secondary bow. Each time light strikes the inner surface of the droplet, some of the light is reflected and the remainder is transmitted through the reflecting surface. The light that is transmitted through the back surface of the droplet does not contribute to the rainbow. Because the rays that form the secondary rainbow experience an additional reflection, they are not as bright as those that form the primary rainbow. Ideally, the secondary rainbow always forms; it is just not often discernible by the observer. In addition, other rainbows result because of three or even more internal reflections. These higher-order rainbows are too dim to be seen.

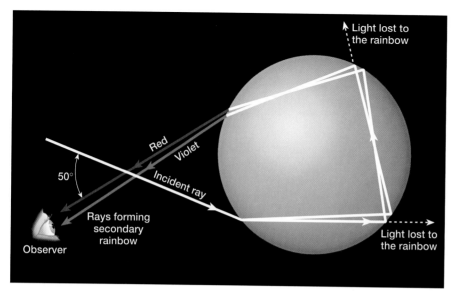

Figure 16–12 Idealized geometry of the rays that contribute to the secondary rainbow. By comparing this figure with Figure 16–11, you will see that the positions of the red and violet rays are reversed, which accounts for the order of the colors observed.

Halos, Sun Dogs, and Solar Pillars

Although a fairly common occurrence, halos are rarely seen by the casual observer. When noticed, the **halo** appears as a narrow whitish ring having a large diameter centered on the Sun (Figure 16–13). Look for halos on days when the sky is covered with a thin layer of cirrus clouds. In addition, this optical phenomenon is generally more often viewed in the morning or late afternoon, when the Sun is near the horizon. Occupants of polar regions, where a low Sun and cirrus clouds are common, are frequently treated to views

of halos and associated phenomena. Occasionally, halos may also be seen around the Moon.

The most common halo is the *22° halo*, so named because its radius subtends an angle of 22° from the observer. Less frequently observed is the larger *46° halo*. Like the rainbow, the halo is produced by dispersion of sunlight. In the case of the halo, however, it is ice crystals, rather than raindrops, that refract light. Thus, as stated earlier, the clouds most often associated with halo formation are cirrus clouds. Because cirrus clouds often form as a result of frontal lifting, which, in turn, is associated with cyclonic storms, halos have been accurately

Figure 16–13 A 22° halo produced by the dispersion of sunlight by cirrostratus clouds. *(Courtesy of Ward's Natural Science Establishment, Inc.)*

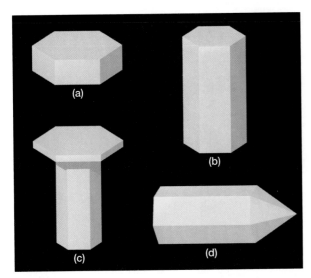

Figure 16–14 Common ice-crystal configurations that contribute to the formation of certain optical phenomena: (a) plate, (b) column, (c) capped column, and (d) bullet.

described as harbingers of foul weather, as the following weather proverb attests:

The moon with a circle brings water in her beak.

Four basic types of ice crystals are believed to contribute to the formation of halos: plates, columns, bullets, and capped columns (Figure 16–14). All these crystals are hexagonal (six sided), as is also the case for snowflakes.

Halos form when the ice crystals that compose cirrus clouds have random orientation. Because sunlight will strike the faces of these crystals at every possible angle, we might expect the scattered light to be dispersed equally in all directions. Yet just as we noted in the formation of the rainbow, as the angle at which the rays strike the surface changes so does the amount of dispersion, but only up to

a point. After this point, a change in the angle of incidence does not appreciably change the direction of scattering. Consequently, a larger portion of the light will be scattered in one direction than in any other. For a six-sided ice crystal, this angle of maximum scattering is 22°; thus, we have the 22° halo.

The primary difference between 22° and 46° halos is the path that the light takes through the ice crystals. The scattering sunlight that is responsible for the 22° halo strikes one of the sides of the ice crystal and exits from an alternating side, as shown in Figure 16–15a. The angle of separation between the alternating faces of an ice crystal is 60°, which is the same as a common glass prism. Consequently, an ice crystal disperses light in a manner similar to a prism in order to produce the 22° halo. The 46° halo, by contrast, is formed from light that passes through one side of the crystal and exits at the base or top (Figure 16–15b). The angle separating these two surfaces is 90°. Light that passes through two ice faces separated by 90° is concentrated at an angle of 46°, which accounts for the latter halo.

Although ice crystals disperse light in the same manner as a raindrop (or prism), halos are generally whitish in color, partly because of the rather imperfect shape and size of ice crystals compared to rain droplets. The colors produced by this dispersion overlap and wash out each other. Occasionally, however, halos will be colored. Most commonly, a reddish band will be seen in the inner portion of the ring. Because red is refracted the least of all colors, we would expect to find it located on the inner edge of the halo, which is nearest the Sun. The other colors, which are refracted more than red, will tend to wash out each other, leaving the red surrounded by a whitish ring.

One of the most spectacular effects associated with a halo is called **sun dogs** or **parhelia**. These two bright regions, or "mock suns" as they are often called, can be seen adjacent to the 22° halo and usually slightly below

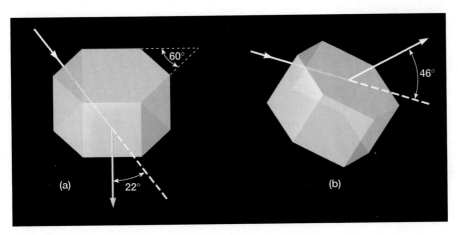

Figure 16–15 The paths that are taken by light to generate (a) the 22° halo and (b) the 46° halo.

Figure 16–16 Sun dogs, or parhelia, produced by the dispersion of sunlight by the ice crystals of cirrus clouds. *(Photo by E. J. Tarbuck)*

the elevation of the sun (Figure 16–16). Sun dogs form under the same conditions as, and in conjunction with, the halo, except that their existence depends on numerous ice crystals oriented vertically (see Figure 16–15). This particular orientation results when ice crystals are allowed to descend slowly. Then a large portion of the striking rays will be concentrated in two areas at a distance usually slightly greater than 22° from the Sun. When the Sun is near the horizon, so that the impact angle is perpendicular to the vertical crystal faces, the mock suns will appear directly on the 22° halo, with the sun positioned between them.

Another optical phenomenon that is related to the halo is the **sun pillar**. These vertical shafts of light are most often viewed near sunset or sunrise, when they appear to extend upward from the Sun (Figure 16–17). These bright pillars of light are created when sunlight is reflected from

the lower sides of descending plates and capped columns, which are oriented like slowly falling leaves. Because direct sunlight is often reddish when the Sun is low in the sky, pillars will appear similarly colored as well. Occasionally, pillars that extend below the Sun can also be viewed.

The Glory

To the Earthbound observer, the **glory** is a spectacle that is rarely witnessed. The next time you are in an airplane and fortunate enough to have a window seat, however, look for the shadow of the aircraft projected on the clouds below. The airplane shadow will often be surrounded by one or more colored rings that constitute the glory. Each ring will be colored in a manner similar to the rainbow, with red being the outermost band and violet the innermost. Generally, however, the colors are not as discernible as those of the primary rainbow. When two or more sets of rings are seen, the inner one will be the brightest and thinnest.

Although the glory is most commonly seen by pilots, its name comes from its appearance when viewed by an observer located on the ground. The glory can be seen if an observer is located so that he or she is above a layer of fog with the Sun at his or her back. Should the observer's shadow be cast on the fog bank, the glory will enshroud the observer's head. When two people witness such a sight simultaneously, only the observer's own head will appear within the glory. Consequently, this type of "halo" has been represented in many ancient artistic works to glorify the wearer.

The glory is formed in a manner not unlike that of the rainbow. But the cloud droplets that are responsible for the glory are much smaller and more uniform in size than the raindrops that scatter the rainbow rays. The light that becomes the glory strikes the very edge of the droplets. These rays then travel to the opposite side of the droplet,

Figure 16–17 A Sun pillar is a shaft of sunlight reflected by ice crystals in high clouds. *(Photo by Frank Zullo/Science Source/Photo Researchers, Inc.)*

partly by one internal reflection, and the remaining distance along the surface of the droplet. This path causes the rays to be backscattered directly toward the Sun. Because the glory always forms opposite the Sun's position, the observer's shadow will always be found within the glory.

The Corona

The only optical phenomenon more commonly witnessed in association with the Moon than the Sun is the corona. Typically, the **corona** appears as a bright whitish disk centered on the Moon or Sun. Whenever colors are discernible, the corona appears as several concentric rings, each with a red outer band and a bluish inner region (Figure 16–18).

Figure 16–18 The corona. Typically, the corona appears as a bright whitish disk centered on the Moon or Sun. Those like the one shown above that display the colors of the rainbow are rare. *(Photo by Henry Lansford)*

The corona is produced when a thin layer of water-laden clouds, usually altostratus, veil the illuminating body. Although water droplets are responsible for scattering the light that produces the corona, the colors are not the result of reflection and refraction, as were the colors of the rainbow. Instead, the corona forms because of a slight bending of light that occurs as light passes near the edges of cloud droplets. Although we described light as traveling in a straight line, which it generally does, light will bend very slightly around sharp edges, a process referred to as **diffraction**. It is because of diffraction that even the sharpest shadow appears blurred at the edges when examined carefully.

Because of their small size, cloud droplets are particularly effective at bending light. From all sides of a cloud droplet, diffracted light will be directed into the "shadow" of the droplet. Here light rays will meet and interfere with each other. It is the **interference** of the various components of white light that generates the colors that make up the corona.

To help understand how interference produces color, we will need to recall our discussion of radiation in Chapter 2. Remember that light exhibits wave motion, much like ocean swells. White light consists of an array of colors, each with a different wavelength (distance from one crest to the next). That portion of visible light with the shortest wavelength appears violet, whereas the portion with the longest wavelength appears red (see Box 16–2).

When a light wave of any type becomes superimposed on another, interference results in a new wave having characteristics different from either of the interacting waves. In the generation of the corona, interference occurs as light waves, which are diffracted around the edge of cloud droplets, converge out of phase. Two waves are out of phase when the crest of one is aligned with the trough of the other. When two similar but out-of-phase waves become superimposed, the result is the cancellation of both waves. When white light is diffracted, such that one color is out of phase whereas the other colors are in phase, the out-of-phase color will be subtracted from the light. When yellow light, for example, is subtracted from white light, the remaining light will appear blue. Because each color is bent a different amount, each color will experience destructive interference at a different angle. The result of the diffraction and interference caused by cloud droplets is an array of colors, each produced by the cancellation and subtraction of a different color.

Box 16–2 Iridescent Clouds

Iridescent clouds are among the more spectacular and elusive of optical phenomena (Figure 16–B). These dramatic clouds contain areas of bright colors, generally violet, pink, and green (Figure 16–B). Like the corona shown in Figure 16–18, the display of colors associated with iridescent clouds is produced by the diffraction of sunlight or moonlight by small cloud droplets or ice crystals. Special conditions must exist before iridescent colors can be observed. It is important that the diffracting particles be sufficiently small and uniform in size. In addition, the cloud must be in the same part of the sky as the Sun or Moon. Most of the time when these conditions occur, the Sun's rays are so intense that the colors go unnoticed.

Furthermore, when clouds of vertical development are present, they generally contain cloud droplets that are too large or too varied in size to generate this play of colors. Such factors contribute to the relatively rare occurrence of these spectacular clouds. The best time to view iridescent clouds is when the Sun is behind a cloud or just after the Sun has set behind a building or topographic barrier.

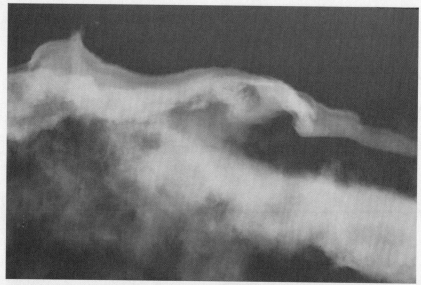

Figure 16–B Iridescent cloud. *(Photo by Tom Schlatter, National Oceanic and Atmospheric Administration/Seattle)*

Chapter Summary

- The four basic properties of light are *reflection, refraction, diffraction,* and *interference.* The *law of reflection* states that when light rays are reflected, they always bounce off the reflecting surface at the same angle (the angle of reflection) at which they meet that surface (the angle of incidence). *Internal reflection* occurs when light that is traveling through a transparent material, such as water, reaches the opposite surface and reflects back into the transparent material.

Internal reflection is an important factor in the formation of optical phenomena, such as rainbows. Refraction is the bending of light due to a change in velocity as it passes obliquely from one transparent medium to another. Furthermore, light will also gradually bend as it traverses a material of varying density. The bending of light by refraction is responsible for such common optical illusions as the apparent displacement of the position of the stars, Moon, and Sun.

- A *mirage* is an optical effect of the atmosphere caused by refraction when light passes from air with one density into air with a different density and the object appears displaced from its true position. An *inferior mirage* occurs when the image appears below the true location of the observed object. During a phenomenon called *looming*, objects sometimes appear to be suspended above the horizon. Looming is considered a *superior mirage* because the image is seen above its true position. A mirage that changes the apparent size of an object is called *towering*. A type of towering, called *Fata Morgana*, is frequently observed in coastal areas as towering castles that appear out of thin air.

- Perhaps the most spectacular and best known atmospheric optical phenomenon is the *rainbow*. Sunlight and water droplets are necessary for the formation of a rainbow. Furthermore, the observer must be between the Sun and rain. When a rainbow forms, the water droplets act as prisms and refraction disperses the sunlight into the spectrum of colors, a process called *dispersion*. The curved shape of the rainbow results because the rainbow rays always travel toward the observer at an angle between 40 and 42° from the path of the sunlight. In any direction (upward, sideward, etc.) at an angle of 42° from the path of the Sun's rays, droplets will be directing the color red toward the observer, thus forming a semicircle of color across the sky. In a *primary rainbow*, sunlight is reflected once within a raindrop. However, in a dimmer, less frequently observed, *secondary rainbow*, light is reflected twice within a raindrop before it exits.

- A *halo* is a narrow whitish ring with a large diameter centered on the Sun. They occur most often when the sky is covered with a thin layer of cirrus clouds. The most common halo is the 22° halo, so named because its radius subtends an angle of 22° from the observer.

Less frequently observed is the larger 46° halo. Halos are produced by dispersion of sunlight from atmospheric ice crystals that refract light. The primary difference between 22° and 46° halos is the path that light takes through the ice crystals. One of the most spectacular effects associated with a halo is called *sun dogs* or *parhelia*. These two bright regions, or mock suns as they are often called, can be seen adjacent to the 22° halo. A *sun pillar*, most often viewed near sunset or sunrise, is a vertical shaft of light that appears to extend upward from the Sun.

- The *glory*, most commonly seen by pilots, consists of one or more colored rings that surround the observer's (airplane's) shadow projected on the clouds below. It forms in a manner not unlike that of a rainbow. The phenomenon can also be viewed by an observer on the ground when he or she is above a layer of fog with the Sun at his or her back. Here, the glory can appear as a "halo" surrounding the shadow of the observer's head cast on the fog.

- The only optical phenomenon more commonly witnessed in association with the Moon than the Sun is the *corona*, a bright whitish disk centered on the Moon or Sun. A corona is produced when water droplets in a thin layer of water-laden clouds, usually altostratus, scatter light from the illuminating body. The colors of the corona are the result of a process called *diffraction*, the slight bending of light as it passes near the edges of cloud droplets. From all sides of a cloud droplet, diffracted light will be directed into the "shadow" of the droplet. Here light rays will meet and interfere with each other. It is the *interference* (interaction of some light frequencies, or colors, which can cause them to be canceled) of the various components of white light that generates the colors that make the corona.

Vocabulary Review

corona (p. 443)
diffraction (p. 443)
dispersion (p. 438)
Fata Morgana (p. 437)
glory (p. 442)
halo (p. 440)
inferior mirage (p. 435)
interference (p. 443)
internal reflection (p. 432)

law of reflection (p. 432)
looming (p. 435)
mirage (p. 435)
rainbow (p. 437)
refraction (p. 433)
sun dogs or parhelia (p. 441)
sun pillar (p. 442)
superior mirage (p. 435)
towering (p. 437)

Review Questions

1. Which of the six colors of the rainbow is refracted at the greatest angle? Why?

2. State the law of reflection.

3. If you have ever been close to a large movie screen, you might have noticed that the screen is made of a number of small glassy particles oriented at slightly different angles rather than one very smooth surface. Why do you think this is so?

4. When light travels from warm air into a region of colder air, its path will curve. Will it curve away from the cold and toward the warm, or vice versa?

5. Why does a mirage always disappear when the observer gets near?

6. What is meant by an inferior mirage? A superior mirage?

7. If you were looking for a rainbow in the morning, which direction would you look? Why that direction?

8. Explain why the secondary rainbow is dimmer than the primary rainbow.

9. How are halos and rainbows similar? How are they different?

10. What is the orientation of the ice crystals that produce a halo? Sun dogs?

11. What gives the glory its name?

12. How are the colors of a corona produced?

13. Describe the relative positions of the observer, the optical phenomenon, and the illuminating body for each of the following:
    a. rainbow
    b. halo
    c. glory
    d. corona
    e. sun dogs

14. At what time of day, if any, can each of the optical phenomena listed in question 13 be best observed?

15. What types of particles (water droplets or ice crystals) are found in the clouds (or fog) that generate each of the optical phenomena listed in question 13?

Atmospheric Science Online

The following are informative and interesting Internet sites that address topics related to those presented in the chapter:

About Rainbows (Unidata Program Center):
• **http://www.unidata.ucar.edu/staff/blynds/ rnbw.html**

Light and Optics (Online Meteorology Guides, University of Illinois):
• **http://ww2010.atmos.uiuc.edu/(Gh)/guides/mtr/ opt/home.rxml**

Refraction of Light (Rick Reed):
• **http://riker.ps.missouri.edu/RicksPage/refract/ refraction.html**

For direct links to these sites and others, chapter objectives and reviews, quiz questions, and topical investigations that utilize Web resources, visit *The Atmosphere, Eighth Edition* Home Page at:
• **http://www.prenhall.com/lutgens**

Metric Units

Table A–1 The international system of units (SI)

I. Basic Units

Quantity	Unit	SI Symbol
Length	meter	m
Mass	kilogram	kg
Time	second	s
Electric current	ampere	A
Thermodynamic temperature	kelvin	K
Amount of substance	mole	mol
Luminous intensity	candela	cd

II. Prefixes

Prefix	Factor by Which Unit is Multiplied	Symbol
tera	10^{12}	T
giga	10^{9}	G
mega	10^{6}	M
kilo	10^{3}	k
hecto	10^{2}	h
deka	10	da
deci	10^{-1}	d
centi	10^{-2}	c
milli	10^{-3}	m
micro	10^{-6}	µ
nano	10^{-9}	n
pico	10^{-12}	p
femto	10^{-15}	f
atto	10^{-18}	a

III. Derived Units

Quantity	Units	Expression
Area	square meter	m^2
Volume	cubic meter	m^3
Frequency	hertz (Hz)	s^{-1}
Density	kilogram per cubic meter	kg/m^3
Velocity	meter per second	m/s
Angular velocity	radian per second	rad/s
Acceleration	meter per second squared	m/s^2
Angular acceleration	radian per second squared	rad/s^2
Force	newton (N)	$kg \cdot ms^2$
Pressure	newton per square meter	N/m^2
Work, energy, quantity of heat	joule (J)	N•m
Power	watt (W)	J/s
Electric charge	coulomb (C)	A•s
Voltage, potential difference, electromotive force	volt (V)	W/A
Luminance	candela per square meter	cd/m^2

Table A–2 Metric–English conversion

When You Want to Convert:	Multiply by:	To Find:
Length		
inches	2.54	centimeters
centimeters	0.39	inches
feet	0.30	meters
meters	3.28	feet
yards	0.91	meters
meters	1.09	yards
miles	1.61	kilometers
kilometers	0.62	miles
Area		
square inches	6.45	square centimeters
square centimeters	0.15	square inches
square feet	0.09	square meters
square meters	10.76	square feet
square miles	2.59	square kilometers
square kilometers	0.39	square miles
Volume		
cubic inches	16.38	cubic centimeters
cubic centimeters	0.06	cubic inches
cubic feet	0.028	cubic meters
cubic meters	35.3	cubic feet
cubic miles	4.17	cubic kilometers
cubic kilometers	0.24	cubic miles
liters	1.06	quarts
liters	0.26	gallons
gallons	3.78	liters
Masses and Weights		
ounces	28.33	grams
grams	0.035	ounces
pounds	0.45	kilograms
kilograms	2.205	pounds

Temperature

When you want to convert degrees Fahrenheit (°F) to degrees Celsius (°C), subtract 32 degrees and divide by 1.8 (also see Table A–3).

When you want to convert degrees Celsius (°C) to degrees Fahrenheit (°F), multiply by 1.8 and add 32 degrees (see also Table A–3).

When you want to convert degrees Celsius (°C) to degrees Kelvin (K), delete the degree symbol and add 273.

When you want to convert degrees Kelvin (K) to degrees Celsius (°C), add the degree symbol and subtract 273.

Table A–3 Temperature conversion table. (To find either the Celsius or the Fahrenheit equivalent, locate the known temperature in the center column. Then read the desired equivalent value from the appropriate column.)

°C		°F	°C		°F	°C		°F	°C		°F
−40.0	−40	−40	−17.2	+1	33.8	5.0	41	105.8	27.2	81	177.8
−39.4	−39	−38.2	−16.7	2	35.6	5.6	42	107.6	27.8	82	179.6
−38.9	−38	−36.4	−16.1	3	37.4	6.1	43	109.4	28.3	83	181.4
−38.3	−37	−34.6	−15.4	4	39.2	6.7	44	111.2	28.9	84	183.2
−37.8	−36	−32.8	−15.0	5	41.0	7.2	45	113.0	29.4	85	185.0
−37.2	−35	−31.0	−14.4	6	42.8	7.8	46	114.8	30.0	86	186.8
−36.7	−34	−29.2	−13.9	7	44.6	8.3	47	116.6	30.6	87	188.6
−36.1	−33	−27.4	−13.3	8	46.4	8.9	48	118.4	31.1	88	190.4
−35.6	−32	−25.6	−12.8	9	48.2	9.4	49	120.2	31.7	89	192.2
−35.0	−31	−23.8	−12.2	10	50.0	10.0	50	122.0	32.2	90	194.0
−34.4	−30	−22.0	−11.7	11	51.8	10.6	51	123.8	32.8	91	195.8
−33.9	−29	−20.2	−11.1	12	53.6	11.1	52	125.6	33.3	92	197.6
−33.3	−28	−18.4	−10.6	13	55.4	11.7	53	127.4	33.9	93	199.4
−32.8	−27	−16.6	−10.0	14	57.2	12.2	54	129.2	34.4	94	201.2
−32.2	−26	−14.8	−9.4	15	59.0	12.8	55	131.0	35.0	95	203.0
−31.7	−25	−13.0	−8.9	16	60.8	13.3	56	132.8	35.6	96	204.8
−31.1	−24	−11.2	−8.3	17	62.6	13.9	57	134.6	36.1	97	206.6
−30.6	−23	−9.4	−7.8	18	64.4	14.4	58	136.4	36.7	98	208.4
−30.0	−22	−7.6	−7.2	19	66.2	15.0	59	138.2	37.2	99	210.2
−29.4	−21	−5.8	−6.7	20	68.0	15.6	60	140.0	37.8	100	212.0
−28.9	−20	−4.0	−6.1	21	69.8	16.1	61	141.8	38.3	101	213.8
−28.3	−19	−2.2	−5.6	22	71.6	16.7	62	143.6	38.9	102	215.6
−27.8	−18	−0.4	−5.0	23	73.4	17.2	63	145.4	39.4	103	217.4
−27.2	−17	+1.4	−4.4	24	75.2	17.8	64	147.2	40.0	104	219.2
−26.7	−16	3.2	−3.9	25	77.0	18.3	65	149.0	40.6	105	221.0
−26.1	−15	5.0	−3.3	26	78.8	18.9	66	150.8	41.1	106	222.8
−25.6	−14	6.8	−2.8	27	80.6	19.4	67	152.6	41.7	107	224.6
−25.0	−13	8.6	−2.2	28	82.4	20.0	68	154.4	42.2	108	226.4
−24.4	−12	10.4	−1.7	29	84.2	20.6	69	156.2	42.8	109	228.2
−23.9	−11	12.2	−1.1	30	86.0	21.1	70	158.0	43.3	110	230.0
−23.3	−10	14.0	−0.6	31	87.8	21.7	71	159.8	43.9	111	231.8
−22.8	−9	15.8	0.0	32	89.6	22.2	72	161.6	44.4	112	233.6
−22.2	−8	17.6	+0.6	33	91.4	22.8	73	163.4	45.0	113	235.4
−21.7	−7	19.4	1.1	34	93.2	23.3	74	165.2	45.6	114	237.2
−21.1	−6	21.2	1.7	35	95.0	23.9	75	167.0	46.1	115	239.0
−20.6	−5	23.0	2.2	36	96.8	24.4	76	168.8	46.7	116	240.8
−20.0	−4	24.8	2.8	37	98.6	25.0	77	170.6	47.2	117	242.6
−19.4	−3	26.6	3.3	38	100.4	25.6	78	172.4	47.8	118	244.4
−18.9	−2	28.4	3.9	39	102.2	26.1	79	174.2	48.3	119	246.2
−18.3	−1	30.2	4.4	40	104.0	26.7	80	176.0	48.9	120	248.0
−17.8	0	32.0									

Table A–4 Wind-conversion table (Wind speed units: 1 mile per hour = 0.868391 knot = 1.609344 km/h = 0.44704 m/s)

Miles per Hour	Knots	Meters per Second	Kilometers per Hour	Miles per Hour	Knots	Meters per Second	Kilometers per Hour
1	0.9	0.4	1.6	51	44.3	22.8	82.1
2	1.7	0.9	3.2	52	45.2	23.2	83.7
3	2.6	1.3	4.8	53	46.0	23.7	85.3
4	3.5	1.8	6.4	54	46.9	24.1	86.9
5	4.3	2.2	8.0	55	47.8	24.6	88.5
6	5.2	2.7	9.7	56	48.6	25.0	90.1
7	6.1	3.1	11.3	57	49.5	25.5	91.7
8	6.9	3.6	12.9	58	50.4	25.9	93.3
9	7.8	4.0	14.5	59	51.2	26.4	95.0
10	8.7	4.5	16.1	60	52.1	26.8	96.6
11	9.6	4.9	17.7	61	53.0	27.3	98.2
12	10.4	5.4	19.3	62	53.8	27.7	99.8
13	11.3	5.8	20.9	63	54.7	28.2	101.4
14	12.2	6.3	22.5	64	55.6	28.6	103.0
15	13.0	6.7	24.1	65	56.4	29.1	104.6
16	13.9	7.2	25.7	66	57.3	29.5	106.2
17	14.8	7.6	27.4	67	58.2	30.0	107.8
18	15.6	8.0	29.0	68	59.1	30.4	109.4
19	16.5	8.5	30.6	69	59.9	30.8	111.0
20	17.4	8.9	32.2	70	60.8	31.3	112.7
21	18.2	9.4	33.8	71	61.7	31.7	114.3
22	19.1	9.8	35.4	72	62.5	32.2	115.9
23	20.0	10.3	37.0	73	63.4	32.6	117.5
24	20.8	10.7	38.6	74	64.3	33.1	119.1
25	21.7	11.2	40.2	75	65.1	33.5	120.7
26	22.6	11.6	41.8	76	66.0	34.0	122.3
27	23.4	12.1	43.5	77	66.9	34.4	123.9
28	24.3	12.5	45.1	78	67.7	34.9	125.5
29	25.2	13.0	46.7	79	68.6	35.3	127.1
30	26.1	13.4	48.3	80	69.5	35.8	128.7
31	26.9	13.9	49.9	81	70.3	36.2	130.4
32	27.8	14.3	51.5	82	71.2	36.7	132.0
33	28.7	14.8	53.1	83	72.1	37.1	133.6
34	29.5	15.2	54.7	84	72.9	37.6	135.2
35	30.4	15.6	56.3	85	73.8	38.0	136.8
36	31.3	16.1	57.9	86	74.7	38.4	138.4
37	32.1	16.5	59.5	87	75.5	38.9	140.0
38	33.0	17.0	61.2	88	76.4	39.3	141.6
39	33.9	17.4	62.8	89	77.3	39.8	143.2
40	34.7	17.9	64.4	90	78.2	40.2	144.8
41	35.6	18.3	66.0	91	79.0	40.7	146.5
42	36.5	18.8	67.6	92	79.9	41.1	148.1
43	37.3	19.2	69.2	93	80.8	41.6	149.7
44	38.2	19.7	70.8	94	81.6	42.0	151.3
45	39.1	20.1	72.4	95	82.5	42.5	152.9
46	39.9	20.6	74.0	96	83.4	42.9	154.5
47	40.8	21.0	75.6	97	84.2	43.4	156.1
48	41.7	21.5	77.2	98	85.1	43.8	157.7
49	42.6	21.9	78.9	99	86.0	44.3	159.3
50	43.4	22.4	80.5	100	86.8	44.7	160.9

APPENDIX B

Explanation and Decoding
of the Daily Weather Map

Weather maps showing the development and movement of weather systems are among the most important tools used by the weather forecaster. Of the several types of maps used, some portray conditions near the surface of Earth and others depict conditions at various heights in the atmosphere. Some cover the entire Northern Hemisphere and others cover only local areas as required for special purposes. The maps used for daily forecasting by the National Weather Service (NWS) are similar in many respects to the printed Daily Weather Map. At NWS offices, maps showing conditions at Earth's surface are drawn four times daily. Maps of upper-level temperature, pressure, and humidity are prepared twice each day.

Principal Surface Weather Map

To prepare the surface map and present the information quickly and pictorially, two actions are necessary: (1) weather observers at many places must go to their posts at regular times each day to observe the weather and send the information to the offices where the maps are drawn; (2) the information must be quickly transcribed to the maps. In order for the necessary speed and economy of space and transmission time to be realized, codes have been devised for sending the information and for plotting it on the maps.

Codes and Map Plotting

A great deal of information is contained in a brief coded weather message. If each item were named and described in plain language, a very lengthy message would be required, one confusing to read and difficult to transfer to a map. A code permits the message to be condensed to a few five-figure numeral groups, each figure of which has a meaning, depending on its position in the message. People trained in the use of the code can read the message as easily as plain language.

The location of the reporting station is printed on the map as a small circle (the station circle). A definite arrangement of the data around the station circle, called the *station model*, is used. When the report is plotted in these fixed positions around the station circle on the weather map, many code figures are transcribed exactly as sent. Entries in the station model that are not made in

code figures or actual values found in the message are usually in the form of symbols that graphically represent the element concerned. In some cases, certain of the data may or may not be reported by the observer, depending on local weather conditions. Precipitation and clouds are examples. In such cases, the absence of an entry on the map is interpreted as nonoccurrence or nonobservance of the phenomena. The letter *M* is entered where data are normally observed but not received.

Both the code and the station model are based on international agreements. These standardized numerals and symbols enable a meteorologist of one country to use the weather reports and weather maps of another country even though that person does not understand the language. Weather codes are, in effect, an international language that permits complete interchange and use of worldwide weather reports so essential in present-day activities.

The boundary between two different air masses is called a *front*. Important changes in weather, temperature, wind direction, and clouds often occur with the passage of a front. Half circles or triangular symbols or both are placed on the lines representing fronts to indicate the kind of front. The side on which the symbols are placed indicates the direction of frontal movement. The boundary of relatively cold air of polar origin advancing into an area occupied by warmer air, often of tropical origin, is called a *cold front*. The boundary of relatively warm air advancing into an area occupied by colder air is called a *warm front*. The line along which a cold front has overtaken a warm front at the ground is called an *occluded front*. A boundary between two air masses, which shows at the time of observation little tendency to advance into either the warm or cold areas, is called a *stationary front*. Air-mass boundaries are known as surface fronts when they intersect the ground and as *upper-air fronts* when they do not. Surface fronts are drawn in solid black, fronts aloft are drawn in outline only. Front symbols are given in Table B–1.

A front that is disappearing or weak and decreasing in intensity is labeled *frontolysis*. A front that is forming is labeled *frontogenesis*. A *squall line* is a line of thunderstorms or squalls usually accompanied by heavy showers and shifting winds (Table B–1).

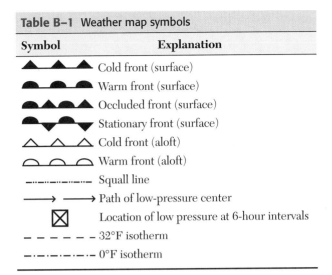

Table B–1 Weather map symbols

Symbol	Explanation
▲▲▲	Cold front (surface)
●●●	Warm front (surface)
●▲●▲	Occluded front (surface)
●▼●▼	Stationary front (surface)
△△△	Cold front (aloft)
⌒⌒⌒	Warm front (aloft)
–··–··–··–	Squall line
⟶ ⟶	Path of low-pressure center
⊠	Location of low pressure at 6-hour intervals
– – – – –	32°F isotherm
–·–·–·–·–	0°F isotherm

The paths followed by individual disturbances are called *storm tracks* and are shown by arrows (Table B–1). A symbol (a box containing an x) indicates past positions of a low-pressure center at six-hour intervals. HIGH (H) and LOW (L) indicate the centers of high and low barometric pressure. Solid lines are isobars and connect points of equal sea-level barometric pressure. The spacing and orientation of these lines on weather maps are indications of speed and direction of windflow. In general, wind direction is parallel to these lines with low pressure to the left of an observer looking downwind. Speed is directly proportional to the closeness of the lines (called *pressure gradient*). Isobars are labeled in millibars.

Isotherms are lines connecting points of equal temperature. Two isotherms are drawn on the large surface weather map when applicable. The freezing, or 32°F, isotherm is drawn as a dashed line, and the 0°F isotherm is drawn as a dash–dot line (Table B–1). Areas where precipitation is occurring at the time of observation are shaded.

Auxiliary Maps

500-Millibar Map

Contour lines, isotherms, and wind arrows are shown on the insert map for the 500-millibar contour level. Solid lines are drawn to show height above sea level and are labeled in feet. Dashed lines are drawn at 5° intervals of temperature and are labeled in degrees Celsius. True wind direction is shown by "arrows" that are plotted as flying with the wind. The wind speed is shown by flags and feathers. Each flag represents 50 knots, each full feather represents 10 knots, and each half-feather represents 5 knots.

Temperature Map (Highest and Lowest)

Temperature data are entered from selected weather stations in the United States. The figure entered above the station dot shows the maximum temperature for the 12-hour period ending 7:00 p.m. EST of the previous day. The figure entered below the station dot shows the minimum temperature during the 12 hours ending at 7:00 a.m. EST. The letter *M* denotes missing data.

Precipitation Map

Precipitation data are entered from selected weather stations in the United States. When precipitation has occurred at any of these stations in the 24-hour period ending at 7:00 a.m. EST, the total amount, in inches and hundredths, is entered above the station dot. When the figures for total precipitation have been compiled from incomplete data and entered on the map, the amount is underlined. *T* indicates a trace of precipitation (less than 0.01 inch) and the letter *M* denotes missing data. The geographical areas where precipitation has fallen during the 24 hours ending at 7:00 a.m. EST are shaded. Dashed lines show depth of snow on ground in inches as of 7:00 a.m. EST.

Explanation of Station Symbols and Map Entries

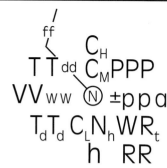

Symbol station model

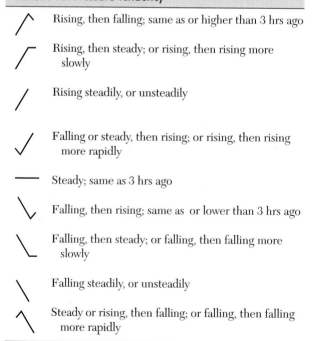

Sample report

N	Total cloud cover—Table E
dd	Wind direction
ff	Wind speed in knots or mi/hr—Table F
VV	Visibility in miles
ww	Present weather—Table H
W	Past weather—Table H
PPP	Barometric pressure reduced to sea level (add an initial 9 or 10 and place a decimal point to the left of last number)
TT	Current air temperature in °F
N_h	Fraction of sky covered by low or middle clouds—Table E (ranges from 0 for no clouds to 9 for sky obscured)
C_L	Low clouds or clouds with vertical development—Table C
h	Height in feet of the base of the lowest clouds—Table D
C_M	Middle clouds—Table C
C_H	High clouds—Table C
T_dT_d	Dewpoint temperature in °F
a	Pressure tendency—Table A
pp	Pressure change in mb in preceding 3 hr (+28 = +2.8)
RR	Amount of precipitation in last 6 hr
R_t	Time precipitation began or ended (0 = none; 1 = <1 hr ago; 2 = 1–2 hr ago; 3 = 2–3 hr ago; 4 = 3–4 hr ago; 5 = 4–5 hr ago; 6 = 5–6 hr ago; 7 = 6–12 hr ago; 8 = >12 hr ago; 9 = unknown)

Table A Air Pressure Tendency

⌃	Rising, then falling; same as or higher than 3 hrs ago
⌐／	Rising, then steady; or rising, then rising more slowly
／	Rising steadily, or unsteadily
✓	Falling or steady, then rising; or rising, then rising more rapidly
—	Steady; same as 3 hrs ago
＼／	Falling, then rising; same as or lower than 3 hrs ago
＼＿	Falling, then steady; or falling, then falling more slowly
＼	Falling steadily, or unsteadily
⌄	Steady or rising, then falling; or falling, then falling more rapidly

Table B Cloud Abbreviations

St	stratus
Fra	fractus
Sc	stratocumulus
Ns	nimbostratus
As	altostratus
Ac	altocumulus
Ci	cirrus
Cs	cirrostratus
Cc	cirrocumulus
Cu	cumulus
Cb	cumulonimbus

Table C Cloud Types

Cu of fair weather, little vertical development and seemingly flattened

Cu of considerable development, generally towering, with or without other Cu or Sc, bases all at same level

Cb with tops lacking clear-cut outlines, but distinctly not cirriform or anvil shaped; with or without Cu, Sc, or St

Sc formed by spreading out of Cu; Cu often present also

Sc not formed by spreading out of Cu

St or StFra, but no StFra of bad weather

StFra and/or CuFra of bad weather (scud)

Cu and Sc (not formed by spreading out of Cu) with bases at different levels

Cb having a clearly fibrous (cirriform) top, often anvil shaped, with or without Cu, Sc, St, or scud

Thin As (most of cloud layer semitransparent)

Thick As, greater part sufficiently dense to hide Sun (or Moon), or Ns

Thin Ac, mostly semitransparent; cloud elements not changing much and at a single level

Thin Ac in patches; cloud elements continually changing and/or occurring at more than one level

Thin Ac in bands or in a layer gradually spreading over sky and usually thickening as a whole

Ac formed by the spreading out of Cu or Cb

Double-layered Ac, or a thick layer of Ac, not increasing; or Ac with As and/or Ns

Ac in the form of Cu-shaped tufts or Ac with turrets

Ac of a chaotic sky, usually at different levels; patches of dense Ci usually present also

Filaments of Ci, or "mares' tails," scattered and not increasing

Dense Ci in patches or twisted sheaves, usually not increasing, sometimes like remains of Cb; or towers or tufts

Dense Ci, often anvil shaped, derived from or associated with Cb

Ci, often hook shaped, gradually spreading over the sky and usually thickening as a whole

Ci and Cs, often in converging bands, or Cs alone; generally overspreading and growing denser; the continuous layer not reaching 45° altitude

Ci and Cs, often in converging bands, or Cs alone; generally overspreading and growing denser; the continuous layer exceeding 45° altitude

Veil of Cs covering the entire sky

Cs not increasing and not covering entire sky

Cc alone or Cc with some Ci or Cs, but the Cc being the main cirriform cloud

Table D Height of Base of Lowest Cloud

Code	Feet	Meters
0	0–149	0–49
1	150–299	50–99
2	300–599	100–199
3	600–999	200–299
4	1000–1999	300–599
5	2000–3499	600–999
6	3500–4999	1000–1499
7	5000–6499	1500–1999
8	6500–7999	2000–2499
9	8000 or above or no clouds	2500 or above or no clouds

Table E Cloud Cover

○	No clouds
◔ (one line)	One-tenth or less
	Two-tenths or three-tenths
	Four-tenths
◐	Five-tenths
	Six-tenths
	Seven-tenths or eight-tenths
	Nine-tenths or overcast with openings
●	Completely overcast (ten-tenths)
⊗	Sky obscured

Table F Wind Speed

Symbol	Knots	Miles per Hour	Kilometers per Hour
◎	1–2	1–2	1–3
	3–7	3–8	4–13
	8–12	9–14	14–19
	13–17	15–20	20–32
	18–22	21–25	33–40
	23–27	26–31	41–50
	28–32	32–37	51–60
	33–37	38–43	61–69
	38–42	44–49	70–79
	43–47	50–54	80–87
	48–52	55–60	88–96
	53–57	61–66	97–106
	58–62	67–71	107–114
	63–67	72–77	115–124
	68–72	78–83	125–134
	73–77	84–89	135–143
	103–107	119–123	192–198

Table G Fronts

Fronts are shown on surface weather maps by the symbols below. (Arrows—not shown on maps—indicate direction of motion of front.)

▼▼▼	Cold front (surface)
●●●	Warm front (surface)
●●▲▲	Occluded front (surface)
▲▲▽▽	Stationary front (surface)
⌒⌒⌒	Warm front (aloft)
▽▽▽	Cold front (aloft)

Table H Weather Conditions

Cloud development NOT observed or NOT observable during past hour

Clouds generally dissolving or becoming less developed during past hour

State of sky on the whole unchanged during past hour

Clouds generally forming or developing during past hour

Visibility reduced by smoke

Light fog (mist)

Patches of shallow fog at station, NOT deeper than 6 feet on land

More or less continuous shallow fog at station, NOT deeper than 6 feet on land

Lightning visible, no thunder heard

Precipitation within sight, but NOT reaching the ground

Drizzle (NOT freezing) or snow grains (NOT falling as showers) during past hour, but NOT at time of observation

Rain (NOT freezing and NOT falling as showers) during past hour, but NOT at time of observation

Snow (NOT falling as showers) during past hour, but NOT at time of observation

Rain and snow or ice pellets (NOT falling as showers) during past hour, but NOT at time of observation

Freezing drizzle or freezing rain (NOT falling as showers) during past hour, but NOT at time of observation

Slight or moderate dust storm or sandstorm has decreased during past hour

Slight or moderate dust storm or sandstorm, no appreciable change during past hour

Slight or moderate dust storm or sandstorm has begun or increased during past hour

Severe dust storm or sandstorm, has decreased during past hour

Severe dust storm or sandstorm, no appreciable change during past hour

Fog or ice fog at distance at time of observation, but NOT at station during past hour

Fog or ice fog in patches

Fog or ice fog, sky discernible, has become thinner during past hour

Fog or ice fog, sky NOT discernible, has become thinner during past hour

Fog or ice fog, sky discernible, no appreciable change during past hour

Intermittent drizzle (NOT freezing), slight at time of observation

Continuous drizzle (NOT freezing), slight at time of observation

Intermittent drizzle (NOT freezing), moderate at time of observation

Continuous drizzle (NOT freezing), moderate at time of observation

Intermittent drizzle (NOT freezing), heavy at time of observation

Intermittent rain (NOT freezing), slight at time of observation

Continuous rain (NOT freezing), slight at time of observation

Intermittent rain (NOT freezing), moderate at time of observation

Continuous rain (NOT freezing), moderate at time of observation

Intermittent rain (NOT freezing), heavy at time of observation

Intermittent fall of snowflakes, slight at time of observation

Continuous fall of snowflakes, slight at time of observation

Intermittent fall of snowflakes, moderate at time of observation

Continuous fall of snowflakes, moderate at time of observation

Intermittent fall of snowflakes, heavy at time of observation

Slight rain shower(s)

Moderate or heavy rain shower(s)

Violent rain shower(s)

Slight shower(s) of rain and snow mixed

Moderate or heavy shower(s) of rain and snow mixed

Moderate or heavy shower(s) of hail, with or without rain, or rain and snow mixed, not associated with thunder

Slight rain at time of observation; thunderstorm during past hour, but NOT at time of observation

Moderate or heavy rain at time of observation; thunderstorm during past hour, but NOT at time of observation

Slight snow, or rain and snow mixed, or hail at time of observation; thunderstorm during past hour, but NOT at time of observation

Moderate or heavy snow, or rain and snow mixed, or hail at time of observation; thunderstorm during past hour, but NOT at time of observation

Table H Weather Conditions

∞ Haze	Widespread dust in suspension in the air, NOT raised by wind, at time of observation	Dust or sand raised by wind at time of observation	Well developed dust whirl(s) within past hour	Dust storm or sandstorm within sight of or at station during past hour
)•(Precipitation within sight, reaching the ground but distant from station	(•) Precipitation within sight, reaching the ground, near to but NOT at station	Thunderstorm, but no precipitation at the station	Squall(s) within sight during past hour or at time of observation	Funnel cloud(s) within sight of station at time of observation
Showers of rain during past hour, but NOT at time of observation	Showers of snow, or of rain and snow, during past hour, but NOT at time of observation	Showers of hail, or of hail and rain, during past hour, but NOT at time of observation	Fog during past hour, but NOT at time of observation	Thunderstorm (with or without precipitation) during past hour, but NOT at time of observation
Severe dust storm or sandstorm has begun or increased during past hour	Slight or moderate drifting snow, generally low (less than 6 ft)	Heavy drifting snow, generally low	Slight or moderate blowing snow, generally high (more than 6 ft)	Heavy blowing snow, generally high
Fog or ice fog, sky NOT discernible, no appreciable change during past hour	Fog or ice fog, sky discernible, has begun or become thicker during past hour	Fog or ice fog, sky NOT discernible, has begun or become thicker during past hour	Fog depositing rime, sky discernible	Fog depositing rime, sky NOT discernible
Continuous drizzle (NOT freezing), heavy at time of observation	Slight freezing drizzle	Moderate or heavy freezing drizzle	Drizzle and rain, slight	Drizzle and rain, moderate or heavy
Continuous rain (NOT freezing), heavy at time of observation	Slight freezing rain	Moderate or heavy freezing rain	Rain or drizzle and snow, slight	Rain or drizzle and snow, moderate or heavy
Continuous fall of snowflakes, heavy at time of observation	Ice prisms (with or without fog)	Snow grains (with or without fog)	Isolated starlike snow crystals (with or without fog)	Ice pellets or snow pellets
Slight snow shower(s)	Moderate or heavy snow showers(s)	Slight shower(s) of snow pellets, or ice pellets with or without rain, or rain and snow mixed	Moderate or heavy shower(s) of snow pellets, or ice pellets, or ice pellets with or without rain or rain and snow mixed	Slight shower(s) of hail, with or without rain or rain and snow mixed, not associated with thunder
Slight or moderate thunderstorm without hail, but with rain, and/or snow at time of observation	Slight or moderate thunderstorm, with hail at time of observation	Heavy thunderstorm, without hail, but with rain and/or snow at time of observation	Thunderstorm combined with dust storm or sandstorm at time of observation	Heavy thunderstorm with hail at time of observation

APPENDIX C
Relative Humidity and Dew Point Tables

Table C–1 Relative humidity (percent)

Dry bulb (°C)	Depression of Wet-Bulb Temperature (Dry-Bulb Temperature Minus Wet-Bulb Temperature = Depression of the Wet Bulb)																					
	1	2	3	4	5	6	7	8	9	10	11	12	13	14	15	16	17	18	19	20	21	22
–20	28																					
–18	40																					
–16	48	0																				
–14	55	11																				
–12	61	23																				
–10	66	33	0																			
–8	71	41	13																			
–6	73	48	20	0																		
–4	77	54	32	11																		
–2	79	58	37	20	1																	
0	81	63	45	28	11																	
2	83	67	51	36	20	6																
4	85	70	56	42	27	14																
6	86	72	59	46	35	22	10	0														
8	87	74	62	51	39	28	17	6														
10	88	76	65	54	43	33	24	13	4													
12	88	78	67	57	48	38	28	19	10	2												
14	89	79	69	60	50	41	33	25	16	8	1											
16	90	80	71	62	54	45	37	29	21	14	7	1										
18	91	81	72	64	56	48	40	33	26	19	12	6	0									
20	91	82	74	66	58	51	44	36	30	23	17	11	5									
22	92	83	75	68	60	53	46	40	33	27	21	15	10	4	0							
24	92	84	76	69	62	55	49	42	36	30	25	20	14	9	4	0						
26	92	85	77	70	64	57	51	45	39	34	28	23	18	13	9	5						
28	93	86	78	71	65	59	53	45	42	36	31	26	21	17	12	8	4					
30	93	86	79	72	66	61	55	49	44	39	34	29	25	20	16	12	8	4				
32	93	86	80	73	68	62	56	51	46	41	36	32	27	22	19	14	11	8	4			
34	93	86	81	74	69	63	58	52	48	43	38	34	30	26	22	18	14	11	8	5		
36	94	87	81	75	69	64	59	54	50	44	40	36	32	28	24	21	17	13	10	7	4	
38	94	87	82	76	70	66	60	55	51	46	42	38	34	30	26	23	20	16	13	10	7	5
40	94	89	82	76	71	67	61	57	52	48	44	40	36	33	29	25	22	19	16	13	10	7

Relative Humidity Values

To determine the relative humidity, find the air (dry-bulb) temperature on the vertical axis (far left) and the depression of the wet bulb on the horizontal axis (top). Where the two meet, the relative humidity is found. For example, when the dry-bulb temperature is 20°C and a wet-bulb temperature is 14°C, then the depression of the wet-bulb is 6°C (20°C – 14°C). From Table C–1, the relative humidity is 51 percent and from Table C–2, the dew point is 10°C.

Table C–2 Dew-point temperature (°C).*

Dry bulb (°C)	Dry-Bulb Temperature Minus Wet-Bulb Temperature = Depression of the Wet Bulb																					
	1	2	3	4	5	6	7	8	9	10	11	12	13	14	15	16	17	18	19	20	21	22
−20	−33																					
−18	−28																					
−16	−24																					
−14	−21	−36																				
−12	−18	−28																				
−10	−14	−22																				
−8	−12	−18	−29																			
−6	−10	−14	−22																			
−4	−7	−12	−17	−29																		
−2	−5	−8	−13	−20																		
0	−3	−6	−9	−15	−24																	
2	−1	−3	−6	−11	−17																	
4	1	−1	−4	−7	−11	−19																
6	4	1	−1	−4	−7	−13	−21															
8	6	3	1	−2	−5	−9	−14															
10	8	6	4	1	−2	−5	−9	−14	−18													
12	10	8	6	4	1	−2	−5	−9	−16													
14	12	11	9	6	4	1	−2	−5	−10	−17												
16	14	13	11	9	7	4	1	−1	−6	−10	−17											
18	16	15	13	11	9	7	4	2	−2	−5	−10	−19										
20	19	17	15	14	12	10	7	4	2	−2	−5	−10	−19									
22	21	19	17	16	14	12	10	8	5	3	−1	−5	−10	−19								
24	23	21	20	18	16	14	12	10	8	6	2	−1	−5	−10	−18							
26	25	23	22	20	18	17	15	13	11	9	6	3	0	−4	−9	−18						
28	27	25	24	22	21	19	17	16	14	11	9	7	4	1	−3	−9	−16					
30	29	27	26	24	23	21	19	18	16	14	12	10	8	5	1	−2	−8	−15				
32	31	29	28	27	25	24	22	21	19	17	15	13	11	8	5	2	−2	−7	−14			
34	33	31	30	29	27	26	24	23	21	20	18	16	14	12	9	6	3	−1	−5	−12	−29	
36	35	33	32	31	29	28	27	25	24	22	20	19	17	15	13	10	7	4	0	−4	−10	
38	37	35	34	33	32	30	29	28	26	25	23	21	19	17	15	13	11	8	5	1	−3	−9
40	39	37	36	35	34	32	31	30	28	27	25	24	22	20	18	16	14	12	9	6	2	−2

Dry-Bulb (Air) Temperature

Dew point temperatures

Correcting Mercurial
Barometer Readings

When barometric readings taken from separate locations are being compared, corrections must be made to ensure that the true differences in atmospheric pressure are shown. In general, this step is done by reducing these barometric readings to normal sea-level conditions. So a correction is made for temperature, elevation, and latitude (gravity correction).

Temperature Correction

This correction accounts for the expansion and contraction of the mercury column and brass scale caused by temperature variations.

Elevation Correction

Because pressure decreases with increases in elevation, barometers not located at sea level will show a reduced pressure. The elevation correction accounts for the position of the barometer as well as for the temperature of the assumed air column extending down to sea level.

Latitude Correction

Earth is not a perfect sphere; it has a shorter polar diameter than an equatorial diameter. Consequently, the pull of gravity on the mercury column is greater at the poles and lesser at the equator. Because the adjustment needed to correct this latitudinal variation in gravity is usually negligible compared with the other corrections, we have not included it in the following procedure for correcting the barometer.

To reduce a mercurial barometer reading to sea level, take the following steps:

1. Read the height of the mercury column on the millimeter scale.
2. Correct this reading for the temperature of the barometer by using Table D–1 and subtracting the obtained value from your reading.
3. Using the out-of-doors air temperature and the elevation (altitude) of the barometer, obtain the temperature–altitude factor from Table D–2.
4. Using the temperature–altitude factor from step 3, the temperature-corrected barometric reading from step 2, and Table D–3, find the sea-level correction.
5. Add the sea-level correction to the temperature-corrected barometer reading. This sum gives the desired barometric reading reduced to sea level.

Table D–1 Temperature correction

Observed Height in Millimeters

Temp. (°C)	620	630	640	650	660	670	680	690	700	710	720	730	740	750	760	770	780	790
0	0.00	0.00	0.00	0.00	0.00	0.00	0.00	0.00	0.00	0.00	0.00	0.00	0.00	0.00	0.00	0.00	0.00	0.00
1	0.10	0.10	0.10	0.11	0.11	0.11	0.11	0.11	0.11	0.12	0.12	0.12	0.12	0.12	0.12	0.13	0.13	0.13
2	0.20	0.21	0.21	0.21	0.22	0.22	0.22	0.23	0.23	0.23	0.24	0.24	0.24	0.25	0.25	0.25	0.25	0.26
3	0.30	0.31	0.31	0.32	0.32	0.33	0.33	0.34	0.34	0.35	0.35	0.36	0.36	0.37	0.37	0.38	0.38	0.39
4	0.40	0.41	0.42	0.42	0.43	0.44	0.44	0.45	0.46	0.46	0.47	0.48	0.48	0.49	0.50	0.50	0.51	0.52
5	0.51	0.51	0.52	0.53	0.54	0.55	0.56	0.56	0.57	0.58	0.59	0.60	0.60	0.61	0.62	0.63	0.64	0.64
6	0.61	0.62	0.63	0.64	0.65	0.66	0.67	0.68	0.69	0.70	0.71	0.71	0.72	0.73	0.74	0.75	0.76	0.77
7	0.71	0.72	0.73	0.74	0.75	0.77	0.78	0.79	0.80	0.81	0.82	0.83	0.85	0.86	0.87	0.88	0.89	0.90
8	0.81	0.82	0.84	0.85	0.86	0.87	0.89	0.90	0.91	0.93	0.94	0.95	0.97	0.98	0.99	1.01	1.02	1.03
9	0.91	0.92	0.94	0.95	0.97	0.98	1.00	1.01	1.03	1.04	1.06	1.07	1.09	1.10	1.12	1.13	1.15	1.16
10	1.01	1.03	1.04	1.06	1.08	1.09	1.11	1.13	1.14	1.16	1.17	1.19	1.21	1.22	1.24	1.26	1.27	1.29
11	1.11	1.13	1.15	1.17	1.18	1.20	1.22	1.24	1.26	1.27	1.29	1.31	1.33	1.35	1.36	1.38	1.40	1.42
12	1.21	1.23	1.25	1.27	1.29	1.31	1.33	1.35	1.37	1.39	1.41	1.43	1.45	1.47	1.49	1.51	1.53	1.55
13	1.31	1.34	1.36	1.38	1.40	1.42	1.44	1.46	1.48	1.50	1.53	1.55	1.57	1.59	1.61	1.63	1.65	1.67
14	1.41	1.44	1.46	1.48	1.51	1.53	1.55	1.57	1.60	1.62	1.64	1.67	1.69	1.71	1.73	1.76	1.78	1.80
15	1.52	1.54	1.56	1.59	1.61	1.64	1.66	1.69	1.71	1.74	1.76	1.78	1.81	1.83	1.86	1.88	1.91	1.93
16	1.62	1.64	1.67	1.69	1.72	1.75	1.77	1.80	1.82	1.85	1.88	1.90	1.93	1.96	1.98	2.01	2.03	2.06
17	1.72	1.74	1.77	1.80	1.83	1.86	1.88	1.91	1.94	1.97	1.99	2.02	2.05	2.08	2.10	2.13	2.16	2.19
18	1.82	1.85	1.88	1.91	1.93	1.96	1.99	2.02	2.05	2.08	2.11	2.14	2.17	2.20	2.23	2.26	2.29	2.32
19	1.92	1.95	1.98	2.01	2.04	2.07	2.10	2.13	2.17	2.20	2.23	2.26	2.29	2.32	2.35	2.38	2.41	2.44
20	2.02	2.05	2.08	2.12	2.15	2.18	2.21	2.25	2.28	2.31	2.34	2.38	2.41	2.44	2.47	2.51	2.54	2.57
21	2.12	2.15	2.19	2.22	2.26	2.29	2.32	2.36	2.39	2.43	2.46	2.50	2.53	2.56	2.60	2.63	2.67	2.70
22	2.22	2.26	2.29	2.33	2.36	2.40	2.43	2.47	2.51	2.54	2.58	2.61	2.65	2.69	2.72	2.76	2.79	2.83
23	2.32	2.36	2.40	2.43	2.47	2.51	2.54	2.58	2.62	2.66	2.69	2.73	2.77	2.81	2.84	2.88	2.92	2.96
24	2.42	2.46	2.50	2.54	2.58	2.62	2.66	2.69	2.73	2.77	2.81	2.85	2.89	2.93	2.97	3.01	3.05	3.08
25	2.52	2.56	2.60	2.64	2.68	2.72	2.77	2.81	2.85	2.89	2.93	2.97	3.01	3.05	3.09	3.13	3.17	3.21
26	2.62	2.66	2.71	2.75	2.79	2.83	2.88	2.92	2.96	3.00	3.04	3.09	3.13	3.17	3.21	3.26	3.30	3.34
27	2.72	2.77	2.81	2.85	2.90	2.94	2.99	3.03	3.07	3.12	3.16	3.20	3.25	3.29	3.34	3.38	3.42	3.47
28	2.82	2.87	2.91	2.96	3.00	3.05	3.10	3.14	3.19	3.23	3.28	3.32	3.37	3.41	3.46	3.51	3.55	3.60
29	2.92	2.97	3.02	3.06	3.11	3.16	3.21	3.25	3.30	3.35	3.39	3.44	3.49	3.54	3.58	3.63	3.68	3.72
30	3.02	3.07	3.12	3.17	3.22	3.27	3.32	3.36	3.41	3.46	3.51	3.56	3.61	3.66	3.71	3.75	3.80	3.85
31	3.12	3.17	3.22	3.27	3.32	3.37	3.43	3.48	3.53	3.58	3.63	3.68	3.73	3.78	3.83	3.88	3.93	3.98
32	3.22	3.28	3.33	3.38	3.43	3.48	3.54	3.59	3.64	3.69	3.74	3.79	3.85	3.90	3.95	4.00	4.05	4.11
33	3.32	3.38	3.43	3.48	3.54	3.59	3.64	3.70	3.75	3.81	3.86	3.91	3.97	4.02	4.07	4.13	4.18	4.23
34	3.42	3.48	3.53	3.59	3.64	3.70	3.75	3.81	3.87	3.92	3.98	4.03	4.09	4.14	4.20	4.25	4.31	4.36
35	3.52	3.58	3.64	3.69	3.75	3.81	3.86	3.92	3.98	4.03	4.09	4.15	4.21	4.26	4.32	4.38	4.43	4.49

Table D–2 Temperature–altitude factor values (enter in Table D–3)

Assumed Temperature of Air Column(°C)

Altitude in meters	–16°	–8°	0°	+4°	+8°	+12°	+16°	+20°	+24°	+28°
10	1.2	1.1	1.1	1.1	1.0	1.0	1.0	1.0	1.0	1.0
50	5.8	5.6	5.4	5.3	5.2	5.2	5.1	5.0	4.9	4.9
100	11.5	11.2	10.8	10.7	10.5	10.3	10.2	10.0	9.9	9.7
150	17.3	16.7	16.2	16.0	15.7	15.5	15.3	15.0	14.8	14.6
200	23.0	22.3	21.6	21.3	21.0	20.7	20.3	20.0	19.7	19.5
250	28.8	27.9	27.0	26.6	26.2	25.8	25.4	25.0	24.7	24.3
300	34.5	33.5	32.5	32.0	31.5	31.0	30.5	30.1	29.6	29.2
350	40.3	39.0	37.9	37.3	36.7	36.2	35.6	35.1	34.6	34.0
400	46.0	44.6	43.3	42.6	42.0	41.3	40.7	40.1	39.5	38.9
450	51.8	50.2	48.7	47.9	47.2	46.5	45.8	45.1	44.4	43.8
500	57.5	55.8	54.1	53.3	52.4	51.6	50.9	50.1	49.4	48.6
550	63.3	61.4	59.5	58.6	57.7	56.8	55.9	55.1	54.3	53.5
600	69.0	66.9	64.9	63.9	62.9	62.0	61.0	60.1	59.2	58.3
650	74.8	72.5	70.3	69.2	68.2	67.1	66.1	65.1	64.2	63.2
700	80.6	78.1	75.7	74.6	73.4	72.3	71.2	70.1	69.1	68.1
750	86.3	83.7	81.1	79.9	78.7	77.5	76.3	75.1	74.0	72.9
800	92.1	89.2	86.5	85.2	83.9	82.6	81.4	80.1	79.0	77.8
850	97.8	94.8	92.0	90.5	89.2	87.8	86.4	85.2	83.9	82.7
900	103.6	100.4	97.4	95.9	94.4	93.0	91.5	90.2	88.8	87.5
950	109.3	106.0	102.8	101.2	99.6	98.1	96.6	95.2	93.8	92.4
1000	115.1	111.5	108.2	106.5	104.9	103.3	101.7	100.2	98.7	97.3
1050	120.8	117.1	113.6	111.8	110.1	108.4	106.8	105.2	103.6	102.1
1100	126.6	122.7	119.0	117.2	115.4	113.6	111.9	110.2	108.6	107.0
1150	132.3	128.3	124.4	122.5	120.6	118.8	117.0	115.2	113.5	111.8
1200	138.1	133.8	129.8	127.8	125.9	123.9	122.0	120.2	118.4	116.7
1250	143.8	139.4	135.2	133.1	131.1	129.1	127.1	125.2	123.4	121.6
1300	149.6	145.0	140.6	138.5	136.3	134.3	132.2	130.2	128.3	126.4
1350	155.3	150.6	146.0	143.8	141.6	139.4	137.3	135.2	133.2	131.3
1400	161.1	156.2	151.4	149.1	146.8	144.6	142.4	140.2	138.2	136.2
1450	166.8	161.7	156.8	154.5	152.1	149.7	147.5	145.3	143.1	141.0
1500	172.6	167.3	162.3	159.8	157.3	154.9	152.5	150.3	148.0	145.9
1550	178.3	172.9	167.7	165.1	162.6	160.1	157.6	155.3	153.0	150.7
1600	184.1	178.5	173.1	170.4	167.8	165.2	162.7	160.3	157.9	155.6
1650	189.8	184.0	178.5	175.7	173.0	170.4	167.8	165.3	162.8	160.5
1700	195.6	189.6	183.9	181.1	178.3	175.6	172.9	170.3	167.8	165.3
1750	201.4	195.2	189.3	186.4	183.5	180.7	178.0	175.3	172.7	170.2
1800	207.1	200.8	194.7	191.7	188.8	185.9	183.1	180.3	177.6	175.0
1850	212.9	206.3	200.1	197.0	194.0	191.0	188.1	185.3	182.6	179.9
1900	218.6	211.9	205.5	202.4	199.3	196.2	193.2	190.3	187.5	184.8
1950	224.4	217.5	210.9	207.7	204.5	201.4	198.3	195.3	192.4	189.6
2000	230.1	223.0	216.3	213.0	209.7	206.5	203.4	200.3	197.4	194.5
2050	235.9	228.6	221.7	218.3	215.0	211.7	208.5	205.3	202.3	199.3
2100	241.6	234.2	227.1	223.7	220.2	216.8	213.5	210.4	207.2	204.2
2150	247.4	239.8	232.5	229.0	225.5	222.0	218.6	215.4	212.2	209.1
2200	253.1	245.4	237.9	234.3	230.7	227.7	223.7	220.4	217.1	213.9
2250	258.9	250.9	243.4	239.6	235.9	232.3	228.8	225.4	222.0	218.8
2300	264.6	256.5	248.8	245.0	241.2	237.5	233.9	230.4	227.0	223.6
2350	270.4	262.1	254.2	250.3	246.4	242.7	239.0	235.4	231.9	228.5
2400	276.1	267.7	259.6	255.6	251.7	247.8	244.0	240.4	236.8	233.4

Table D–2 Continued

Assumed Temperature of Air Column(°C)

Altitude in meters	–16°	–8°	0°	+4°	+8°	+12°	+16°	+20°	+24°	+28°
2450	281.9	273.2	265.0	260.9	256.9	253.0	249.1	245.4	241.8	238.2
2500	287.6	278.8	270.4	266.2	262.2	258.1	254.2	250.4	246.7	243.1
2550	293.4	284.4	275.8	271.6	267.4	263.3	259.3	255.4	251.6	247.9
2600	299.1	290.0	281.2	276.9	272.6	268.5	264.4	260.4	256.6	252.8
2650	304.9	295.5	286.6	282.2	277.9	273.6	269.5	265.4	261.5	257.7
2700	310.6	201.1	292.0	287.5	283.1	278.8	274.5	270.4	266.4	262.5
2750	316.4	306.7	297.4	292.9	288.4	283.9	279.6	275.4	271.4	267.4
2800	322.1	312.3	302.8	298.2	293.6	289.1	284.7	280.4	276.3	272.2
2850	327.9	317.8	308.2	303.5	298.8	294.3	289.8	285.4	281.2	277.1
2900	333.6	323.4	313.6	308.8	304.1	299.4	294.9	290.4	286.2	282.0
2950	339.4	329.0	319.0	314.2	309.3	304.6	299.9	295.5	291.1	286.8
3000	345.1	334.5	324.4	319.5	314.6	309.7	305.0	300.5	296.0	291.7

Table D–3 Barometric correction values

Temp-Alt. Factor	Barometer Reading				
	780 MM	760 MM	740 MM	720 MM	700 MM
10.9	0.9	0.9	0.8	0.8	
5	4.5	4.4	4.3	4.2	4.0
10	9.0	8.8	8.6	8.3	8.1
15	13.6	13.2	12.9	12.5	12.2
20	18.2	17.7	17.2	16.8	16.3
25	22.8	22.2	21.6	21.0	20.4
30	27.4	26.7	26.0	25.3	24.6
35	...	31.2	30.4	29.6	28.8

	760 MM	740 MM	720 MM	700 MM	680 MM	660 MM
40	35.8	34.9	33.9	33.0	32.0	31.1
45	40.4	39.3	38.3	37.2	36.2	35.1
50	45.0	43.8	42.7	41.5	40.3	39.1
55	49.7	48.4	47.1	45.8	44.5	43.1
60	...	52.9	51.5	50.1	48.6	47.2
65	...	57.5	55.9	54.4	52.8	51.3
70	...	62.1	60.4	58.7	57.1	55.4
75	...	66.7	64.9	63.1	61.3	59.5

	720 MM	700 MM	680 MM	660 MM	640 MM
80	69.5	67.5	65.6	63.7	61.7
85	74.0	72.0	69.9	67.9	65.8
90	78.6	76.4	74.2	72.1	69.9
95	83.2	80.9	78.6	76.3	74.0
100	87.9	85.4	83.0	80.5	78.1
105	...	89.9	87.4	84.8	82.2
110	...	94.5	91.8	89.1	86.4
115	...	99.1	96.3	93.4	90.6
120	...	103.7	100.7	97.8	94.8
125	...	108.3	105.3	102.2	99.1

	680 MM	660 MM	640 MM	620 MM	600 MM
125	105.3	102.2	99.1	96.0	92.9
130	109.8	106.6	103.3	100.1	96.9
135	114.3	111.0	107.6	104.3	100.9
140	118.9	115.4	111.9	108.4	104.9
145	123.5	119.9	116.3	112.6	109.0
150	128.2	124.4	120.6	116.9	113.1
155	...	128.9	125.0	121.1	117.2
160	...	133.5	129.4	125.4	121.4
165	...	138.1	133.9	129.7	125.5
170	...	142.7	138.4	134.0	129.7

Temp-Alt. Factor	Barometer Reading				
	640 MM	620 MM	600 MM	580 MM	560 MM
170	138.4	134.0	129.7	125.4	121.1
175	142.9	138.4	133.9	129.5	125.0
180	147.4	142.8	138.2	133.6	129.0
185	151.9	147.2	142.4	137.7	132.9
190	156.5	151.6	146.7	141.8	136.9
195	161.1	156.1	151.0	146.0	141.0
200	165.7	160.5	155.4	150.2	145.0
205	170.4	165.0	159.7	154.4	149.1
210	...	169.6	164.1	158.6	153.2
215	...	174.1	168.5	162.9	157.3

	620 MM	600 MM	580 MM	560 MM	540 MM
215	174.1	168.5	162.9	157.3	151.7
220	178.7	172.9	167.2	161.4	155.7
225	183.3	177.4	171.5	165.6	159.7
230	188.0	181.9	175.8	169.8	163.7
235	192.6	186.4	180.2	174.0	167.8
240	...	191.0	184.6	178.2	171.9
245	...	195.5	189.0	182.5	176.0
250	...	200.1	193.4	186.8	180.1
255	...	204.7	197.9	191.1	184.3
260	...	209.4	202.4	195.4	188.4

	580 MM	560 MM	540 MM	520 MM
260	202.4	195.4	188.4	181.5
265	206.9	199.8	192.6	185.5
270	211.5	204.2	196.9	189.6
275	216.0	208.6	201.1	193.7
280	220.6	213.0	205.4	197.8
285	225.2	217.5	209.7	201.9
290	229.9	222.0	214.0	206.1
295	...	226.5	218.4	210.3
300	...	231.0	222.8	214.5

Temp-Alt. Factor	560 MM	540 MM	520 MM	500 MM	480 MM
305	235.6	227.2	218.8	210.3	201.9
310	240.2	231.6	223.0	214.4	205.9
315	244.8	236.0	227.3	218.6	209.8
320	249.4	240.5	231.6	222.7	213.8
325	254.1	245.0	236.0	226.9	217.8
330	...	249.6	240.3	231.1	221.8
335	...	254.1	244.7	235.3	225.9
340	...	258.7	249.1	239.6	230.0
345	...	263.3	253.6	243.8	234.1

APPENDIX E
Laws Relating to Gases

Kinetic Energy

All moving objects, by virtue of their motion, are capable of doing work. We call this energy of motion, or *kinetic energy*. The kinetic energy of a moving object is equal to one-half its mass (M) multiplied by its velocity (v) squared. Stated mathematically:

$$\text{Kinetic energy} = \tfrac{1}{2}Mv^2$$

Therefore, by doubling the velocity of a moving object, the object's kinetic energy will increase four times.

First Law of Thermodynamics

The first law of thermodynamics is simply the thermal version of the law of conservation of energy, which states that energy cannot be created or destroyed, only transformed from one form to another. Meteorologists use the first law of thermodynamics along with the principles of kinetic energy extensively in analyzing atmospheric phenomena. According to the kinetic theory, the temperature of a gas is proportional to the kinetic energy of the moving molecules. When a gas is heated, its kinetic energy increases because of an increase in molecular motion. Further, when a gas is compressed, the kinetic energy will also be increased and the temperature of the gas will rise. These relationships are expressed in the first law of thermodynamics, as follows: The temperature of a gas may be changed by the addition or subtraction of heat, or by changing the pressure (compression or expansion), or by a combination of both. It is easy to understand how the atmosphere is heated or cooled by the gain or loss of heat. However, when we consider rising and sinking air, the relationships between temperature and pressure become more important. Here an increase in temperature is brought about by performing work on the gas and not by the addition of heat. This phenomenon is called the *adiabatic form* of the first law of thermodynamics.

Boyle's Law

About 1600, the Englishman Robert Boyle showed that if the temperature is kept constant when the pressure exerted on a gas is increased, the volume decreases. This principle, called *Boyle's law*, states: At a constant temperature, the volume of a given mass of gas varies inversely with the pressure. Stated mathematically:

$$P_1V_1 = P_2V_2$$

The symbols P_1 and V_1 refer to the original pressure and volume, respectively, and P_2 and V_2 indicate the new pressure and volume, respectively, after a change occurs. Boyle's law shows that if a given volume of gas is compressed so that the volume is reduced by one-half, the pressure exerted by the gas is doubled. This increase in pressure can be explained by the kinetic theory, which predicts that when the volume of the gas is reduced by one-half, the molecules collide with the walls of the container twice as often. Because density is defined as the mass per unit volume, an increase in pressure results in increased density.

Charles's Law

The relationships between temperature and volume (hence, density) of a gas were recognized about 1787 by the French scientist Jacques Charles, and were stated formally by J. Gay-Lussac in 1802. *Charles's law* states: At a constant pressure, the volume of a given mass is directly proportional to the absolute temperature. In other words, when a quantity of gas is kept at a constant pressure, an increase in temperature results in an increase in volume and vice versa. Stated mathematically:

$$\frac{V_1}{V_2} = \frac{T_1}{T_2}$$

where V_1 and T_1 represent the original volume and temperature, respectively, and V_2 and T_2 represent the final volume and temperature, respectively. This law explains the fact that a gas expands when it is heated. According to the kinetic theory, when heated, particles move more rapidly and therefore collide more often.

The Ideal Gas Law or Equation of State

In describing the atmosphere, three variable quantities must be considered: pressure, temperature, and density (mass per unit volume). The relationships among these variables can be found by combining in a single statement the laws of Boyle and Charles as follows:

$$PV = RT \text{ or } P = \rho RT$$

where P is pressure, V volume, R the constant of proportionality, T absolute temperature, and ρ density. This law, called the *ideal gas law*, states:

1. When the volume is kept constant, the pressure of a gas is directly proportional to its absolute temperature.

2. When the temperature is kept constant, the pressure of a gas is proportional to its density and inversely proportional to its volume.

3. When the pressure is kept constant, the absolute temperature of a gas is proportional to its volume and inversely proportional to its density.

APPENDIX F

Climate Data

The following climate data are for 51 representative stations (Table F–1) from around the world. Temperatures are given in degrees Celsius and precipitation in millimeters. Names and locations are given in Table F–2 along with the elevation (in meters) of each station and its Kppen classification. This format was selected so that you can use the data in exercises to reinforce your understanding of climatic controls and classification. For example, after classifying a station using Table 15–1, write out a likely location based on such items as mean annual temperature, annual temperature range, total precipitation, perature, annual temperature range, total precipitation,

and seasonal precipitation distribution. Your location need not be a specific city; it could be a description of the station's setting, such as "middle-latitude continental" or "subtropical with a strong monsoon influence." It would also be a good idea to list the reasons for your selection. You may then check your answer by examining the list of stations in Table F–2.

If you simply wish to examine the data for a specific place or data for a specific climatic type, consult the list at the end of this appendix.

Table F–1 Selected climate data for the world

	J	F	M	A	M	J	J	A	S	O	N	D	YR.
1.	1.7	4.4	7.9	13.2	18.4	23.8	25.8	24.8	21.4	14.7	6.7	2.8	13.8
	10	10	13	13	20	15	30	32	23	18	10	13	207
2.	−10.4	−8.3	−4.6	3.4	9.4	12.8	16.6	14.9	10.8	5.5	−2.3	−6.4	3.5
	18	25	25	30	51	89	64	71	33	20	18	15	459
3.	10.2	10.8	13.7	17.9	22.2	25.7	26.7	26.5	24.2	19.0	13.3	10.0	18.4
	66	84	99	74	91	127	196	168	147	71	53	71	1247
4.	−23.9	−17.5	−12.5	−2.7	8.4	14.8	15.6	12.8	6.4	−3.1	−15.8	−21.9	−3.3
	23	13	18	8	15	33	48	53	33	20	18	15	297
5.	−17.8	−15.3	−9.2	−4.4	4.7	10.9	16.4	14.4	10.3	3.3	−3.6	−12.8	−0.3
	58	58	61	48	53	61	81	71	58	61	64	64	738
6.	−8.2	−7.1	−2.4	5.4	11.2	14.7	18.9	17.4	12.7	7.4	−0.4	−4.6	5.4
	21	23	29	35	52	74	39	40	35	29	26	21	424
7.	12.8	13.9	15.0	15.0	17.8	20.0	21.1	22.8	22.2	18.3	17.2	15.0	17.6
	69	74	46	28	3	3	0	0	5	10	28	61	327
8.	18.9	20.0	21.1	22.8	25.0	26.7	27.2	27.8	27.2	25.0	21.1	20.0	23.6
	51	48	58	99	163	188	172	178	241	208	71	43	1520
9.	−4.4	−2.2	4.4	10.6	16.7	21.7	23.9	22.7	18.3	11.7	3.8	−2.2	10.4
	46	51	69	84	99	97	97	81	97	61	61	51	894
10.	−5.6	−4.4	0.0	6.1	11.7	16.7	20.0	18.9	15.5	10.0	3.3	−2.2	7.5
	112	96	109	94	86	81	74	61	89	81	107	99	1089
11.	−2.1	0.9	4.7	9.9	14.7	19.4	24.7	23.6	18.3	11.5	3.4	−0.2	10.7
	34	30	40	45	36	25	15	22	13	29	33	31	353
12.	12.8	13.9	15.0	16.1	17.2	18.8	19.4	22.2	21.1	18.8	16.1	13.9	15.9
	53	56	41	20	5	0	0	2	5	13	23	51	269
13.	−0.1	1.8	6.2	13.0	18.7	24.2	26.4	25.4	21.1	14.9	6.7	1.6	13.3
	50	52	78	94	95	109	84	77	70	73	65	50	897
14.	2.7	3.2	7.1	13.2	18.8	23.4	25.7	24.7	20.9	15.0	8.7	3.4	13.9
	77	63	82	80	105	82	105	124	97	78	72	71	1036
15.	12.8	15.0	18.9	21.1	26.1	31.1	32.7	33.9	31.1	22.2	17.7	13.9	23.0
	10	9	6	2	0	0	6	13	10	10	3	8	77

Table F–1 Continued

	J	F	M	A	M	J	J	A	S	O	N	D	YR.
16.	25.6	25.6	24.4	25.0	24.4	23.3	23.3	24.4	24.4	25.0	25.6	25.6	24.7
	259	249	310	165	254	188	168	117	221	183	213	292	2619
17.	25.9	25.8	25.8	25.9	26.4	26.6	26.9	27.5	27.9	27.7	27.3	26.7	26.7
	365	326	383	404	185	132	68	43	96	99	189	143	2433
18.	13.3	13.3	13.3	13.3	13.9	13.3	13.3	13.3	13.9	13.3	13.3	13.9	13.5
	99	112	142	175	137	43	20	30	69	112	97	79	1115
19.	25.9	26.1	25.2	23.9	22.3	21.3	20.8	21.1	21.5	22.3	23.1	24.4	23.2
	137	137	143	116	73	43	43	43	53	74	97	127	1086
20.	13.8	13.5	11.4	8.0	3.7	1.2	1.4	2.9	5.5	9.2	11.4	12.9	7.9
	21	16	18	13	25	15	15	17	12	7	15	18	171
21.	1.5	1.3	3.1	5.8	10.2	12.6	15.0	14.7	12.0	8.3	5.5	3.3	7.8
	179	139	109	140	83	126	141	167	228	236	207	203	1958
22.	−0.5	0.2	3.9	9.0	14.3	17.7	19.4	18.8	15.0	9.6	4.7	1.2	9.5
	41	37	30	39	44	60	67	65	45	45	44	39	556
23.	6.1	5.8	7.8	9.2	11.6	14.4	15.6	16.0	14.7	12.0	9.0	7.0	10.8
	133	96	83	69	68	56	62	80	87	104	138	150	1126
24.	10.8	11.6	13.6	15.6	17.2	20.1	22.2	22.5	21.2	18.2	14.4	11.5	16.6
	111	76	109	54	44	16	3	4	33	62	93	103	708
25.	−9.9	−9.5	−4.2	4.7	11.9	16.8	19.0	17.1	11.2	4.5	−1.9	−6.8	4.4
	31	28	33	35	52	67	74	74	58	51	36	36	575
26.	8.0	9.0	10.9	13.7	17.5	21.6	24.4	24.2	21.5	17.2	12.7	9.5	15.9
	83	73	52	50	48	18	9	18	70	110	113	105	749
27.	−9.0	−9.0	−6.6	−4.1	0.4	3.6	5.6	5.5	3.5	−0.6	−4.5	−7.6	−1.9
	202	180	164	166	197	249	302	278	208	183	190	169	2488
28.	−2.9	−3.1	−0.7	4.4	10.1	14.9	17.8	16.6	12.2	7.1	2.8	0.1	6.6
	43	30	26	31	34	45	61	76	60	48	53	48	555
29.	12.8	13.9	17.2	18.9	22.2	23.9	25.5	26.1	25.5	23.9	18.9	15.0	20.3
	66	41	20	5	3	0	0	0	3	18	46	66	268
30.	24.6	24.9	25.0	24.9	25.0	24.2	23.7	23.8	23.9	24.2	24.2	24.7	24.4
	81	102	155	140	133	119	99	109	206	213	196	122	1675
31.	21.1	20.4	20.9	21.7	23.0	26.0	27.3	27.3	27.5	27.5	26.0	25.2	24.3
	0	2	0	0	1	15	88	249	163	49	5	6	578
32.	20.4	22.7	27.0	30.6	33.8	34.2	33.6	32.7	32.6	30.5	25.5	21.3	28.7
	0	0	0	0	0	2	1	11	2	0	0	0	16
33.	17.8	18.1	18.8	18.8	17.8	16.2	14.9	15.5	16.8	18.6	18.3	17.8	17.5
	46	51	102	206	160	46	18	25	25	53	109	81	922
34.	20.6	20.7	19.9	19.2	16.7	13.9	13.9	16.3	19.1	21.8	21.4	20.9	18.7
	236	168	86	46	13	8	0	3	8	38	94	201	901
35.	11.7	13.3	16.7	18.6	19.2	20.0	20.3	20.5	20.5	19.1	15.9	12.9	17.4
	20	41	179	605	1705	2875	2455	1827	1231	447	47	5	11437
36.	26.2	26.3	27.1	27.2	27.3	27.0	26.7	27.0	27.4	27.4	26.9	26.6	26.9
	335	241	201	141	116	97	61	50	78	91	151	193	1755
37.	−0.8	2.6	5.3	8.5	13.1	17.0	17.2	17.3	15.3	11.5	5.7	0.3	9.4
	0	0	1	1	18	72	157	151	68	4	1	0	473
38.	24.5	25.8	27.9	30.5	32.7	32.5	30.7	30.1	29.7	28.1	25.9	24.6	28.6
	24	7	15	25	52	53	83	124	118	267	308	157	1233
39.	−18.7	−18.1	−16.7	−11.7	−5.0	0.6	5.3	5.8	1.4	−4.2	−12.3	−15.8	−7.5
	8	8	8	8	15	20	36	43	43	33	13	12	247
40.	−21.9	−18.6	−12.5	−5.0	9.7	15.6	18.3	16.1	10.3	0.8	−10.6	−18.4	−1.4
	15	8	8	13	30	51	51	51	28	25	18	20	318
41.	−4.7	−1.9	4.8	13.7	20.1	24.7	26.1	24.9	19.9	12.8	3.8	−2.7	11.8
	4	5	8	17	35	78	243	141	58	16	10	3	623

Table F–1 Continued

	J	F	M	A	M	J	J	A	S	O	N	D	YR.
42.	24.3	25.2	27.2	29.8	29.5	27.8	27.6	27.1	27.6	28.3	27.7	25.0	27.3
	8	5	6	17	260	524	492	574	398	208	34	3	2530
43.	25.8	26.3	27.8	28.8	28.2	27.4	27.1	27.1	26.7	26.5	26.1	25.7	27.0
	6	13	12	65	196	285	242	277	292	259	122	37	1808
44.	3.7	4.3	7.6	13.1	17.6	21.1	25.1	26.4	22.8	16.7	11.3	6.1	14.7
	48	73	101	135	131	182	146	147	217	220	101	60	1563
45.	−15.8	−13.6	−4.0	8.5	17.7	21.5	23.9	21.9	16.7	6.1	−6.2	−13.0	5.3
	8	15	15	33	25	33	16	35	15	47	22	11	276
46.	−46.8	−43.1	−30.2	−13.5	2.7	12.9	15.7	11.4	2.7	−14.3	−35.7	−44.5	−15.2
	7	5	5	4	5	25	33	30	13	11	10	7	155
47.	19.2	19.6	18.4	16.4	13.8	11.8	10.8	11.3	12.6	16.3	15.9	17.7	15.2
	84	104	71	109	122	140	140	109	97	106	81	79	1242
48.	28.2	27.9	28.3	28.2	26.8	25.4	25.1	25.8	27.7	29.1	29.2	28.7	27.6
	341	338	274	121	9	1	2	5	17	66	156	233	1562
49.	21.9	21.9	21.2	18.3	15.7	13.1	12.3	13.4	15.3	17.6	19.4	21.0	17.6
	104	125	129	101	115	141	94	83	72	80	77	86	1205
50.	−7.2	−7.2	−4.4	−0.6	4.4	8.3	10.0	8.3	5.0	1.1	−3.3	−6.1	0.7
	84	66	86	62	89	81	79	94	150	145	117	79	1132
51.	−4.4	−8.9	−15.5	−22.8	−23.9	−24.4	−26.1	−26.1	−24.4	−18.8	−10.0	−3.9	−17.4
	13	18	10	10	10	8	5	8	10	5	5	8	110

Table F–2 Locations and climate classifications for Table F–1

Station No.	City	Location	Elevation (m)	Köppen Classification
		North America		
1	Albuquerque, N.M.	lat. 35°05′N long. 106°40′W	1593	BWk
2	Calgary, Canada	lat. 51°03′N long. 114°05′W	1062	Dfb
3	Charleston, S.C.	lat. 32°47′N long. 79°56′W	18	Cfa
4	Fairbanks, Alaska	lat. 64°50′N long. 147°48′W	134	Dfc
5	Goose Bay, Canada	lat. 53°19′N long. 60°33′W	45	Dfb
6	Lethbridge, Canada	lat. 49°40′N long. 112°39′W	920	Dfb
7	Los Angeles, Calif.	lat. 34°00′N long. 118°15′W	29	BSk
8	Miami, Fla.	lat. 25°45′N long. 80°11′W	2	Am
9	Peoria, Ill.	lat. 40°45′N long. 89°35′W	180	Dfa
10	Portland, Me.	lat. 43°40′N long. 70°16′W	14	Dfb
11	Salt Lake City, Utah	lat. 40°46′N long. 111°52′W	1288	BSk
12	San Diego, Calif.	lat. 32°43′N long. 117°10′W	26	BSk

Table F–2 Continued

Station No.	City	Location	Elevation (m)	Köppen Classification
13	St. Louis, Mo.	lat. 38°39′N long. 90°15′W	172	Cfa
14	Washington, D.C.	lat. 38°50′N long. 77°00′W	20	Cfa
15	Yuma, Ariz.	lat. 32°40′N long. 114°40′W	62	BWh
		South America		
16	Iquitos, Peru	lat. 3°39′S long. 73°18′W	115	Af
17	Manaus, Brazil	lat. 3°01′S long. 60°00′W	60	Am
18	Quito, Ecuador	lat. 0°17′S long. 78°32′W	2766	Cfb
19	Rio de Janeiro, Brazil	lat. 22°50′S long. 43°20′W	26	Aw
20	Santa Cruz, Argentina	lat. 50°01′S long. 60°30′W	111	BSk
		Europe		
21	Bergen, Norway	lat. 60°24′N long. 5°20′E	44	Cfb
22	Berlin, Germany	lat. 52°28′N long. 13°26′E	50	Cfb
23	Brest, France	lat. 48°24′N long. 4°30′W	103	Cfb
24	Lisbon, Portugal	lat. 38°43′N long. 9°05′W	93	Csa
25	Moscow, Russia	lat. 55°45′N long. 37°37′E	156	Dfb
26	Rome, Italy	lat. 41°52′N long. 12°37′E	3	Csa
27	Santis, Switzerland	lat. 47°15′N long. 9°21′E	2496	ET
28	Stockholm, Sweden	lat. 59°21′N long. 18°00′E	52	Dfb
		Africa		
29	Benghazi, Libya	lat. 32°06′N long. 20°06′E	25	BSh
30	Coquilhatville, Zaire	lat. 0°01′N long. 18°17′E	21	Af
31	Dakar, Senegal	lat. 14°40′N long. 17°28′W	23	BSh
32	Faya, Chad	lat. 18°00′N long. 21°18′E	251	BWh
33	Nairobi, Kenya	lat. 1°16′S long. 36°47′E	1791	Csb
34	Harare, Zimbabwe	lat. 17°50′S long. 30°52′E	1449	Cwb

Table F–2 Continued

Station No.	City	Location	Elevation (m)	Köppen Classification
		Asia		
35	Cherrapunji, India	lat. 25°15′N long. 91°44′E	1313	Cwb
36	Djakarta, Indonesia	lat. 6°11′S long. 106°45′E	8	Am
37	Lhasa, Tibet	lat. 29°40′N long. 91°07′E	3685	Cwb
38	Madras, India	lat. 13°00′N long. 80°11′E	16	Aw
39	Novaya Zemlya, Russia	lat. 72°23′N long. 54°46′E	15	ET
40	Omsk, Russia	lat. 54°48′N long. 73°19′E	85	Dfb
41	Beijing, China	lat. 39°57′N long. 116°23′E	52	Dwa
42	Rangoon, Myanmar	lat. 16°46′N long. 96°10′E	23	Am
43	Ho Chi Minh City, Viet Nam	lat. 10°49′N long. 106°40′E	10	Aw
44	Tokyo, Japan	lat. 35°41′N long. 139°46′E	6	Cfa
45	Urumchi, China	lat. 43°47′N long. 87°43′E	912	Dfa
46	Verkhoyansk, Russia	lat. 67°33′N long. 133°23′E	137	Dfd
		Australia and New Zealand		
47	Auckland, New Zealand	lat. 37°43′S long. 174°53′E	49	Csb
48	Darwin, Australia	lat. 12°26′S long. 131°00′E	27	Aw
49	Sydney, Australia long. 151°179E	lat. 33°52′S	42	Cfb
		Greenland		
50	Ivigtut, Greenland long. 48°109W	lat. 61°12′N	29	ET
		Antarctica		
51	McMurdo Station, Antarctica	lat. 77°53′S long. 167°00′E	2	EF

Glossary

Absolute Humidity The mass of water vapor per volume of air (usually expressed as grams of water vapor per cubic meter of air).

Absolute Instability The condition of air that has an environmental lapse rate that is greater than the dry adiabatic rate (1°C per 100 meters).

Absolute Stability The condition of air that has an environmental lapse rate that is less than the wet adiabatic rate.

Absolute Zero The zero point on the Kelvin temperature scale, representing the temperature at which all molecular motion is presumed to cease.

Acid Precipitation Rain or snow with a pH value that is less than the value for uncontaminated rain.

Adiabatic Temperature Change The cooling or warming of air caused when air is allowed to expand or is compressed, not because heat is added or subtracted.

Advection Horizontal convective motion, such as wind.

Advection Fog Fog formed when warm moist air is blown over a cool surface and chilled below the dew point.

Aerosols Tiny solid and liquid particles suspended in the atmosphere.

Aerovane A device that resembles a wind vane with a propeller at one end. Used to indicate wind speed and direction.

Air A mixture of many discrete gases, of which nitrogen and oxygen are most abundant, in which varying quantities of tiny solid and liquid particles are suspended.

Air Mass A large body of air, usually 1600 kilometers or more across, that is characterized by homogeneous physical properties at any given altitude.

Air-Mass Thunderstorm A localized thunderstorm that forms in a warm, moist, unstable air mass. Most frequent in the afternoon in spring and summer.

Air-Mass Weather The conditions experienced in an area as an air mass passes over it. Because air masses are large and relatively homogeneous, air-mass weather will be fairly constant and may last for several days.

Air Pollutants Airborne particles and gases occurring in concentrations that endanger the health and well-being of organisms or disrupt the orderly functioning of the environment.

Air Pressure The force exerted by the weight of a column of air above a given point.

Albedo The reflectivity of a substance, usually expressed as a percentage of the incident radiation reflected.

Aleutian Low A large cell of low pressure centered over the Aleutian Islands of the North Pacific during the winter.

Altimeter An aneroid barometer calibrated to indicate altitude instead of pressure.

Altitude (of the Sun) The angle of the Sun above the horizon.

Analog Method A statistical approach to weather forecasting in which current conditions are matched with records of similar past weather events with the idea that the succession of events in the past will be paralleled by current conditions.

Anemometer An instrument used to determine wind speed.

Aneroid Barometer An instrument for measuring air pressure; it consists of evacuated metal chambers that are very sensitive to variations in air pressure.

Annual Mean Temperature An average of the 12 monthly means.

Annual Temperature Range The difference between the warmest and coldest monthly means.

Anticyclone An area of high atmospheric pressure characterized by diverging and rotating winds and subsiding air aloft.

Anticyclonic Flow Winds blow out and flow clockwise about an anticyclone (high) in the Northern Hemisphere, and they blow out and flow counterclockwise about an anticyclone in the Southern Hemisphere.

Aphelion The point in the orbit of a planet that is farthest from the Sun.

Apparent Temperature The air temperature perceived by a person.

Arctic (A) Air Mass A bitterly cold air mass that forms over the frozen Arctic Ocean.

Arctic Sea Smoke A dense and often extensive steam fog occurring over high-latitude ocean areas in winter.

Arid *See* Desert.

Atmosphere The gaseous portion of a planet, the planet's envelope of air; one of the traditional subdivisions of Earth's physical environment.

Atmospheric Window Refers to the fact that the troposphere is transparent (i.e., does not absorb) to terrestrial radiation between 8 and 11 micrometers in length.

Aurora A bright and ever-changing display of light caused by solar radiation interacting with the upper atmosphere in the region of the poles. It is called *aurora borealis* in the Northern Hemisphere and *aurora australis* in the Southern Hemisphere.

Automated Surface Observing System (ASOS) A widely used, standardized set of automated weather instruments that provide routine surface observations .

Autumnal Equinox *See* Equinox.

Azores High Name given the subtropical anticyclone when it is situated over the eastern part of the North Atlantic Ocean.

Backing Wind Shift A wind shift in a counterclockwise direction, such as a shift from east to north.

Barograph A recording barometer.

Barometric Tendency *See* Pressure Tendency.

Beaufort Scale A scale that can be used for estimating wind speed when an anemometer is not available.

Bergeron Process A theory that relates the formation of precipitation to supercooled clouds, freezing nuclei, and the different saturation levels of ice and liquid water.

Bermuda High Name given the subtropical high in the North Atlantic during the summer when it is centered near the island of Bermuda.

Bimetal Strip A thermometer consisting of two thin strips of metal welded together and that have widely different coefficients of thermal expansion. When temperature changes, the two metals expand or contract unequally and cause changes in the curvature of the element. Commonly used in thermographs.

Biosphere The totality of life-forms on Earth.

Blackbody A material that is able to absorb 100 percent of the radiation that strikes it.

Blizzard A violent and extremely cold wind laden with dry snow picked up from the ground.

Bora In the region of the eastern shore of the Adriatic Sea, a cold, dry northeasterly wind that blows down from the mountains.

Buys Ballot's Law With your back to the wind in the Northern Hemisphere, low pressure will be to your left and high pressure to your right. The reverse is true in the Southern Hemisphere.

Calorie The amount of heat required to raise the temperature of 1 gram of water 1°C.

Ceiling The height ascribed to the lowest layer of clouds or obscuring phenomena when the sky is reported as broken, overcast, or obscured and the clouds are not classified "thin" or "partial." The ceiling is termed *unlimited* when the foregoing conditions are not present.

Celsius Scale A temperature scale (at one time called the centigrade scale) devised by Anders Celsius in 1742 and used where the metric system is in use. For water at sea level, 0° is designated the ice point and 100° the steam point.

Chinook The name applied to a foehn wind in the Rocky Mountains.

Circle of Illumination The line (great circle) separating daylight from darkness on Earth.

Cirrus One of three basic cloud forms; also one of the three high cloud types. They are thin, delicate ice-crystal clouds often appearing as veil-like patches or thin, wispy fibers.

Climate A description of aggregate weather conditions; the sum of all statistical weather information that helps describe a place or region.

Climate Change A study dealing with variations in climate on many different time scales from decades to millions of years, and the possible causes of such variations.

Climate-Feedback Mechanisms Because the atmosphere is a complex interactive physical system, several different possible outcomes may result when one of the system's elements is altered. These various possibilities are called *climate-feedback mechanisms*.

Climate System The exchanges of energy and moisture occurring among the atmosphere, hydrosphere, lithosphere, biosphere, and cryosphere.

Cloud A form of condensation best described as a dense concentration of suspended water droplets or tiny ice crystals.

Cloud Seeding The introduction into clouds of particles (most commonly dry ice or silver iodide) for the purpose of altering the cloud's natural development.

Clouds of Vertical Development A cloud that has its base in the low height range but extends upward into the middle or high altitudes.

Cold Front The discontinuity at the forward edge of an advancing cold air mass that is displacing warmer air in its path.

Cold-Type Occluded Front A front that forms when the air behind the cold front is colder than the air underlying the warm front it is overtaking.

Cold Wave A rapid and marked fall of temperature. The National Weather Service applies this term to a fall of temperature in 24 hours equaling or exceeding a specified number of degrees and reaching a specified minimum temperature or lower. These specifications vary for different parts of the country and for different periods of the year.

Collision–Coalescence Process A theory of raindrop formation in warm clouds (above 0°C) in which large cloud droplets ("giants") collide and join together with smaller droplets to form a raindrop. Opposite electrical charges may bind the cloud droplets together.

Condensation The change of state from a gas to a liquid.

Condensation Nuclei Microscopic particles that serve as surfaces on which water vapor condenses.

Conditional Instability The condition of moist air with an environmental lapse rate between the dry and wet adiabatic rates.

Conduction The transfer of heat through matter by molecular activity. Energy is transferred during collisions among molecules.

Constant-Pressure Surface A surface along which the atmospheric pressure is everywhere equal at any given moment.

Continental (c) Air Mass An air mass that forms over land; it is normally relatively dry.

Continental Climate A climate lacking marine influence and characterized by more extreme temperatures than in marine climates: therefore, it has a relatively high annual temperature range for its latitude.

Contrail A cloud-like streamer frequently observed behind aircraft flying in clear, cold, and humid air and caused by the addition to the atmosphere of water vapor from engine exhaust gases.

Controls of Temperature Those factors that cause variations in temperature from place to place, such as latitude and altitude.

Convection The transfer of heat by the movement of a mass or substance. It can only take place in fluids.

Convection Cell Circulation that results from the uneven heating of a fluid; the warmer parts of the fluid expand and rise because of their buoyancy and the cooler parts sink.

Convergence The condition that exists when the wind distribution within a given region results in a net horizontal inflow of air into the area. Because convergence at lower levels is associated with an upward movement of air, areas of convergent winds are regions favorable to cloud formation and precipitation.

Cooling Degree-Days Each degree of temperature of the daily mean above 65°F is counted as one cooling degree-day. The amount of energy required to maintain a certain temperature in a building is proportional to the cooling degree-days total.

Coriolis Effect The deflective effect of Earth's rotation on all free-moving objects, including the atmosphere and oceans. Deflection is to the right in the Northern Hemisphere and to the left in the Southern Hemisphere.

Corona A bright, whitish disk centered on the Moon or Sun that results from diffraction when the objects are veiled by a thin cloud layer.

Country Breeze A circulation pattern characterized by a light wind blowing into a city from the surrounding countryside. It is best developed on clear and otherwise calm nights when the urban heat island is most pronounced.

Cryosphere Collective term for the ice and snow that exist on Earth. One of the *spheres* of the climate system.

Cumulus One of three basic cloud forms; also the name given one of the clouds of vertical development. Cumulus are billowy, individual cloud masses that often have flat bases.

Cumulus Stage The initial stage in thunderstorm development in which the growing cumulonimbus is dominated by strong updrafts.

Cup Anemometer *See* Anemometer.

Cyclogenesis The process that creates or develops a new cyclone; also the process that produces an intensification of a preexisting cyclone.

Cyclone An area of low atmospheric pressure characterized by rotating and converging winds and ascending air.

Cyclonic Flow Winds blow in and counterclockwise about a cyclone (low) in the Northern Hemisphere and in and clockwise about a cyclone in the Southern Hemisphere.

Daily Mean Temperature The mean temperature for a day that is determined by averaging the hourly readings or, more commonly, by averaging the maximum and minimum temperatures for a day.

Daily Temperature Range The difference between the maximum and minimum temperatures for a day.

Dart Leader *See* Leader.

Deposition The process whereby water vapor changes directly to ice without going through the liquid state.

Desert One of the two types of dry climate—the driest of the dry climates.

Dew A form of condensation consisting of small water drops on grass or other objects near the ground that forms when the surface temperature drops below the dew point. Usually associated with radiation cooling on clear, calm nights.

Dew Point The temperature to which air has to be cooled in order to reach saturation.

Diffraction The slight bending of light as it passes sharp edges.

Diffused Light Solar energy is scattered and reflected in the atmosphere and reaches Earth's surface in the form of diffuse blue light from the sky.

Discontinuity A zone characterized by a comparatively rapid transition of meteorological elements.

Dispersion The separation of colors by refraction.

Dissipating Stage The final stage of a thunderstorm that is dominated by downdrafts and entrainment leading to the evaporation of the cloud structure.

Diurnal Daily, especially pertaining to actions that are completed within 24 hours and that recur every 24 hours.

Divergence The condition that exists when the distribution of winds within a given area results in a net horizontal outflow of air from the region. In divergence at lower levels, the resulting deficit is compensated for by a downward movement of air from aloft; hence, areas of divergent winds are unfavorable to cloud formation and precipitation.

Doldrums The equatorial belt of calms or light variable winds lying between the two trade-wind belts.

Doppler Radar A type of radar that has the capacity of detecting motion directly.

Drizzle Precipitation from stratus clouds consisting of tiny droplets.

Dry Adiabatic Rate The rate of adiabatic cooling or warming in unsaturated air. The rate of temperature change is 1°C per 100 meters.

Dry Climate A climate in which yearly precipitation is not as great as the potential loss of water by evaporation.

Dryline A narrow zone in the atmosphere along which there is an abrupt change in moisture as when dry continental tropical air converges with humid maritime tropical air. The denser cT air acts to lift the less dense mT air producing clouds and storms.

Dry-Summer Subtropical Climate A climate located on the west sides of continents between latitudes 30° and 45°. It is the only humid climate with a strong winter precipitation maximum.

Dynamic Seeding A type of cloud seeding that uses massive seeding, a process resulting in an increase of the release of latent heat and causing the cloud to grow larger.

Easterly Wave A large migratory wavelike disturbance in the trade winds that sometimes triggers the formation of a hurricane.

Eccentricity The variation of an ellipse from a circle.

Electromagnetic Radiation *See* Radiation.

Elements (atmospheric) Those quantities or properties of the atmosphere that are measured regularly and that are used to express the nature of weather and climate.

El Niño The name given to the periodic warming of the ocean that occurs in the central and eastern Pacific. A major El Niño episode can cause extreme weather in many parts of the world.

Entrainment The infiltration of surrounding air into a vertically-moving air column. For example, the influx of cool, dry air into the downdraft of a cumulonimbus cloud; a process that acts to intensify the downdraft.

Environmental Lapse Rate The rate of temperature decrease with height in the troposphere.

Equatorial Low A quasi-continuous belt of low pressure lying near the equator and between the subtropical highs.

Equinox The point in time when the vertical rays of the Sun are striking the equator. In the Northern Hemisphere, March 20 or 21 is the *vernal* or *spring equinox* and

September 22 or 23 is the *autumnal equinox*. Lengths of daylight and darkness are equal at all latitudes at equinox.

Evaporation The process by which a liquid is transformed into gas.

Eye A roughly circular area of relatively light winds and fair weather at the center of a hurricane.

Eye Wall The doughnut-shaped area of intensive cumulonimbus development and very strong winds that surrounds the eye of a hurricane.

Fahrenheit Scale A temperature scale devised by Gabriel Daniel Fahrenheit in 1714 and used in the English system. For water at sea level, 32° is designated the ice point and 212° the steam point.

Fall Wind *See* Katabatic Wind.

Fata Morgana A mirage most frequently observed in coastal areas in which extreme towering occurs.

Fixed Points Reference points, such as the steam point and the ice point, used in the construction of temperature scales.

Flash The total discharge of lightning, which is usually perceived as a single flash of light but which actually consists of several flashes (*see* Stroke).

Foehn A warm, dry wind on the ice side of a mountain range that owes its relatively high temperature largely to adiabatic heating during descent down mountain slopes.

Fog A cloud with its base at or very near Earth's surface.

Forecasting Skill *See* Skill.

Freezing The change of state from a liquid to a solid.

Freezing Nuclei Solid particles that have a crystal form resembling that of ice; they serve as cores for the formation of ice crystals.

Freezing Rain *See* Glaze.

Front A boundary (discontinuity) separating air masses of different densities, one warmer and often higher in moisture content than the other.

Frontal Fog Fog formed when rain evaporates as it falls through a layer of cool air.

Frontal Wedging The lifting of air resulting when cool air acts as a barrier over which warmer, lighter air will rise.

Frontogenesis The beginning or creation of a front.

Frontolysis The destruction and dying of a front.

Frost Occurs when the temperature falls to 0°C or below (**See** White Frost)

Fujita Intensity Scale (F-scale) A scale developed by T. Theodore Fujita for classifying the severity of a tornado, based on the correlation of wind speed with the degree of destruction.

Geostationary Satellite A satellite that remains over a fixed point because its rate of travel corresponds to Earth's rate of rotation. Because the satellite must orbit at distances of about 35,000 kilometers, images from this type of satellite are not as detailed as those from polar satellites.

Geostrophic Wind A wind, usually above a height of 600 meters, that blows parallel to the isobars.

Glaze A coating of ice on objects formed when supercooled rain freezes on contact. A storm that produces glaze is termed an "icing storm."

Global Circulation The general circulation of the atmosphere; the average flow of air over the entire globe.

Glory A series of rings of colored light most commonly appearing around the shadow of an airplane that is projected on clouds below.

Gradient Wind The curved airflow pattern around a pressure center resulting from a balance among pressure-gradient force, Coriolis force, and centrifugal force.

Greenhouse Effect The transmission of shortwave solar radiation by the atmosphere coupled with the selective absorption of longer-wavelength terrestrial radiation, especially by water vapor and carbon dioxide, resulting in warming of the atmosphere.

Growing Degree-Days A practical application of temperature data for determining the approximate date when crops will be ready for harvest.

Gust Front The boundary separating the cold downdraft from a thunderstorm and the relatively warm, moist surface air. Lifting along this boundary may initiate the development of thunderstorms.

Hadley Cell The thermally driven circulation system of equatorial and tropical latitudes consisting of two convection cells, one in each hemisphere. The existence of this circulation system was first proposed by George Hadley in 1735 as an explanation for the trade winds.

Hail Precipitation in the form of hard, round pellets or irregular lumps of ice that may have concentric shells formed by the successive freezing of layers of water.

Halo A narrow whitish ring of large diameter centered around the Sun. The commonly observed 22° *halo* subtends an angle of 22° from the observer.

Heat The kinetic energy of random molecular motion.

Heat Budget The balance of incoming and outgoing radiation.

Heating Degree-Day Each degree of temperature of the daily mean below 65°F is counted as one heating degree-day. The amount of heat required to maintain a certain temperature in a building is proportional to the heating degree-days total.

Heterosphere A zone of the atmosphere beyond about 80 kilometers where the gases are arranged into four roughly spherical shells, each with a distinctive composition.

High Cloud A cloud that normally has its base above 6000 meters; the base may be lower in winter and at high-latitude locations.

Highland Climate Complex pattern of climate conditions associated with mountains. Highland climates are characterized by large differences that occur over short distances.

Homosphere A zone of atmosphere extending from Earth's surface to about 80 kilometers that is uniform in terms of the proportions of its component gases.

Horse Latitudes A belt of calms or light variable winds and subsiding air located near the center of the subtropical high.

Humid Continental Climate A relatively severe climate characteristic of broad continents in the middle latitudes between approximately 40° and 50° north latitude. This climate is not found in the Southern Hemisphere, where the middle latitudes are dominated by the oceans.

Humidity A general term referring to water vapor in the air.

Humid Subtropical Climate A climate generally located on the eastern side of a continent and characterized by hot, sultry summers and cool winters.

Hurricane A tropical cyclonic storm having minimum winds of 119 kilometers per hour; also known as *typhoon* (western Pacific) and *cyclone* (Indian Ocean).

Hurricane Warning A warning issued when sustained winds of 119 kilometers per hour or higher are expected within a specified coastal area in 24 hours or less.

Hurricane Watch An announcement aimed at specific coastal areas that a hurricane poses a possible threat, generally within 36 hours.

Hydrologic Cycle The continuous movement of water from the oceans to the atmosphere (by evaporation), from the atmosphere to the land (by condensation and precipitation), and from the land back to the sea (via stream flow).

Hydrosphere The water portion of our planet; one of the traditional subdivisions of Earth's physical environment.

Hydrostatic Equilibrium The balance maintained between the force of gravity and the vertical pressure gradient that does not allow air to escape to space.

Hygrometer An instrument designed to measure relative humidity.

Hygroscopic Nuclei Condensation nuclei having a high affinity for water, such as salt particles.

Ice Cap Climate A climate that has no monthly means above freezing and supports no vegetative cover except in a few scattered high mountain areas. This climate, with its perpetual ice and snow, is confined largely to the ice sheets of Greenland and Antarctica.

Icelandic Low A large cell of low pressure centered over Iceland and southern Greenland in the North Atlantic during the winter.

Ice Point The temperature at which ice melts.

Ideal Gas Law The pressure exerted by a gas is proportional to its density and absolute temperature.

Inclination of the Axis The tilt of Earth's axis from the perpendicular to the plane of Earth's orbit (plane of the ecliptic). Currently, the inclination is about 23 1/2° away from the perpendicular.

Inferior Mirage A mirage in which the image appears below the true location of the object.

Infrared Radiation with a wavelength from 0.7 to 200 micrometers.

Interference Occurs when light rays of different frequencies (i.e., colors) meet. Such interference results in the cancellation or subtraction of some frequencies, which is responsible for the colors associated with coronas.

Internal Reflection Occurs when light that is traveling through a transparent material, such as water, reaches the opposite surface and is reflected back into the material. This is an important factor in the formation of such optical phenomena as rainbows.

Intertropical Convergence Zone (ITCZ) The zone of general convergence between the Northern and Southern Hemisphere trade winds.

Ionosphere A complex atmospheric zone of ionized gases extending between 80 and 400 kilometers, thus coinciding with the lower thermosphere and heterosphere.

Isobar A line drawn on a map connecting points of equal barometric pressure, usually corrected to sea level.

Isohyet A line connecting places having equal rainfall.

Isotherm A line connecting points of equal air temperature.

ITCZ *See* Intertropical Convergence Zone.

Jet Stream Swift geostrophic airstreams in the upper troposphere that meander in relatively narrow belts.

Jungle An almost impenetrable growth of tangle vines, shrubs, and short trees characterizing areas where the tropical rain forest has been cleared.

Katabatic Wind The flow of cold, dense air downslope under the influence of gravity; the direction of flow is controlled largely by topography.

Kelvin Scale A temperature scale (also called the *absolute scale*) used primarily for scientific purposes and having intervals equivalent to those on the Celsius scale but beginning at absolute zero.

Köppen Classification Devised by Wladimir Köppen, a system for classifying climates that is based on mean monthly and annual values of temperature and precipitation.

Lake-Effect Snow Snow showers associated with a *cP* air mass to which moisture and heat are added from below as it traverses a large and relatively warm lake (such as one of the Great Lakes), rendering the air mass humid and unstable.

Land Breeze A local wind blowing from the land toward the sea during the night in coastal areas.

La Niña An episode of strong trade winds and unusually low sea–surface temperatures in the central and eastern Pacific. The opposite of *El Niño*.

Lapse Rate *See* Environmental Lapse Rate; Normal Lapse Rate.

Latent Heat The energy absorbed or released during a change of state.

Law of Conservation of Angular Momentum The product of the velocity of an object around a center of rotation (axis) and the distance of the object from the axis is constant.

Leader The conductive path of ionized air that forms near a cloud base prior to a lightning stroke. The initial conductive path is referred to as a *step leader* because it extends itself earthward in short, nearly invisible bursts. A *dart leader*, which is continuous and less branched than a step leader, precedes each subsequent stroke along the same path.

Lifting Condensation Level The height at which rising air that is cooling at the dry adiabatic rate becomes saturated and condensation begins.

Lightning A sudden flash of light generated by the flow of electrons between oppositely charged parts of a cumulonimbus cloud or between the cloud and the ground.

Liquid-in-Glass Thermometer A device for measuring temperature that consists of a tube with a liquid-filled bulb at one end. The expansion or contraction of the fluid indicates temperature.

Lithosphere Rigid outer layer of Earth that includes the crust and uppermost part of the mantle.

Long-Range Forecasting Estimating rainfall and temperatures for a period beyond 3 to 5 days, usually for 30-day periods. Such forecasts are not as detailed or reliable as those for shorter periods.

Looming A mirage that allows objects that are below the horizon to be seen.

Low Cloud A cloud that forms below a height of about 2000 meters.

Macroscale Winds Such phenomena as cyclones and anticyclones, which persist for days or weeks and have a horizontal dimension of hundreds to several thousands of kilometers; also, features of the atmospheric circulation that persist for weeks or months and have horizontal dimensions of up to 10,000 kilometers.

Marine Climate A climate dominated by the ocean; because of the moderating effect of water, sites having this climate are considered relatively mild.

Marine West Coast Climate A climate found on windward coasts from latitudes 40° to 65° and dominated by maritime air masses. Winters are mild and summers are cool.

Maritime (m) Air Mass An air mass that originates over the ocean. These air masses are relatively humid.

Mature Stage The second of the three stages of a thunderstorm. This stage is characterized by violent weather as downdrafts exist side by side with updrafts.

Maximum Thermometer A thermometer that measures the maximum temperature for a given period in time, usually 24 hours. A constriction in the base of the glass tube allows mercury to rise but prevents it from returning to the bulb until the thermometer is shaken or whirled.

Mediterranean Climate A common name applied to the dry-summer subtropical climate.

Melting The change of state from a solid to a liquid.

Mercury Barometer A mercury-filled glass tube in which the height of the column of mercury is a measure of air pressure.

Mesocyclone A vertical cylinder of cyclonically rotating air (3 to 10 kilometers in diameter) that develops in the updraft of a severe thunderstorm and that often precedes the development of damaging hail or tornadoes.

Mesopause The boundary between the mesosphere and the thermosphere.

Mesoscale Convective Complex (MCC) A slow-moving roughly circular cluster of interacting thunderstorm cells covering an area of thousands of square kilometers that may persist for 12 hours or more.

Mesoscale Winds Small convective cells that exist for minutes or hours, such as thunderstorms, tornadoes, and land and sea breezes. Typical horizontal dimensions range from 1 to 100 kilometers.

Microscale Winds Phenomena such as turbulence, with life spans of less than a few minutes that affect small areas and are strongly influenced by local conditions of temperature and terrain.

Middle Cloud A cloud occupying the height range from 2000 to 6000 meters.

Middle-Latitude (Midlatitude) Cyclone Large low-pressure center with diameter often exceeding 1000 kilometers that moves from west to east and may last from a few days to more than a week and usually has a cold front and a warm front extending from the central area of low pressure.

Midlatitude Jet Stream A jet stream that migrates between the latitudes of 30° and 70°.

Milankovitch Cycles Systematic changes in three elements of Earth's orbit—eccentricity, obliquity, and precession. Named for Yugoslavian astronomer Milutin Milankovitch.

Millibar The standard unit of pressure measurement used by the National Weather Service. One millibar (mb) equals 100 newtons per square meter.

Minimum Thermometer A thermometer that measures the minimum temperature for a given period of time, usually 24 hours. By checking the small dumbbell-shaped index, the minimum temperature can be read.

Mirage An optical effect of the atmosphere caused by refraction in which the image of an object appears displaced from its true position.

Mistral A cold northwest wind that blows into the western Mediterranean basin from higher elevations to the north.

Mixing Depth The height to which convectional movements extend above Earth's surface. The greater the mixing depths, the better the air quality.

Mixing Ratio The mass of water vapor in a unit mass of dry air; commonly expressed as grams of water vapor per kilogram of dry air.

Monsoon The seasonal reversal of wind direction associated with large continents, especially Asia. In winter, the wind blows from land to sea; in summer, it blows from sea to land.

Monthly Mean Temperature The mean temperature for a month that is calculated by averaging the daily means.

Mountain Breeze The nightly downslope winds commonly encountered in mountain valleys.

National Weather Service (NWS) The federal agency responsible for gathering and disseminating weather-related information.

Negative-Feedback Mechanism As used in climatic change, any effect that is opposite of the initial change and tends to offset it.

Newton A unit of force used in physics. One newton is the force necessary to accelerate 1 kilogram of mass 1 meter per second per second.

Normal Lapse Rate The average drop in temperature with increasing height in the troposphere; about 6.5°C per kilometer.

Nor'easter The term used to describe the weather associated with an incursion of mP air into the Northeast from the North Atlantic; strong northeast winds, freezing or near-freezing temperatures, and the possibility of precipitation make this an unwelcome weather event.

Nowcasting Short-term weather forecasting techniques that are generally applied to predicting severe weather.

Numerical Weather Prediction (NWP) Forecasting the behavior of atmospheric disturbances based upon the solution of the governing fundamental equations of hydrodynamics, subject to the observed initial conditions.

Because of the vast number of calculations involved, high-speed computers are always used.

Obliquity The angle between the planes of Earth's equator and orbit.

Occluded Front A front formed when a cold front overtakes a warm front.

Occlusion The overtaking of one front by another.

Ocean Current The mass movement of ocean water that is either wind-driven or initiated by temperature and salinity conditions that alter the density of seawater.

Orographic Lifting Mountains or highlands acting as barriers to the flow of air force the air to ascend. The air cools, adiabatically, and clouds and precipitation may result.

Outgassing The release of gases dissolved in molten rock.

Overrunning Warm air gliding up a retreating cold air mass.

Oxygen-Isotope Analysis A method of deciphering past temperatures based on the precise measurement of the ratio between two isotopes of oxygen, ^{16}O and ^{18}O. Analysis is commonly made of seafloor sediments and cores from ice sheets.

Ozone A molecule of oxygen containing three oxygen atoms.

Paleosol Old, buried soil that may furnish some evidence of the nature of past climates because climate is the most important factor in soil formation.

Parcel An imaginary volume of air enclosed in a thin elastic cover. Typically it is considered to be a few hundred cubic meters in volume and is assumed to act independently of the surrounding air.

Parhelia *See* Sun dogs.

Perihelion The point in the orbit of a planet closest to the Sun.

Permafrost The permanent freezing of the subsoil in tundra regions.

Persistence Forecast A forecast that assumes that the weather occurring upstream will persist and move on and will affect the areas in its path in much the same way. Persistence forecasts do not account for changes that might occur in the weather system.

pH Scale A 0 to 14 scale that is used for expressing the exact degree of acidity or alkalinity of a solution. A pH of 7 signifies a neutral solution. Values below 7 signify an acid solution, and values above 7 signify an alkaline solution.

Photochemical Reaction A chemical reaction in the atmosphere triggered by sunlight, often yielding a secondary pollutant.

Photosynthesis The production of sugars and starches by plants using air, water, sunlight, and chlorophyll. In the process, atmospheric carbon dioxide is changed to organic matter and oxygen is released.

Photosynthesis The production of sugars and starches by plants using air, water, sunlight, and chlorophyll. In the process, atmospheric carbon dioxide is changed to organic matter and oxygen is released.

Plane of the Ecliptic Plane of Earth's orbit around the Sun.

Plate Tectonics Theory A theory that states the outer portion of Earth is made up of several individual pieces, called *plates*, which move in relation to one another upon a partially molten zone below. As plates move, so do continents, which explains some climatic changes in the geologic past.

Polar (P) Air Mass A cold air mass that forms in a high-latitude source region.

Polar Climates Climates in which the mean temperature of the warmest month is below 10°C; climates that are too cold to support the growth of trees.

Polar Easterlies In the global pattern of prevailing winds, winds that blow from the polar high toward the subpolar low. These winds, however, should not be thought of as persistent winds, such as the trade winds.

Polar Front The stormy frontal zone separating air masses of polar origin from air masses of tropical origin.

Polar Front Theory A theory developed by J. Bjerknes and other Scandinavian meteorologists in which the polar front, separating polar and tropical air masses, gives rise to cyclonic disturbances that intensify and move along the front and pass through a succession of stages.

Polar High Anticyclones that are assumed to occupy the inner polar regions and are believed to be thermally induced, at least in part.

Polar Satellite Satellites that orbit the poles at rather low altitudes of a few hundred kilometers and require only 100 minutes per orbit.

Positive-Feedback Mechanism As used in climatic change, any effect that acts to reinforce the initial change.

Potential Energy Energy that exists by virtue of a body's position with respect to gravitation.

Precession The slow migration of Earth's axis that traces out a cone over a period of 26,000 years.

Precipitation Fog *See* Frontal Fog.

Pressure Gradient The amount of pressure change occurring over a given distance.

Pressure Tendency The nature of the change in atmospheric pressure over the past several hours. It can be a useful aid in short-range weather prediction.

Prevailing Westerlies The dominant west-to-east motion of the atmosphere that characterizes the regions on the poleward side of the subtropical highs.

Prevailing Wind A wind that consistently blows from one direction more than from any other.

Primary Pollutant A pollutant emitted directly from an identifiable source.

Prognostic Chart A computer-generated forecast showing the expected pressure pattern at a specified future time. Anticipated positions of fronts are also included. They usually represent the graphical output associated with a numerical weather prediction model.

Psychrometer Device consisting of two thermometers (wet bulb and dry bulb) that is rapidly whirled and, with the use of tables, yields the relative humidity and dew point.

Radiation The wavelike energy emitted by any substance that possesses heat. This energy travels through space at 300,000 kilometers per second (the speed of light).

Radiation Fog Fog resulting from radiation cooling of the ground and adjacent air; primarily a nighttime and early morning phenomenon.

Radiosonde A lightweight package of weather instruments fitted with a radio transmitter and carried aloft by a balloon.

Rainbow A luminous arc formed by the refraction and reflection of light in drops of water.

Rain Shadow Desert A dry area on the lee side of a mountain range.

Rawinsonde A radiosonde that is tracked by radio-location devices in order to obtain data on upper-air winds.

Reflection, Law of The angle of incidence (incoming ray) is equal to the angle of reflection (outgoing ray).

Refraction The bending of light as it passes obliquely from one transparent medium to another.

Relative Humidity The ratio of the air's water vapor content to its water vapor capacity.

Return Stroke Term applied to the electric discharge resulting from the downward (earthward) movement of electrons from successively higher levels along the conductive path of lightning

Revolution The motion of one body about another, as Earth about the Sun.

Ridge An elongate region of high atmospheric pressure.

Rime A delicate accumulation of ice crystals formed when supercooled fog or cloud droplets freeze on contact with objects.

Rossby Waves Upper-air waves in the middle and upper troposphere of the middle latitudes with wavelengths of from 4000 to 6000 kilometers; named for C. G. Rossby, the meteorologist who developed the equations for parameters governing the waves.

Rotation The spinning of a body, such as Earth, about its axis.

Saffir–Simpson Scale A scale, from 1 to 5, used to rank the relative intensities of hurricanes.

Santa Ana The local name given a foehn wind in southern California.

Saturation The maximum possible quantity of water vapor that the air can hold at any given temperature and pressure.

Saturation Vapor Pressure The vapor pressure, at a given temperature, wherein the water vapor is in equilibrium with a surface of pure water or ice.

Savanna A tropical grassland, usually with scattered trees and shrubs.

Sea Breeze A local wind blowing from the sea during the afternoon in coastal areas.

Secondary Pollutant A pollutant that is produced in the atmosphere by chemical reactions occurring among primary pollutants.

Semiarid *See* Steppe.

Severe Thunderstorm A thunderstorm that produces frequent lightning, locally damaging wind, or hail that is 2 centimeters or more in diameter. In the middle latitudes, most thunderstorms form along or ahead of cold fronts.

Siberian High The high-pressure center that forms over the Asian interior in January and produces the dry winter monsoon for much of the continent.

Skill An index of the degree of accuracy of a set of forecasts as compared to forecasts based on some standard, such as chance or climatic data.

Sleet Frozen or semifrozen rain formed when raindrops pass through a subfreezing layer of air.

Smog A word currently used as a synonym for general air pollution. It was originally created by combining the words "smoke" and "fog."

Snow Precipitation in the form of white or translucent ice crystals, chiefly in complex branched hexagonal form and often clustered into snowflakes.

Solstice The point in time when the vertical rays of the Sun are striking either the Tropic of Cancer (summer solstice in the Northern Hemisphere) or the Tropic of Capricorn (winter solstice in the Northern Hemisphere). Solstice represents the longest or shortest day (length of daylight) of the year.

Source Region The area where an air mass acquires its characteristic properties of temperature and moisture.

Southern Oscillation The seesaw pattern of atmospheric pressure change that occurs between the eastern and western Pacific. The interaction of this effect and that of El Niño can cause extreme weather events in many parts of the world.

Specific Heat The amount of heat needed to raise 1 gram of a substance 1°C at sea-level atmospheric pressure.

Specific Humidity The mass of water vapor per unit mass of air, including the water vapor (usually expressed as grams of water vapor per kilogram of air).

Speed Divergence The divergence of air aloft that results from the variations in velocity occurring along the axis of a jet stream. On passing from a zone of lower wind speed to one of faster speed, air accelerates and therefore experiences divergence.

Squall Line Any nonfrontal line or narrow band of active thunderstorms.

Stable Air Air that resists vertical displacement. If it is lifted, adiabatic cooling will cause its temperature to be lower than the surrounding environment; and if it is allowed, it will sink to its original position.

Standard Atmosphere The idealized vertical distribution of atmospheric pressure (as well as temperature and density), which is taken to represent average conditions in the real atmosphere.

Standard Rain Gauge Having a diameter of about 20 centimeters, this gauge funnels rain into a cylinder that magnifies precipitation amounts by a factor of 10, allowing for accurate measurement of small amounts.

Static Seeding The most commonly used technique of cloud seeding; based on the assumption that cumulus clouds are deficient in freezing nuclei and that the addition of nuclei will spur additional precipitation formation.

Stationary Front A situation in which the surface position of a front does not move; the flow on either side of such a boundary is nearly parallel to the position of the front.

Statistical Methods (in forecasting) Methods in which tables or graphs are prepared from a long series of observations to show the probability of certain weather events under certain conditions of pressure, temperature, or wind direction.

Steam Fog Fog having the appearance of steam; produced by evaporation from a warm water surface into the cool air above.

Steam Point The temperature at which water boils.

Step Leader *See* Leader.

Steppe One of the two types of dry climate; a marginal and more humid variant of the desert that separates it from bordering humid climates. Steppe also refers to the short-grass vegetation associated with this semiarid climate.

Storm Surge The abnormal rise of the sea along a shore as a result of strong winds.

Stratopause The boundary between the stratosphere and the mesosphere.

Stratosphere The zone of the atmosphere above the troposphere characterized at first by isothermal conditions and then a gradual temperature increase. Earth's ozone is concentrated here.

Stratus One of three basic cloud forms; also the name given one of the low clouds. Stratus clouds are sheets or layers that cover much or all of the sky.

Stroke One of the individual components that make up a flash of lightning. There are usually three to four strokes per flash, roughly 50 milliseconds apart.

Subarctic Climate A climate found north of the humid continental climate and south of the polar climate and characterized by bitterly cold winters and short cool summers. Places within this climatic realm experience the highest annual temperature ranges on Earth.

Sublimation The process whereby a solid changes directly to a gas without going through the liquid state.

Subpolar Low Low pressure located at about the latitudes of the Arctic and Antarctic circles. In the Northern Hemisphere, the low takes the form of individual oceanic cells; in the Southern Hemisphere, there is a deep and continuous trough of low pressure.

Subsidence An extensive sinking motion of air, most frequently occurring in anticyclones. The subsiding air is warmed by compression and becomes more stable.

Subtropical High Not a continuous belt of high pressure, but rather several semipermanent anticyclonic centers characterized by subsidence and divergence located roughly between latitudes 25° and 35°.

Summer Solstice *See* Solstice.

Sun Dogs Two bright spots of light, sometimes called "mock suns," that sit at a distance of 22° on either side of the Sun.

Sun Pillar Shafts of light caused by reflection from ice crystals that extend upward or, less commonly, downward from the Sun when the Sun is near the horizon.

Sunspot A dark area on the Sun associated with powerful magnetic storms that extend from the Sun's surface deep into the interior.

Supercell A type of thunderstorm that consists of a single, persistent, and very powerful cell (updraft and downdraft) and that often produces severe weather, including hail and tornadoes.

Supercooled The condition of water droplets that remain in the liquid state at temperatures well below 0°C.

Superior Mirage A mirage in which the image appears above the true position of the object.

Synoptic Weather Forecasting A system of forecasting based on careful studies of synoptic weather charts over a period of years; from such studies a set of empirical rules is established to aid the forecaster in estimating the rate and direction of weather-system movements.

Synoptic Weather Map A weather map describing the state of the atmosphere over a large area at a given moment.

Taiga The northern coniferous forest; also a name applied to the subarctic climate.

Temperature A measure of the degree of hotness or coldness of a substance.

Temperature Inversion A layer in the atmosphere of limited depth where the temperature increases rather than decreases with height.

Thermal Low An area of low atmospheric pressure created by abnormal surface heating.

Thermistor An electric thermometer consisting of a conductor whose resistance to the flow of current is temperature-dependent; commonly used in radiosondes.

Thermocouple An electric thermometer that operates on the principle that differences in temperature between the junction of two unlike metal wires in a circuit will induce a current to flow.

Thermograph An instrument that continuously records temperature.

Thermometer An instrument for measuring temperature; in meteorology, generally used to measure the temperature of the air.

Thermosphere The zone of the atmosphere beyond the mesosphere in which there is a rapid rise in temperature with height.

Thunder The sound emitted by rapidly expanding gases along the channel of a lightning discharge.

Thunderstorm A storm produced by a cumulonimbus cloud and always accompanied by lightning and thunder. It is of relatively short duration and usually accompanied by strong wind gusts, heavy rain, and sometimes hail.

Tipping-Bucket Gauge A recording rain gauge consisting of two compartments ("buckets"), each capable of holding 0.025 centimeter of water. When one compartment fills, it tips and the other compartment takes its place.

Tornado A violently rotating column of air attended by a funnel-shaped or tubular cloud extending downward from a cumulonimbus cloud.

Tornado Warning A warning issued when a tornado has actually been sighted in an area or is indicated by radar.

Tornado Watch A forecast issued for areas of about 65,000 square kilometers, indicating that conditions are such that tornadoes may develop; they are intended to alert people to the possibility of tornadoes.

Towering A mirage in which the size of an object is magnified.

Trace of Precipitation An amount less than 0.025 centimeter.

Trade Winds Two belts of winds that blow almost constantly from easterly directions and are located on the equatorward sides of the subtropical highs.

Transpiration The release of water vapor to the atmosphere by plants.

Trend Forecast A short-range forecasting technique that assumes that the weather occurring upstream will persist and move on to affect the area in its path.

Tropic of Cancer The parallel of latitude, 23-1/2° north latitude, marking the northern limit of the Sun's vertical rays.

Tropic of Capricorn The parallel of latitude, 23-1/2° south latitude, marking the southern limit of the Sun's vertical rays.

Tropical (T) Air Mass A warm-to-hot air mass that forms in the subtropics.

Tropical Depression By international agreement, a tropical cyclone with maximum winds that do not exceed 61 kilometers per hour.

Tropical Disturbance A term used by the National Weather Service for a cyclonic wind system in the tropics that is in its formative stages.

Tropical Rain Forest A luxuriant broadleaf evergreen forest; also the name given the climate associated with this vegetation.

Tropical Storm By international agreement, a tropical cyclone with maximum winds between 61 and 115 kilometers per hour.

Tropical Wet and Dry A climate that is transitional between the wet tropics and the subtropical steppes.

Tropopause The boundary between the troposphere and the stratosphere.

Troposphere The lowermost layer of the atmosphere marked by considerable turbulence and, in general, a decrease in temperature with increasing height.

Trough An elongate region of low atmospheric pressure.

Tundra Climate Found almost exclusively in the Northern Hemisphere or at high altitudes in many mountainous regions. A treeless climatic realm of sedges, grasses, mosses, and lichens dominated by a long, bitterly cold winter.

Ultraviolet Radiation with a wavelength from 0.2 to 0.4 micrometer.

Unstable Air Air that does not resist vertical displacement. If it is lifted, its temperature will not cool as rapidly as the surrounding environment and so it will continue to rise on its own.

Upslope Fog Fog created when air moves up a slope and cools adiabatically.

Upwelling The process by which deep, cold, nutrient-rich water is brought to the surface, usually by coastal currents that move water away from the coast.

Urban Heat Island Refers to the fact that temperatures within a city are generally higher than in surrounding rural areas.

Valley Breeze The daily upslope winds commonly encountered in a mountain valley.

Vapor Pressure That part of the total atmospheric pressure attributable to its water-vapor content.

Veering Wind Shift A wind shift in a clockwise direction, such as a shift from east to south.

Vernal Equinox *See* Equinox.

Virga Wisps or streaks of water or ice particles falling out of a cloud but evaporating before reaching Earth's surface.

Visibility The greatest distance that prominent objects can be seen and identified by unaided, normal eyes.

Visible Light Radiation with a wavelength from 0.4 to 0.7 micrometer.

Warm Front The discontinuity at the forward edge of an advancing warm air mass that is displacing cooler air in its path.

Warm-Type Occluded Front A front that forms when the air behind the cold front is warmer than the air underlying the warm front it is overtaking.

Water Hemisphere A term used to refer to the Southern Hemisphere, where the oceans cover 81 percent of the surface (compared to 61 percent in the Northern Hemisphere).

Weather The state of the atmosphere at any given time.

Weather Analysis The stage prior to developing a weather forecast. This stage involves collecting, compiling, and transmitting observational data.

Weather Forecasting Predicting the future state of the atmosphere.

Weather Modification Deliberate human intervention to influence and improve atmospheric processes.

Weighting Gauge A recording precipitation gauge consisting of a cylinder that rests on a spring balance.

Westerlies *See* Prevailing Westerlies.

Wet Adiabatic Rate The rate of adiabatic temperature change in saturated air. The rate of temperature change is variable, but it is always less than the dry adiabatic rate.

White Frost Ice crystals that form on surfaces instead of dew when the dew point is below freezing.

Wind Air flowing horizontally with respect to Earth's surface.

Windchill A measure of apparent temperature that uses the effects of wind and temperature on the cooling rate of the human body. The windchill chart translates the cooling power of the atmosphere with the wind to a temperature under nearly calm conditions.

Wind Vane An instrument used to determine wind direction.

Winter Solstice *See* Solstice.

World Meteorological Organization (WMO) Established by the United Nations, the WMO consists of more than 130 nations and is responsible for gathering needed observational data and compiling some general prognostic charts.

Index